# CO: TWENTY-FIVE YEARS OF MILLIMETER-WAVE SPECTROSCOPY

*Cover figure:* Contour map with 1" resolution of CO (7-6) from the gravitationally lensed Cloverleaf quasar (H1413+117) at $z$ = 2.6 overlaid on the grey scale HST image.

INTERNATIONAL ASTRONOMICAL UNION

UNION ASTRONOMIQUE INTERNATIONALE

# CO: TWENTY-FIVE YEARS OF MILLIMETER-WAVE SPECTROSCOPY

PROCEEDINGS OF THE 170TH SYMPOSIUM OF THE INTERNATIONAL ASTRONOMICAL UNION, HELD IN TUCSON, ARIZONA, MAY 29–JUNE 5, 1995

EDITED BY

WILLIAM B. LATTER

*NASA Ames Research Center,*
*Moffett Field, CA, USA*

SIMON J.E. RADFORD

*National Radio Astronomy Observatory,*
*Tucson, AZ, USA*

PHILIP R. JEWELL

*National Radio Astronomy Observatory,*
*Tucson, AZ, USA*

JEFFREY G. MANGUM

*National Radio Astronomy Observatory,*
*Tucson, AZ, USA*

and

JOHN BALLY

*CASA, University of Colorado,*
*Boulder, USA*

KLUWER ACADEMIC PUBLISHERS

DORDRECHT / BOSTON / LONDON

A C.I.P. Catalogue record for this book is available from the Library of Congress.

ISBN 0-7923-4283-6

---

*Published on behalf of*
*the International Astronomical Union*
*by*
*Kluwer Academic Publishers, P.O. Box 17, 3300 AA Dordrecht, The Netherlands.*

*Kluwer Academic Publishers incorporates*
*the publishing programmes of*
*D. Reidel, Martinus Nijhoff, Dr W. Junk and MTP Press.*

*Sold and distributed in the U.S.A. and Canada*
*by Kluwer Academic Publishers,*
*101 Philip Drive, Norwell, MA 02061, U.S.A.*

*In all other countries, sold and distributed*
*by Kluwer Academic Publishers Group,*
*P.O. Box 322, 3300 AH Dordrecht, The Netherlands.*

*Printed on acid-free paper*



*Printed in the Netherlands*

# CONTENTS

## STAR FORMATION

## INSTRUMENTATION

## GALAXIES AND GALACTIC NUCLEI

## CO IN PLANETARY SYSTEMS AND IN STELLAR DEATH

# PREFACE

Interstellar carbon monoxide (CO) was first detected in 1970 with the 36 foot diameter telescope of the National Radio Astronomy Observatory[1] on Kitt Peak in Southern Arizona. R. W. Wilson, K. B. Jefferts, and A. A. Penzias of Bell Labs reported, "We have found intense 2.6 mm line radiation from nine Galactic sources which we attribute to carbon monoxide."[2] Soon afterward, several other basic molecules were also observed in space. IAU Symposium 170, *CO: Twenty Five Years of Millimeter Wave Spectroscopy,* was organized to commemorate those discoveries. The Symposium reviewed the accomplishments of a quarter century of research on interstellar molecular gas, surveyed the current state of millimeter-wave spectroscopy, and gave a glimpse of what the next 25 years might hold.

Studies of interstellar CO have revolutionized our understanding of the phases and dynamics of the interstellar medium, the initial and final stages of stellar evolution, the chemistry of dense and diffuse interstellar matter and of the solar system, the structure of the Milky Way galaxy, and the content and structure of other galaxies, some very distant. Spectroscopic studies of CO and other molecules are primary tools for investigating all these topics, which are among the most fundamental and active research areas in astrophysics. New developments in instrumentation, including several powerful new telescopes, continue to keep millimeter and submillimeter wavelength radio astronomy at the forefront of research.

Held from May 29 to June 2, 1995, the Symposium drew more than 225 researchers from 19 countries to Tucson. Together with many pioneers of the field, a large group of young astronomers and students participated in the presentations and discussions.

[1]NRAO is a facility of the National Science Foundation, operated under cooperative agreement by Associated Universities, Inc.

[2]*Astrophysical Journal*, **161**, L43.

A highlight of the Symposium banquet was Robert Wilson's account of the CO discovery observations, illustrated with historical slides. To commemorate the occasion, plaques bearing small sections of the original surface of the 36 ft telescope were presented to Wilson, Jefferts, and Penzias.

A Symposium only occurs through the generous contributions of many people. The IAU (through Commissions 28, 33, 34, and 40), URSI, NRAO, SMTO, and AUI all provided sponsorship and support. Our colleagues on the Scientific and Local Organizing Committees provided many hard hours of service. For their work behind the scenes in making the meeting comfortable and enjoyable, we are pleased to thank the NRAO and SMTO staffs, including Nancy and Jeff Clarke, Lisa Engel, Bill Hale, Susan Lake, Inger Neighbours, Matt Waddel, and Dale Webb. We also thank Susanne Aalto, Per Bergman, Chad Englebracht, Valyo Ivanov, Gopal Narayanan, Julio Sancedo, and Milagros Ruiz. Jennifer Neighbours deserves special recognition for her tireless efforts coordinating the Symposium and producing these Proceedings.

A goal of the Symposium was to gather researchers from all branches of millimeter-wave astronomy together to discuss the state of the field. The reviews and contributions recorded in these Proceedings are one aspect of the Symposium. Equally important was the chance for direct, personal interaction between those attending. The Symposium's success will only be measured by the questions and ideas taken away by the participants, perhaps to inspire new discoveries and new insights.

*The Editors*

| *Scientific Organizing Committee* | *Local Organizing Committee* |
|---|---|
| John Bally, Chair | William B. Latter, Co-Chair |
| Jaap Baars | Simon Radford, Co-Chair |
| Leo Blitz | Jaap Baars |
| Dennis Downes | Darrel Emerson |
| Edith Falgarone | Philip Jewell |
| Tetsuo Hasegawa | Jeffery Mangum |
| Harvey Liszt | Robert Martin |
| Nicholas Scoville | Connstance Walker |
| Ewine van Dishoeck | |
| Glenn White | |

# MOLECULAR CLOUDS IN THE MILKY WAY

# PRESENT AND FUTURE CO SURVEYS

P. THADDEUS
*Harvard-Smithsonian Center for Astrophysics*
*60 Garden Street*
*Cambridge, MA 02138*

*Surveys aren't the most important thing in astronomy — they're the only thing.* — J. Ostriker (with apologies to V. Lombardi)

**Abstract.** Much of what is known about molecular clouds and their relation to star formation and galactic structure has been obtained from the large scale CO surveys. For several reasons, however, these have been done less systematically and less completely than the early 21 cm surveys; a great deal remains to be done. After a very brief general summary of what has been learned, I will emphasize the work now underway on the outer Galaxy and at high and intermediate latitudes. I will also briefly describe work in progress on M31. It will be argued in conclusion that a complete Galactic CO survey at an angular resolution of $1'$ is quite feasible with a dedicated multiple-feed telescope.

## 1. Introduction

A large part of what is now known about molecular clouds and their bearing on star formation and galactic structure in a spiral system like ours has been obtained from the large scale CO surveys. Among the most important general findings are:

- Molecular clouds are CO clouds. CO is the most readily observed molecule in all or nearly all molecular clouds and the best general tracer of the largely invisible $H_2$ and He that constitute the overwhelming bulk of the mass.
- Much of the total interstellar gas resides in molecular clouds.
- Essentially all star formation in a spiral galaxy like ours occurs in molecular clouds and the progeny of young stars (H II regions, infrared



W.B. Latter et al. (eds.), CO: Twenty-Five Years of Millimeter-Wave Spectroscopy, 3–10.

sources, masers, type II supernovae) are generally found in or near molecular clouds—the large ones especially.
- Molecular clouds are excellent tracers of galactic spiral structure.
- The mass spectrum of molecular clouds is steep. Much of the mass is contained in the largest clouds or cloud complexes, the GMCs, which may exceed $1 \times 10^6 M_\odot$ in total mass and represent the largest localized objects in a galaxy like ours.
- The mass distribution of molecular clouds peaks fairly sharply in the crowded spiral arms 4–6 kpc from the Galactic center—the so-called molecular ring.
- In $z$, the distribution of molecular clouds is approximately Gaussian, with a full width at half maximum of about 120 pc at the molecular ring and increasing slightly as one proceeds outwards to the solar circle. The thickness of the molecular disk is only about half that of the atomic gas, but (not coincidentally) is almost the same as that of OB stars.
- The GMCs are the source of much of the diffuse Galactic high energy $\gamma$-ray emission observed by the SAS-2, COS-B, and Compton observatories.

And if space permitted one could add a number of other general findings heavily dependent on the large scale surveys, some now part of the conventional wisdom regarding molecular clouds and so taken for granted.

All of the major CO surveys are direct outgrowths of the first detection of CO with the 12 m telescope—which given the occasion I will call the 36 foot—and the early molecular observations that were done with that pioneering instrument when it was about the only telescope equipped with adequate spectroscopic equipment in the millimeter-wave band. Struck by the ease with which CO could be observed, even with the rather primitive uncooled Schottky receivers available 25 years ago, and by the wide angular extent of the molecular emission in the vicinity of H II regions like the Orion Nebula and Ori B, we quickly realized that smaller telescopes that could be largely dedicated to CO work were needed. It is with instruments of this kind that much of the survey work has been done.

Rather than survey the surveys—hardly possible in the space available here—let me simply describe some of the CO survey work that remains to be done or is underway. Since a rather complete and uniform inventory of local gas was done by the early 21 cm astronomers in less than a dozen years, it might seem strange that after 25 years our Galaxy and its close neighbors would be incompletely studied in the molecule universally accepted as the best tracer of molecular gas, but that is indeed the case. With the development of ultra-sensitive SIS receivers, approaching the quantum limit in the millimeter band, our early expectation that the essential survey work would require only a few years and several Ph. D dissertations has turned out to be exceedingly naive. Our 1.2 m telescopes in Cambridge and Chile have now turned out over 20 dissertations at Columbia, Harvard, and other universities, and there may be that many yet to come.

Although the entire Galactic plane has now been studied in CO, much of it at fairly high resolution with the Massachusetts and Bell telescopes, one has only to step a short distance off the plane—a few degrees—to find large areas that have been studied only at low resolution. Our whole-Galaxy survey at a resolution of 0.5° (Dame et al. 1987), a composite of a number of smaller scale studies done in the U. S. and Chile with our two telescopes, covered an irregular band 10–20° wide with several extensions to include nearby objects like the Orion and Taurus clouds. Many of the off-plane molecular clouds within this large region remain unstudied at higher resolution, and in the vast regions above and below our survey to the Galactic poles—about 80% of the sky or ~30,000 square degrees—only limited surveys have been undertaken.

Here are some CO surveys underway, planned, or possible that I think are particularly needed and of major interest.

## 2. A Complete Survey of the Second Quadrant and the Perseus Arm

The Perseus Arm, starting in the upper first quadrant of the Milky Way and crossing the entire second quadrant into the third, is one of the most impressive and easily observed spiral features of our Galaxy, optically visible over much of its length and containing some of the best studied Population I objects (e.g., Cas A, NGC 7538, W3). It is prominent in our 0.5° whole-Galaxy survey, well separated kinematically from the Local Arm over at least 80° in longitude. From the first rather primitive CO study of a large longitude range in this direction (Cohen et al. 1980), it was immediately apparent that molecular clouds are excellent tracers of spiral structure—not at all axially symmetric in their Galactic distribution, as some undersampled studies toward the inner Galaxy had suggested.

A full survey of the Perseus Arm at a resolution much better than 0.5° is a difficult undertaking, because the arm is at least 4° thick in $b$ and is a good deal fainter in CO generally than the inner arms of the Milky Way.

Two ambitious large scale CO surveys of the outer Galaxy towards the Perseus Arm are underway. Heyer in the present volume is reporting on a survey being done at the University of Massachusetts with their excellent array receiver. The other is being done with the CfA telescope. Starting a year ago with a large amount of data taken by previous observers, Dap Hartmann will complete within one year a fully sampled deep survey from about 75° to 206° in $l$, covering a strip 4° thick, with several extensions to include prominent clouds. The Massachusetts and CfA surveys taken together should shed light on a variety of interesting questions, including the mass spectrum of molecular clouds, the structure of the Local Arm (toward Perseus bifurcated at several places, possibly as a result of multiple supernova explosions), the molecular arm-interarm contrast, star formation beyond the solar circle, and the properties of the distant molecular clouds beyond the Perseus Arm—and much more.

## 3. A Complete High Latitude Survey

It is only very recently that anyone has had the persistence to attempt a reasonably complete CO survey covering the tens of thousands of square degrees at high Galactic latitude. There have been extensive studies of the molecular gas associated with the infrared cirrus discovered by IRAS, and well sampled and sensitive studies of selected areas, leading to the discovery of quite extensive molecular clouds well off the Galactic plane (e.g., Heithausen et al. 1993).

A year ago Hartmann and Magnani began an unbiased CO survey with the CfA telescope of the entire high-Galactic-latitude sky ($|b| > 30°$) readily accessible from Cambridge ($\delta > -17.5°$). Observations are being made on a locally Cartesian $1° \times 1°$ grid of nearly 16,000 points to a $3\sigma$ sensitivity of 0.3 K, as described in a poster contributed to the present meeting. As of April 1996, over 75% of the planned observations are in hand. One of the main goals of this project is to determine rigorously the local mass density represented by the high latitude molecular clouds for comparison with what has already been determined for the normal in-plane population by Dame et al. (1987).

## 4. High $z$ Gas in the First Quadrant

Unfortunately, even a complete inventory of the molecular cirrus at high latitudes can tell little about the distribution of molecular clouds far from the Galactic plane. Because the cirrus is almost always quiescent and devoid of star formation, distances are difficult or impossible to obtain. There is little evidence to date that the molecular component of the cirrus is anything but the expected small fraction of normal in-plane molecular clouds that happen to lie above or below the Sun in $z$. For the time being, this is clearly the simplest and most reasonable hypothesis to adopt.

The distribution with $z$, however, can readily be determined by observing molecular clouds near the terminal velocity several kiloparsecs away in the first Galactic quadrant, which is what Dame has accomplished with the CfA telescope over the past several years (the same, of course, could be done in the fourth quadrant from the southern hemisphere, but to my knowledge no one is attempting that). Some of Dame's recent findings are briefly described elsewhere in this volume. His investigation began with an attempt to see whether there was a molecular counterpart to the thick 21 cm disk observed by Heiles (1984) and Lockman (1984) or to the very high-$z$ CO emission claimed by Garcia-Burillo et al. (1992) in the edge-on spiral NGC 891. If, as widely believed, supernovae and stellar winds are responsible for ejecting atomic gas to high $z$, one might argue that this mechanism would readily dissociate the comparatively fragile molecular gas and no thick molecular disk would be expected. On the other hand, there is a good deal of information to indicate that the disruption of giant molecular clouds by star formation is a ragged, incomplete process and that substantial molecular fragments are sometimes ejected intact, some at a velocity sufficient (i.e., >15 km $s^{-1}$) to carry them well above the Galactic plane (e.g., Nyman et al. 1987). We decided that it was best to keep an open mind and simply to look for high-$z$ CO.

Briefly, what has been found consistently between Galactic longitudes 23° and 51° are thick wings to the normal distribution of molecular clouds, reminiscent of the Heiles-Lockman thick 21 cm disk, but, as one might suspect, on a reduced scale (Dame & Thaddeus 1994, 1995). This thick molecular disk is also, to within the observational uncertainties, approximately Gaussian in shape, with a thickness that is typically 2–3 times that of the normal in-plane disk. The total molecular mass locked up in this component is far from trivial: about $50 \times 10^6$ "$M_\odot$ within the solar circle (under certain assumptions, the most questionable being that the in-plane mass calibration is applicable). There is also distinct structure in the thick-disk gas, but there is not yet enough data to determine the mass spectrum of the putative "clouds" (if that is the right term). The questions raised by this discovery are intriguing: Is the thick-disk component of molecular gas really ejected from nearer the plane, or is it somehow made *in situ*? If ejected, just how—by what specific mechanism that avoids dissociation? And what are the dynamical effects of this component on the halo and disk of the Galaxy?

## 5. The Outer Galaxy

In the distant second quadrant, Digel and collaborators have shown, from limited CO surveys toward slight ripples observed in the distribution of H I, that there are molecular clouds at or beyond the claimed edge of the optical disk at 18 kpc (Digel et al. 1994). Star formation seems to be associated with at least one of these—paradoxically, the most kinematically distant at 28 kpc (de Geus et al. 1993). There may be at least 100 of these objects north and south, so a more systematic survey than has been done so far is very much in order. Given the sensitivity required, and the flare and warp in the Galactic disk far from the center, that is not an easy task. An array receiver of the kind discussed in §8 below would be of great help.

## 6. M31

As the nearest external spiral, M31 ought to be one of the best places to study the relationship of molecular clouds to spiral arms, but it has been difficult to come to grips with this system because it is so large in angular size and is on average faint in CO relative to the Milky Way (by a factor of about five). There have been a number of studies of particular clouds and small segments of arms with the big millimeter-wave telescopes (including the 36 foot: e.g., Combes et al. 1977, Blitz 1985) and several large-area studies with the Bell 7 m telescope (Stark et al. 1981; Ryden & Stark 1986; Casoli et al. 1987; see Hodge 1992 for a review), but the only complete CO survey is one done several years ago with our 1.2 m telescope in Cambridge (Koper et al. 1991; Dame et al. 1993).

A deep high resolution CO survey of M31 is now underway, however, with the University of Massachusetts 15 m telescope and its 15 element QUARRY array receiver as part of the doctoral dissertation of Laurent Loinard at the University of Grenoble. As of April 1996, much of the southwest side of the galaxy has been uniformly surveyed at a linear resolution of 150 pc and the results are of considerable interest. Baade's spiral arm S4 near the peak of the M31 molecular ring is delineated with great clarity over at least 10 kpc. Over much of that length, the CO arm-interarm contrast is large: at least 10:1. The highest contrast observed anywhere is only two and one half times larger (in the Milky Way between the Local and Perseus Arms, Digel et al. 1996). Loinard has shown that the GMCs along S4 are about as large and as luminous in CO as those along the Carina Arm of the Milky Way at about the same distance from the Galactic center and that they are spaced along S4 at about the same interval as those along Carina—roughly every 600 pc (Loinard et al. 1996). M31 is apparently a weak molecular galaxy relative to the Milky Way only because its molecular ring is large and there is little CO emission 4-6 kpc from the center where the spiral arms comprising our molecular ring are especially intense. Beyond the radius of the solar circle, the structure of the two galaxies appears to be remarkably similar in molecular clouds.

## 7. The J = 2–1 Line

Professor Hasegawa and his group at the University of Tokyo have constructed a 60 cm telescope to survey the 2–1 CO line, the aperture chosen to match the 8.7′ beam of our northern and southern 1.2 m telescopes at the

1-0 line. After extensive survey work in Japan (e.g., of the Orion molecular clouds: Sakamoto et al. 1994; the first quadrant Galactic plane: Sakamoto et al. 1995), a copy of this instrument has recently been transported to Chile to study the molecular clouds in the third and fourth quadrants. Papers by Hasegawa and Oka in the present volume describe some of their recent results. These telescopes should prove two of the most useful and active survey instruments during the years ahead.

## 8. A High-Resolution All-Sky CO Survey

The CO surveys, with their precise kinematic information, have proven repeatedly to be indispensable to the analysis of spacecraft continuum surveys, particularly those in the infrared and at high energy. Molecular clouds are the common denominator. They are the source of much of the Galactic high-energy diffuse $\gamma$-rays observed by the COS-B and Compton gamma ray observatories, much of the Galactic far-infrared emission observed by IRAS and COBE, and much of the X-ray absorption (more-or-less local) observed by ROSAT. The CO surveys have the great merit of providing the precise kinematic information that all these continuum surveys require to place objects along the line of sight and for a rounded interpretation generally. Even the lowest resolution CO surveys are higher resolution than any of the gamma ray surveys. It is hardly necessary to add that all the CO surveys have been done at a very small fraction of the cost of any one of the spacecraft surveys.

The spacecraft surveys do have the virtue, however, of very large scale uniform coverage of much or all of the sky. It is clear that a similar all-sky CO survey done at moderately high angular resolution ($\sim 1'$) and at fairly high sensitivity would be an astronomical resource of enormous value to a large number of astronomers—a veritable Palomar Survey of the molecular clouds here and in our nearest extragalactic neighbors. Like the Palomar survey, an all-sky CO survey would be useful for many years. Even with the fairly sophisticated new electronics required, its cost effectiveness would be high.

How could this best be done? One strategy might be to build a single portable instrument capable of surveying the northern sky in about 3 years, after which it could be transported to Chile to do the southern sky in the same amount of time. If the CO 1–0 line is the chosen vehicle, the antenna requirement is not particularly daunting—an aperture in the 7–15 m range, costing say \$1–2M, which would probably represent a minor fraction of the cost of the project. Alternately, one might start with the Bell 7 m antenna, an excellent, thoroughly tested system well able to carry the complicated receivers required for the project. This is not a very portable telescope, however, and shipping it south and installing it there would be a major undertaking.

The main expense would be in receivers and spectrometers. The survey would require a cryogenic multi-feed receiver like the QUARRY array now operating so effectively at the University of Massachusetts, but with many more beams—probably at least 30—and with a spectrometer array to match, either digital or acousto-optical. It would clearly be a mistake to undertake the labor and expense of an all-sky survey without adequate spectral resolution or spectral range, so the spectrometer requirements are formidable but capable of being met: say 5000 resolution elements per beam, each 0.1 km $s^{-1}$ wide, for a fairly generous spectral range of 500 km $s^{-1}$.

The total survey could probably be done for less than seven million dollars, or less than 0.1% of the runout cost of the Space Telescope, but still a lot of money for a ground-based project of this kind. It is not clear at all in the present climate how such an amount could be found. Too often the catchwords in the funding process are terms like *biggest*, *unique*, and *dramatic*, rather than *useful*, *cost effective*, or *judicious*. Jerry Ostriker's wise dictum quoted at the beginning of this article has yet to sink in.

## References


Blitz, L. 1985, ApJ, 296, 481

Casoli, F., Combes, F., & Stark, A. A. 1987, A&A, 173, 43

Cohen, R. S., Cong, H., Dame, T. M., & Thaddeus, P. 1980, ApJ, 239, L53

Combes, F., Encrenez, P. J., Lucas, R., & Weliachew, L. 1977, A&A, 61, L7

Dame, T. M. et al. 1987, ApJ, 322, 706

Dame, T. M., Koper, E., Israel, F. P., & Thaddeus, P. 1993, ApJ, 418, 730

Dame, T. M., & Thaddeus, P. 1994, ApJ, 436, L173

Dame, T. M., & Thaddeus, P. 1995, in The Physics of the Interstellar Medium and the Intergalactic Medium, eds. A. Ferrara, C. F. McKee, C. Heiles, & P. Shapiro (Amer. Soc. Phys. Conf. Series. 80), p. 15

de Geus, E., Vogel, S. N., Digel, S. W., & Gruendl, R. A. 1993, ApJ, 413, L97

Digel, S., Lyder, D., Philbrick, A., Puche, D., & Thaddeus, P. 1996, ApJ, 458, 561

Digel, S. W., de Geus, E., & Thaddeus, P. 1994, ApJ, 422, 92

Garcia-Burillo, S., Guelin, M., Cernicharo, J., & Dahlem, M. 1992, A&A, 266, 21

Heiles, C. 1984, ApJS, 55, 585

Heithausen, A., Stacy, J. G., de Vries, H. W., Mebold, U., & Thaddeus, P. 1993, A&A, 268, 265

Hodge, P. 1992, The Andromeda Galaxy (Dordrecht: Klewer)

Koper, E., Dame, T. M., Israel, F. P., & Thaddeus, P. 1991, ApJ, 383, L11

Lockman, F. J. 1984, ApJ, 283, 90

Loinard, L., Dame, T. M., Koper, E., Lequeux, J., Thaddeus, P., & Young, J. S. 1996, ApJL, submitted

Nyman, L. A., Thaddeus, P., Bronfman, L., & Cohen, R. S. 1987, ApJ, 314, 374

Ryden, B. S., & Stark, A. A. 1986, ApJ, 305, 823

Sakamoto, S., Hayashi, M., Hasegawa, T., Handa, T., & Oka, T. 1994, ApJ, 425, 641

Sakamoto, S., Hasegawa, T., Hayashi, M., Handa, T., & Oka, T. 1995, ApJS, 100, 125

Stark, A. A., Linke, R. A., & Frerking, M. A. 1981, BAAS, 13, 535

# CO IN THE MILKY WAY

LEO BLITZ
*Astronomy Department*
*University of Maryland*
*College Park, MD 20742 USA*

**Abstract.** If the CO distribution of the Milky Way is described as a truncated exponential rather than as a molecular ring with some gas at large radii, it becomes easier to understand the evolution of the disk of stars. The star formation rate per unit molecular gas mass is constant as a function of radius, and the $H_2$ depletion time turns out to be only a few percent of the Hubble time. This very short timescale requires that the atomic gas act as a reservoir for the active star forming gas. Because the HI has such a different radial distribution, there must either be infall from outside the Galaxy, an efficient way for the atomic gas in the disk to lose angular momentum, or both, leading to measurable infall or inflow velocities. The truncation radius of CO is probably due to the recently identified stellar bar.

## 1. Introduction

The ubiquity of CO in the Milky Way was noticed soon after the discovery paper by Wilson *et al.* (1970) and it was shortly thereafter that CO surveys of different kinds of objects in the Milky Way (Wilson *et al.* 1974) and the Milky Way itself (Schwartz *et al.* 1972) were attempted. The difficulty of making an accurate CO survey of the Galactic plane soon became apparent, however, and it was not until the late 1970s that the first truly successful CO surveys were made (*e.g.*, Cohen 1977; Burton & Gordon 1978, Sanders *et al.* 1984). Figure 1 below shows the radial CO distribution (plotted as $H_2$ mass surface density) from the most completely sampled data set, which was made with the 1.2 m sky survey telescope now located at the Center for Astrophysics (from Dame 1993). Also shown is the radial distribution of HI using two different outer Galaxy rotation curves. The figure succinctly points out what has been known since the mid-1970s: the distribution of



*W.B. Latter et al. (eds.), CO: Twenty-Five Years of Millimeter-Wave Spectroscopy*, 11–18.

molecules is very different from that of the atomic gas. The surface density ratio between the two phases varies by more than an order of magnitude between 0.5 and 2 $R_\odot$. Figure 1 also shows the large $H_2$ surface density between 3 and 7 kpc known as the "molecular ring." $R_\odot$ is taken to be 8.5 kpc.

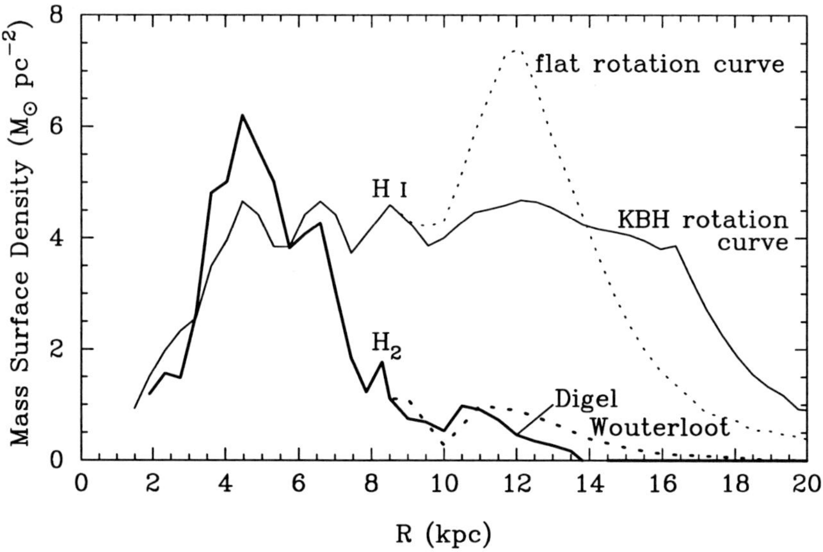


*Figure 1.* Plot of $H_2$ and HI surface density as a function of Galactic radius from Dame (1993). The $H_2$ surface density is measured as CO surface density multiplied by a constant conversion factor as described in the text; a recent recalibration lowers the $H_2$ curve from that shown here by ~25%. See Dame 1993 for further details.

The conversion from CO emissivity to $H_2$ surface density uses the well-known "X–factor" (Bloemen *et al.* 1986), which relates the integrated CO line strength $T_A dv$ to $N(H_2)$. The value of $X$ used in Figure 1 is $2.3 \times 10^{20}$ cm$^{-2}$ (K km s$^{-1}$)$^{-1}$; recent recalibration of the X–factor using GRO data suggests that a better value is $1.7 - 1.9 \times 10^{20}$ cm$^{-2}$ (K km s$^{-1}$)$^{-1}$ (Strong & Mattox 1996). The lower value would decrease the amplitude of the $H_2$ curve in Figure 1 by 25%, implying that the $H_2$ surface density is everywhere less than that of the atomic gas. The total mass of $H_2$ in the Milky Way is $1.3 \times 10^9$ $M_\odot$ if the larger X-factor is used (Dame 1993) and $1.0 \times 10^9$ $M_\odot$ using the lower value. This compares to a total HI mass of about $5 \times 10^9$ $M_\odot$ for HI integrated out as far as it is detected (Henderson *et al.* 1982; Wouterloot *et al.* 1990). Neither value includes the associated helium.

## 2. A Truncated Exponential

Why does the CO distribution have the form that it does? Why is it so profoundly different from the HI? The difference is mirrored in the CO and HI distributions in all spiral galaxies and its origin provides a key to understanding the structure and evolution of disk galaxies in general. Part of the key is in recognizing that "molecular ring" is a misnomer. If one plots the CO distribution on a semi-log scale (see Figure 2), it is apparent that an exponential distribution is a good description of the radial dependence. The CO distribution is then more appropriately described as a truncated exponential rather than a ring; there is indeed nothing special about any part of the distribution outside the truncation radius of ~3 kpc. The question then is what has removed the gas from within a radius of 4 kpc, rather than why is there so much gas between 3 and 7 kpc.

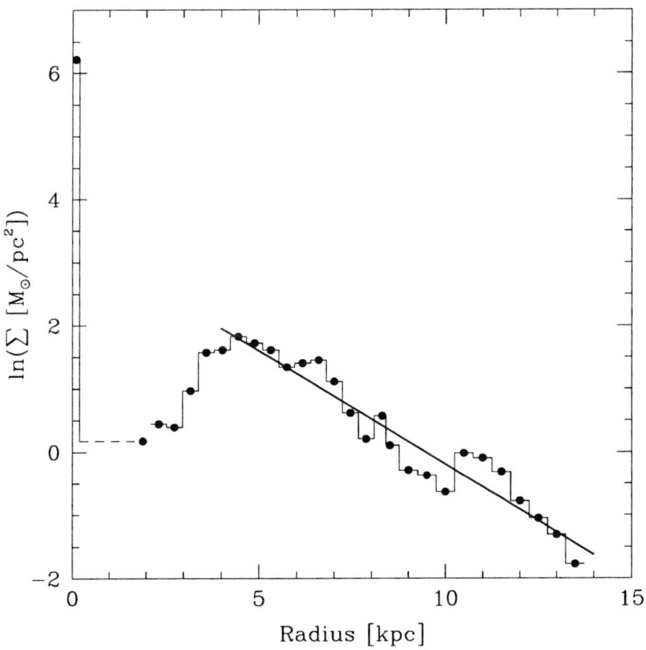


*Figure 2.* The $H_2$ data from the curve marked Digel in Figure 1 with a least squares fit to the data at R > 4 kpc. The innermost point is based on data from Dame *et al.* 1987.

The scale length of the CO distribution in Figure 2 is 2.8 kpc. This is in good agreement with the scale length determined for all of the other tracers of current and past star formation. In a recent paper summarizing the distribution of various components of the disk (Prantzos & Aubert 1995), the distributions of radio HII regions (Güsten & Mezger 1983) and pulsars (Lyne *et al.* 1985) are found to have an exponential radial distribution with a scale length of about 3 kpc, similar to that of the CO. This result is not surprising; because the CO bearing gas is the progenitor of these population I objects, one would expect that they would all have the same radial distribution. However, an analysis of the COBE near infrared data (Spergel *et al.* 1996), which traces primarily much older stars, also obtains a good fit with an exponential disk of 3 kpc scale length. The clear implication is that the Galactic distribution of currently forming stars echoes what has been taking place for the past several Gyr.

However, this raises an intriguing question. Does the distribution of star–forming gas that we see in the Milky Way reflect the gravitational potential determined by the radial stellar component, or is it the other way around? Is it the distribution of gas that determines the stellar distribution in the Milky Way and consequently in the disks of all spiral galaxies?

## 3. Gas Depletion Times

Some insight into this question comes from looking at the star formation rates in the Milky Way and the rate at which gas is returned to the ISM. Let us take the $H_2$ mass of the Milky Way as $10^9$ $M_\odot$. There have been a number of estimates of the star formation rate, but a reasonably reliable one comes from Miller & Scalo (1979) as part of their determination of the IMF. They obtain a local star formation rate of $3 - 7 \times 10^{-9}$ $M_\odot$ $pc^{-2}$ $yr^{-1}$, which translates to a global star formation rate in the Milky Way of $2 - 4$ $M_\odot$ $yr^{-1}$. Because all star formation results from the conversion of molecular gas into stars, the molecular gas would be depleted in $2.5 - 5 \times 10^8$ yr, only a few percent of the Hubble time! Furthermore, because the length scale of the star formation is the same as that of the molecular gas, the depletion timescale is independent of Galactic radius. Could it be, on the other hand, that the return of gas to the ISM from dying stars substantially makes up for the molecular depletion? Since the stellar distribution is already exponential, if the gas return could efficiently be transformed into molecules, it could explain the exponential gas distribution.

There are four broad classes of stars that return gas to the ISM: ordinary evolved stars in the AGB phase, massive stars giving rise to high velocity winds, supernovae, and main sequence stars producing winds similar to the solar wind. The return to the ISM has been estimated by various authors and is summarized in the table below. These numbers are for total mass returned to the ISM, rather than just hydrogen and helium mass, and are thus upper limits to the mass available for future star formation. Even if any one of these values has been significantly underestimated, the rate at which existing molecular gas is turned into new stars is 1 – 3 $M_\odot$ $yr^{-1}$, insubstantially changing the depletion time for the molecular gas.

TABLE 1. Gas Replenishment Rates

| Source Type | $M_\odot$ $yr^{-1}$ | Ref. |
|---|---|---|
| Evolved Stars | 0.55 | Knapp (1991) |
| Massive Steller Winds | $\sim$ 0.2 | Bieging (1989) |
| Supernovae | $\sim$ 0.02 | van Buren (1985) |
| Main Sequence Stars | negligible | |
| Total | 0.8 | |

## 3.1. ATOMIC HYDROGEN

Although the reservoir of molecular gas is relatively small, the reservoir of atomic gas is about 5–10 times larger. If all of the atomic gas were available for conversion into molecular gas, the gas depletion time could be increased to 1.5 – 5 $\times$ $10^9$ yr, about 10–50% of the Hubble time, a value that is no longer implausibly short. However, because the distribution of atomic hydrogen is so different from that of the molecular gas (both in the Milky Way and most normal spirals), the gas depletion time near the truncation radius of the molecular gas remains disturbingly short, only about twice the value obtained from consideration of the molecular gas alone, or about 5 – 10 $\times$ $10^8$ yr, even if all of the atomic gas in that region were converted into stars. It seems that the only way around this dilemma is to require that there be a net inflow or infall of atomic gas, a conclusion reached some years back by Lacey and Fall (1985) from an analysis of the star formation history of the disk of the Milky Way. But if the molecular gas is replenished by inflow through the disk, there must be a process that efficiently transfers angular momentum of the gas outward. Furthermore, this process must produce a *relative* motion between the atomic and molecular gas to account for their different radial distributions.

A rough estimate of the inflow velocity can be obtained by requiring $10^9$ $M_\odot$ of atomic gas to replenish the molecular gas inward of 2 $R_\odot$ in 5 $\times$ $10^8$ yr. In this case, gas must travel a mean distance of 4 kpc, the distance outside the solar circle within which there is about $10^9$ $M_\odot$ of atomic gas, at a mean velocity of about 7 $km\,s^{-1}$. This is a measurable velocity, and would be present in existing data, though its effects are rather subtle in most cases. Lacey and Fall (1985) obtain somewhat lower velocities, some of which is the result of making up for the molecular gas depletion by infall as well as inflow and by having significantly higher star formation rates in the past. When account is made of the different assumptions used in their work and the estimates presented here, the estimated flow velocities are similar.

In any event, in order that the molecular gas not be depleted in a small fraction of a Hubble time, the molecular gas be replenished by inflow from the disk or infall from ouside the Galaxy if the current estimates of the star formation rate and gas return rates from stars are approximately correct. In addition, any inflow must be accompanied by angular momentum transfer and the size of the effects should be measurable.

### 3.2. EXPONENTIAL DISKS

We may now ask how long, at the current rate of star formation, would it take to make a disk with the current mass? If the local star formation rate is 3–7 $\times$ $10^{-9}$ $M_\odot$ $yr^{-1}$, and if the local disk surface density is $\sim$ 35 $M_\odot$ $pc^{-2}$ Gould (1996), then with a constant star formation rate, one can build up the disk of the Milky Way in 7–12 Gyr, a plausible amount of time. In other words, if the gas flows and conversion from atomic to molecular gas have both been in a steady state until today and if the efficiency of star formation has been unchanged, then the present exponential stellar disk of stars would have been built up over a Hubble time. The observational evidence is thus consistent with the idea that the stellar disk is exponential because the gas is exponential. On the other hand, the current observational evidence allows a range of past star formation rates and it is desirable to have tighter constraints on the star formation history of the disk. New data on stellar metallicities that might come from large scale photometric surveys, such as the OGLE, MACHO, EROS, and DUO microlensing searches as well as as the Sloan Digital Sky Survey, could produce valuable new data on this important question.

## 4. Evidence for Inward Gas Motions

### 4.1. THE BAR

Bars have long been argued to produce shocks in the gas within them causing inflows (*e.g.,* Sanders and Huntley 1976). The supposition that the Milky Way is a barred galaxy has been around since 1964 (de Vaucouleurs) and has received strong confirmation in the last five years (see Blitz 1994 for a review). Dwek *et al.* (1994) have analyzed the COBE data, have made the first reliable modeling of the bar, and have concluded that the axial ratio of the bar is about $3\frac{1}{2}$:1. Evidently, the Milky Way is a strongly barred galaxy.

Binney *et al.* (1991) have made dynamical arguments for the existence of a bar based on the kinematics of the gas in the central regions of the Milky Way and the recognition that stable gas orbits must be non-intersecting. They showed that the observed gas distributions could be reasonably explained if there were a stellar bar with an extent of about 3–3.5 kpc. Inward of this radius, there are two families of non self-intersecting orbits that correspond to locations where gas could exist for relatively long periods of time. At other locations, the gas would rapidly shock and be driven inward, piling up at stable locations until other effects further change the radial gas distribution (Blitz *et al.* 1993).

If the bar is then responsible for the truncation of the exponential CO distribution in the Milky Way, similar effects should be seen in other barred galaxies. Indeed, millimeter interferometry has shown that barred galaxies always have a resolved central maximum similar to that seen in the Milky Way, often with complex structure, and a relatively empty region out to a location near the ends of the bar, which may or may not show secondary maxima (*e.g.*, Kenney *et al.* 1992; Regan 1996). But if bars redistribute the angular momentum of the gas in the central regions of a galaxy, what would redistribute the angular momentum of the atomic gas in the outer regions of the disk, and what would produce a relative difference in HI vs. $H_2$ flow velocities?

A related question that may be asked is whether the molecular gas seen at the Galactic center is simply the result of the the redistribution of the molecular gas inside the truncation radius by the bar. This is equivalent to asking whether the area under the point closest to the center in Figure 2 is equal to that under the fitted line inside 4 kpc extrapolated to R = 0 kpc. The answer is no. The $H_2$ mass one would expect inside of 4 kpc is $7 \times 10^8$ $M_\odot$. The mass of the nuclear CO disk is $8 \times 10^7$ $M_\odot$ if the X-factor is the same both in the center and in the disk. If anything, changes in the X-factor would imply lower masses in the center than what one infers from the CO emissivity alone (Blitz *et al.* 1986; Sodrowski *et al.* 1994).

### 4.2 EVIDENCE FOR INFALL AND INFLOW

If the star formation in the disk of the Milky Way is in or near a steady state, then one may ask, is there evidence for infall or inflow of HI to make up for the depletion of the molecular gas? There is in fact evidence for both, but it is ambiguous and may be attributable to other phenomena. The high velocity hydrogen clouds, for example (see Wakker 1993 for a review), and the low negative velocity gas observed toward both northern and southern galactic polar caps (Kulkarni & Fich 1985) might both be manifestations of infalling gas. Evidence for inflow might be the negative HI and CO velocities relative to the stars seen in the Galactic anticenter (*e.g.*, Metzger & Schechter 1994). Further work is needed to disentangle a clear infall or inflow signature from other effects.

## References

Bloemen, J.B.G.M., *et al.*1986, A&A, 154, 25
Bieging, J. 1989, in *The Evolution of the Interstellar Medium*, L. Blitz, ed., ASP: San Francisco, 137
Binney, J. *et al.* 1991, MNRAS, 252, 210
Blitz, L. 1994, in *Back to the Galaxy*, S.S. Holt & F. Verter, eds., AIP Press: New York, 98
Blitz, L., Bloemen, J.B.G.M., Hermsen, W., & Bania, T.M. 1985, A&A, 143, 267
Blitz, L., *et al.* 1993, *Nature*, 361, 417
Burton, W.B. & Gordon, M.A. 1978, A&A, 63, 7
Cohen, R.S. 1977, Ph.D. Dissertation, Columbia University
Dame, T. *et al.* 1987, ApJ, 322, 706
Dame, T. 1993, in *Back to the Galaxy*, S.S. Holt & F. Verter, eds., AIP Press: New York, 267.
de Vaucouleurs, G. 1964, in *The Galaxy and the Magellanic Clouds*, eds. F.J. Kerr & Rodgers, A.W., Australian Academy of Science: Sydney, 195
Güsten R., & Mezger, P. 1983, Vistas Astr., 26, 159
Henderson, A.P., Jackson, P.D., & Kerr, F.J. 1982, ApJ, 263 116
Kenney, J., *et al.* 1992, ApJ, 395, L79
Knapp, G.R. 1991 in *Frontiers of Stellar Evolution*, D.L. Lambert ed., ASP: San Francisco, 229
Kulkarni, S.R., & Fich, M. 1985, ApJ, 290, 792
Lacey, C.G. & Fall, S.M. 1985 ApJ, 290, 154
Lyne, A., Manchester, R. & Taylor J. 1985, MNRAS, 213, 613
Spergel, D.N., Malhotra, S., & Blitz, L. 1996, in *Spiral Galaxies in the Near-IR*, ESO-Garching, D. Minnitti & H. Rix, eds., ESO-Garching, in press
Metzger, M.R., & Schechter, P.L. 1994, ApJ, 420 , 177
Miller, G., & Scalo, J. 1979, ApJS, 41, 513
Prantzos, N., & Aubert O. 1995, A&A, 302, 69
Regan, M. 1996, Ph. D. Dissertation, University of Maryland
Sanders, D.B., Solomon, P.M., & Scoville, N.Z. 1984, ApJ, 276, 182.
Sanders, R.H. & Huntley, J.M. 1976, ApJ, 209, 53
Schwartz, P.R., Wilson, W.J., & Epstein, E.E. 1972, ApJ, 186, 529
Sodroski, T., *et al.* 1994, ApJ, 428, 638
Spergel, D.N., Malhotra, S., & Blitz, L. 1996, in *Spiral Galaxies in the Near-IR*, D. Minnitti & H. Rix, eds. ESO-Garching, in press
Strong, A.W., & Mattox, J. 1996, A&A (Lett.), 308, 21
van Buren, D. 1985, ApJ, 294, 567
Wakker, B. 1993, Ph. D. Dissertation, University of Leiden
Wouterloot, J.G.A., Brand, J., Burton, W.B., & Kwee, K.K. 1990, A&A, 230, 21
Wilson, R.W., Jefferts, K.B. & Penzias, A.A. 1970, ApJ, 161, L43
Wilson, W.J., *et al.* 1974, ApJ, 191, 357

# A MOLECULAR WORM IN SCUTUM

T. M. DAME
*Harvard-Smithsonian Center for Astrophysics*
*60 Garden Street*
*Cambridge, MA 02138*

Over the past four years, Pat Thaddeus and I have been using the CfA 1.2 m telescope to carry out a new CO survey of the first Galactic quadrant with a sensitivity per solid angle 3–10 times higher than previous CO surveys of the plane, and 16 times higher than the only previous extensive wide-latitude survey (Dame et al. 1987). Since a preliminary purpose of this survey is the study of molecular gas out of the plane in the inner Galaxy, the observations have so far been confined mainly to several vertical strips across the plane covering a few degrees in Galactic longitude and 6°–8° in Galactic latitude; details of this on-going survey are given in Dame & Thaddeus (1995).

Our survey has revealed faint, filamentary CO emission extending over 200 pc from the plane in many directions; this emission is evidence for a thick, low-density molecular disk in the inner Galaxy ~3 times as wide as that of the dense central $H_2$ layer and comparable in width to the central H I layer (Dame & Thaddeus 1994). We have also found a number of examples of well-defined molecular clouds at high $z$ which appear to have been ejected from active star-forming regions in the plane. This brief contribution discusses one of the more interesting of these clouds, which appears to lie at the top of a molecular "worm" of the sort long observed in H I (e.g., Koo, Heiles, & Reach 1992), but not observed previously in CO.

Figure 1*a* is a 3-dimensional rendering of a surface of constant $T_{mb}$ in a portion of our CO survey below the plane near $l$ = 24°. Although this unusual representation of the data is of little use for quantitative analysis, it displays the large-scale structure and coherence of the high-$z$ feature of interest here better than any single slice or integration through the survey. That feature appears to originate from an intense region near the plane at $l$ = 23.4°, $v$ = 100 km s$^{-1}$ (arrow) and extends at roughly constant longitude to a well-defined cloud at $b = -1.3°$, $v$ = 115 km s$^{-1}$ (labeled). Since this cloud lies ~6 km s$^{-1}$ above the terminal velocity in this direction (Clemens



*W.B. Latter et al. (eds.), CO: Twenty-Five Years of Millimeter-Wave Spectroscopy*, 19–21.

1985), we adopt the subcentral distance of 7.8 kpc. That implies a vertical displacement of 177 pc. Figure 1*a* suggests that this high-$z$ cloud may have been ejected from a region of active star formation near the plane. The region marked by the white spot in Figure 1*a* is one of the most intense in CO in the first Galactic quadrant, lying near the peak of the molecular ring and the tangent of the 4-kpc spiral arm. There is also a pronounced concentration of H II regions in this direction, most falling in the range 70-115 km s$^{-1}$ (see Fig. 2 of Lockman 1989). If our cloud has been ejected, then its $H_2$ mass of $\sim 6 \times 10^4 M_\odot$ and radial velocity $\sim$ 15 km s$^{-1}$ with respect to its parent cloud imply a kinetic energy of $\sim 2 \times 10^{50}$ ergs; its vertical displacement implies a comparable amount of potential energy. With an initial vertical speed of 15 km s$^{-1}$, the cloud could have reached its present height in $\sim 2 \times 10^7$ yrs. In terms of its mass, vertical displacement, and energetics, this high-$z$ cloud is quite similar to the one discussed by Nyman et al. (1987), but is three times further away.

As Figure 1*b* shows, the H I emission in this region displays a similar vertical structure, one which Koo et al. (1992) have identified as a prominent worm (GW23.0-1.6+100 in their catalog). Although the high-$z$ end of the H I worm is displaced by $\sim 1.5°$ in longitude from the CO feature and extends more than twice as high, the H I worm is inclined to the Galactic plane (see Fig. 13 of Koo et al.) and seems to envelope the CO feature at its base. Further, since the CO and H I structures have nearly identical (forbidden) velocities of $\sim$ 115 km s$^{-1}$, there seems little doubt that the two are related, and that what we have detected in CO is the molecular counterpart or core of this H I worm.

## References

Clemens, D. P. 1985, ApJ, **295**, 422

Dame, T. M. et al. 1987, ApJ, **322**, 706

Dame, T. M. & Thaddeus, P. 1994, ApJ, **436**, L173

Dame, T. M. & Thaddeus, P. 1995, Physics of the Interstellar Medium and the Intergalactic Medium, PASP Conference Series, ed. A. Ferrara, C. Heiles, C. F. Mc Kee, & P. Shapiro

Hartmann, D., & Burton, W. B., 1996, "Atlas of Galactic H I Emission" Cambridge University Press (in press)

Koo, B.-C., Heiles, C., & Reach, W. T. 1992, ApJ, **390**, 108

Lockman, F. J. 1989, ApJS, **71**, 469

Nyman, L.-A, Thaddeus, P., Bronfman, L., & Cohen, R. S. 1987, ApJ, **314**, 374

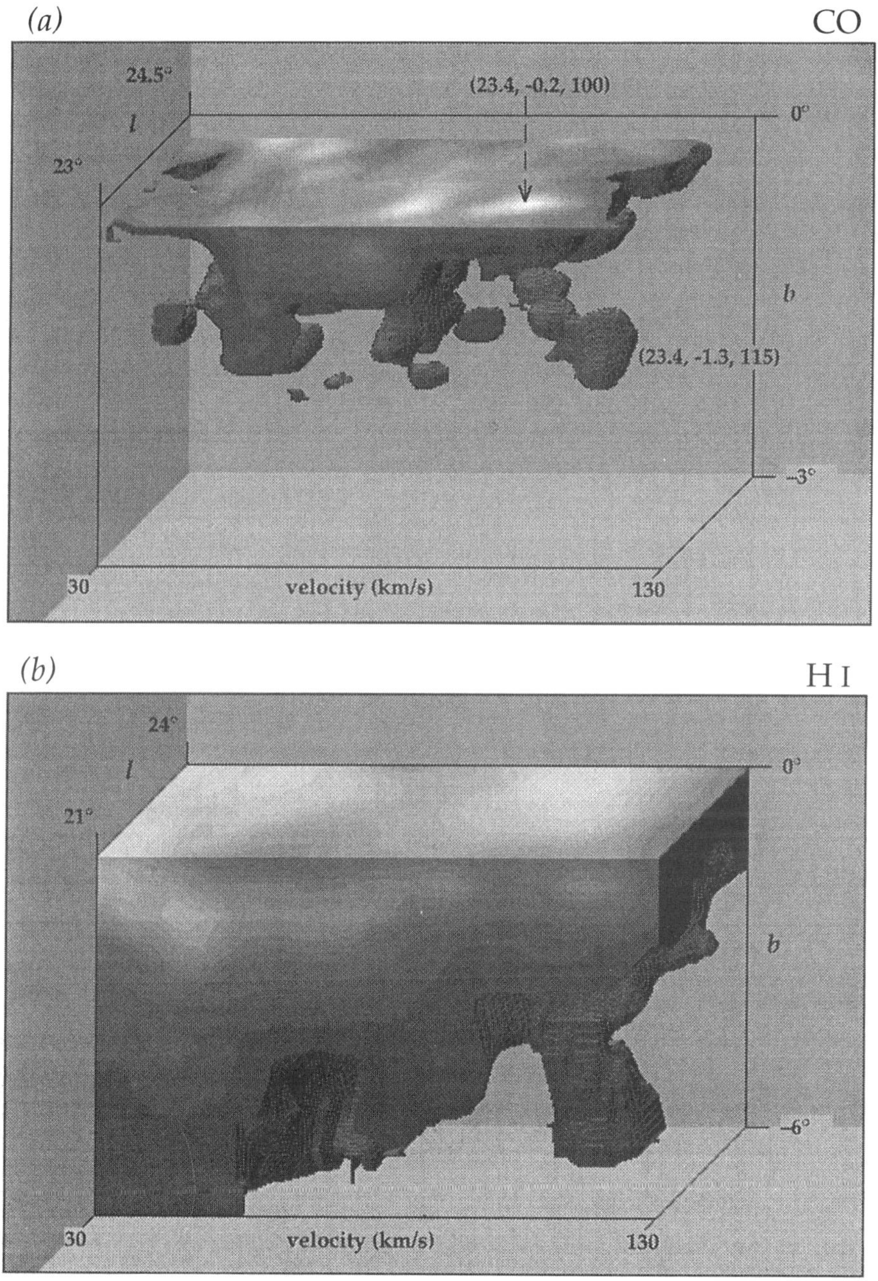


*Figure 1.* *(a)* A shaded isosurface at $T_{mb}$ = 0.28 K from a CO survey of the region $l$ = 23° to 24.5°, $b$ = −3° to −0.2°, and $v$ = 30 to 130 km s$^{-1}$. At the edges of the survey, higher intensities inside the isosurface are represented by a black-to-white grayscale, the brightest point (*white*) at 6.5 K. An intense region near the plane and the high-$z$ cloud are labeled with ($l$, $b$, $v$). *(b)* A similar isosurface at $T_b$ = 2.35 K from the H I survey of Hartmann & Burton (1995), but covering about twice the range of Galactic longitude and latitude.

# MOLECULAR CLOUDS OBSERVED WITH THE EGRET GAMMA-RAY TELESCOPE

SETH. W. DIGEL, STANLEY D. HUNTER,
AND RESHMI MUKHERJEE
*NASA/Goddard Space Flight Center*

EUGÉNE J. DE GEUS
*California Institute of Technology*

ISABELLE A. GRENIER
*Centre d'Etudes Saclay*

ANDREAS HEITHAUSEN
*Universität zu Köln*

GOTTFRIED KANBACH
*Max-Planck-Institut für Extraterrestrische Physik*

AND

PATRICK THADDEUS
*Harvard-Smithsonian Center for Astrophysics*

## 1. Introduction

EGRET, the high-energy $\gamma$-ray telescope on the Compton Gamma-Ray Observatory, has the sensitivity, angular resolution, and background rejection necessary to study diffuse $\gamma$-ray emission from the interstellar medium (ISM). High-energy $\gamma$ rays produced in cosmic-ray (CR) interactions in the ISM can be used to determine the CR density and calibrate the CO line as a tracer of molecular mass. Dominant production mechanisms for $\gamma$ rays of energies ~30 MeV–30 GeV are the decay of pions produced in collisions of CR protons with ambient matter and Bremsstrahlung scattering of CR electrons.

Studies of diffuse $\gamma$-ray emission with EGRET are perhaps the best way to derive $X = N(H_2)/W_{CO}$. Methods for determining $X$ require indirect determinations of molecular column density, such as dust optical depth measurements or virial mass calculations. These methods require assump-



*W.B. Latter et al. (eds.), CO: Twenty-Five Years of Millimeter-Wave Spectroscopy, 22–24.*

tions, such as single, uniform dust temperatures, or virialized clouds, often of questionable validity. The assumptions involved in adopting $\gamma$-ray intensity as a tracer of column density, that high-energy CRs uniformly penetrate atomic and molecular gas and that molecular clouds do not contain $\gamma$-ray sources distributed like the molecular gas, are less serious and more widely applicable. The validity of the latter assumption cannot be proved, but at the resolution of EGRET, observational evidence indicates that embedded sources in nearby molecular clouds would have to be distributed like the gas on the scale of a few parsecs to be unresolved; such sources would be too numerous to be understood.

Interstellar cloud complexes best suited for study with EGRET are nearby, therefore subtending large solid angles, and either far from the plane or toward the outer Galaxy, limiting confusion from diffuse emission at greater distances. To date, the Ophiuchus (Hunter et al. 1994), Orion (Digel, Hunter, and Mukherjee 1995a), and Cepheus Flare complexes, as well as the complexes behind the Flare in the Perseus arm (Digel et al. 1995b), have been studied. A summary of the findings is presented here; the reader is referred to the references cited for important details of the analysis, which is founded on maximum likelihood analyses of diffuse emission with COS-B (e.g., Bloemen et al. 1984; Strong et al. 1988; Grenier and Lebrun 1990), and discussions of the $\gamma$-ray emissivities and CR densities.

## 2. Results and Discussion

Table 1 compares the values of $X$ determined by various methods for the molecular clouds studied with EGRET. The EGRET findings, all for energies $E > 100$ MeV, are in general agreement with non-$\gamma$-ray determinations. The origin of the difference with the COS-B study of Orion by Bloemen et al. (1984) is not clear; however, the model used for COS-B's instrumental background had not been finalized at the time of that study. For COS-B it is an important consideration; the instrumental background intensity is at least as great as that due to the molecular gas, has structure on about the same angular scale, and if incompletely subtracted would increase the value of $X$ derived. The disagreement with the COS-B study of Cepheus is less serious and may be due to the presence of two $\gamma$-ray point sources included in the EGRET analysis that were not resolved by COS-B. The near-agreement of $X$ for the three regions is remarkable on consideration of their large range of masses and different star-formation histories; the best average value for local clouds is $X = (1.0 \pm 0.2) \times 10^{20}$ molecules cm$^{-2}$ (K km s$^{-}$1)$^{-1}$, on scales of $\sim$3 pc and larger.

The EGRET results suggest that $X$ in the Perseus arm may be greater than at the Solar circle. A trend of increasing $X$ with Galactocentric dis-

**Table 1. Determinations of the $X$-ratio**

| | Method | $X^a$ | |
|---|---|---|---|
| *Ophiuchus* | Opt. extinction | $1.1 \pm 0.5$ | Lebrun & Huang (1984) |
| | Dust opt. depth | $> 0.5^{+0.5}_{-0.2}$ | de Geus & Burton (1991) |
| | EGRET | $1.1 \pm 0.2$ | Hunter et al. (1994) |
| *Orion* | COS-B | $2.2 \pm 0.6$ | Bloemen et al. (1984) |
| | $^{13}CO/^{12}CO$ LTE | ~0.9 | Maddalena et al. (1986) |
| | EGRET | $1.06 \pm 0.14$ | Digel et al. (1995a) |
| *Cepheus* | COS-B | $1.9 \pm 1.0$ | Grenier & Lebrun (1990) |
| | EGRET | $0.92 \pm 0.14$ | Digel et al. (1995b) |
| *Outer Galaxy* | CO ($R$ ~12–13 kpc) | ~4 | Mead & Kutner (1988) |
| | CO (12–13 kpc) | $8 \pm 4$ | Digel et al. (1990) |
| | CO (~11 kpc) | 4–6 | Sodroski (1991) |
| | EGRET (~11 kpc) | $2.5 \pm 0.9$ | Digel et al. (1995b) |

[a] Units: $10^{20}$ molecules cm$^{-2}$ (K km s$^{-1}$)$^{-1}$; scaled to the corrected CO radiation temperature scale of Bronfman et al. (1988) if necessary.

tance in the outer Galaxy has been suggested by several CO-only studies (Table 1). The segment of the Perseus arm behind the Cepheus flare contains the most massive molecular clouds in the Perseus arm, the complexes associated with Cas A and NGC 7538. That the $\gamma$-ray $X$-ratio is only marginally determined even there suggests that a far larger segment of the outer Galaxy will have to be studied in order to significantly reduce the uncertainty.

## References


Bloemen, J. B. G. M., Caraveo, P. A., Hermsen, W., Lebrun, F., Maddalena, R. J., Strong, A. W., Thaddeus, P. 1984, *A&A*, **139**, 37

Bronfman, L., Cohen, R. S., Alvarez, H., May, J., Thaddeus, P. 1988, *ApJ*, **324**, 248

de Geus, E. J., Burton, W. B. 1991, *A&A*, **246**, 559

Digel, S., Bally, J., Thaddeus, P. 1990, *ApJ*, **357**, L29

Digel, S. W., Hunter, S. D., Mukherjee, R. 1995a, *ApJ*, **441**, 270

Digel, S. W., Grenier, I. A., Heithausen, A., Hunter, S. D., Thaddeus, P. 1995b, *ApJ*, submitted

Grenier, I. A., Lebrun, F. 1990, *ApJ*, **360**, 129

Hunter, S. D., Digel, S. W., de Geus, E. J., Kanbach G. 1994, *ApJ*, **436**,216

Lebrun, F., Huang, Y. 1984, *ApJ*, **281**, 634

Maddalena, R. J., Morris, M., Moscowitz, J., Thaddeus, P. 1986, *ApJ*, **303**, 375

Mead, K. N., Kutner, M. L. 1988, *ApJ*, **330**, 399

Sodroski, T. J. 1991, *ApJ*, **366**, 95

Strong, A. W., et al. 1988, *A&A*, **207**, 1

# CO AND THE MULTIPHASE ISM

CHRISTOPHER F. MCKEE
*Departments of Physics and Astronomy*
*University of California*
*Berkeley CA 94720*

**Abstract.** CO observations indicate that molecular clouds have a complex multiphase structure, and this is compared with the multiphase structure of the diffuse interstellar medium. The trace ionization within the molecular gas is governed primarily by UV photoionization. Magnetic fields contribute a significantly larger fraction of the pressure in molecular clouds than in the diffuse interstellar medium. Observations suggest that the total Alfvén Mach number, $m_{A\,\mathrm{tot}}$, of the turbulence in the diffuse ISM exceeds unity; Zeeman observations are consistent with $m_{A\,\mathrm{tot}} \lesssim 1$ in molecular clouds, but more data are needed to verify this. Most molecular clouds are self-gravitating, and they can be modeled as multi-pressure polytropes with thermal, magnetic, and wave pressure. The pressure and density within self-gravitating clouds is regulated by the pressure in the surrounding diffuse ISM.

## 1. Introduction

The molecular gas traced by CO is ubiquitous in the Galaxy. In order to understand how this gas fits into the interstellar medium (ISM) and is related to star formation, one must consider a wide range of scales, from the Galaxy as a whole down to small, dense cores that are undergoing gravitational collapse. Large scale phenomena, particularly spiral density waves, are crucial for the formation of molecular clouds (e.g., Elmegreen 1991), whereas the small scale phenomena determine the properties of individual young stellar objects (Shu, Adams, & Lizano 1987). However, phenomena at intermediate scales determine which gas will form stars, the mass of the resulting stars (the initial mass function), and the star formation rate. It is also at these intermediate scales that CO observations provide the most



*W.B. Latter et al. (eds.), CO: Twenty-Five Years of Millimeter-Wave Spectroscopy*, 25–32.

information. These observations have shown that most of the CO is organized into giant molecular clouds (GMCs) that occupy a small fraction of the volume of the ISM (e.g., Scoville & Sanders 1987). Furthermore, the molecular gas within GMCs is confined to clumps that occupy a small fraction of the volume of the GMC (Zuckerman & Palmer 1974).

In this paper, I shall discuss several topics related to CO on intermediate scales: How are molecular clouds related to the multiphase ISM? In particular, what is the relative importance of the magnetic field in molecular gas compared to that in the diffuse ISM? How can molecular clouds and the clumps within them be modeled when several types of pressure—thermal, magnetic, and turbulent—are present? Space precludes discussion of another topic, star formation in a clumpy GMC; the interested reader is referred to Bertoldi & McKee (1996).

## 2. GMCs and the Multiphase ISM

The diffuse interstellar medium exhibits a complex structure that is generally described in terms of a multiphase medium (Field, Goldsmith, & Habing 1969; McKee & Ostriker 1977; McKee 1995). Cold H I clouds ($T \sim 10^2$ K–the CNM) occupy a small fraction of the volume, whereas warm H I ($T \sim 10^4$ K), both neutral (WNM) and ionized (WIM), occupies a significant fraction of the volume. In the three-phase model of the ISM, hot gas ($T \sim 10^6$ K) fills somewhat over half the volume.

GMCs have a complex structure as well, although it is less well understood than that of the diffuse ISM. Much of the mass in GMCs is concentrated in clumps that are apparent in $^{13}$CO and that occupy a few percent of the volume (e.g., Blitz & Shu 1980; Bertoldi & McKee 1992; Williams, deGeus, & Blitz 1994). The nature of the interclump medium is uncertain: both cold H I (Blitz 1991) and warm H I (Falgarone & Puget 1986; McKee 1989) have been suggested. It is thus quite possible that molecular clouds are themselves three phase media, composed of $H_2$ , CNM, and WNM (McKee 1995). Just as there is a substantial range in density observed in the diffuse ISM (Jenkins et al 1983), so too does the density of molecular gas extend over a considerable range. In part, this is due to the effects of self gravity, but very dense molecular gas is also observed on small scales where the effects of self-gravity are negligible (Falgarone et al 1991).

To continue the comparison between the diffuse, atomic ISM and the dense, molecular ISM, consider the energy sources. In the diffuse gas, UV starlight provides the energy for the photoelectric heating that is dominant for both the cold and warm phases (see Wolfire et al 1995 for a recent discussion). The same is true for the outer parts of molecular clouds; in the inner parts, however, which are shielded from the UV radiation, cosmic ray heating dominates. The primary energy source for nonthermal motions in the diffuse ISM is supernovae; in molecular clouds, it is outflows from young stellar objects (Norman & Silk 1980).

Next, consider the source of the ionization. In the diffuse ISM, cosmic rays and X-rays ionize the H I, whereas EUV radiation, primarily from hot stars, maintains the ionization of the H II. In the molecular gas, the ionization is very weak; FUV radiation from stars ionizes the outer parts of molecular clouds, whereas cosmic rays ionize the cores. Most molecular gas in the ISM is in the outer, photoionized regions (McKee 1989).

### 2.1. MAGNETIC FIELDS IN THE DIFFUSE ISM AND IN GMCS

Finally, we consider magnetic fields. For the diffuse ISM, Boulares & Cox (1990) have determined that the total pressure in the midplane of the Galaxy in the solar vicinity is $P_{\rm tot}/k \simeq 28000$ K cm$^{-3}$. Of this, they estimated that cosmic rays contribute about 7000 K cm$^{-3}$ and thermal plus turbulent pressures about 11000 K cm$^{-3}$. As a result, they concluded that the remaining pressure of 10000 K cm$^{-3}$ is due to the magnetic field, corresponding to $B = 5.8$ $\mu$G. (They also considered a model with a higher total pressure and a yet stronger field.) Subsequent analysis has failed to confirm such large magnetic pressures, however. In a recent review, Heiles (1996) concludes that the magnetic field in the solar vicinity consists of a uniform component $B_u \simeq 2.2$ $\mu$G and a random component $B_r \simeq 3.6$ $\mu$G, corresponding to a total field $B_{\rm tot} = (B_u^2 + B_r^2)^{1/2} \simeq 4.2$ $\mu$G. Pulsar observations suggest that the correlation length of the random component of the field is in the range 10–100 pc in the Galactic plane (Ohno & Shibata 1993). Little is known about the structure of $B_r$ out of the Galactic plane; Boulares & Cox suggested that the vertical component of the field $B_z$ (which has a random sign) increases systematically with height so as to account for observations of the Galactic synchrotron radiation.

Because the field has a random component, it exerts a smaller force than a uniform field. The force due to a uniform component of the field $B_u$ is simply $-\nabla B_u^2/8\pi$. However, the large-scale force exerted in the $z$ direction, say, by the random component $B_r$ is $-\partial/\partial z(B_{r\,\rm eff}^2/8\pi)$, where

$$B_{r\,\rm eff}^2 = \langle B_r^2 - 2B_{rz}^2\rangle \tag{1}$$

(Parker 1969; Boulares & Cox 1990). In deriving this result, it has been assumed that the force is averaged over a volume large enough that the averages $\langle B_{rx}B_{rz}\rangle$ and $\langle B_{ry}B_{rz}\rangle$ both vanish. The effective magnetic pressure on large scales is then

$$P_{B\rm eff} = \frac{1}{8\pi}\left(B_u^2 + \frac{1}{3}B_r^2\right), \tag{2}$$

if $\vec{B}_r$ is isotropic (which is reasonable near the Galactic plane at least), so that $B^2_{r\,\rm eff} = B^2_r/3$. For the field strengths found by Heiles (1996), this means that the large-scale effective field pressure is only $P_{B\,\rm eff}/k = 2640$ K cm$^{-3}$, less than 10% of the total pressure. The full magnetic pressure $P_{B\,\rm tot} = B^2_{\rm tot}/8\pi \simeq 5000$ K cm$^{-3}$ is effective only on scales smaller than the correlation length of the field; it is about 18% of the total pressure. In either case, the magnetic pressure is substantially smaller than that inferred by Boulares & Cox.

If the magnetic field does not support the ISM, what does? Most likely, turbulent motions in the interstellar gas provide the necessary support (McKee 1990; Lockman & Gehman 1991). With a thermal pressure of 3600 K cm$^{-3}$ (Jura 1975), the turbulent pressure required is 15000 K cm$^{-3}$. The Alfvén Mach number $m_A$ defined in terms of $B_u$ and the 1-D nonthermal velocity dispersion $\sigma_{\rm nt}$,

$$m_A^2 = \frac{3\sigma^2_{\rm nt}}{v_A^2} = \frac{3}{2}\left(\frac{\rho\sigma^2}{B_u^2/8\pi}\right), \tag{3}$$

is then quite large, $m_A \simeq 4$. It is often stated that the Alfvén Mach number should be less than or of order unity since the dissipation becomes very large for super-Alfvénic motions. However, it is important to distinguish between $m_A$, which is defined in terms of the uniform field $B_u$, and $m_{A\,\rm tot} \propto 1/B_{\rm tot}$, which is defined in terms of the total field strength $B_{\rm tot}$. Once a wave steepens to the point that its wavelength is less than a correlation length in the field, it will feel the full magnetic pressure $B^2_{\rm tot}/8\pi$. Shocks will not form until $m_{A\,\rm tot}$ exceeds unity; for the ISM as a whole, we infer $m_{A\,\rm tot} \simeq 2$. Although $m_{A\,\rm tot}$ exceeds unity, it is not clear that shocks will form, since a significant fraction of the turbulent pressure is in clouds that are moving sub-Alfvénically through the intercloud medium.

In GMCs, the total pressure is much greater than in the diffuse ISM (Scoville & Sanders 1987; Blitz 1991). As a result, the thermal pressure, which appears to be comparable to that in the diffuse medium (Falgarone & Puget 1986), and the cosmic ray pressure represent a smaller fraction of the total pressure in GMCs than in the diffuse ISM. Furthermore, the cosmic ray pressure is approximately uniform, so that cosmic rays exert almost no force. As a result, the total pressure in GMCs is dominated by dynamical pressure and magnetic pressure,

$$P \simeq \rho\sigma^2_{\rm nt} + \frac{B_u^2}{8\pi} + \frac{1}{3}\,\frac{B_r^2}{8\pi}. \tag{4}$$

Here we have assumed that $B_r$ is isotropic; note that $B_r$ may include a time dependent part $\delta B$ associated with wave motions as well as a static part.

Only the uniform component of the field $B_u$ is accessible to observation through the Zeeman effect (Zweibel & McKee 1995).

Analysis of the observations of a number of molecular clouds by Myers & Goodman (1988) showed that typically $m_A \sim 1$; the uncertainties inherent in this result were discussed by Heiles et al (1993). More recently, however, Crutcher et al (1993) searched for the Zeeman effect in OH in a number of molecular clouds and for the most part found only upper limits. Their data show that the harmonic mean $m_A$ in 10 clouds with $1\sigma$ errors less than 4 $\mu$G is 1.5, so that the mean $m_A$ is somewhat higher, perhaps about 2. If the total field is about twice the uniform field, as it is in the diffuse ISM, then the data of Crutcher et al are consistent with $m_{A\,\mathrm{tot}} \lesssim 1$, so that the observed motions are sub-Alfvénic with respect to $B_{\mathrm{tot}}$. In this case, the magnetic pressure must be a substantial fraction of the total pressure in molecular clouds: equation (4) implies that in GMCs

$$\frac{P_{B\,\mathrm{tot}}}{P} \gtrsim \frac{1}{1+\frac{2}{3}m_{A\,\mathrm{tot}}^2}, \tag{5}$$

so that the total magnetic pressure would be at least 60% of the total pressure. It would be very valuable to make more sensitive observations of the magnetic field in molecular clouds in order to verify this.

We conclude that the total magnetic pressure in the diffuse ISM is less than about 20% of the total pressure, whereas in molecular clouds it is typically of order 60%. The increased role of magnetic pressure in molecular clouds is due to two factors: (1) the total pressure is much larger, so the thermal and cosmic ray pressure constitute a smaller fraction; and (2) the Alfvén Mach number associated with the turbulent motions in the clouds appears to be smaller than in the diffuse ISM. This latter effect can be understood in terms of the equipartition arguments developed by Zweibel & McKee (1995; see also Zweibel 1995). They showed that in the absence of damping the kinetic energy in the waves, which is $(3/2)\rho\sigma_{\mathrm{nt}}^2$ after averaging over direction, must equal the potential energy associated with the restoring forces that drive the waves. For transverse waves, which suffer less damping than longitudinal waves, the potential energy consists of the fluctuating field energy, $\delta B^2/8\pi$, and the tidal gravitational energy, which they wrote as $\delta_2 w_G(\mathrm{tidal})$. Equipartition then implies that $m_{A\,\mathrm{tot}}$ is given by

$$m_{A\,\mathrm{tot}}^2 \simeq \frac{\delta B^2}{B_{\mathrm{tot}}^2} + \frac{\delta_2 w_G(\mathrm{tidal})}{B_{\mathrm{tot}}^2/8\pi}. \tag{6}$$

In the diffuse ISM, stars, not gas, dominate the gravitational field. As a result, the tidal term can be larger in the diffuse ISM than in molecular clouds, enabling $m_{A\,\mathrm{tot}}$ to be larger there as well.

## 3. Multi-Pressure Polytrope Models of Molecular Clouds

Most molecular gas in the Galaxy is self-gravitating (Larson 1981), and as a result its pressure is significantly greater than that of the diffuse ISM. Just as in the case of stars, it is therefore convenient to make polytropic models of molecular clouds and the clumps within them by assuming that the pressure scales as some power of the density, $P \propto \rho^{\gamma_p}$ (Dickman & Clemens 1983; Chieze 1987; Maloney 1988; Elmegreen 1989). Previous polytropic models of molecular clouds suffered from the problem that it is difficult to determine the appropriate value of the polytropic index $\gamma_p$. The simplest approach is to assume that the cloud is isothermal ($\gamma_p = 1$); in that case, one finds that the maximum mass that the cloud can have is the Bonnor-Ebert mass, $M_{BE} = 1.18\sigma^4/(G^3P_0)^{1/2}$ (Bonnor 1956; Ebert 1955). Such a model makes no allowance for the effect of magnetic fields or turbulent motions, however, and is in conflict with the observation that the velocity dispersion in a cloud increases with scale (Larson 1981). In response to this problem, Maloney (1988) suggested that $\gamma_p$ should be less than unity, so that the clouds would be hotter on the outside than in the center. The difficulty with this type of model is that the mean pressure in the clouds is then not much greater than at the surface, which again is in conflict with observation.

To overcome these problems, Holliman & McKee (1996; hereafter HM) have introduced multi-pressure polytropes that treat all the important sources of pressure in molecular clouds: thermal motions ($P_{th}$), static magnetic fields ($P_B$), and fluctuating fields and velocities ($P_w$). This last pressure is modeled as being due to a superposition of Alfvén waves (Arons & Max 1975; McKee & Zweibel 1995). The main limitation of this polytropic model is the assumption of spherical symmetry. However, as shown by Mouschovias & Spitzer (1976), spherical models can be used to determine the stability of self-gravitating, magnetized clouds with reasonable accuracy. HM show that in the absence of ambipolar diffusion, static fields can be represented by $\gamma_p = 4/3$; ambipolar diffusion leads to $\gamma_p < 4/3$. The pressure due to the Alfvén waves, $P_w = (3/2)\rho\sigma_{nt}^2$, is isotropic, and it includes both the dynamical pressure and the pressure due to the fluctuating magnetic field (McKee & Zweibel 1995). The polytropic index for the waves is 1/2. At any point, the pressure in the cloud is

$$P(r) = \Sigma K_i \rho(r)^{\gamma_{p,i}}, \tag{7}$$

where the sum extends over the three components of the pressure.

The stability of the polytropes is governed by the adiabatic index $\gamma$, which describes the response of the gas to a compression, as well as by the polytropic index $\gamma_p$. For an isothermal gas, $\gamma = 1$; for the static field,

$\gamma = 4/3$, which ensures that the relative contribution of the field to the support of the cloud is independent of compression; and, for Alfvén waves, $\gamma = 3/2$ (McKee & Zweibel 1995). HM show that a central result from the study of single-pressure polytropes (Chieze 1987, Elmegreen 1989) remains valid for multi-pressure polytropes: the pressure and density within a cloud are regulated by the pressure at its surface. Only magnetically subcritical clouds, which remain stable for arbitrarily high densities, can escape this conclusion. Observed clouds, however, derive a significant amount of support from thermal pressure and waves, and are therefore magnetically supercritical (McKee 1989). Thermal pressure and wave pressure each have $\gamma < 4/3$ and $\gamma_p < 1.2$, and HM show that this places an upper bound on the ratio of central to surface pressures for stable molecular clouds. Molecular clouds are thus the opposite of stars, which can exist in a zero pressure environment.

Observations show that GMCs have relatively large pressure ratios. For self-gravitating clouds, the mean pressure is $\bar{P}/k \simeq 5000\bar{A}_V^2$ K cm$^{-3}$, where $\bar{A}_V$ is the mean extinction through the cloud (McKee 1989). Solomon et al (1987) inferred $\bar{A}_V \simeq 7.5$ in the sample of GMCs they observed. This corresponds to a mean pressure of about $3 \times 10^5$ K cm$^{-3}$, ten times the local interstellar pressure. By contrast, an isothermal sphere has a mean pressure that is at most 2.5 times the surface value. HM show that when the shielding layer of atomic gas around the molecular cloud is taken into account (Elmegreen 1989), multi-pressure polytropes can account for the large observed pressure ratio.

Multi-pressure polytropes have also been used to model low-mass cores. These models improve upon the phenomenological "TNT" models of Myers & Fuller (1992) by including the effects of the static magnetic field and by integrating the equation of hydrostatic equilibrium. HM find models that span a factor of about 100 in density from the center of the core to the surface, considerably greater than the factor 14 possible for stable isothermal clouds that are supported by gas pressure alone. The density profile in the envelope is somewhat steeper than $r^{-2}$, the profile for a singular isothermal sphere. The results for $\sigma_{\rm nt}(r)$ are consistent with the observations of Fuller & Myers (1992) except at the center, where the omission of wave damping in the model leads to a higher value of $\sigma_{\rm nt}$ than observed. Application of the model to high-mass cores gives a flatter $\sigma_{\rm nt}(r)$ relation than for low-mass cores, as observed by Caselli & Myers (1995).

## Acknowledgements

My research is supported by NSF grant AST 92-21289, and, for research on star formation, by a NASA grant to the Center for Star Formation Studies.

## References


Arons, J., & Max, C. E. 1975, ApJ, **196**, L77.

Bertoldi, F., and McKee, C.F. 1992, ApJ, **395**, 140.

Bertoldi, F., and McKee, C.F. 1996, to be submitted to ApJ.

Blitz, L. 1991, in *The Physics of Star Formation and Early Stellar Evolution*, eds. C.J. Lada and N.D. Kylafis (Dordrecht: Kluwer), p.3.

Blitz, L., & Shu, F. H. 1980, ApJ, **238**, 148.

Bonnor, W. B. 1956, MNRAS, **116**, 351.

Boulares, A., & Cox, D. P. 1990, ApJ, **365**, 544.

Caselli, P., & Myers, P. C. 1995, ApJ, 446, 665.

Chieze, J. P. 1987, A&A, **171**, 225.

Crutcher, R. M. 1993, ApJ, **407**, 175.

Dickman, R. L., & Clemens, D. P. 1983, ApJ, **271**, 143.

Ebert, R. 1955, *Z. Astrophys.*, **37**, 222.

Elmegreen, B. G. 1989, ApJ, **338**, 178.

Elmegreen, B. G. 1991, ApJ, **378**, 139.

Falgarone, E., & Puget, J. L. 1986, A&A, **162**, 235.

Falgarone, E., Phillips, T. G., & Walker, C. K. 1991, ApJ, **378**, 186.

Field, G. B., Goldsmith, D. W., & Habing, H. J. 1969, ApJ, **155**, L149.

Fuller, G. A., & Myers, P. C. 1992, ApJ, **384**, 523.

Heiles, C. 1996, in *Polarimetry of the Interstellar Medium*, ed. W. G. Roberge & D. C. B. Whittet (San Francisco: ASP), in press.

Heiles, C., Goodman, A., McKee, C. F., & Zweibel, E. 1993, in *Protostars and Planets III*, eds. E. Levy and J. Lunine (Tucson: University of Arizona Press), p.279.

Holliman, J. H., & McKee, C. F. 1996, ApJ, to be submitted (HM).

Jenkins, E. B., Jura, M., & Loewenstein, M. 1983, ApJ, **270**, 88.

Jura, M. 1975, ApJ, **197**, 581.

Larson, R. B. 1981, MNRAS, **194**, 809.

Lockman, F. J., & Gehman, C. S. 1991, ApJ, **382**, 182.

Maloney, P. 1988, ApJ, **334**, 761.

McKee, C. F. 1989, ApJ, **345**, 782.

McKee, C. F. 1990, in *The Evolution of the Interstellar Medium*, ed. L. Blitz (San Francisco: ASP), p. 3.

McKee, C. F. 1995, in *The Physics of the Interstellar Medium and Intergalactic Medium*, ed. A. Ferrara, C. F. McKee, C. Heiles, & P. Shapiro (San Francisco: ASP), p. 292.

McKee, C. F., & Ostriker, J. P. 1977, ApJ, **218**, 148.

McKee, C. F., & Zweibel, E. G. 1995, ApJ, **440**, 686.

Mouschovias, T., & Spitzer, L. 1976, ApJ, **210**, 326.

Myers, P. C., & Fuller, G. A. 1992, ApJ, **392**, 631.

Myers, P. C., & Goodman, A. A. 1988, ApJ, **326**, L27.

Norman, C. A., & Silk, J., 1980, ApJ, **238**, 158.

Ohno, H., & Shibata, S. 1993, MNRAS, **262**, 953.

Parker, E. N. 1969, *Space Sci. Rev.*, **9**, 651.

Scoville, N. Z., & Sanders, D. B. 1987, in *Interstellar Processes*, eds. D. Hollenbach and H. Thronson (Dordrecht: Reidel), p.21.

Shu, F. H.. Adams, F. C., & Lizano, S. 1987, ARA&A, **25**, 23.

Solomon, P. M., Rivolo, A. R., Barrett, J. W., & Yahil, A. 1987, ApJ, **319**, 730.

Williams, J. P., deGeus, E. J., & Blitz, L. 1995, ApJ, **428**, 693.

Wolfire, M. G., Hollenbach, D. J., McKee, C. F., Tielens, A. G. G. M., & Bakes, E. L. O. 1995, ApJ, **443**, 152.

Zuckerman, B., & Palmer, P. 1974, ARA&A, **12**, 279.

Zweibel, E. G. 1995, in *The Physics of the Interstellar Medium and Intergalactic Medium*, ed. A. Ferrara, C. F. McKee, C. Heiles, & P. Shapiro (San Francisco: ASP), p. 524.

Zweibel, E. G., & McKee, C. F. 1995, ApJ, **439**, 779.

# THE URSA MAJOR MOLECULAR CLOUDS

MARC W. POUND
*Radio Astronomy Laboratory, University of California*
*Berkeley, CA 94720*

AND

ALYSSA A. GOODMAN
*Harvard-Smithsonian Center for Astrophysics*
*Cambridge, MA 02138*

## 1. Introduction

The Ursa Major molecular cloud complex lies in the direction of an expanding HI shell known as the North Celestial Pole loop. The NCP loop, which is centered at $(l, b) \sim (138°, 30°)$ and easily seen in IRAS 100 $\mu$m emission, is some 60 pc across and 150 pc distant (Meyerdierks et. al 1991). At 100 $\mu$m, the Ursa Major clouds appear in projection as "finger" ($l \sim 140, b \sim 38$) which "hangs down" towards the center of the loop and the plane of the Galaxy. Distance estimates to the molecular clouds (Penprase 1993) are consistent with that of the NCP loop, indicating that the clouds are physically associated with the loop.

Figure 1 shows the CO integrated intensity map, ($I$(CO)), of the Ursa Major complex. The CO spectra ($\sim$ 22000) were obtained with the AT&T Bell Labs 7 m telescope (HPBW = $1'.7$), and were spaced at $1'$. The $H_2$ mass of the region mapped in CO is about 220 solar masses assuming a relation between $I$(CO) and $H_2$ column density $N(H_2)/I(CO) = 2.1 \times 10^{20}$ (K km/s)$^{-1}$ cm$^{-2}$. The structure of the Ursa Major complex is dominated by several long ($> 2$ pc) filaments, some of which are both extremely straight and extremely narrow ($< 0.2$ pc across). The filaments, clumps, and "bow shocks" seen in the CO maps correlate in detail with features in the IRAS 100 micron flux and opacity maps and, coupled with the general appearance of the atomic gas, suggest a recent shock event. In channel maps (not shown) the lower tip of the CO filaments have a bow shape,



*W.B. Latter et al. (eds.), CO: Twenty-Five Years of Millimeter-Wave Spectroscopy,* 33–35.

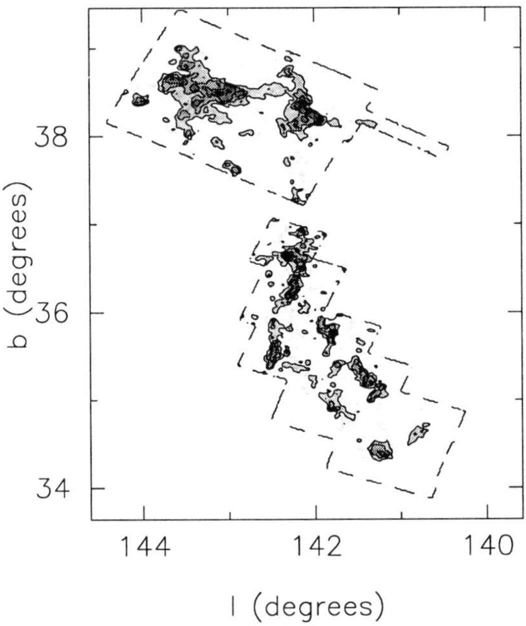


*Figure 1.* CO integrated intensity I(CO) from $V_{LSR}$ = −2 to −3.5 km/s. Contours are 1.5, 3.0, ..., 7.5 K km/s. The data have been smoothed to 4′.

suggesting the gas is being compressed by a wind from the direction of the Galactic plane.

In HI maps made with the Hat Creek (HPBW = 36′) and Effelsberg (HPBW = 9′) telescopes, the atomic gas appears to form a sheath around the CO filaments (see Myers et. al 1995). Figure 2 summarizes the HI and CO kinematic data. In the atomic component (upper plots in Fig. 2), there is a clear correlation of *both* FWHM line width $\Delta V$ and LSR velocity with $b$. In the CO data (lower plot in Fig. 2), the correlations are also present but with much more scatter. These data suggest that the filamentary portion of the Ursa Major complex is arc-shaped when viewed perpendicular to our line of sight, and that this arc is slowly tumbling back down toward the Galactic plane (see Figure 3). A filament rotating as a "solid body" (in the direction of the curved arrows in Fig. 3) would give rise to the observed increase in $V_{LSR}$ with $b$. Two differing components of the line width, along the filament ("tangential") and across the filament ("radial"), would explain the increase of $\Delta V$ with $b$. Depending on the orientation of the filament with respect to the observer, $\Delta V_{\rm tan} >> \Delta V_{\rm rad}$ (upper picture in Fig. 3) or $\Delta V_{\rm tan} << \Delta V_{\rm rad}$ (lower picture in Fig. 3). Alternatively, the correlation of $V_{\rm LSR}$ with $b$ could be explained by an expanding filament rather than a rotating one. However, the loci of temperature in position-velocity diagrams do not show the curvature that signify expansion.

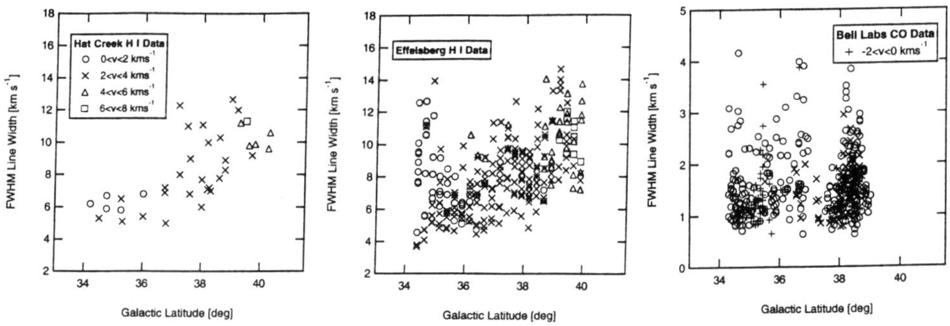


*Figure 2.* Summary of HI and CO kinematic data. Symbols represent $V_{LSR}$ range and have same value in each plot. See text for explanation.

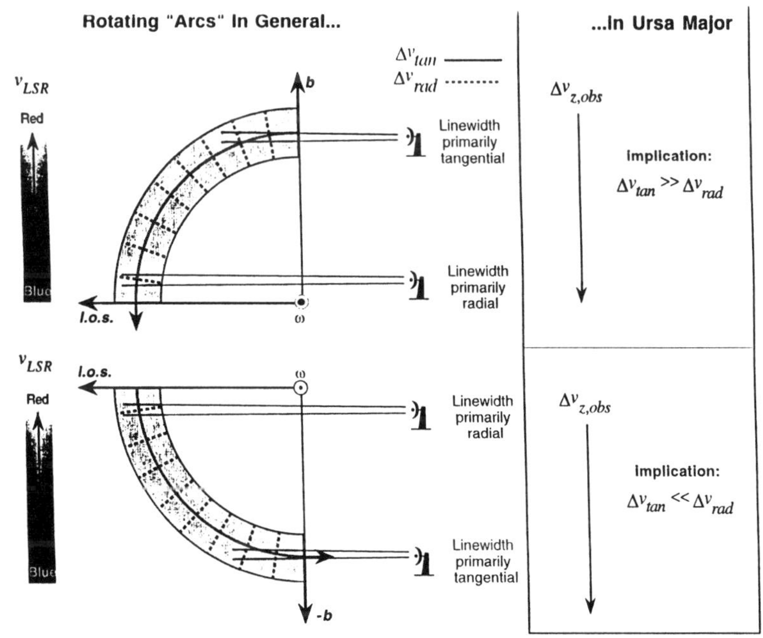


*Figure 3.* Two possible geometries of the HI filament consistent with the data in Figure 2. See text for explanation.

# FCRAO CO SURVEY OF THE OUTER GALAXY

MARK H. HEYER
*Five College Radio Astronomy Observatory*
*Lederle Research Building*
*University of Massachusetts at Amherst*
*Amherst, MA 01003*

## 1. Introduction

FCRAO has initiated an imaging survey of CO $J$=1–0 emission from the outer Galaxy to better understand the physical properties of the molecular gas component of the interstellar medium and to determine the relationship of this component to spiral arm features. When completed in Spring 1996, the CO Survey will cover $103° < l_{II} < 145°$, $-3° < b_{II} < 5°$, and $-150 < V_{LSR} < 50$ km s$^{-1}$ with angular sampling of 50″ and velocity resolution of 0.8 km s$^{-1}$. During the first year of observations, we have focused upon two fields with longitude ranges of $103° < l_{II} < 112°$ and $131° < l_{II} < 139°$. The first field corresponds to the Cepheus region while the second field contains the W3, W4, and W5 complex of HII regions. To date, 700,000 CO spectra have been collected. In this contribution, we present preliminary results from the first year of observations.

## 2. Results

An image of peak antenna temperature ($T_A^*$) of CO $J$=1–0 emission from the Cepheus region over the $V_{LSR}$ range of $-65$ to $+5$ km s$^{-1}$ is shown in Figure 1. It includes emission from molecular clouds within the solar neighborhood (distance < 1 kpc) in addition to clouds associated with the Perseus spiral arm located at a distance of 3.5–5 kpc. The bright, high contrast emitting regions within the image are exclusively associated with luminous far infrared sources and optical HII regions. In most cases, the far infrared sources are not the exciting stars of the optical nebulae but rather, newborn stars still deeply embedded within the cloud cores.



*W.B. Latter et al. (eds.), CO: Twenty-Five Years of Millimeter-Wave Spectroscopy*, 36–38.

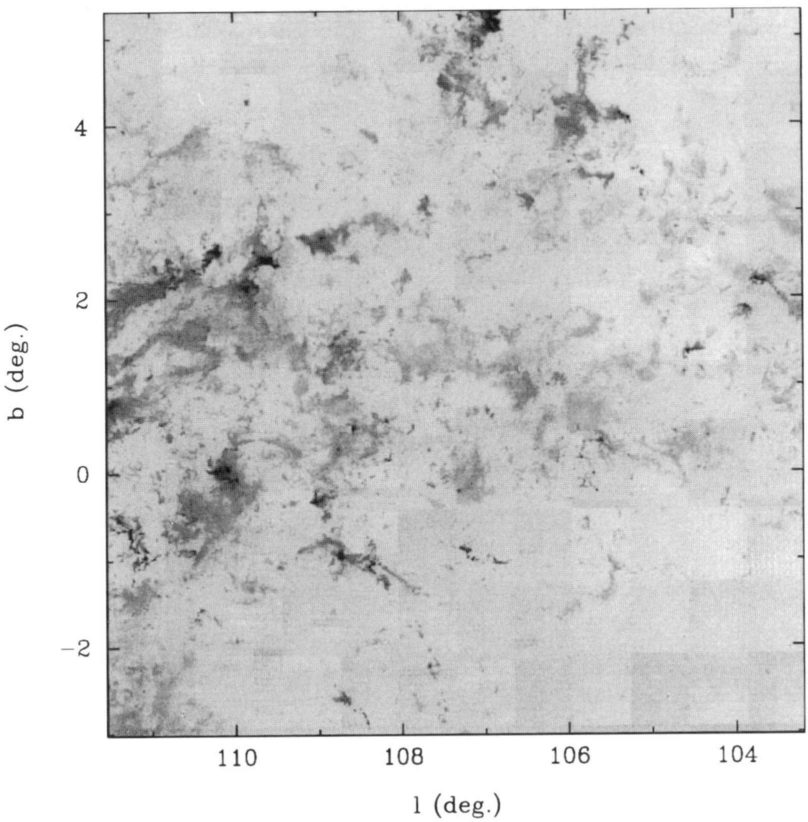


*Figure 1.* Peak antenna temperature of CO $J$=1–0 emission from the Cep region of the plane of the Galaxy. The image is constructed from 360,000 spectra.

The images provide an unprecedented view into the internal structure and kinematics of the molecular interstellar medium. We have identified 1140 "clouds" which are defined as closed surfaces of antenna temperature within the $l$-$b$-$v$ data cube at a threshold of 2 K ($T_R^*$). This is similar to the cloud definitions used by Solomon et al. (1987) and Scoville et al. (1987) but at a much lower threshold of antenna temperature. The distributions of velocities for the identified objects within the two regions are shown in Figure 2. There are strong enhancements in the number of objects at the velocities −10 and −45 km s$^{-1}$ which correspond to the local spiral arm and the Perseus spiral arm respectively. Within the spiral arms, there is no evidence for large scale coherent velocity structure although radial streaming motions of Perseus spiral arm clouds have been inferred by many studies (see Brand and Blitz 1993). We estimate the one dimensional cloud-cloud velocity dispersion to be 4–6 km $s^{-1}$ . This includes both random and streaming, non-circular motions within the spiral arms.

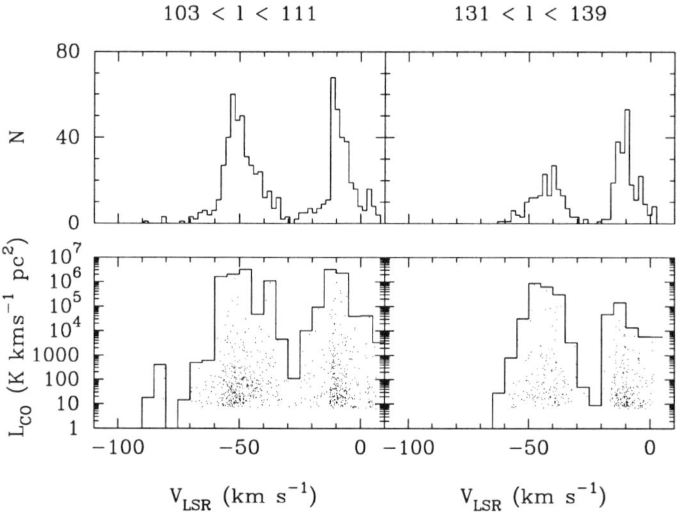


*Figure 2.* The distribution of velocities for the objects identified within the 2 observed fields. (bottom) CO luminosity (points) as function of $V_{LSR}$ . The solid lines are the sum of all CO luminosities within 5 km $s^{-1}$ bin.

Assuming a flat rotation curve in the outer Galaxy, $\Theta_0$=220 km $s^{-1}$ and $R_0$=8.5 kpc, kinematic distances and CO luminosities are calculated for each identified object. The bottom panels of Figure 2 show the CO luminosity $L_{CO}$, for each object as a function of $V_{LSR}$ and the sum within each velocity bin. There is a similar strong enhancement of $L_{CO}$ at the nominal spiral arm velocities. Although we detect clouds at velocities which correspond to interarm regions, the arm-interarm contrast is large in the outer Galaxy. These figures also show that within a given $V_{LSR}$ bin, the largest contribution to the total CO luminosity is from several large clouds. Only a small fraction of the total CO luminosity originates from the low level "chaff" of the molecular interstellar medium. Thus, most of the mass of the molecular interstellar medium resides within the small number of very massive clouds.

## References

Brand, J. & Blitz, L. (1993) The velocity field of the outer Galaxy, *AA*, **275**, pp. 67–90.

Scoville, N. Z., Yun, M. S., Clemens, D. P., Sanders, D. B., & Waller, W. H. (1987) Molecular Clouds and Cloud Cores in the Inner Galaxy, *ApJ Suppl*, **63**, pp. 821–915.

Solomon, P. M, Rivolo, A. R., Barret, J., & Yahil, A. (1987) Mass, Luminosity, and Line Width Relations of Galactic Molecular Clouds, *ApJ*, **319**, pp. 730–741.

# THE CO 2–1/1–0 RATIO

TETSUO HASEGAWA
*Institute of Astronomy, The University of Tokyo,*
*Osawa 2-21-1, Mitaka, Tokyo 181, Japan*

**Abstract.** A Galactic CO $J$=2–1 survey compatible with the Columbia CO $J$=1–0 survey has been conducted with a 60-cm telescope at Nobeyama. Molecular clouds in the solar neighborhood exhibit variations of the $J$=2–1/$J$=1–0 intensity ratio from place to place in a cloud and from one cloud to another. This mainly reflects a variation of the hydrogen density near the *CO photospheres*, where the opacity of the lines reaches $\sim 1$ at cloud surfaces. The ratio often exceeds unity at molecular cloud-HII region interfaces. The Milky Way survey has shown that the average $J$=2–1/$J$=1–0 ratio in volume emissivity increases gradually from 0.6 in the solar neighborhood to 0.75 at 4 kpc from the Galactic center. This trend is due to an increased contribution of the gas with higher $J$=2–1/$J$=1–0 ratio ($\geq 0.7$, *High Ratio Gas*) that concentrates in spiral arms in the inner Galaxy. The amount of *High Ratio Gas* there is too large to be explained as molecular gas under direct influence of HII regions; rather, it may be a manifestation of dense molecular gas that is ready to form massive stars.

## 1. Introduction — Basic Problems

From millimeter wave spectroscopy, we have learned the amount of molecular gas in the Galaxy, its radial distribution, and its spatial structure, especially its concentration to giant molecular clouds (Scoville & Solomon 1975; Burton et al. 1975; Sanders, Solomon, & Scoville 1984; for a recent review, see, e.g., Dame 1993). A natural next step to take is a study of its physical conditions, i.e., its density, temperature, etc. To understand star formation that varies from place to place, we need a knowledge of gas conditions and their variation on spatial scales ranging from dense molecular cloud cores ($\leq 1pc$) to a galaxy (over 10 kpc).



*W.B. Latter et al. (eds.), CO: Twenty-Five Years of Millimeter-Wave Spectroscopy,* 39–46.

The physical conditions of gas in selected regions of molecular clouds have been studied (e.g., Goldsmith 1987). On a larger scale, gas conditions in the central kiloparsec regions in galaxies are being investigated (e.g., Braine & Combes 1992; Aalto-Bergman in these proceedings). In the IRAS luminous galaxies, the massive central concentrations of dense molecular gas may be responsible for the extremely intense star formation. However, limited spatial resolution often makes it difficult to convert the observed information to the physics of molecular gas.

Our knowledge of the change of the gas condition on a scale range that links cloud internal structures and galactic structures is poor. To fill this gap, we have conducted a survey of Galactic CO $J$=2–1 emission at 230 GHz with a dedicated 60-cm radio telescope at Nobeyama (Hayashi et al. 1990; Handa et al. 1993). The survey was designed to be compatible with the Columbia survey of the $J$=1–0 transition (Cohen, Dame, & Thaddeus 1986; Dame et al. 1987); the telescopes have the same beamsize (HPBW = 9′), and in most cases we observed on the same grid as the 1–0 observations. These cases enable us to measure the 2–1/1–0 line intensity ratio accurately.

In this paper, with the data taken in the last a few years at hand, we address the following questions:

- What is the CO 2–1/1–0 ratio in various clouds with and without massive star formation?
- What makes the ratio high or low?
- How can we understand its galaxy scale variation in terms of formation of clouds and stars?

## 2. Lessons from Nearby Molecular Clouds

It is often stated that CO lines are optically thick and are in local thermodynamical equilibrium (LTE). If this is the case, the line brightness of any low-$J$ rotational transition becomes $B_\nu(T_\mathrm{k})$, the Planck function at a gas kinetic temperature $T_\mathrm{k}$. However, this textbook understanding is correct only partially. High resolution maps of molecular clouds reveal many clumps and filaments within them. The CO lines get optically thick at small depths at their surfaces. We call these regions *CO photospheres* where most of the CO line photons reaching us are emitted. The gas density there is not necessarily high enough to excite the levels to the LTE limit.

What do we probe by measuring the CO 2–1/1–0 line intensity ratio? Although intensities of both lines roughly scale with the column density of gas, the $J$=2–1 emission samples warmer and denser regions compared with the $J$=1–0 emission. By taking their ratio, we can cancel the strong dependence on the *quantity* of gas to examine its *quality*, i.e., the conditions of the gas near the photosphere.

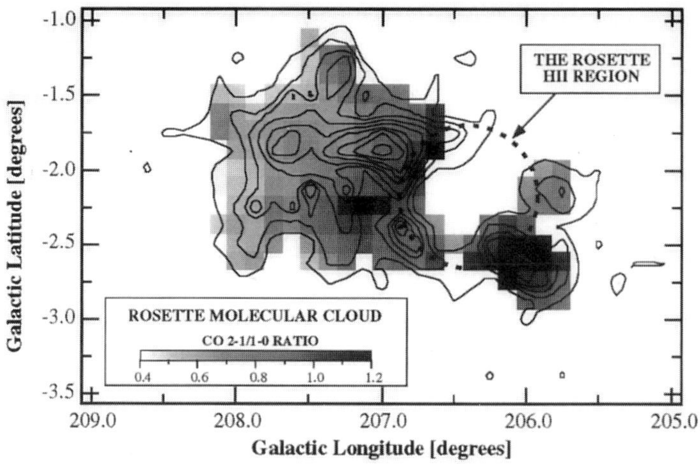


*Figure 1.* The Rosette Molecular Cloud in CO $J$=2–1 integrated intensity (*contours*) and the 2–1/1–0 ratio (*grayscale*). The $J$=1–0 data from Blitz & Thaddeus (1980) and May et al. (1993) were used to derive the 2–1/1–0 ratio. (Morino et al. 1996)

We can get insight into these problems by visiting nearby molecular clouds. Sakamoto et al. (1994) mapped the Orion A and B clouds, the nearest examples of massive ($> 10^5$ M$_\odot$), giant molecular clouds (GMCs) with ongoing massive star formation. The map of the CO 2–1/1–0 intensity ratio formed from the $J$=1–0 data by Maddalena et al. (1986) shows its global variation over the surface of the cloud (Figs. 2*a*, *b* in Sakamoto et al. 1994). In the central ridges of the clouds, the ratio is close to, but smaller than, unity (0.9 typically), and becomes smaller ($\approx 0.5$) in their peripheral regions.

Figure 1 shows another GMC associated with an HII region, the Rosette molecular cloud (Morino et al. 1996). The CO 2–1/1–0 intensity ratio exceeds unity in the molecular shell immediately outside the HII region. The thickness of the gas shell with this very high ratio ($> 1.0$) is less than 10 pc. Away from this shell, the ratio is 0.7–0.6 in the region of high CO intensity and gets lower ($< 0.5$) in the cloud boundary to the southeast (to the left in Fig. 1).

As a prototype of less massive ($\sim 10^4$ M$_\odot$) dark clouds with low-mass star formation, the Taurus molecular cloud complex has been observed (Hayashi et al. 1996). The CO 2–1/1–0 ratio derived with the $J$=1-0 data by Ungerechts & Thaddeus (1987) shows the common systematic variation that the ratio is higher (0.6) in the ridges of the clouds while it is lower (0.4) in their outskirts. It is interesting to note that known protostars and T Tauri stars are located in the regions of higher 2–1/1–0 ratio.

Figure 2 shows the distribution of the 2–1/1–0 intensity ratio in the

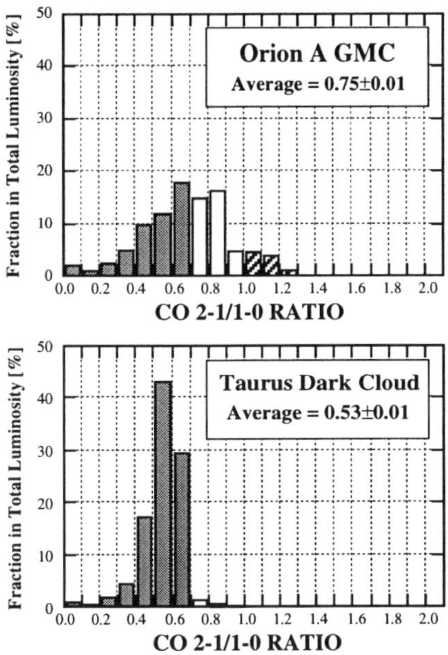


*Figure 2.* The distribution of the CO 2–1/1–0 ratio in the Orion A giant molecular cloud (*upper*) and the Taurus dark cloud (*lower*). Spatial resolution is 1.2 pc in both cases. Gas with *Low Ratio*, *High Ratio*, and *Very High Ratio* explained in the text are shown shaded, open, and hatched, respectively. CO 2–1/1–0 luminosity ratios of the entire clouds are shown as average ratios; the quoted errors do not include common antenna calibration errors ($\sim 10$ %).

clouds in Orion and Taurus. As the CO luminosity scales with the mass of molecular gas to first order, the Figure shows the mass fraction of molecular gas at certain conditions indicated by the CO 2–1/1–0 ratio. Differences between the two clouds are clear: In the Taurus dark cloud, most emission comes from the gas with a ratio smaller than 0.7. We call it *Low Ratio Gas*. In the Orion A GMC, on the other hand, a significant fraction of the luminosity comes from the gas with the ratio 0.7–1.0 (*High Ratio Gas*) and that with the ratio larger than 1.0 (*Very High Ratio Gas*). The 2–1/1–0 CO luminosity ratios in these clouds are differ accordingly; 0.53 average for Taurus and 0.75 for Orion.

What do these observations mean? A variation in the temperature of LTE gas cannot explain *Low Ratio Gas* unless the gas temperature gets significantly below 10 K. Sakamoto et al. (1994) modeled the variation of the 2–1/1–0 intensity ratio as due mainly to a variation of density at the CO photospheres of clumps that compose the clouds. The photon trapping effect places the CO $J$=1–0 and $J$=2–1 photospheres near each other (especially at low densities and temperatures), and this justifies our limited use of one-zone large velocity gradient (LVG) model calculations in excitation analyses of the CO lines there. In this scheme, *Low Ratio Gas* is in the form of clumps with extended envelopes, whose CO photospheres occur at low density. On the contrary, *High Ratio Gas* is in tightly confined clumps with a steep density gradient, which have CO photospheres at higher density.

The above model cannot explain *Very High Ratio Gas* often accompanied by relatively bright $^{13}$CO emission. This may reflect small-scale clumping and/or a steep temperature gradient at their surfaces that are expected in photodissociation regions (Gierens, Stutzki, & Winnewisser 1992).

## 3. The Milky Way Survey

With the understanding of the CO 2–1/1–0 ratio acquired from nearby molecular clouds described above, we can now analyze the Milky Way survey data to see how the conditions of molecular gas vary with location in the Galaxy.

The CO $J$=2–1 survey of the first Galactic quadrant was made in three parts and covered up to 40% of the inner galactic disk between 2 kpc and 8 kpc from the Galactic center:

- The pilot survey (Handa et al. 1993): $l = 2° - 88°$ sampling at $\Delta l = 1°$ along the Galactic plane ($b = 0°$)
- Two dimensional survey, Part I (Sakamoto et al. 1995): $l = 20° - 60°$, $|b| \leq 1°$, at $\Delta l = \Delta b = 0.25°$ sampling
- Two dimensional survey, Part II (Sorai et al. 1996): $l = 6° - 20°$, $|b| \leq 1°$, at $\Delta l = \Delta b = 0.25°$ sampling ($\Delta l = \Delta b = 0.125°$ for $|b| \leq 0.25°$)

One of the first questions to answer is whether the average condition of the molecular gas in the Galactic disk varies with Galactocentric radius $R_{\rm G}$. Handa et al. (1993) analyzed the pilot survey data with the assumption of axisymmetry and circular rotation at a constant speed. The result reproduced by filled circles in Figure 3 showed a gradient of the CO 2–1/1–0 ratio with $R_{\rm G}$ for the first time.

Similar analyses of the two dimensional surveys improved the spatial sampling by a large factor and confirmed the variation (open symbols in Fig. 3). In these cases we are limited not by observational errors but by intrinsic fluctuations of molecular gas in the disk; we can see this in the discrepancy at $R_{\rm G} > 7$ kpc, where a small number of clouds in the near side dominate an average at constant $R_{\rm G}$. Nevertheless, we see the trend of the CO 2–1/1–0 ratio increasing gradually from 0.6 ($\pm$0.1) in the solar neighborhood to 0.75 ($\pm$0.05) at $R_{\rm G} = 4$ kpc.

In working with small telescopes, beam dilution is a serious problem. Indeed, if we were to identify molecular clouds from our data, incompleteness posed by our spatial resolution and sampling of a few $\times$10 pc at $R_{\rm G} \approx 4$

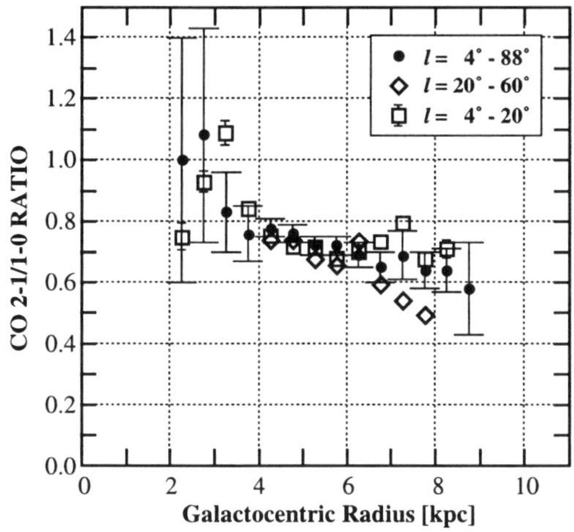


*Figure 3.* The 2–1/1–0 ratio of the CO volume emissivity as a function of Galactocentric radius, $R_G$. We adopt $R_G = 8.5$ kpc for the Sun. The three marks correspond to the three surveys explained in the text. The $J$=1–0 data from Cohen, Dame, & Thaddeus (1986) were used to derive the 2–1/1–0 ratio. Indicated errors do not include common antenna calibration errors ($\sim$ 10 %).

kpc would be a large problem. We can avoid this difficulty by measuring the CO line intensities in volume emissivity, the luminosity in unit volume of space in the Galactic disk. The ratios shown in Figure 3 are in volume emissivity, and we are free from these problems.

If we take these 2–1/1–0 ratios superficially and apply the one-zone LVG analysis described in the preceding section, the gradient could be interpreted as an increase of the "mean" photospheric density of molecular clouds in the inner Galaxy. However, molecular gas contributing to the CO luminosity in the Galactic disk is highly heterogeneous.

As the first approximation, we view the gas as composed of *High Ratio Gas* (here we include *Very High Ratio Gas* in it) and *Low Ratio Gas*. The former is found in bright ridges in GMCs, while the latter is in dark clouds and in the peripheries of GMCs. Sakamoto et al. (1996) binned the data with a $\Delta V_{\rm LSR} = 5$ km s$^{-1}$ resolution, and classified volume elements with detected CO emission in the $l$-$b$-$V_{\rm LSR}$ datacube into these two categories. This analysis showed that the higher 2–1/1–0 CO volume emissivity ratio at smaller $R_G$ is a reflection of an increased contribution of *High Ratio Gas* that is concentrated in spiral arms of the inner Galaxy.

It is interesting to ask whether the gradient in the CO 2–1/1–0 ratio continues into the Galactic center region. The observations show that, although abundant *Very High Ratio Gas* is found in the midplane, a large amount of *Low Ratio Gas* off the plane counteracts to make the total 2–

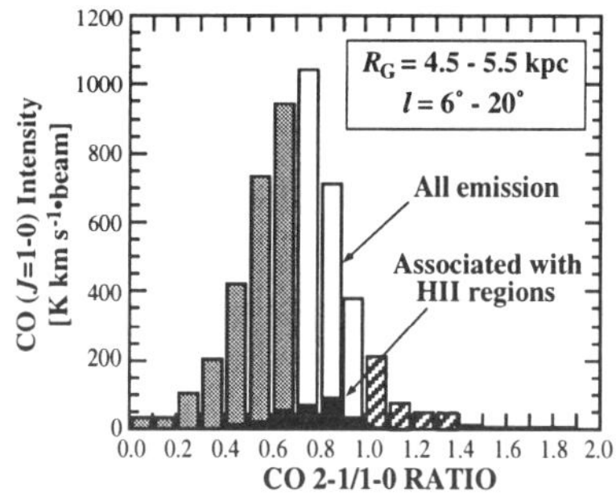


*Figure 4.* The distribution of the CO 2–1/1–0 ratio in the gas located in the Galactic disk at $R_G = 4.5 - 5.5$ kpc and $l = 6° - 20°$. Filled bars show the CO $J$=1–0 emission coincident in position (within $\sim$ 10 pc) and velocity (within $\pm 7.5$ km s$^{-1}$) with radio HII regions catalogued by Lockman (1989). (Sorai et al. 1996)

1/1–0 CO luminosity ratio as low as 0.65 for all emission within $|l| \leq 2.5°$, $|b| \leq 1.0°$, and $|V_{LSR}| \leq 150$ km s$^{-1}$ (Oka et al. 1996a, b).

A survey of the outer Galaxy is in progress. In the Perseus arm at $l = 130° - 151°$, the volume emissivity of *Low Ratio Gas* is extremely low (Sato et al. in these proceedings). This may be related to the fact that the $H_2$/HI abundance ratio turns over at this $R_G$.

## 4. Massive Star Formation and High Ratio Gas

In the solar neighborhood, *High Ratio Gas* has been found only in GMCs so far. It may show surfaces of tightly bound clumps in bright ridges of molecular clouds, where we see many young stars and protostars of intermediate mass. In the Galactic disk, we found that *High Ratio Gas* concentrated along spiral arms. Because massive star formation highlights spiral arms, they may have a causal relationship. Is *High Ratio Gas* a cause or an effect of massive star formation?

As we have seen in Orion and Rosette GMCs, massive star formation results in *Very High Ratio Gas* around HII regions. However, it is limited to $<$ 10 pc from the molecular cloud-HII region interface, and most of the *High Ratio Gas* lies beyond this range of direct effect. It appears that *High Ratio Gas* is *not* a direct effect of OB star formation in GMCs but is related to the massive star formation expected in them. A comparison between the CO 2–1/1–0 ratio and masses of dense cores in various types of clouds would be interesting, because a theory predicts a close relationship between the core mass and the mass of a star that forms in it (Nakano, Hasegawa, & Norman 1995).

For the molecular gas in the Galactic disk we can test the one-to-one correspondence between *High Ratio Gas* and HII regions observed in radio wavelengths in the $l$-$b$-$V_{\rm LSR}$ datacube. In general, HII regions are located in or near *High Ratio Gas*, but there is much more *High Ratio Gas* without coincident HII regions (Sakamoto et al. 1996; Sorai et al. 1996). Figure 4 shows this quantitatively for the gas located near the peak of the molecular ring in the Galactic disk. The CO emission coincident with HII regions in position and velocity are indicated by solid bars (we double-count the near- and far-side emission). Although molecular gas directly associated with HII regions exhibits systematically higher CO 2–1/1–0 ratio, it has only a minor ($\sim$ 10 %) fraction in the total CO luminosity from *High Ratio Gas*; most of the *High Ratio Gas* resides outside the range of direct effect of O stars.

The above argument leads us to a working hypothesis that GMCs containing *High Ratio Gas* are formed by some mechanism(s) related to the spiral arms, and they provide sites of massive star formation; *High Ratio Gas* may be a manifestation of dense molecular gas that is ready for massive star formation.

*Acknowledgment*: The author thanks his collaborators whose works formed basic ingredients of this paper. We express special thanks to Pat Thaddeus and Tom Dame for kindly providing their CO $J$=1–0 survey data.

## References

Blitz, L. & Thaddeus, P. 1980, ApJ, **241**, 676
Braine, J. & Combes, F. 1992, A&A, **264**, 433
Burton, W. B., Gordon, M. A., Bania, T. M., & Lockman, F. J. 1975, ApJ, **202**, 30
Cohen, R. S., Dame, T. M., & Thaddeus, P. 1986, ApJS, **60**, 695
Dame, T. M. 1993, in *Back to the Galaxy*, ed. Holt & Verter (New York: AIP), 262
Dame, T. M., et al. 1987, ApJ, **322**, 706
Gierens, K. M., Stutzki, J., & Winnewisser, G. 1992, A&A, **259**, 271
Goldsmith, P. F. 1987, in *Interstellar processes*, ed. Hollenbach & Thronson (Dordrecht: Kluwer), 51
Handa, T., et al. 1993, in *Back to the Galaxy*, ed. Holt & Verter (New York: AIP), 315
Hayashi, M., Hasegawa, T., Sunada, K., & Kaifu, N. 1990, in *Submillimetre Astronomy*, ed. Watt & Webster (Dordrecht: Kluwer), 63
Hayashi, M. et al. 1996, in preparation
Lockman, F. 1989, ApJS, **71**, 469
Maddalena, R. J., Morris, M., Moscowitz, J., & Thaddeus, P. 1986, ApJ, **303**, 375
May, J., et al. 1993, A&AS, **99**, 105
Morino, J. et al. 1996, in preparation
Nakano, T., Hasegawa, T., & Norman, C. 1995, ApJ, **450**, 183
Oka,T., Hasegawa, T., Handa, T., Hayashi, M.,& Sakamoto, S. 1996a, ApJ, **460**, 334
Oka,T., Hasegawa, T., Hayashi, M., Handa, T., & Sakamoto, S. 1996b, ApJ, submitted
Sakamoto, S., Hasegawa, T., Hayashi, M., Handa, T., & Oka, T. 1995, ApJS, **100**, 125
Sakamoto, S., Hasegawa, T., Hayashi, M., Handa, T., & Oka, T. 1996, ApJ, submitted
Sakamoto, S., Hayashi, M., Hasegawa, T., Handa, T., & Oka, T. 1994, ApJ, **425**, 641
Sanders, D. B., Solomon, P. M., & Scoville, N. Z. 1984, ApJ, **276**, 182
Scoville, N. Z. & Solomon, P. M. 1975, ApJ, **199**, L105
Sorai, K. et al. 1996, in Proc. East Asian Meeting on Astronomy, ed. Kaifu, in press
Ungerechts, H. & Thaddeus, P. 1987, ApJS, **63**, 645

# MAPPING DUST EXTINCTION IN MOLECULAR CLOUDS

JOÃO ALVES AND CHARLES J. LADA

*Smithsonian Astrophysical Observatory*
*60 Garden Street*
*Cambridge, MA 02138*

**Abstract.** Infrared array cameras have enabled the measurement and mapping of dust extinction within molecular clouds to significantly greater optical depths and at significantly higher angular resolution than previously possible. In this paper we present a map of the dust extinction in the dark cloud L978 in Cygnus derived from a large-scale multi-wavelength infrared survey. Analysis of our observations reveals significant variations in cloud structure on scales smaller than our $1'.5$ resolution, thus confirming trends previously observed in IC5146 dark cloud.

## 1. Introduction

Although molecular clouds are composed almost entirely of molecular hydrogen their most basic physical properties (e.g., size, mass, density and temperature) have been derived almost exclusively from observations of rare, trace molecules such as CO, CS and $NH_3$. However, the most direct way to trace the hydrogen content of a molecular cloud is to measure the distribution of dust through it since the gas-to-dust ratio in interstellar clouds appears to be constant (e.g., Lilley 1955; Bohlin, Savage, & Drake 1978).

Recently, Lada et al. (1994) demonstrated that large scale imaging surveys of dark clouds can be used to directly trace the distribution of dust extinction and column density through a molecular cloud at considerably higher angular resolution and greater optical depth than previously possible from star counts (Bok 1937).



*W.B. Latter et al. (eds.), CO: Twenty-Five Years of Millimeter-Wave Spectroscopy*, 47–49.

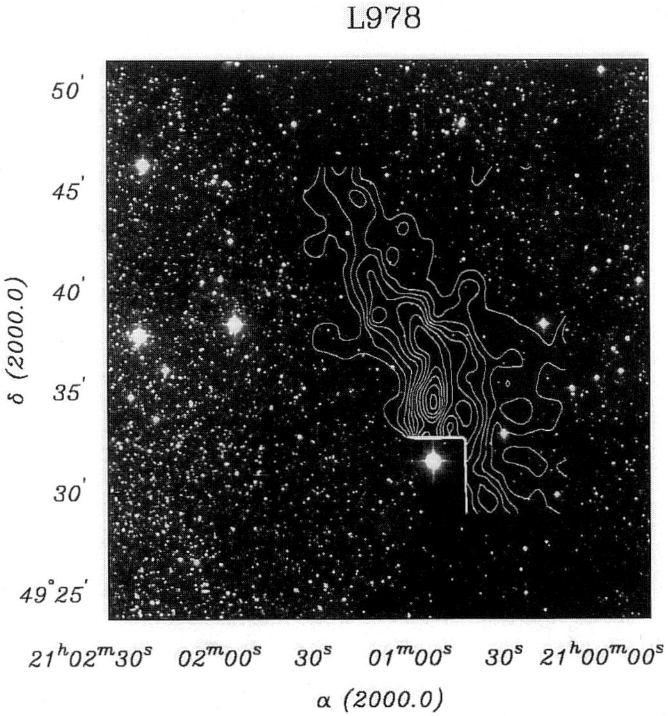


*Figure 1.* Map of the dust extinction in L978. Contours start at 5 magnitudes of visual extinction and increase by steps of 2 magnitudes. The map is overlayed on the Palomar red plate.

## 2. The Color Excess Method

Briefly, the dust column density is measured from the H-K colors of thousands of background stars observed through a cloud using the color excess: $E(H-K) \equiv (H-K)_{observed} - (H-K)_{intrinsic}$ where the intrinsic color is determined from the observation of unextincted background field stars. The color excess is directly related to the extinction at any wavelength via the reddening law (e.g., $A_v = 15.9 \times E(H-K)$; Rieke & Lebofsky 1985). The measured color excesses or extinctions then can be spatially binned to produce uniform sampled maps at angular resolutions comparable to mm wave observations for direct comparison with molecular line data. Observations of the IC5146 molecular cloud done by Lada et al. (1994) enabled very precise mass determinations of this dark cloud as well as direct abundances measurements of $C^{13}O$, $C^{18}O$, and CS, and permitted a quantitative description of the cloud structure at small angular scales.

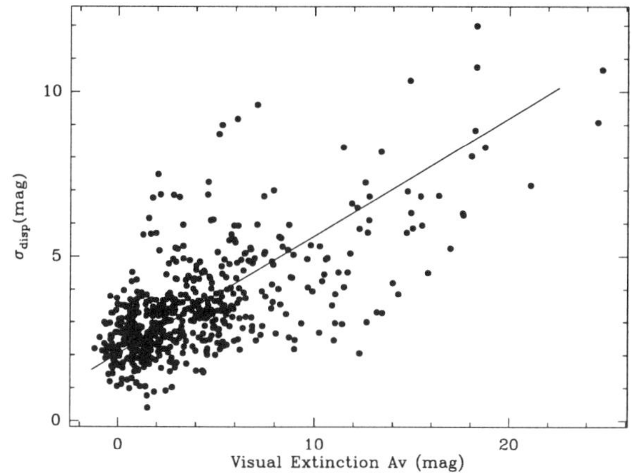


*Figure 2.* The variation of $\sigma_{\rm disp}$, the dispersion in the extinction measurements, with $A_v$ the derived mean extinction for L978. Also plotted is the least-squares linear fit to the data.

## 3. Application to L978 and Concluding Remarks

Figure 1 displays the extinction map we derived for the L978 molecular cloud using this method (Alves et al. 1996). Almost 2000 stars were observed in the J, H, and K near-infrared bands to derive the dust extinction with a dynamic range of $5<A_v<25$ mags. The distribution of extinction agrees well with the shape of the cloud from the optical plate. Presented in Figure 2 is the relation between the dispersion in each (binned) pixel of the map versus the mean visual extinction in the same pixel. This behaviour, $\sigma_{\rm disp}$ increasing in a systematic fashion with increasing $A_v$, first discovered by Lada et al. in IC5146, demonstrates the existence of significant variations in cloud structure on scales smaller than our $1.5'$ binned resolution. Our observations appear to confirm the trends found in IC5146 suggesting that analogous distributions of dust characterize both clouds on small spatial scales.

## References

Alves, J., Kenyon, S., Phelps, R., Lada, E. A., Lada, C. J. 1996, in preparation
Bohlin, R. C., Savage, B. D. and Drake, J. F. 1978, ApJ, **224**, 132
Bok, B. J. 1937 The Distribution of Stars in Space, (Chicago, Univ. of Chicago Press)
Lada, C. J., Lada, E. A., Clemens, D. P., Bally, J. 1994, ApJ, **429**, 694
Lilley, A. E. 1955, ApJ, **121**, 559
Rieke, G. H., & Lebofsky, M. J. 1985, ApJ, **288**, 618

# ON THE FRACTAL STRUCTURE OF MOLECULAR CLOUDS

J. STUTZKI, A. HEITHAUSEN AND F. BENSCH
*I. Physikalisches Institut der Universität zu Köln*
*Zülpicher Straße 77*
*D-50937 Köln*

Molecular clouds show structure on all scales observed and many attempts have been made to characterize the structure which is linked to important aspects of star formation. Being generated predominantly by the turbulent internal cloud motion, the structure of molecular clouds is likely to be random to a large degree. As the information on the structural properties of random function is basically contained in its power spectrum, it is obviously important to study the power spectrum of observed cloud images. For HI clouds (Green 1993), the azimuthally averaged power spectra derived directly from interferometer observations are well characterized by a power law with a spectral index close to 2.8.

A very powerful tool to study the power spectrum of a random time series has been introduced by Allan (1966) within the context of characterizing the stability of atomic clocks. For a random time series, the Allan variance on a given time scale $T$ is basically the variance of the differences of subsequent averages over time $T$. In Fourier space, the Allan variance turns out to be the weighted average of the power spectrum of the random function with the weighting $\sin(\pi f T)^4/(\pi f T)^2$. This filtering function peaks approximately at $1/(\pi T)$. If the random function is purely white noise, i.e., has a flat power spectrum, the Allan variance goes as $\sigma_A^2 \propto 1/T$; if the random function has drift behaviour, i.e., $< (f(t+T) - f(t))^2 > \propto T^{2H}$, the power spectrum is $PS \propto f^{-\beta}$ with $\beta = 1 + 2H$ and the Allan variance is $\sigma_A^2 \propto T^{\beta-1}$ for $0 < \beta < 3$, and $\sigma_A^2 \propto T^2$ for $\sigma > 3$. Note that the discrimination at $\beta = 3$ implies that the Allan variance actually contains information different from the power spectrum itself.

Extension to higher dimensions is pretty straightforward. For the analysis of a 2-dim spatial structure, such as a molecular cloud image, we have to take the difference between adjacent averages of size $L$. With no azimuthal direction preferred, this is naturally done in cylindrical symmetry, i.e., using a function with a value of +1 on a circle of diameter $L$ enclosed



*W.B. Latter et al. (eds.), CO: Twenty-Five Years of Millimeter-Wave Spectroscopy,* 50–53.

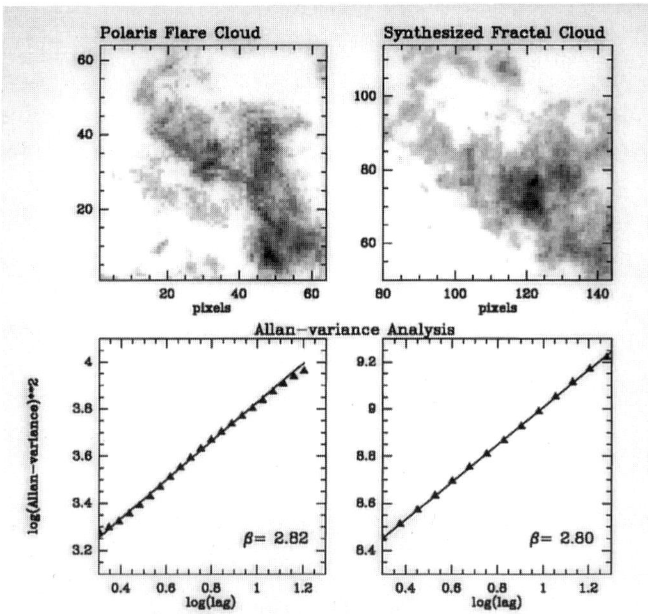


*Figure 1.* Allan variance of (left) an observed image of the Polaris flare cloud (Heithausen et al. 1990) showing a power law power spectrum with a spectral index $\beta = 2.8$ and (right) the same for a synthesized fBm-image.

by a region with value $-1/8$ (for equal weight of the two areas) on a ring of outer radius $3L/2$ as the spatial filter function. One then calculates the 2-dim Allan variance as the average of the 2-dim power spectrum weighted with the corresponding power spectrum of the 2-dim filter function. Figure 1 (left) shows the Allan variance as defined above for an integrated intensity image of the Polaris Flare to closely follow a power law with a power law spectral index of $\beta = 2.8$, similar to values for other observed molecular cloud images.

It is well established within the work that has been done over the last years in analysing and synthesizing fractal images (e.g., in order to generate artificial landscapes for computer animations, etc.) that such structures can easily be generated as so-called "fractal Brownian motion" (fBm)-images. These are generated by specifying a power law power spectrum with a given index $\beta = E + 2H$, where $E$ is the Euclidian dimension of the image and $H$ ranges between 0 and 1, and completely random phases of the image. Such structures can be shown to have a drift behaviour (autocorrelation function)$< (f(\rightarrow r + \rightarrow D) - f(\rightarrow r))^2 > \propto D^{2H}$, and their iso-intensity contours follow an area/perimeter relation with a fractal dimension $d_H = (E-1)+1-H$ (for a short introduction, see the book by Peitgen & Saupe, 1988). Figure 1 (right) shows that such a structure actually gives a good representation of the structure observed for molecular cloud images. In fact, eye inspection of a series of synthesized fBm images covering the full range of $H = 0...1$ allow to identify the spectral index for observed molecular clouds to be close to 2.8 (i.e., $H$ around 0.4). This value is in good agreement with the value derived above from the Allan variance analysis, and with the range of 1.4–1.6 derived for the area/perimeter fractal dimension.

Observed clouds have a power law clump mass spectrum $dN/dM \propto M^{-\alpha}$, with $\alpha$ in the range 1.5 to 1.7 for most clouds (Stutzki & Güsten 1990; Williams et al. 1994). Although the concept of dividing the observed structure into discrete subentities seems to be almost orthogonal to analysing the structure in terms of fractal dimensions or power spectra, it can be shown that $\alpha$ is connected to the power law spectral index for the 2-dim projected image by $\beta = 6 - 2\alpha$. The derivation assumes that the clumps have, on average, constant column density (following one of Larson's relation). Note that the regime of observed values for $\alpha$ and $\beta$ is consistent with the above relation.

It can be shown (see Stutzki et al. 1995) that the 3-dim structure resulting in the observed 2-dim image with a power law power spectrum and random phases also has a power law spectrum with identical spectral index, i.e., $\beta = 2.8$. In addition, under the reasonable assumption that the 3-dim structure has no correlations introducing some preferential 3-dim shapes that are hidden in the 2-dim projection, the 3-dim structure should also have random phases. This is a very important result, because it implies that the 3-dim structure is completely dominated by surface. A 3-dim fBm structure is limited to the range $\beta = E + 2H = 3...5$, corresponding to $H = 0...1$. The respective surface/volume relation for the fractal structure of the iso-density surfaces (the 3-dim analogon to the perimeter/area relation for the iso-intensity contours), then has a fractal dimension in the range $d_H = E + 1 - H = 4...3$, i.e., $\log(A) \propto \log(V)^{d_H/3}$. The extreme value of $H = 1$, i.e., $d_H = 3$, already corresponds to the surface increasing proportionally to the volume. The value of $\beta = 2.8$ for a 3-dim structure derived above then implies an even faster increase of surface with volume.

The result that the 3-dim cloud structure is dominated by surface is in accordance with the fact that even $^{12}CO$, though being a completely optically thick tracer, measures cloud mass, as well as the recently emerging view, that most line emission from molecular cloud tracers is largely dominated by surface effects. From the results presented, it seems very important to try to understand the derived value for the power law power spectrum index as a unique structure parameter from theoretical arguments on the turbulent structure of molecular clouds. For a full understanding of the 3-dim cloud structure, a link to the velocity structure is crucial. Velocity structure becomes very important as a tool to tell overlapping 3-dim spatial structures apart, and to keep the emission of individual clumps from mutual shielding in the radiative transfer.

## References

Allan, D.W., 1966, Proc. IEEE, **54**, 221
Heithausen, A. & Thaddeus, P. 1990, ApJL, **353**, L49
Green, D.A., 1993, MNRAS, **262**, 327
Peitgen, H.O., & Saupe, D. (editors) 1988, "The Science of Fractal Images," Springer
Stutzki J. & Güsten, R., 1990, ApJ, **356**, 513
Stutzki, J., et al. 1995 (in prep.)
Williams, J.P., de Geus, E.J. & Blitz, L. 1994, ApJ, **428**, 693

# THE ORIGIN AND STRUCTURE OF MOLECULAR CLOUDS

JOHN BALLY
*Dept. of Astrophysical, Planetary, and Atmospheric Sciences*
*Center for Astrophysics and Space Astronomy*
*Campus Box 389, University of Colorado*
*Boulder, CO 80309 USA*

Although we have known about giant molecular clouds (GMCs) for nearly 25 years, their origin, evolution, and destruction is still poorly understood. Formation mechanisms can be grouped into two general categories; *galactic scale processes* including global gravitational instabilities of the galactic disk, Jeans-Parker type magnetically regulated instabilities, and spiral density waves, and *local processes* based on self-propagating star formation or supershell fragmentation (cf. Elmegreen 1994).

The Solar vicinity lying within about 0.5 to 1 kpc is the only place where we have reasonably complete knowledge of the location and motion of stars and associated gas. The radial velocities and proper motions of stars, their locations in 3 dimensions, and association with gas, make it possible to reconstruct the history of recent ($< 60$ Myr) star formation. HII regions and low mass T-Tauri stars trace the youngest ($< 3$ Myr old) massive stars and sites of on-going star formation. Nearby OB associations (Orion OB1, Per OB2, Sco-Cen, and Lac OB1) trace sites of star formation active within the last 15 Myr. GMCs and giant shells of expanding HI are associated with these OB associations. Late B and A stars and proper motion groups, such as the Cas-Tau "fossil" OB association, trace star formation over the last 60 Myr (Blaauw 1991).

After correction for the effect of Galactic differential rotation, most of the ISM in the Solar vicinity (GMCs and HI) is found to be expanding with a mean velocity of about 5 km s$^{-1}$ from a point located near $l =$ 150, $d = 200$ pc, the approximate centroid of the Cas-Tau group. This is the 'Lindblad ring' recognized in early 21 cm maps and is traced by local molecular gas (see Dame et al. 1987; Poppel et al. 1994). Associated B and A stars trace Gould's Belt of nearby intermediate age and young stars.

The Lindblad ring appears to be a 30 to 60 Myr old fossil supershell driven into the local ISM by a massive OB association whose remnants



*W.B. Latter et al. (eds.), CO: Twenty-Five Years of Millimeter-Wave Spectroscopy,* 54–56.

are recognized as the Cas-Tau group. Support for this view comes from the agreement between the observed radial velocity of HI and nearby CO clouds with models of an expanding and tidally shearing 30 to 60 Myr old superbubble (see Poppel et al. 1994). The location of holes in the distribution of high galactic latitude HI (the Lockman Hole) and dust (Baade's Hole) above and below the Cas-Tau group centroid indicates that the ancient bubble burst out of the Galactic plane. Excess low negative velocity HI at high latitudes indicates that dense clumps accelerated out of the plane 25 to 50 million years ago are now falling back due to the gravity of the disk.

Most nearby molecular clouds may have formed by gravitational fragmentation of the Cas-Tau fossil supershell. Gravitational instabilities in a swept-up shell lead to a fragment mass of order $10^5 M_\odot$, and to a global velocity structure similar to what is observed in the local ISM.

Massive stars in OB associations dominate the energy injected into the ISM. Over the 5 to 15 Myr formation time of an OB association, between 10 to 100 stars with masses in excess of $8 M_\odot$ are produced (in addition to more than $10^4$ lower mass stars). Their Lyman continuum (LyC) radiation dissociates and ionizes surrounding molecular gas, their powerful winds shock and expel this plasma, and their demise in supernova (SN) explosions contribute to the $10^{52}$ to $10^{54}$ ergs of energy available to inflate superbubbles in the ISM.

The energy injected into the ISM by OB stars can be approximately represented as a time variable OB association driven wind with $L_w \approx 10^{38}$ erg s$^{-1}$ that first evolves through an adiabatic (energy conserving) phase, which after a bubble cooling time or after the bubble blows out of the Galactic disk and loses pressure, enters an isothermal (momentum conserving) phase. In the first phase ($t < 1$ to 10 Myr), the shell radius evolves according to $R_s = 70(pc) L_{38}^{1/5} n_0^{-1/5} t^{3/5}$ where $L_{38}$ is the wind mechanical luminosity in $10^{38}$ erg s$^{-1}$, $n_0$ is the mean ISM density, and $t_6$ is the bubble age in Myr. In the second phase, $R_s = (L_w/\alpha V_0 n_0)^{1/4} t^{1/2}$ where $\alpha$ is of order unity and $V_0$ is the shell velocity when it enters the isothermal phase. After the last massive star dies (at $t > 30$ Myr), the shell coasts, with $R_s = (E_0/\alpha' V_0 n_0)^{1/4} t^{1/4}$ where $E_0$ is the shell kinetic energy when the shell enters this third phase.

Bubble evolution is modified (with respect to uniform density, spherical models) by *non-coeval star formation*; by *poisoning* resulting from the evaporation of clumps that have been over run, adding coolants to the bubble; *blowout* when the bubble bursts out of the galactic disk; *shear* induced by galactic differential rotation; *magnetic fields*; *clumps* in the ISM; *density waves* which modulate the ISM pressure; and *gravitational instabilities.*

Gravitational instabilities cause the shell walls to *curdle* when the escape velocity from a piece of the shell exceeds the effective velocity dispersion

($v_{\mathrm{eff}}$) in the same volume ($v_{\mathrm{eff}}$ is the sum in quadrature of the sound speed, turbulent velocity, alfven speed, and shell divergence velocity). As shown by McCray & Kafatos (1987), supershells tend to fragment into $10^5 M_\odot$ clumps at a characteristic shell radius of 200 pc and a shell age of about 30 Myr. Condensation on this mass, size, and time scale is enhanced in the Solar vicinity by the blowout scale and the time taken for ballistic objects to start to fall back onto the plane (1/4 to 1/2 of the $z$-oscillation time scale of 80 Myr). Galactic differential rotation (shear) elongates fragmenting fossil supershells into ovoids with a major axis of about 500 - 800 pc and a minor axis of about 200 - 400 pc in this time scale.

The above model and data can be used to reconstruct the 30 to 60 Myr history of the Solar vicinity. A large OB association formed between 30 and 60 Myr ago at the present location of the Cas-Tau group, possibly triggered by the previous passage of the Sun through a major spiral arm. The resulting superbubble overran preexisting clouds, swept-up low density ISM into a $10^6 M_\odot$ supershell, and blew out of the disk. About 15 to 25 Myr ago, the first GMC-sized condensations formed near the center of the present day Sco-Cen, Orion, Lac OB1, and Per OB2 associations. These associations started forming about 15 Myr ago, and are presently driving their own superbubbles into the remnants of the ancient Cas-Tau fossil supershell. The expansion velocity of the Cas-Tau fossil shell is imprinted on the currently forming OB groups, the Great Rift clouds, the Lindblad ring, and other clouds lying within 500 pc of the Sun.

The model is self-sustaining. The nearby younger supershells will in turn fossilize, and curdle into new GMCs that will produce future generations of OB associations. In this model, the nearby 'hot cavity' may be an outlying portion of the Sco-Cen superbubble which is presently over-running the Solar system as evidenced by Solar Lyman-$\alpha$ back-scatter and the streaming of the ISM through the Solar system (Crutcher 1982) which indicates outflow from the general direction of the Sco-Cen OB association.

## References

Blaauw, A. 1991, in *The Physics of Star Formation ...*, ed. Lada & Kylafis, p. 125.
Crutcher, R. M. 1982, ApJ, 254, 82.
Dame, T. M., et al. 1987, ApJ, 322, 702.
Elmegreen, B. G. 1994, in *Protostars and Planets III*, ed. Levy & Lunine, p. 97.
McCray, R. & Kafatos, M. 1987, ApJ, 317, 190.
Poppel, W. G. L., Marronetti, P., & Benaglia, P. 1994, A&A, 287, 601.

# MOLECULAR GAS NEAR THE GALACTIC CENTER

MARK MORRIS
*Department of Physics & Astronomy, UCLA*
*Los Angeles, CA 90095-1562, USA*

## 1. Molecular Surveys

Surveys of molecular gas in the central kiloparsec of the Galaxy have been underway throughout the 25 years since the discovery of CO. Table 1 lists the large-scale surveys published during this interval, including absorption-line studies of $H_2CO$ and OH and emission-line studies of CO, $NH_3$, CS, and, to a lesser extent, HCN and $H_2$. With its several accessible isotopes, and because of its ubiquity, CO has provided perhaps the most complete picture of the large-scale distribution and kinematics of molecular gas, although the denser material ($n(H_2) > 10^4$ cm$^{-3}$) is better probed by $NH_3$, CS, and HCN.

## 2. Contents of the Molecular Arena

The region within 2° (300 pc) of the Galactic center contains the most prominent concentration of molecular gas in the Galaxy, with a mean surface density of $\sim 200$ $M_\odot$ pc$^{-2}$. This stands in contrast to the much smaller surface density in the HI nuclear disk beyond a few hundred parsecs from the center: $\sim 5$ $M_\odot$ pc$^{-2}$. Figure 1 shows this "central molecular zone" or CMZ, as viewed in $^{12}$CO. This large reservoir of gas is probably attributable to the inexorable inflow of gas from the rest of the Galaxy, occuring as a result of the combination of several angular momentum loss processes (Morris & Serabyn 1996), not the least of which are those attributed to the pronounced stellar bar which dominates the central few kiloparsecs of the Galaxy. The residence time for gas in the CMZ is on the order of 0.4–$1 \times 10^9$ years, far shorter than the age of the Galaxy, so the CMZ is in a quasi-steady state in which the inflow of gas from the Galaxy is balanced primarily by star formation and by evaporation of clouds into the surrounding hot ($10^7$ - $10^8$K) medium implied by X-ray observations (Koyama *et*



*W.B. Latter et al. (eds.), CO: Twenty-Five Years of Millimeter-Wave Spectroscopy*, 57–64.

TABLE 1. Large-Scale Molecular Line Surveys of the Galactic Center

| Year | Molecule & Line | Reference |
|---|---|---|
| 1970 | OH 1667 MHz | Robinson, B.J., McGee, R.X.: Aust. J. Phys. 23, 405. see also p. 541 & p. 777 |
| 1972 | $H_2CO$ $1_{11} \leftarrow 1_{10}$ | Scoville, N.Z., *et al.*: ApJ 172, 335 |
| 1972 | CO 1→0 | Solomon, P.M., *et al.*: ApJ 178, 125 |
| 1973 | $H_2CO$ $1_{11} \leftarrow 1_{10}$ | Scoville, N.Z., Solomon, P.M.: ApJ 180, 55 |
| 1974 | CO 1→0 | Scoville, N.Z., *et al.*: ApJL 187, L63 |
| 1974 | CO 1→0 | Sanders, R.H., Wrixon, G.T.: A&A 33, 9 |
| 1975 | $NH_3$ (1,1) | Kaifu, N., *et al.*: ApJ 201, 98 |
| 1976 | OH 1665/7 MHz | Cohen, R.J., Few, R.W.: MNRAS 176, 495 |
| 1977 | CO 1→0 | Liszt, H.S., *et al.*: ApJ 213, 38 |
| 1977 | CO 1→0 | Bania, T.M.: ApJ 216, 381 |
| 1978 | CO 1→0 | Liszt, H.S., Burton, W.B.: ApJ 226, 790 |
| 1979 | $H_2CO$ $1_{11} \leftarrow 1_{10}$ | Whiteoak, J.B., Gardner, F.F.: MNRAS 188, 445 |
| 1980 | $H_2CO$ $1_{11} \leftarrow 1_{10}$ | Bieging, J.H., *et al.*: A&AS, 42, 163 |
| 1980 | CO 1→0 | Bania, T.M.: ApJ 242, 95 |
| 1981 | $H_2CO$ $1_{11} \leftarrow 1_{10}$ | Cohen, R.J., Few, R.W.: MNRAS 194, 711 |
| 1982 | CO 1→0 | Inatani, J.: PASJ 34, 515 |
| 1983 | $NH_3$ (1,1), (2,2), (3,3) | Morris, M., *et al.*: AJ 88, 1228 |
| 1984 | CO 1→0 | Sanders, D.B., *et al.*: ApJ 276, 182 |
| 1986 | CO 1→0 | Bania, T.: ApJ 308, 868 |
| 1987 | $^{13}$CO 1→0 | Heiligman, G.M.: ApJ 314, 747 |
| 1987 | CO 1→0 | Dame, T.M., Ungerechts, H., Cohen, R.S.: ApJ 322, 706 |
| 1987 | $^{13}$CO 1→0 & CS 2→1 | Bally, J., *et al.*: ApJS 65, 13 |
| 1988 | CO 1→0 | Bally, J., *et al.*: ApJ 324, 223 |
| 1989 | CS 1→0, 2→1 | Tsuboi, M., *et al.*: in *The Center of the Galaxy*, ed: M. Morris, Dordrecht: Kluwer, p. 121 |
| 1992 | CO 1→0 | Burton, W.B., Liszt, H.S.: A&AS 95, 9 |
| 1992 | $H_2CO$ $1_{11} \leftarrow 1_{10}$, $2_{11} \leftarrow 2_{12}$ | Zylka, R., *et al.*: A&AS 96, 525 |
| 1994 | OH 1665/7 MHz | Boyce, P.J., Cohen, R.J.: A&AS 107, 563 |
| 1996 | HCN, CO 1→0 | Jackson, J.M., *et al.*: ApJL 456, L91 |
| 1996 | $H_2$ | Pak, S., Jaffe, D.T., Keller, L.D.: ApJL 457, L43 |
| 1996 | CO 2→1 | Oka, T., *et al.*: ApJ 460, 334 |
| 1996 | $C^{18}O$ 1→0 | Dahmen, G. *et al.*: in *4th ESO/CTIO Workshop: The Galactic Center,* ed: R. Gredel, ASP Conf. Ser., in press |
| 1996 | CO 1→0 | Uchida, K., Bally, J., Morris, M.: in preparation |

*al.* 1989, Yamauchi *et al.* 1990, Sunyaev *et al.* 1993, Koyama *et al.* 1996). This hot medium, which may lie at the base of a thermally-driven Galactic wind, has approximately the same extent as the CMZ.

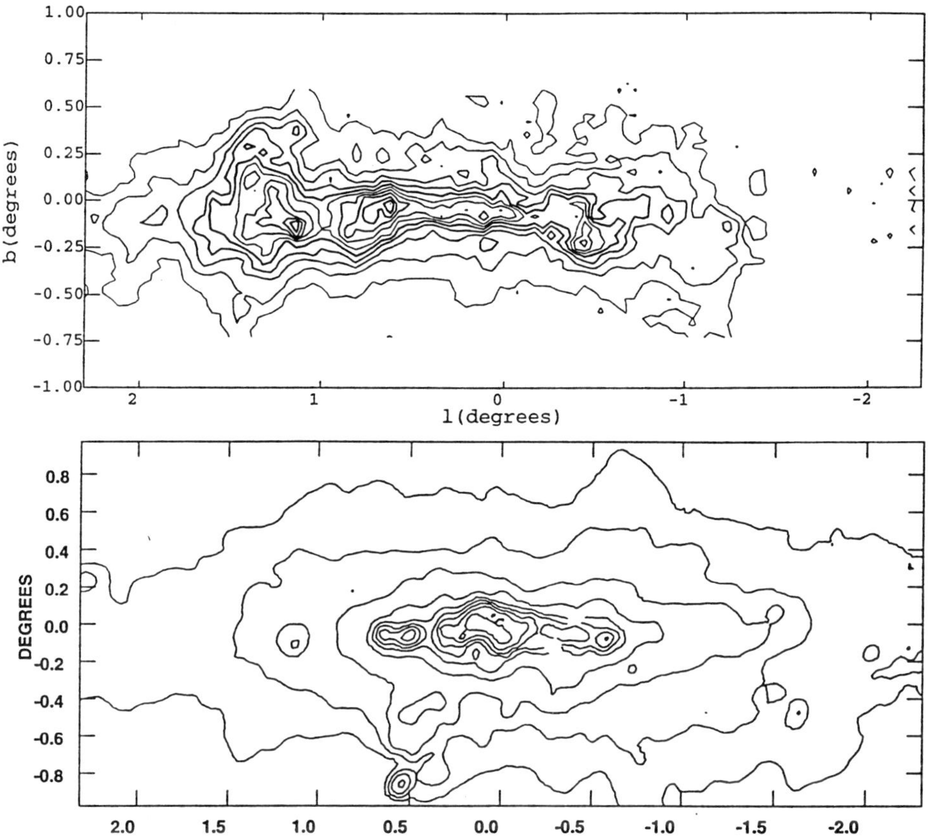


*Figure 1.* The Galaxy's Central Molecular Zone. Top: integrated intensity of $^{12}CO$ $J = 1 \rightarrow 0$ emission, from the AT&T Bell Labs survey (Uchida *et al.* 1996a). Bottom: IRAS 60 $\mu$m emission map, showing the distribution of relatively warm dust.

## 3. Molecular Streams and other Continous Features

Much of the molecular gas in the CMZ is arranged in well-defined, continuous "streams" which can be followed through longitude-latitude-velocity ($l$-$b$-$v$) data cubes for distances on the order of $10^2$ pc (Uchida *et al.* 1996a). These streams probably result from the tidal shear to which the gas clouds are subjected (Güsten & Downes 1980). The most imposing of these features appears as a parallelogram, or ellipse, in longitude-velocity plots (figure 2), and appears to be a quasi-continuous ring of gas completely surrounding the Galactic center at a projected radius of 180 pc. This feature, which approximately coincides with the outer boundary of the CMZ, was originally hypothesized to be an expanding molecular ring (Scoville 1972; Kaifu

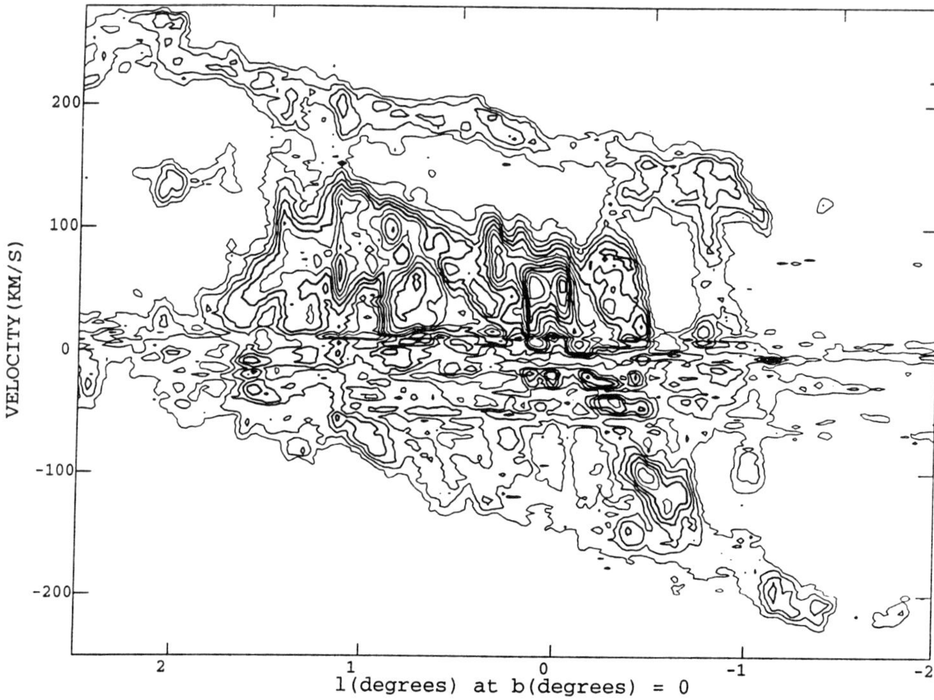


*Figure 2.* $^{12}$CO $J = 1 \rightarrow 0$ emission contours in a position-velocity plane in which position extends along the line from $l, b = 2.5°, -0.333°$ to $-2.0°, 0.1333°$, from the AT&T Bell Labs survey (Uchida *et al.* 1996a). This line, tilted by $\sim 6°$ with respect to the Galactic plane, was chosen to reflect the tilt of the plane containing the 180 pc molecular ring (Uchida *et al.* 1994a). Contour units are 1.4, 2.8, 4.4, 6.2, 8.2, 10.4, ... K.

*et al.* 1972) provoked by an explosion at or near the Galactic center. Various implications of this hypothesis have been pursued recently, including the shocks which radially expanding gas should produce (Uchida *et al.* 1994a,b) and the recognition that the expanding molecular ring should really be an expanding molecular shell (Sofue 1995b). However, the 180-pc ring has more recently been modelled as the response of gas to a central bar potential (Binney *et al.* 1991). In this latter model, which has been widely adopted, the 180 pc molecular ring coincides with the innermost of the non-self-intersecting "$X_1$" family of orbits (Contopoulos & Mertzanides 1977): elongated oval orbits having their major axes aligned with the bar. This kinematics of this ring, and its tilt by about 20° with respect to the Galactic plane, are continuous with the properties of HI located further out in the "HI nuclear disk", indicating that the molecular ring likely marks the location of an HI/$H_2$ transition, presumably induced by shocks.

The bulk of the gas in the Galactic center, however, about $5 \times 10^7$ $M_\odot$ (Güsten 1989, Cox & Laureijs 1989, Sodroski *et al.* 1995, Sofue 1995a), resides inside the 180 pc molecular ring and is said to be on $X_2$ orbits: oval orbits lying deeper within the central gravitational potential well than the $X_1$ orbits and having their major axes oriented perpendicular to the bar. Sofue (1995a) has examined the continous gas streams evident in $^{13}CO$ plotted in the $l$-$v$ plane and has concluded that most of the gas interior to the 180 pc ring is organized into two "arms" which form large arcs upon a ring of galactocentric radius 120 pc. (A similar picture had earlier been suggested by Scoville *et al.* 1974.) Such structures could arise as gas builds up in the outermost occupied $X_2$ orbits as a result of the inevitable shock dissipation occuring in the innermost occupied $X_1$ orbit (Binney *et al.* 1991, Morris & Serabyn 1996).

## 4. Transience of the Molecular Gas Reservoir

The most massive clouds in the CMZ can reside there for time scales of only a few times $10^8$ years, because they are doomed to spiral inwards under the influence of dynamical friction with the stars in the central bulge (Stark *et al.* 1991). Less massive clouds may lose angular momentum on similar time scales, because they are subject to magnetic viscosity and relatively frequent dissipative cloud collisions. Assuming a steady-state inflow, Morris & Serabyn (1996) estimate a mean residence time for gas in $X_2$ orbits of $0.4 \times 10^8$–$10^9$ years. For a star formation rate of a few tenths of a solar mass per year in the CMZ (Güsten 1989), this is sufficient time for almost all of the gas in the CMZ to be converted to stars. Some gas eventually finds its way into the central circumnuclear disk, which has an inner radius of about 1 pc. However, this feature is likely to be transient, and may represent a recent chance encounter of a cloud with the central gravitational potential well. The reader is referred to Morris & Serabyn (1996) for a more thorough discussion of the structure and evolution of the circumnuclear disk.

## 5. Tilts and Asymmetries

The CMZ shows a strong asymmetry about the Galactic center, with about 3/4 of the gas lying at positive longitudes. The 180 pc molecular ring is also asymmetric (fig. 2), its centroid being displaced by several tens of parsecs from the Galaxy's dynamical center and the core of the central stellar cluster (Uchida *et al.* 1994a, Blitz 1994). Furthermore, the ring appears to be displaced by about 30 or 40 km s$^{-1}$ toward positive velocity, relative to the expected LSR velocity of 0 km s$^{-1}$. These displacements and asymmetries are suggestive of a "sloshing" mode (or $m = 1$ wave) in the gas layer, possibly coupled to the strong stellar bar (an $m = 2$ mode), and possibly related to the pronounced tilt of the gas layer (another manifestation of an $m = 1$ mode). Indeed, organized vertical gas motions appear to be an important element of the kinematical picture, and progress in analyzing the dynamics of the CMZ will require accounting for the fact that some of the continuous molecular streams extend well out of the Galactic plane in projection. In at least one location, a strong, large-scale (100 pc) vertical shock is implied by the CO data (Uchida *et al.* 1994b).

## 6. Temperatures and Internal Motions in Galactic Center Clouds

Several groups, mostly using the ammonia molecule, have shown that the temperatures of clouds in the CMZ are relatively high (25–200 K, typically $\sim$ 70 K, Morris *et al.* 1983, Güsten *et al.* 1985, Armstrong & Barrett 1985, Mauersberger *et al.* 1986, Hüttemeister *et al.* 1993) compared to clouds in the outer Galactic disk. A widespread gas heating mechanism is thus called for, although there is currently no consensus on the nature of that mechanism. Highly supersonic internal velocity dispersions ($\sim$ 15–30 km s$^{-1}$), comparable to the local intercloud velocity dispersion, are also prevalent throughout the Galactic center. The elevated temperatures and linewidths may be related, as dissipation of turbulent energy is a prime candidate for direct heating of the gas. This remains true if the cloud velocity dispersions represent magnetosonic disturbances, as seems likely in this highly magnetized environment, although the dissipative heating in that case is likely to be dominated by magnetically induced drift of ions and electrons through the viscous, neutral medium (ambipolar diffusion; Roberge *et al.* 1995), rather than by shocks or viscous shear. The tidal field has been suggested as a source of energy to account for the large linewidths (Fleck 1980); Das & Jog (1995) conclude that this mechanism is efficient only for a central bar potential, in which tidal fields vary markedly along elongated orbits. This requirement appears to be satisfied in the Galactic center.

## 7. Magnetic Fields

As the previous section indicated, there are several lines of evidence that the magnetic field near the Galactic center is quite strong, both inside and outside of clouds. Outside clouds, the field is evidenced by the large-scale synchrotron filaments observed in the radio. The strength of the field is estimated to be in the milligauss range on the basis of the apparent rigidity of the filaments in the face of their clear interactions with the high-velocity-dispersion clouds (Morris 1994, 1996, and references therein). The magnetic field is perpendicular to the Galactic plane wherever it is probed by the presence of a filament; considerations of pressure confinement of the filamentary structures have been used to argue that this poloidal field is uniform and pervasive throughout the inner 100–150 pc.

Inside molecular clouds, the field has been probed using measurements of the polarization of the thermal far-IR emission from dust grains which have presumably been aligned by the field (*e.g.*, Hildebrand *et al.* 1993, Morris *et al.* 1995). Although this technique does not yield the strength of the field directly, the large polarizations measured (up to about 10%) and the uniformity of the polarization vectors are both consistent with a strong and highly organized magnetic field within clouds. Interestingly, among the four Galactic center clouds which have so far been studied with far-IR polarization, the predominant magnetic field direction within three of them is roughly parallel to the galactic plane, while the other (Sgr B2) does not quite conform to this trend, although its interpretation is complicated by the fact that the polarization in that case arises primarily from far-IR absorption rather than from emission (Hildebrand *et al.* 1995). It appears, therefore, that the magnetic field orientation within Galactic center clouds is quite different from that outside of clouds, giving rise to the possibility that interesting MHD phenomena can take place at cloud surfaces. Indeed, Serabyn & Morris (1994) suggested that the nonthermal radio filaments are generated at the surfaces of molecular clumps as a result of particle acceleration accompanying magnetic field line reconnection. The other requisite element of that hypothesis is a source of ionization to provide the needed electrons at the cloud surface. Existing observations are consistent with this hypothesis for all of the radio filaments which have been sufficiently well studied (*e.g.*, Uchida *et al.* 1996b). Zeeman measurements of magnetic fields in Galactic center clouds have been far less revealing than the polarization measurements, in part because of the large intrinsic linewidths there. Beyond the circumnuclear disk, only upper limits have so far been reported (Uchida & Güsten 1995), and the results seem oddly discrepant with the polarization data.

## References

Armstrong, J.T. and Barrett, A.H. (1985) *ApJ*, **57**, pp. 535-70.

Binney, J., Gerhard, O.E., Stark, A.A., Bally, J. and Uchida, K.I. (1991) *MNRAS,* **252**, pp. 210-18.

Blitz, L. (1994) in *Physics of the Gaseous and Stellar Disks of the Galaxy*, ed: I. King, San Francisco: ASP, pp. 1-13.

Contopoulos, G. and Mertzanides, C. (1977) *A&A*, **61**, pp. 477-85.

Cox, P. and Laureijs, R. (1989) in *The Center of the Galaxy*, ed: M. Morris, Kluwer Academic Publishers, Dordrecht, pp. 121-128.

Das, M. and Jog, C. (1995) *ApJ*, **451**, pp. 167-75.

Fleck, R.C. (1980) *ApJ*, **242**, pp. 1019-22.

Güsten, R. (1989) in *The Center of the Galaxy*, ed: M. Morris, Kluwer Academic Publishers, Dordrecht, pp. 89-106.

Güsten, R. and Downes, D. (1980) *A&A*, **87**, pp. 6-19.

Güsten, R., Walmsley, C.M., Ungerechts, H. and Churchwell, E. (1985) *A&A*, **142**, pp. 381-87.

Hildebrand, R.H., Davidson, J., Dotson, J., Figer, D.F., Novak, G., Platt, S.R. and Tao, L. (1993) *ApJ*, **417**, pp. 565-71.

Hildebrand, R.H., Dotson, J.L., Dowell, C.D., Platt, S.R., Schleuning, D., Davidson, J.A. and Novak, G. (1995) in *Airborne Astronomy Symposium on the Galactic Ecosystem*, eds: M.R. Haas, J.A. Davidson, E.F. Erickson, San Francisco: ASP, ASP Conf. Ser. **73**, pp. 97-103.

Hüttemeister, S., Wilson, T.L., Bania, T.M. and Martin-Pintado, J. (1993) *A&A*, **280**, p. 255-67.

Kaifu, N., Kato, T. and Iguchi, T. (1972) *Nature Phys. Sci*, **238**, pp. 105-07.

Koyama, K., Awaki, H., Kunieda, H., Takano, S., Tawara, Y., et al. (1989) *Nature*, **339**, pp. 603-05.

Koyama, K., Maeda, Y., Sonobe, T., Takeshima, T., Tanaka, Y. and Yamauchi, S. (1996) *PASJ*, in press.

Mauersberger, R., Henkel, C., Wilson, T.L. and Walmsley, C.M. (1986) *A&A*, **162**, pp. 199-210.

Morris, M. (1994),in *The Nuclei of Normal Galaxies: Lessons from the Galactic Center*, eds: R. Genzel and A.I. Harris, Dordrecht: Kluwer, pp. 185-98.

Morris, M. (1996) in IAU Symp. No.169: *Unsolved Problems of the Milky Way*, eds: L. Blitz & P.J. Teuben, Dordrecht: Kluwer, in press.

Morris, M., Davidson, J.A. and Werner, M.W. (1995) in *Airborne Astronomy Symposium on the Galactic Ecosystem*, eds: M.R. Haas, J.A. Davidson, E.F. Erickson, San Francisco: ASP, ASP Conf. Ser. **73**, pp. 477-88.

Morris, M., Polish, N., Zuckerman, B. and Kaifu, N. (1983) *AJ*, **88**, pp. 1228-35.

Morris, M. and Serabyn, E. (1996) *Ann. Rev. Astron. Ap.*, in press.

Roberge, W.G., Hanany, S. and Messinger, D.W. (1995) *ApJ*, **453**, pp. 238-55.

Scoville, N.Z. (1972) *ApJ*, **175**, pp. L127-32.

Scoville, N.Z., Solomon, P.M. and Jefferts, K.B. (1974), *ApJL*, **187**, pp. L63-66.

Serabyn, E. and Morris, M. (1994) *ApJL*, **424**, pp. L91-94.

Sodroski, T.J., Odegard, N., Dwek, E., Hauser, M.G., Franz, B.A., Freedman, I., Kelsall, T., Wall, W.F., Berriman, G.B., Odenwald, S.F., Bennett, C., Reach, W.T. and Weiland, J.L. (1995) *ApJ*, **452**, pp. 262-68.

Sofue, Y. (1995a) *PASJ*, **47**, pp. 527-549.

Sofue, Y. (1995b) *PASJ*, **47**, pp. 551-559.

Stark, A.A., Gerhard, O.E., Binney, J. and Bally, J. (1991), *MNRAS*, **248**, pp. 14p-17p.

Sunyaev, R., Markevitch, M. and Pavlinsky, M. (1993) *ApJ*, **407**, pp. 606-10.

Uchida, K.I. and Güsten, R. (1995) *A&A*, **298**, pp. 473-81.

Uchida, K.I., Morris, M. and Bally, J. (1994a), in *The Nuclei of Normal Galaxies: Lessons from the Galactic Center*, eds: R. Genzel, A.I. Harris, NATO ASI Series C: **445**, pp. 99-103.

Uchida, K.I., Morris, M., Serabyn, E. and Bally, J. (1994b) *ApJ*, **421**, pp. 505-16.

Uchida, K.I., Morris, M., Serabyn, E. and Güsten, R. (1996b) *ApJ*, in press.

Uchida, K.U., Bally, J. and Morris, M. (1996a) in preparation.

Yamauchi, S., Kawada, M., Koyama, K., Kunieda, H., Tawara, Y. and Hatsukade, I. (1990) *ApJ* **365**, pp. 532-38.

# CO MAPPING OF THE INNER FEW HUNDRED PARSECS OF THE GALAXY

TOMOHARU OKA, TETSUO HASEGAWA,
AND TOSHIHIRO HANDA
*Institute of Astronomy, Faculty of Science,
University of Tokyo,
Osawa 2-21-1, Mitaka, Tokyo 181, Japan*

FUMIO SATO
*Department of Astronomy and Earth Sciences,
Tokyo Gakugei University,
Nukuikitamachi 4-1-1, Koganei, Tokyo 184, Japan*

AND

MASATO TSUBOI
*Department of Physics, Faculty of Science, Ibaraki University,
Bunkyo 2-1-1, Mito, Ibaraki 310, Japan*

**Abstract.** We have made high resolution mapping observations of $^{12}$CO ($J$=1–0) and $^{13}$CO ($J$=1–0) from the central few hundred parsecs of the Galaxy using the NRO 45m telescope (HPBW=16″). The high resolution CO maps reveal many molecular arc and/or shell structures which could have resulted from interactions with supernovae. We also found a large (radius $\sim$ 50 pc) ring structure surrounding the region of diffuse X-ray emission, expanding at $V_{\rm EXP} \simeq 200$ km s$^{-1}$. This and other evidence suggest that a galactic superwind following a burst of star formation in the recent past is sweeping up the ambient gas. The Galactic center may have experienced a burst of star formation $\sim 5 \times 10^7$ years ago, and it may be in post-starburst phase now.

## 1. Introduction and Observations

Molecular gas is strongly concentrated in the inner few hundred parsecs of the Galaxy (see Morris in this volume), where nearly 10% of the molecular



*W.B. Latter et al. (eds.), CO: Twenty-Five Years of Millimeter-Wave Spectroscopy*, 65–67.

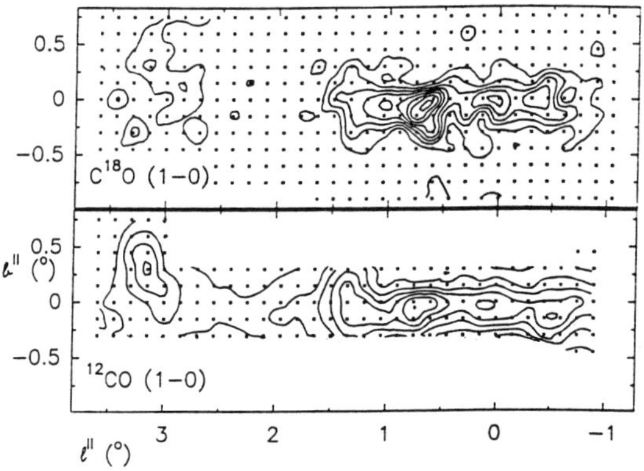


*Figure 1.* $^{12}$CO channel maps. a) Molecular arcs, b) Filaments

mass in the inner Galaxy is found. Toward these galactic center molecular clouds, the lines that trace the high density region ($n(H_2) \geq 10^{4-5}cm^{-3}$) are strongly detected, and this suggests that this region contains a large amount of dense gas which naturally leads to active star formation. However, current "star formation efficiency" (SFE = SFR/$M_g$) is not more than a half of that in the molecular annulus in the Galaxy (Güsten 1989). This moderate star formation activity appears inconsistent with the abundance of dense molecular gas in a picture with steady star formation.

We have made high resolution mapping observations of $^{12}$CO ($J$=1–0) and $^{13}$CO ($J$=1–0) lines with the NRO 45m telescope (beamwidth 16″) using the newly equipped $2 \times 2$ multi-beam SIS receiver. We have collected about 25000 $^{12}$CO and over 9000 $^{13}$CO spectra so far with 34″ grid spacing. The $^{12}$CO data cover almost the full extent of the Galactic center molecular cloud complex. We used wide-band AOSs with 250 MHz bandwidth (650 km s$^{-1}$ @ 115 GHz), which is sufficiently wide to cover the broad line profile from the Galactic center molecular clouds.

## 2. Parsec-Scale Structure of Galactic Center Molecular Clouds

In our large-scale and high-resolution CO maps, we found quite a few, molecular arc structures in the Galactic center molecular clouds (Figure 1*a*). Some of them show clear kinematics of an expanding shell that may have been formed by interactions with supernova remnants and/or stellar wind. The clouds within $|l| \leq 1°$ are bright in CO and show very sharp emission edges compared with the ones in the outer region. The sharp edge could be understood as a result of high external pressure (Spergel & Blitz 1992), due to hot gas ($T_k \geq 10^{7-8}$ K) indicated by X-ray observations (Yamauchi et al. 1990). We also find a number of molecular filaments in the CO emission within $|l| \leq 1°$ at positive velocity side (Figure 1*b*). They

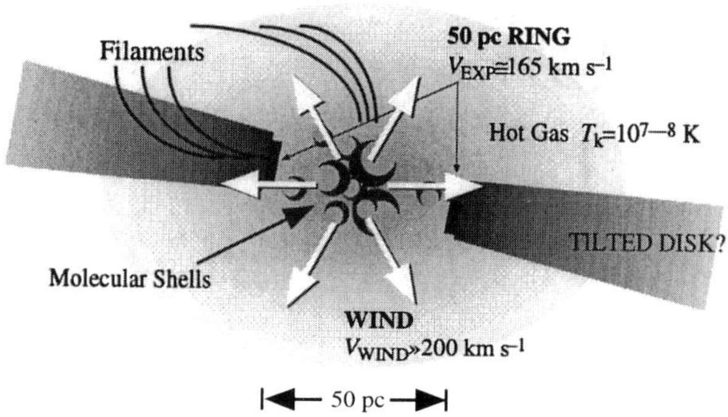


*Figure 2.* Schematic view of the Galactic center

rise up out of the plane as the velocity becomes higher as if being blown off by strong wind from the inner region.

We also find a large (radius $\sim 50$ pc) ring structure expanding at $V_{\rm EXP} \sim 200$ km s$^{-1}$. Its tight anti-correlation with X-ray (Watson et al. 1981), as well as its enhanced $^{13}$CO/$^{12}$CO intensity ratio at the highest velocities, suggests that a hot gas supplied by a number of supernovae and stellar winds is sweeping up the ambient gas. Kinetic energy of this expanding ring is about $10^{54}$ ergs, corresponding to the kinetic energy released by a thousand supernovae. This value is consistent with the estimation from X-ray observations with the Ginga satellite (Yamauchi et al. 1990).

We suggest these results may give the key to solve the inconsistency raised in the introduction. A burst of star formation with a short duration time will have peak in the mechanical energy released as superwind about $5 \times 10^7$ years after a starburst (Heckman et al. 1993). If our Galactic center is now in this post-starburst phase, we can consistently understand the observed features described above and the concentration of intermediate mass (4–9 $M_\odot$) AGB stars in the central 100 pc (Lindqvist et al. 1991). These AGB stars seem to be remnants of previous starburst.

## References

Güsten, R. 1989, in IAU Symposium 136, *The Center of the Galaxy*, ed. M. Morris (Dordrecht: Kluwer), 89

Heckman, T. M., Lehnert, M. D., Armus, L. 1993, in *The Environment and Evolution of Galaxies*, eds. J. M. Shull and H. A. Thronson, Jr. (Dordrecht: Kluwer), 455

Lindqvist, M. , Winnberg, A. , Habing, H. J. , and Matthews, H. E. 1991, A&AS, **92**, 43

Morris, M. 1995, these proceedings

Spergel, N., and Blitz, L. 1992, Nature, **357**, 665

Watson, M. G., Willingale, R., Grindlay, J. E., and Hertz, P. 1981, ApJ, **250**, 142

Yamauchi, S. , et al. 1990, ApJ, **365**, 532

# CHEMISTRY

# THEORETICAL CHEMISTRY

ERIC HERBST
*Departments of Physics and Astronomy*
*The Ohio State University*
*Columbus, Ohio 43210-1106*
*USA*

## 1. Introduction: CO and CI

Theoretical astrochemistry is concerned with the explanation and prediction of the abundances of the many likely and detected interstellar and circumstellar molecules. An updated list of the more than 100 detected gas-phase molecules is given by Herbst (1995). Carbon monoxide, by virtue of its ubiquity and high abundance, occupies a special place of importance.

As a guidepost for understanding the chemistry, quiescent dense interstellar clouds can be divided into shells or slabs of material of increasing density and visual extinction as one proceeds towards the cloud interior. Unlike ultra-violet photons, cosmic rays penetrate throughout the cloud and power a rich ion-molecule chemistry, especially in the high density interior. In outer layers, the dominant form of gaseous carbon is $C^+$ (CII) because of the diffuseness of the gas and the strength of the interstellar radiation field (Tielens & Hollenbach 1985). In intermediate layers ($A_V \approx 1 - 5$), a significant abundance of neutral atomic carbon (C or CI) exists. In the interior, with gas densities in the range $10^{3-5}$ $cm^{-3}$ and large extinction, CO dominates. Indeed, gas-phase models show that CO is produced from precursor atoms in a relatively short time (Herbst & Leung 1989). By $10^5$ yr, the CO fractional abundance has reached its asymptotic value, which depends both on metallicity and on the elemental depletions assumed. For cloud interiors in our galaxy, a fractional abundance of $10^{-4}$ is typically calculated, whereas for analogous cloud interiors in other galaxies, the value can be reduced by an order of magnitude or even more (Millar & Herbst 1990).

Once CO becomes the dominant carbon-containing molecule in the cloud interior, a small residual abundance of C (C/CO $\approx$ 0.001 - 0.01)



*W.B. Latter et al. (eds.), CO: Twenty-Five Years of Millimeter-Wave Spectroscopy,* 71–78.

is maintained via an indirect photodissociation of CO caused by cosmic rays (Prasad & Tarafdar 1983; Gredel et al. 1989). The small residual C abundance derives from the standard, or LIP (low ionization phase), solution to the differential equations governing the gas-phase chemistry; a more exotic solution, labeled the HIP (high ionization phase), exists in which the C/CO abundance ratio is considerably higher, approaching 0.1 (Flower et al. 1994). In the normal solution, the overall column density of C through the cloud is dominated by intermediate shells at times past $10^5$ yr. This column density can be as large as $10^{-5}$ that of $H_2$ for times through $10^6$ yr although it declines thereafter (Lee et al. 1995). The large C abundance calculated at times before steady state is achieved ($10^{7-8}$ yr) may well explain C abundances towards relatively quiescent sources.

The shell-like $C^+$/C/CO structure discussed here is an idealization. Even if nearly spherical shells exist, there is the possibility of diffusion among them (Xie et al. 1995). To the extent that the interior of the cloud is clumpy, both diffusion between diffuse and dense gas and photon penetration from external photons will occur. Star formation and young stellar activity lead to still more photons. The net effect of all of these complications is to increase the internal abundances of C and $C^+$.

## 2. Gas-Phase Interstellar Chemistry

The ion-molecule picture (Herbst 1995; Herbst & Leung 1989; Millar et al. 1991; Bergin et al. 1995) of interstellar chemistry in cloud interiors leads to five successful general predictions, one of which, the high $CO/H_2$ abundance ratio, has already been mentioned. The second prediction concerns the abundances of protonated ions, which are calculated to be sufficiently large for various species to be observed. The detection of the protonated ions $HCO^+$, $HN_2^+$, $HOCO^+$, $HCNH^+$, and $HC_3NH^+$ is in good accord with the ion-molecule picture. The third prediction concerns metastable species, which are calculated to have abundances comparable to normal species. Known to chemists as isomers, metastable species have geometrical structures corresponding to local but not to global potential minima. The best known interstellar example is probably HNC. A more exotic example, l-$C_3H$, is calculated to lie above c-$C_3H$, where the designations 'l' and 'c' refer to linear and cyclic structures, respectively. Although metastable species are often especially reactive in the laboratory, their abundances are enhanced in ion-molecule chemistry for two main reasons. First, both stable and metastable species are typically produced via the dissociative recombination of precursor positive ions and electrons. The reactions are thought to lead to the breaking of differing chemical bonds; for example:

$$HCNH^+ + e^- \longrightarrow HCN + H; HNC + H. \tag{1}$$

Secondly, once produced, stable and metastable species react equally efficiently with abundant ions, typically on every strong collision.

Isotopic fractionation, leading to abundance ratios between so-called isotopomers and normal species very different from the abundance ratios of the isotopes themselves, represents the fourth strong success for the ion-molecule picture. Fractionation in the interstellar medium occurs at low temperatures because the direction of isotopic exchange reactions is critical if the exothermicity is large compared with the temperature. The fractionation of carbon is discussed by Langer in this volume. Here we consider deuterium fractionation. Consider, as an example, the exchange reactions

$$H_3^+ + HD \rightleftharpoons H_2D^+ + H_2. \tag{2}$$

The left-to-right reaction is exothermic by 230 K, so that at temperatures considerably below this value, the right-to-left (endothermic) reaction occurs only slowly. Since, in the interiors of dense clouds, HD is the major repository of deuterium ($HD/H_2 \approx 3 \times 10^{-5}$), the above reaction system can produce a large $H_2D^+/H_3^+$ abundance ratio ($\leq 0.10$) which can then propagate throughout the ion-molecule network to produce large fractional abundances of an assortment of deuterated isotopomers, both ionized and neutral. This propagation is important because most ion-HD exchange reactions do not occur due to energy barriers. One ion produced from $H_2D^+$ is $DCO^+$, generated by the reaction

$$H_2D^+ + CO \longrightarrow DCO^+ + H_2. \tag{3}$$

The observed abundance ratio between this ion and its normal isotopomer is reproduced well by theory. The dissociative recombination of $DCO^+$ leads to a large abundance of atomic deuterium which, in turn, leads to another sequence of fractionation reactions. Detailed model calculations which include a variety of deuterium fractionation reactions show reasonable agreement with observation for many species in cold cloud cores (Millar et al. 1989).

The fifth successful prediction concerns the unsaturated (hydrogen-poor) nature of the chemistry. Many ions with more than 4–5 atoms do not react with $H_2$ because the reactions are endothermic. The result is that the ions cannot incorporate much hydrogen into their structures. Dissociative recombination then leads preferentially to very unsaturated neutral species such as bare carbon clusters and hydrocarbons with only one or two hydrogen atoms.

Despite these general areas of agreement between theory and observation, predictions of individual molecular abundances can be incorrect. This is not surprising; there are numerous chemical, physical, and astronomical uncertainties afflicting the model calculations. Among the chemical uncertainties is a lack of knowledge of the rates of many neutral-neutral reactions at low temperature. Among the physical uncertainties is a poor understanding of how species can be removed from grains once they stick to their surfaces. Among the astronomical uncertainties are the extent of clumpiness in the gas and the role of star formation in influencing the physical conditions and creating 'active' regions, discussed below.

One major uncertainty in ion-molecule model predictions is the extent of large molecule formation. Large molecules, here defined to be more than 4–5 atoms in size, are produced by long sequences of reactions, so that their predicted abundances are especially sensitive to uncertainties in the rates of individual reactions, many of which have not been measured in the laboratory. In particular, the uncertain role of neutral-neutral reactions strongly affects the calculated abundances (Bettens et al. 1995). In addition to this uncertainty, the calculated abundances of complex molecules show a time dependence which is quite distinct from that of smaller species. The latter reach their steady-state abundances by $10^4 - 10^5$ yr; the former reach peak abundances at this time but their abundances decrease sharply at still later times. For successful comparison with the observed abundances in molecule factories such as TMC-1 and L134N, calculated abundances at or near their peak values must be utilized. This peak at so-called 'early time' derives from the oxygen-rich nature of the elemental abundances chosen. The larger molecules are mainly carbon-containing in nature, and at late times virtually all of the carbon is consumed to produce CO. The time dependence may not be critical since it is difficult to see how large molecules can be kept off grains for more than $10^6$ yr. Even so, investigators have increased the predicted abundances of complex species at late times by assuming that diffusion of atomic carbon occurs from outer shells or from diffuse gas in a clumpy interior.

Bettens & Herbst (1995) have recently extended the maximum size of molecules in gas-phase models from $\approx 10$ atoms to upwards of sixty atoms. The synthesis of hydrocarbons has been taken from recent laboratory work, which shows that the growth of linear chains leads not to PAH's but, via isomerization, to large monocyclic rings, which then condense to form tricyclic rings, which finally isomerize to form fullerenes, including the famous $C_{60}$. Growth can even occur under diffuse cloud conditions (Thaddeus 1994). The production of PAH's occurs preferentially in C-rich circumstellar shells (Frenklach & Feigelson 1989).

## 3. Dust Observations and Chemistry

Along with other condensed-phase species such as water and methanol, CO is detected in grain mantles via broad absorption features in the infrared (Whittet 1993). A detailed examination of the CO feature at 4.67 $\mu$m in many sources shows that most CO is gaseous, with at most 25% in the condensed phase. In addition, the line does not correspond well in frequency with laboratory CO spectra taken in a water-ice-dominated environment. Rather, it is necessary to have a considerable amount of CO in a non-polar environment such as itself, $O_2$, and $N_2$ (Tielens et al. 1991). This implies a schizophrenic nature for grain mantles since water ice is known to dominate them, with $H_2O$ (solid)/$H_2 \approx 10^{-4}$.

For dust temperatures under 20 K, the low abundance of non-gaseous CO cannot be explained by evaporation, which does not occur competitively. A variety of non-thermal desorption mechanisms for CO and other mantle molecules has been proposed (Williams 1993). We (Dzegilenko & Herbst 1995) have shown that CO can be desorbed from ice surfaces via an indirect mechanism in which infrared radiation at 3.05 $\mu$m excites an OH stretch, which relaxes via several channels, among which is the transfer of energy into the so-called librational (hindered rotational) mode of a neighboring CO molecule. The librational energy is strongly coupled to the physisorption mode between CO and the grain surface, and the sudden infusion of librational energy permits desorption of the CO in a time scale of picoseconds. Interestingly, the direct absorption of infrared radiation to excite the CO stretch has been measured not to lead to desorption since the two modes appear to be almost completely uncoupled.

The need for two environments for condensed-phase CO can be explained by time-dependent chemical effects. In the initial stages of a cloud, the gas is rich in atomic hydrogen, which is uniquely reactive on grain surfaces because of its rapid random (tunneling) motion from site to site. The formation of molecular hydrogen occurs when two surface H atoms collide with one another and deposit the energy released in chemical bond formation into the energy modes of the grain. In addition, H atoms can coalesce with slower moving heavy surface atoms such as O and N to produce the radicals OH and NH, which in turn react with additional H atoms. The process of association continues until saturation is reached, and species such as water and ammonia ($NH_3$) formed. Later in the cloud history, the gas-phase hydrogen becomes predominantly molecular, and the nature of the grain chemistry changes. Polar saturated species no longer dominate and frosting of molecules such as CO produced in the gas occurs, although the stability of these molecules is not guaranteed. Assuming that CO can be maintained on grain surfaces, the condensed-phase CO observations are consistent with a picture in which the inner mantle layers are rich in water and ammonia, while the outer layers contain the bulk of the CO.

A number of gas-grain chemical models, in which both gas and grain chemistry are followed along with adsorption and desorption processes, have been carried out (Hasegawa & Herbst 1993; Willacy et al. 1994). Most of these models use differential rate equations to describe the grain chemistry; however, it has just been realized by Tielens that these equations can be in error because they inappropriately average the discrete nature of the chemistry that occurs on small particles. We have recently begun to correct the rate equations; an alternative procedure, the Monte Carlo method, is appropriate for steady-state models (Tielens & Hagen 1982) but harder to implement for time-dependent calculations. With some current rate equation approaches, it is difficult to account for the non-polar environment of CO although others are successful (Willacy et al. 1994).

## 4. Active Regions

Star formation affects the chemistry in a variety of ways, which have been investigated in some detail. In the vicinity of low mass star formation, the dominant effect is an increase in the gas and dust density, which results in an increase in the adsorption rate on grains. In the absence of any large increase in the temperature, the grain mantles grow rapidly and the gas phase becomes depleted in molecules containing heavy elements, especially those molecules with relatively large surface binding energies. In the vicinity of high mass star formation, an increase in the temperature results in almost the opposite behavior - the grain mantles are almost totally desorbed and the gas is enriched in molecules produced previously via grain chemistry (Brown et al. 1988). These so-called 'Hot Cores' are discussed by Walmsley in this volume.

Photon-dominated regions (PDR's) represent another active area of extreme current interest (see Tielens & Hollenbach 1985; Sternberg & Dalgarno 1995; Hollenbach et al. 1991) since they represent the theoretical paradigm for explaining $C^+$ and C abundances in cloud interiors. In these regions, the radiation field is enhanced over the typical interstellar value, either by the presence of an external star near to the source, or, more likely, by internal newly-born stars. If clumpiness is neglected, the enhanced radiation field leads to shells similar to those discussed earlier, in which material near the radiation source contains carbon mainly in the form of $C^+$, material somewhat farther removed from the radiation source is rich in neutral atomic carbon, and material still farther removed contains excited CO. Models of PDR's also consider the thermal balance in these sources as well as their unusual chemistry. An example of the latter is the high abundance of the non-protonated ion $CO^+$. The addition of clumpiness to the models makes for a more complex picture with the external layers of clumps being strongly affected by the radiation. Most chemical models of PDR's have been steady-state in nature; the role of time-dependent effects remains largely unexplored. Somewhat akin to PDR's are cosmic ray-dominated regions (Farquhar et al. 1994), in which the effective cosmic ray ionization rate $\zeta$ is larger than the standard value, and X-ray dominated regions. These last two active sources may well be important in models of clouds in external galaxies.

Regions of high temperature possess perhaps the most distinct chemistry from standard quiescent sources. As the temperature rises, exothermic chemical reactions with activation energy barriers as well as endothermic reactions begin to occur. Regions beset intermittently by turbulence constitute one type of high temperature source; these are discussed by Falgarone in this volume. Shocked regions have been studied for some time, initially to explain the high abundance of $CH^+$ in diffuse clouds. These are now divided into two classes depending on magnetic field. In J-type shocks, the temperature rises rapidly to thousands of degrees and then relaxes slowly (Mitchell 1988; Neufeld & Dalgarno 1989). In C-type shocks, the rise in temperature is less dramatic but a two-fluid structure arises in which ions can stream with respect to neutrals, enhancing the importance of endothermic ion-molecule reactions (Roberge & Draine 1990; Pineau des Forêts et al. 1993). Modellers of high temperature regions should be cautious in utilizing thermal rate coefficients for high temperature processes measured in the laboratory; although excitation of internal degrees of freedom can preferentially power endothermic reactions, the available energy in shocks is mainly translational.

## 5. Summary and Future Prospects

During its first twenty-five years, interstellar chemistry has made much progress, especially in the area of gas-phase processes. The general nature of the chemistry in dense cloud interiors is understood. Large models now exist to predict the gas-phase abundances of hundreds of species in quiescent sources as well as in a variety of active sources such as Hot Cores, PDR's, and shocks. Although the abundances of many molecules in different types of sources are predicted accurately by the models, many questions remain due to lingering chemical, physical, and astronomical uncertainties. Grain chemistry remains less well understood although progress is being made.

The future should see more progress in understanding the rates of chemical reactions and physical processes such as desorption. As the input into models becomes more secure, modellers will use their models to describe more complex physical situations, taking account of heterogeneities as well as their time dependence. Some dynamic models are already available (Prasad et al. 1991; Rawlings et al. 1992). The era of calculations with a single spatial point may soon end. Given the range of problems remaining, the next twenty-five years should witness substantial progress.

## 6. Acknowledgments

The support of the National Science Foundation is gratefully acknowledged.

## References

Bergin, E. A., Langer, W. D., & Goldsmith, P. F. 1995, ApJ, 441, 222
Bettens, R. P. A., & Herbst, E. 1995, Int. J. Mass Spec. Ion Proc., in press
Bettens, R. P. A., Lee, H.-H., & Herbst, E. 1995, ApJ, 443, 664
Brown, P. D., Charnley, S. B., & Millar, T. J. 1988, MNRAS, 231, 409
Dzegilenko, F., & Herbst, E. 1995, ApJ, 443, L81
Farquhar, P. R. A., Millar, T. J., & Herbst, E. 1994, MNRAS, 269, 641
Flower, D. R., Le Bourlot, J., Pineau des Forêts, G., & Roueff, E. 1994, A&A, 282, 225
Frenklach, M., & Feigelson, E. D. 1989, ApJ, 341, 372
Gredel, R., Lepp, S., Dalgarno, A., & Herbst, E. 1989, ApJ, 347, 289
Hasegawa, T. I., & Herbst, E. 1993, MNRAS, 263, 589
Herbst, E. 1995, Ann. Rev. Phys. Chem., 46, 27
Herbst, E., & Leung, C. M. 1989, ApJS, 69, 271
Hollenbach, D. J., Takahashi, T., & Tielens, A. G. G. M. 1991, ApJ, 377, 192
Lee, H.-H., Pineau des Forêts, G., Roueff, E., & Herbst, E., in preparation
Millar, T. J., Bennett, A., & Herbst, E. 1989, ApJ, 340, 906
Millar, T. J., & Herbst, E. 1990, MNRAS, 242, 92
Millar, T. J., Rawlings, J. M. C., Bennett, A., Brown, P. D., & Charnley, S. B. 1991, A&AS, 87, 585
Mitchell, G. F. 1988, in Astrochemistry, ed. M. S. Vardya & S. P. Tarafdar (Dordrecht: Reidel), 87
Neufeld, D. A., & Dalgarno, A. 1989, ApJ, 344, 251
Pineau des Forêts, G., Roueff, E., Schilke, P., & Flower, D. R. 1993, MNRAS, 262, 915
Prasad, S. S., Heere, K. R., & Tarafdar, S. P. 1991, ApJ, 373, 123
Prasad, S. S. , & Tarafdar, S. P. 1983, ApJ, 267, 603
Rawlings, J. M. C., Hartquist, T. W., Menten, K. M., & Williams, D. A . 1992, MNRAS, 255, 471
Roberge, W. G. & Draine, B. T. 1990, ApJ, 350, 700
Sternberg, A., & Dalgarno, A. 1995, ApJS, 99, 565
Thaddeus, P. 1994, in Molecules and Grains in Space, ed. I. Nenner & L. Trojanowski (New York: Am. Inst. of Physics), 711
Tielens, A. G. G. M., & Hagen, W. 1982, A&A, 114, 245
Tielens, A. G. G. M., & Hollenbach, D. J. 1985, ApJ, 291, 722
Tielens, A. G. G. M., Tokunaga, A. T., Geballe, T. R., & Baas, F. 1991, ApJ, 381, 181
Whittet, D. C. B. 1993, in Dust and Chemistry in Astronomy, ed. T. J. Millar & D. J. Williams (Bristol: Inst. of Physics), 9
Willacy, K., Williams, D. A., & Duley, W. W. 1994, MNRAS, 267, 949 (1994).
Williams, D. A. 1993, in Dust and Chemistry in Astronomy, ed. T. J. Millar & D. J. Williams (Bristol: Inst. of Physics), 143
Xie, T., Allen, M., & Langer, W. D. 1995, ApJ, 440, 674

# OBSERVATIONAL CHEMISTRY

C. M. WALMSLEY
*University of Cologne*
*I Physikalisches Institut, Universität zu Köln*
*Zülpicherstrasse 77*
*D-50937 Köln*
*Germany*

## 1. Introduction

Observational Chemistry is a vast topic and I will not attempt to review it in this brief contribution. There are, moreover, several excellent reviews already available (e.g., Irvine et al. 1987; van Dishoeck et al. 1993) which give an overview of molecular abundance determinations. Other discussions of specific aspects of the problem are given by Walmsley and Schilke (1992, 1993), by Mundy et al. (1995), and by van Dishoeck and Blake (1995). In this review, I will consider the specific question of gas–grain interactions in molecular clouds. Specifically, I will review some recent work discussing evidence for depletion of heavy elements onto dust grain surfaces as well as the inverse problem of the mechanisms which cause material deposited onto dust grains to be recycled into the gas phase.

## 2. Evidence for Depletion of C, N, O, and Si in Molecular Clouds

There is a long standing problem arising from the fact that carbon, nitrogen, and oxygen do not show convincing evidence for depletion onto dust grains in the high density clumps in molecular clouds. For a normal distribution of dust grain sizes, one can estimate the timescale for depletion $\tau_d$ (in yrs) as:

$$\tau_d = 10^9/(s\,n(\mathrm{H_2})) \qquad (1)$$

where $n(\mathrm{H_2})$ is the molecular hydrogen density in $\mathrm{cm}^{-3}$ and $s$ is the sticking coefficient for collisions of molecules with grain surfaces. It has been demonstrated that $s$ is likely to be close to unity for likely physical conditions and grain surface materials (e.g., Leitch-Devlin and Williams 1985).



*W.B. Latter et al. (eds.), CO: Twenty-Five Years of Millimeter-Wave Spectroscopy,* 79–86.

From equation 1, one derives timescales for depletion of order $10^5$ years at typical densities in molecular cloud cores. This is less than current estimates for the dynamical lifetimes of such clumps and it, therefore, seems surprising that there has been little evidence for substantial depletion. Léger et al. (1985) consider explanations for this.

It is important to note that CO, unlike most other observed interstellar molecules, is a majority gas phase species in the sense that it is believed to be the repository of most gas–phase carbon. One finds typically $[CO]/[H_2]$ of order $10^{-4}$ which is a factor of roughly 6 less than the interstellar carbon abundance. The usual assumption is that the missing 5/6 of the carbon is in the solid state either in the form of graphite or as some refractory organic species. While it is possible that a lot of carbon is locked up in some as yet unobserved gas phase form, it seems unlikely. Theory suggests very strongly that, where dust extinction is sufficiently great that UV radiation is excluded from cloud interiors, most gas–phase carbon will become transformed into CO.

On the other hand, molecules such as CS, $NH_3$, etc., only contain a small fraction of their constituent elements and hence their abundance does *not* depend linearly on the amount of C, S or N in the gas phase. Their (gas phase) destruction rates may be reduced as a consequence of depletion of (for example) oxygen onto grain surfaces and, hence, they may retain high abundances even when a large fraction of C, N, and O has disappeared into ice (see Pineau des Forêts et al. 1991). Thus measurements of $[CO]/[H_2]$ are a better guide to depletion in dense cores than abundances of other species. Of course in situations where, say, $10^{-4}$ or less of solar C, N, and O is left in the gas phase, essentially all species will show reduced abundances relative to standard molecular cloud values. But, in general, results from species containing a small fraction of their constituent elements are a bad guide to depletion.

It is in this context that recent observations by Wilson et al. (1995) using the BIMA array to observe $C^{18}O(1\text{-}0)$ towards NGC 2024 are interesting (see also poster by Mehringer et al., this conference). These authors have a similar synthesised beam to the single dish angular resolution used by Mezger et al. (1988, 1992) to observe the 1.3 mm and 870 $\mu$m dust emission. They can thus compare rather well the CO column density with the dust (and by inference hydrogen) column density of Mezger et al. (see also Mauersberger et al. 1992). The BIMA results show rather convincingly that *on the small (0.01 pc) scale*, there is an anti-correlation between $C^{18}O$ emission and dust. The inferred $[CO]/[H_2]$ ratio is of the order of $10^{-5}$ or lower which is an order of magnitude less than that typically observed in molecular clouds. Although there is considerable uncertainty in hydrogen column densities derived from dust emission and although the kinetic temperature (and hence the $C^{18}O$ partition function) in these high density clumps is also uncertain, the evidence that CO is underabundant seems strong.

Thus, the contention of Mauersberger et al. (1992) and earlier of Mezger et al. (1988) that molecular depletion was occurring in the high density cores in NGC 2024 is confirmed by the new BIMA data from Wilson et al. (1995). It is also interesting to note that a recent $C^{34}S$ map of NGC 2024 by Wiesemeyer et al. (1995) made with the Plateau de Bure Interferometer shows reasonable agreement between the distribution of high density clumps seen in $C^{34}S$ and $C^{18}O$. This suggests that in spite of depletion seen in CO, the [CS]/CO] abundance ratio remains approximately at the canonical molecular cloud value of around $3\,10^{-5}$.

The situation is slightly different for ammonia which has been mapped using the VLA by Gaume et al. (1992) and by Ho et al. (1993) and the $[NH_3]/[CO]$ abundance ratio varies over two orders of magnitude (from $10^{-5}$ to $10^{-3}$) from one position to another in NGC 2024. In at least one case (FIR3), ammonia seems to be coincident with the high density cold clump. Thus ammonia appears to behave differently from CO and CS. One possibility is that ammonia is present in an outer layer of the clump (but see Walmsley and Schilke 1992 for another interpretation).

The NGC 2024 high density clumps are not unique in showing signs of heavy element depletion onto dust grain surfaces. Blake et al. (1995) have recently completed a detailed study of the gas surrounding the "extreme class I source" ( i.e., embedded submm–continuum sources) NGC1333/IRAS 4A and B. They find depletion factors of 10–20 for all species including CO in regions where the beam–averaged density exceeds $10^7$ cm$^{-3}$. The molecules may only be present in an outer skin inside of which molecules containing heavy elements have been frozen out (see model calculations by Rawlings et al. 1992).

On the basis of the above studies, one concludes that at densities above $10^7$ cm$^{-3}$ depletion becomes important in the sense that $[CO]/[H_2]$ descends to $10^{-5}$ or smaller values and other molecules appear to behave in analogous fashion. Some depletion is occurring at much lower densities as witnessed by the solid–state CO and $H_2O$ features which are observed to contain a few percent of carbon and oxygen along sight–lines passing through "normal molecular cloud gas" (densities of a few thousand cm$^{-3}$). Nevertheless, one observes gas whose density is of orders $10^5 - 10^6$ cm$^{-3}$ where the $[CO]/[H_2]$ ratio appears normal. The most impressive evidence for this, perhaps, is the direct measurement of $[CO]/[H_2] = 2.5\,10^{-4}$ towards NGC 2024 IRS2 by Lacy et al.(1994) who observed the near infrared vibration-rotation transitions of both species. The column density $N(H_2)$ in this line of sight is $3.5\,10^{22}$ cm$^{-2}$ and the density must be at least $10^5$

$cm^{-3}$. The dust temperature (which is a critical parameter) is unknown. It would be of considerable interest to obtain abundances for a variety of species towards IRS2.

There is one species, however, which appears to be depleted even at dark cloud densities ($10^4 - 10^5$ $cm^{-3}$). That is SiO for which the upper limits to the abundance $[SiO]/[H_2]$ in sources such as TMC1 and L183 are of order $3\,10^{-12}$ (Ziurys et al. 1989). This is surprising because SiO is thought to be the major gas–phase form of silicon (Herbst et al. 1989, Langer and Glassgold 1989) and because the solar silicon abundance is $[Si/H] = 3.5\,10^{-5}$. Thus, it appears that silicon is depleted in dark clouds by a factor of at least $10^7$ whereas carbon (as one deduces from the CO abundance) is depleted by only a factor of a few. While silicon is depleted more than carbon in diffuse clouds (Sofia et al. 1994), the low upper limit to SiO in TMC1 and L183 suggests that *within the dense core*, depletion is much more effective for silicon than for carbon, nitrogen, etc. One possibility is simply that the sticking coefficient is considerably larger for SiO than for other species. This would then imply a relatively low sticking coefficient for species such as CO and would provide a partial explanation for the low depletion rates of carbon–rich species in general. Another difference between SiO and CO, however, is that the latter can be evaporated at very modest grain temperatures.

An interesting new development in the context of silicon depletion is the detection by Greaves et al. (1995) of SiO in absorption in the spiral arm clouds along the line of sight towards W49 and SgrB2 which are thought to have densities of a few thousand $cm^{-3}$(i.e., intermediate between the $\zeta$Oph cloud where silicon is depleted by a factor 20, and sources like TMC1 and L183). Greaves et al. find $[SiO]/[H_2]$ of order $10^{-10}$ consistent with the idea that silicon depletion increases rapidly with increasing density.

## 3. Molecules in hot cores and outflows

A considerable amount of work has gone into abundance determinations in the Orion hot core and other regions with analogous properties. One motivation for this has been the idea that there is an early phase in the evolution of a massive star (prior to the switch–on of a compact HII region) where the embedded object heats up the surrounding molecular gas to temperatures at which icy mantles on dust grain surfaces can evaporate. One then observes objects such as the Orion hot core with temperatures well above 100 K, densities above $10^7$ $cm^{-3}$ and column densities in the range $10^{24} - 10^{25}$ $cm^{-2}$. Example are the hot cores in SgrB2-N, G10.47, G31.4, G34.26, and of course Orion (see Walmsley and Schilke 1993 for a more detailed discussion). These objects have the characteristic that relatively saturated (hydrogenated) species such as $NH_3$, $CH_3OH$, $CH_3CN$, $C_2H_5CN$, and $CH_2CHCN$ are found to have abundances an order of magnitude or more larger than those determined in cold dark dust clouds.

The interpretation has been that these molecules are manufactured on grain surfaces due mainly to reactions with atomic hydrogen which is expected to be both mobile and reactive. Then (due to the formation of an embedded O–B star), the grains heat up causing evaporation of the icy mantles. The constituents of the mantles are ejected into the gas phase and stay there for more than $10^4$ years until gas phase chemistry reasserts itself and establishes new steady state abundances. Model calculations of the evolving chemistry for a variety of scenarios have been carried out by Brown et al. (1988), Charnley et al. (1992), and by Caselli et al. (1993).

A characteristic of their results is that one expects water and ammonia in hot cores to contain a considerable fraction of the available oxygen and nitrogen respectively. Thus one expects abundances for these species of order $10^{-4}$ relative to $H_2$. It has been known for some time, however, that the ammonia abundance in hot cores is more like $10^{-5} - 10^{-6}$ and it is becoming clear (Gensheimer et al. 1995; Zmuidzinas (this Symposium) that the situation for water is similar. This creates a serious problem for the case of water whose solid state abundance is directly measured (see van Dishoeck et al. 1993) and seems to be an order of magnitude larger than the $H_2O$ abundance measured in hot cores. Thus, water in the solid state does not seem to be simply transformed into gas phase water in the hot core.

There are many possible explanations for this including the comforting one that the uncertainties in the various measurements are such that the "discrepancy" should not be taken seriously. While this may be true, one should perhaps consider alternate possibilities. One is that in the core which precedes the "turn-on" of the embedded O–B star, the grain mantle composition is fundamentally different from that seen in regions which one can detect in the 3 $\mu$m ice feature. If for example, most oxygen in the pre–hot–core–mantle was in the form of $O_2$, the low gas phase water abundance would be comprehensible. But then presumably, one would expect to see evidence for $O_2$ which does not seem to be forthcoming. All this underlines the fact that we are still unsure what form most of the oxygen is in, in molecular clouds in general and in hot cores in particular.

Another type of explanation for the low observed water abundance in hot cores is that the process of evaporation itself needs to be considered. The process of warm–up is presumably not instantaneous. While warm–up is taking place, the mobility and reactivity of a variety of species on the grain surfaces is enhanced. The less polar portions of the mantle should evaporate first and will be mobile for a short time prior to evaporation. If there are "stored radical sites" on the grain such as envisaged by Greenberg

TABLE 1. Comparison of Abundances (rel. to $H_2$) in the L1157 Outflow and the Orion Hot Core

| | Orion | L1157 |
|---|---|---|
| $NH_3$ | $6\,10^{-7}$ | $1.3\,10^{-6}$ |
| $CH_3OH$ | $3\,10^{-7}$ | $7\,10^{-6}$ |
| SiO | $2\,10^{-7}$ | $10^{-7}$ |

and collaborators (e.g., Greenberg 1989), these may be capable of powering reactions which, for example, destroy water. Then the start-up conditions for hot core chemistry may be different than envisaged in the model calculations.

Yet another possibility is that the ices come off the grains in shocks. There have been a series of recent studies showing evidence for enhanced abundances of particular molecules in outflows from young stars. The surprise was that the list of molecules found to have enhanced abundances in outflows is rather similar to that of the molecules with enhanced abundances in hot cores. For example, there has been considerable recent work on the L1157 flow (see e.g. Bachiller et al. 1995; Mikami et al. 1992; Umemoto et al. 1992; Tafalla and Bachiller 1995). These results show that in the blue lobe of the L1157 flow, ammonia and methanol are enhanced by rather similar amounts to that estimated for the Orion hot core. Table 1 gives a comparison of the different abundance estimates from the literature. While more data for more species are clearly needed, there appears to be a similarity between Orion hot core abundances and L1157 blue lobe abundances. This suggests that in both cases, the contents of grain mantles have been returned recently to the gas phase. However, this speculation needs to be tested by obtaining more detailed information on the L1157 outflow as well as towards similar sources.

In the case of L1157, presumably shocks are responsible for the ejection of grain mantles since the dust temperature is thought to be low. Moreover, one observes SiO in the L1157 flow and SiO is thought to be a shock indicator (see above and Martin–Pintado et al. 1992). Hence, one may have conditions similar to those envisaged by Flower and Pineau des Forêts (1995) who find that their C-shock models can explain a small amount of SiO ejection with shock velocities as low as 25 km $s^{-1}$. Another possibility (see Mackay 1995) is that there is a small amount of silane ($SiH_4$) mixed in to the icy mantle which, subsequent to ejection in the shock, gets processed by gas phase reactions to SiO. Testing this would require a search for other silicon–containing species towards outflow sources such as L1157.

A completely different approach to studying abundances in hot outflow material is that adopted by Carr et al. (1995) who have observed mid–infrared absorption lines of a variety of species towards the bright ($2\,10^4$ $L_\odot$) outflow source GL2591. They compare their data to an earlier CO study by Mitchell et al. (1989) and find an enhancement by two orders of magnitude in the HCN abundance close to the hot young star relative to that found in the surroundings (using mm lines). This is reminiscent of the Orion hot core where HCN is also enhanced (see Schilke et al. 1992) and again suggests that icy mantle material is returning to the gas phase in this source.

Finally, I note that there are some interesting puzzles to do with sulfur chemistry whose solution is almost certainly relevant to understanding the chemistry of shocks and outflows. SO and $SO_2$ are known to show enhanced abundances in outflows (see e.g., Serabyn in this proceedings; Helmich et al. 1994) but it is unclear whether the explanation lies in surface or in gas phase processes. The phenomenon does *not* seem to correlate well with high abundances for SiO. Moreover, McMullin et al. (1993) present maps of the Orion South concentration showing different distributions for $CH_3OH$, $SO_2$, and SiO. All of these three molecules are thought to be formed in shocks but the data suggest that we are going to be forced to distinguish between different types of shocks.

## 4. Conclusions

The aim of this short review has been to show that there is considerable progress being made both in determining in which localities matter is disappearing onto grain surfaces and where it is reappearing. In both cases, a lot of work is still needed to understand the physical processes involved. Clearly, direct measurements of the species of interest in the solid state will help also. It seems likely that results from ISO may help here but one can already conclude that even in the post-ISO era, we will be to some extent dependent upon gas phase abundance measurements to know what is happening in the solid state.

## References

Bachiller, R., Liechti, S., Walmsley, C.M., Colomer, F. 1995, *Astron. Astrophys.* **295**, L51.

Blake, G.A., Sandell, G., van Dishoeck, E.F., Groesbeck, T.D., Mundy, L.G., Aspin, C. 1995, *Astrophys. J.* **441**, 689.

Brown, P.D., Charnley, S., Millar, T.J. 1988, *Monthly Notices Roy. Astron. Soc.* **231**, 409. Keenan, F.P., Dufton,P.L. 1994, *Astrophys. J.* **420**, L29.

Carr, J.S., Evans, N.J. II, Lacy, J.H., Zhou, S. 1995 *Astrophys. J.* (in press).
Caselli, P., Hasegawa, T., Herbst, E. 1993, *Astrophys. J.* **408**, 548.
Charnley, S., Tielens, A.G.G.M., Millar, T.J. 1992, *Astrophys. J.* **399**, L71.
Flower, D.R., Pineau des Forêts, G. 1995, *Monthly Notices Roy. Astron. Soc.*(in press).
Gaume, R.M., Johnston, K.J., Wilson, T.L. 1992, *Astrophys. J.* **388**, 489.
Gensheimer, P.D. et al. 1995 *Astron. Astrophys.* (in press).
Greaves, J.S., Ohishi, M., Nyman, L.-A, 1995, *Astron. Astrophys.* (in press).
Greenberg, J.M. 1989 pp7-52 in *Evolution of Interstellar Dust and related topics*, edited A. Bonetti, J.M. Greenberg, S. Aiello, publ. North Holland, Proceedings of the "International School of Physics Enrico Fermi".
Helmich, F.P, Jansen, D.J., de Graauw, Th., van Dishoeck, E.F. 1994 *Astron. Astrophys.* **283**, 626.
Herbst, E., Millar, T.J., Wlodek, S., Bohme, D.K. 1989, *Astron. Astrophys.* **222**, 205.
Ho, P.T.P., Peng, Y.-L, Torrelles, J.M., Gómez, J.F., Rodríguez, L.F., Cantó, J. 1993, *Astrophys. J.* **408**, 565.
Irvine W.M., Goldsmith, P.F., and Hjalmarson, A. (1987), p 561-611 in *Interstellar Processes*, edited D.J.Hollenbach and H.A.Thronson, publ. D.Reidel.
Lacy, J.H., Knacke R., Geballe, T.R., Tokunaga A.T. 1994, *Astrophys. J.* **428**, L69.
Langer,W.D., Glassgold,A. 1990 *Astrophys. J.* **352**, 43.
Léger,A., Jura,M., Omont,A. 1985 *Astron. Astrophys.* **144**. 147.
Leitch–Devlin, M.A., Williams, D.A. 1985, *Monthly Notices Roy. Astron. Soc.* **213**, 295.
Mackay, D. 1995 *Monthly Notices Roy. Astron. Soc.* **274**, 694.
Martín-Pintado J., Bachiller R., Fuente A. 1992, *Astron. Astrophys.* **254**, 315.
Mauersberger R., Wilson T.L., Mezger P.G., Gaume R., Johnston K.J. 1992, *Astron. Astrophys.* **256**, 640.
McMullin,J.P., Mundy, L.G., Blake, G.A. 1993, *Astrophys. J.* **405**, 599.
Mezger, P., Chini, R., Kreysa, E., Wink, J. and Salter, C.J. 1988, *Astron. Astrophys.* **191**, 44.
Mezger, P. et al. 1992 *Astron. Astrophys.* **256**, 631.
Mikami, H., Umemoto, T., Yamamoto, S., and Saito, S. 1992, *Astrophys. J.* **392**, L87.
Mitchell, G.F., Curry,C., Maillard, J.-P, and Allen, M. 1989, *Astrophys. J.* **341**, 1020.
Mundy L.G., McMullin J.P., and Blake G.A. 1995, *Astrophys. Space Science*, **224**, 81
Pineau des Forêts, G., Flower, D.R., and Millar, T.J. 1991, *Monthly Notices Roy. Astron. Soc.* **253**, 217.
Rawlings J.M.C., Hartquist,T.W., Menten,K.M., and Williams, D.A. 1992, *Monthly Notices Roy. Astron. Soc.* **255**, 471.
Schilke P., Walmsley, C.M., Pineau des Forêts, G., Roueff, E., Flower, D.R. and Guilloteau, S. 1992, *Astron. Astrophys.* **256**, 618.
Sofia, U.J., Cardelli, J.A., Savage, B.D. 1994, *Astrophys. J.* **430**, 650.
Tafalla M. and Bachiller R. 1995 *Astrophys. J.* **443**, L37.
Umemoto T. et al. 1992 *Astrophys. J.* **392**, L83.
van Dishoeck E.F., Blake G.A., and Lunine J.I. (1993), in *Protostars and Planets III*, edited by E.H. Levy, J.I. Lunine, M.S. Matthews, publ. U. of Arizona press, Tucson.
van Dishoeck E.F., and Blake G.A. 1995, *Astrophys. and Space Science*, **224**, 237.
Walmsley C.M. and Schilke P. 1992 p251-8 in *Astrochemistry of Cosmic Phenomena*, IAU Symposium 150, edited P.D. Singh, published Kluwer Academic.
Walmsley C.M. and Schilke,P. 1993, p37 in *Dust and Chemistry in Astronomy*, edited T.J.Millar and D.A.Williams, IOP Publishing Ltd.
Wiesemeyer, H., Güsten R., Wink, J., and Yorke, H. 1995 (in preparation).
Wilson T.L., Mehringer D.M., and Dickel H.R. 1995 *Astron. Astrophys.*(in press).
Ziurys, L.M., Friberg, P., and Irvine, W.M. 1989, *Astrophys. J.* **343**, 201.

# CO AT OTHER WAVELENGTHS

JOHN H. BLACK
*Steward Observatory*
*University of Arizona*
*Tucson, AZ 85721 USA*

## 1. Introduction

Although carbon monoxide is most widely observed in interstellar and circumstellar matter through its pure rotational transitions at radio wavelengths, it can be observed in the ultraviolet and infrared regions of the spectrum, too. These observations complement the mm-wave data in important ways. Ultraviolet measurements probe photon-dominated regions where CO co-exists with C, $C^+$, and O. Infrared absorption spectroscopy makes possible the direct measurement of the CO abundance both in solid and gaseous form where cold $H_2$ is now measurable as well. Ultraviolet absorption lines of CO and $H_2$ probe the amount of molecular gas in QSO absorbers at high redshift ($z \approx 2-4$).

The first report of mm-wave line emission of interstellar CO (Wilson, Jefferts, & Penzias 1970) was followed rapidly by a detection in ultraviolet absorption toward $\zeta$ Oph (Smith & Stecher 1971), which also yielded an estimate of the $^{12}CO/^{13}CO$ abundance ratio: 105. The initial measurements with rocket-borne spectrometers were supplanted by observations with the orbiting *Copernicus* (Spitzer & Jenkins 1975) and *IUE* instruments. Although Hubble Space Telescope (*HST*) has superior resolution and sensitivity in the ultraviolet, it is virtually blind at wavelengths between the H I Lyman limit (912 Å) and the threshold for absorption lines of ground-state $H_2$ (near 1110 Å); therefore, *Copernicus* remains the source of almost all direct measurements of molecular hydrogen in the same regions where CO is seen. Although CO has been observed in the photospheres of the Sun and other stars for a long time (Goldberg et al 1965, Kuiper 1962, Sinton 1962), interstellar studies in the infrared ($2-4$ $\mu$m) had to await improvements in the sensitivity and resolution of IR spectrometers (Kleinmann et al. 1978, Hall et al. 1978).



*W.B. Latter et al. (eds.), CO: Twenty-Five Years of Millimeter-Wave Spectroscopy*, 87–94.

## 2. Interstellar Absorption Lines

Absorption line spectroscopy at infrared and ultraviolet wavelengths has several important advantages over mm-wave observations of CO. It is often possible to observe lines arising in a number of rotational levels simultaneously, with the same observing conditions. All measurements toward a common background continuum source usually refer to exactly the same absorbing column; therefore, the determination of relative abundances is rather straightforward. This is especially important where cold, interstellar $H_2$ can be observed *directly*. The strong electronic transitions at ultraviolet wavelengths are sensitive to small amounts of CO: e.g., limits on column density of $N(\mathrm{CO}) < 2 \times 10^{12}$ cm$^{-2}$ have been reached even for absorbers at high redshift, $z \approx 1.7 - 2.3$ (Levshakov et al. 1989). On the other hand, absorption-line measurements will always be fundamentally limited to regions where suitable background sources are available, whereas mm-wave emission-line observations have the advantage in determining the distribution of molecular gas. Because the extinction per unit column density of interstellar matter rises very steeply to shorter wavelengths, hot stars that lie behind chemically interesting amounts of interstellar matter are often too faint for high-resolution spectroscopy in the ultraviolet.

To compare the detectability of small amounts of cold, dilute CO by various techniques, consider a uniform cloud of $H_2$ density $n(\mathrm{H_2}) = 300$ cm$^{-3}$ at a kinetic temperature $T_k = 15$ K with a CO linewidth of $\Delta V = 1.0$ km s$^{-1}$. When $N(\mathrm{CO}) = 1.75 \times 10^{14}$ cm$^{-2}$, the $J = 0$ and $J = 1$ rotational levels have fractional populations $X_0 = 0.541$ and $X_1 = 0.417$, respectively (i.e. the excitation temperature is $T_{01} = 4.07$ K). In the $J = 1 \rightarrow 0$ rotational transition, the line-center optical depth is $\tau = 0.10$ and the peak Rayleigh-Jeans brightness temperature is $T_R = 0.10$ K, in excess of the 2.726 K cosmic background radiation. Such a column density is detectable at mm wavelengths in emission, and in absorption toward a suitable background source (e.g. Hogerheijde et al. 1995). For high-resolution absorption-line spectroscopy, we can express the minimum detectable column density as

$$N_{\mathrm{min}}(\mathrm{CO}) = \frac{6.26 \times 10^{13}}{\lambda f} \left(\frac{0.541}{X_0}\right) \left(\frac{10^5}{R}\right) \left(\frac{100}{S/N}\right) \quad \mathrm{cm}^{-2}$$

where $\lambda$ is the wavelength in Å, $f$ is the oscillator strength, and $R = \lambda/\delta\lambda$ is the resolving power. The rms signal-to-noise ratio is $S/N$, and the detection level is set at 3 rms. For the indicated scalings, $N_{\mathrm{min}}(\mathrm{CO}) = 1.1 \times 10^{14}$ cm$^{-2}$ for the (1,0) R(0) vibration-rotation line at $\lambda = 46575$ Å ($f = 1.2 \times 10^{-5}$), and $1.1 \times 10^{12}$ cm$^{-2}$ for the $\mathrm{A}^1\Pi$-$\mathrm{X}^1\Sigma^+$ (2,0) R(0) line at 1477 Å ($f = 0.04$).

### 2.1. ULTRAVIOLET OBSERVATIONS OF CO

The ultraviolet spectrum of CO longward of the H I Lyman limit ($\lambda >$ 912 Å) consists of many electronic transitions, each with vibrational and rotational fine structure. The transitions of interstellar interest (Morton & Noreau 1994) that arise in the ground vibronic state, X $^1\Sigma^+(v=0)$, include the 4th-positive system, A $^1\Pi$ – X $^1\Sigma^+$, which extends to $\lambda = 1544$ Å with bands that span a large range in oscillator strength: from $f_{14,0} = 1.5\times10^{-5}$ at $\lambda$1214 Å to $f_{2,0} = 4.0\times10^{-2}$ at 1478 Å. Absorption in the predissociated transitions at $\lambda < 1100$ Å destroys CO and determines its lifetime against photodissociation (van Dishoeck & Black 1988).

In translucent molecular clouds, $^{12}$CO, $^{13}$CO, C, $C^+$, H, and $H_2$ are all measurable in the same absorbing gas, along with a variety of other simple molecules such as CH, $CH^+$, $C_2$, CN, and OH. Thus it is possible to investigate the chemistry and fractionation of CO exactly where it is just beginning to dominate the gas-phase abundances of carbon and oxygen. The rotational structure is extremely compact in all ultraviolet bands of CO, so that a resolving power $R > 50,000$ is required to disentangle the individual lines and to yield accurate column densities (Black & van Dishoeck 1988).

Although earlier studies of interstellar CO with the *Copernicus* and *IUE* satellites (e.g. Federman et al. 1980; Black 1980; Wannier et al. 1982) were hampered by insufficient resolution, observations with the Goddard High-Resolution Spectrograph on *HST* have recently advanced the state of this art. *HST* observations of the obsessively well studied diffuse clouds toward $\zeta$ Oph have provided a variety of recent values for the column density $N(\mathrm{CO}) = (1.8\pm0.2)\times10^{15}$ cm$^{-2}$ (Lyu et al. 1994), $2.2\times10^{15}$ cm$^{-2}$ (Sheffer et al. 1992), and $2.5\times10^{15}$ cm$^{-2}$ (Lambert et al. 1994). More significant is the disagreement over the abundance ratio of isotopomers, $^{12}$CO/$^{13}$CO $= 82\pm25$, $150\pm27$, and 167, in the same three studies, respectively: the smallest value is in harmony with no fractionation, while the larger values suggest fractionation controlled by isotope-selective photodissociation. Federman et al. (1994) have identified 7 weak intersystem (singlet-triplet) transitions of CO in the spectrum of $\zeta$ Oph, and their assessment of the oscillator strengths for all observed features favors the interpretation of Lambert et al. (1994): $N(\mathrm{CO}) = 2.5\times10^{15}$ cm$^{-2}$ with an uncertainty of 20%. The case for the oscillator strengths preferred by Federman et al. has received considerable support from the recent laboratory measurements of Smith et al. (1994). The $\zeta$ Oph clouds are a pathological case with two narrow velocity components separated by only $\approx 1$ km s$^{-1}$; therefore, the analysis is extremely sensitive to uncertainties in the oscillator strengths and to the (unresolved) line broadening.

The visual extinction toward $\zeta$ Oph is $A_V = 1.0$ mag. With *HST* it is now possible to observe thicker foreground clouds. For example, toward HD 154368, where $A_V = 2.5$ mag, the CO column density is a factor of 10 higher, $N(\text{CO}) \approx 2.5 \times 10^{16}$ cm$^{-2}$ (Snow et al. 1996; Black et al. 1996). Comparison of the fine-structure excitation of atomic coolants like C and $C^+$ with the rotational excitation of CO shows that CO accounts for only a small fraction, 1–2%, of the total cooling.

### 2.2. INFRARED OBSERVATIONS OF CO

Both the (1,0) and (2,0) vibration-rotation bands of CO are observable in absorption in the interstellar medium, at 4.67 and 2.34 $\mu$m, respectively. The line positions and oscillator strengths are accurately determined (Goorvitch & Chackerian 1994a,b; Goorvitch 1994). The infrared technique is also valuable as a means of observing non-polar molecules that have no allowed rotational transitions at radio wavelengths, such as $CH_4$ (Knacke et al. 1985, Lacy et al. 1991), $C_2H_2$ (Lacy et al. 1989), $H_3^+$ (Oka 1981, Geballe & Oka 1989, Black et al. 1990), and potentially $CO_2$. The first direct detection of cold $H_2$ toward NGC 2024 IRS2 and possibly NGC 2264 IRS by Lacy et al. (1994) is a great triumph. In these lines of sight, the CO is also well observed, so that direct determinations of the CO/$H_2$ abundance ratio can be made. Toward NGC 2024 IRS2, the CO is thermalized at $T = 44$ K and has a total column density of $N(\text{CO}) = 7 \times 10^{18}$ cm$^{-2}$ (Black & Willner 1984, corrected for improved oscillator strengths), while the column density of $H_2$ is $N(\text{H}_2) = 3.5 \times 10^{22}$ cm$^{-2}$, including an allowance for the undetected molecules in the $J = 1$ level at the CO temperature (Lacy et al. 1994). This yields $N(\text{CO})/N(\text{H}_2) = 2 \times 10^{-4}$ in the gas phase. The total column density of CO in the solid phase is $2 \times 10^{17}$ cm$^{-2}$, based on the results of Tielens et al. (1991) corrected for an improved absorption strength (see §3). The carbon monoxide accounts for 25% of a solar abundance of carbon, the remainder probably residing in atoms (C and $C^+$), other molecules, and the cores of dust grains. The detection of $H_2$ is difficult because its vibration-rotation lines are weak, electric quadrupole transitions: the product of population fraction and oscillator strength for the (1,0) S(0) line at 2.2 $\mu$m is $X_0 f = 7.2 \times 10^{-14}$ for gas at $T_k = 50$ K and $n(\text{H}_2) = 10^4$ cm$^{-3}$. The corresponding factor for the *permitted* CO (2,0) R(2) line at 2.3 $\mu$m is $X_2 f = 1.1 \times 10^{-8}$; therefore, the strongest 2-$\mu$m CO lines are only 15 times stronger than the $H_2$ line for CO/$H_2 \approx 10^{-4}$. Furthermore, where the $H_2$ is barely measurable and the $^{12}$CO lines are barely saturated in the (2,0) band, $\tau \leq 1$, the $^{13}$CO lines of the (1,0) band will have modest line-center optical depths ($\tau \lesssim 5$) for $^{12}$CO/$^{13}$CO $= 60$. This means that all three species can often be measured in weak or mildly saturated lines, as demonstrated by Scoville et al. (1983) toward

Orion BN where $^{12}CO/^{13}CO = 96 \pm 5$. Similarly, Mitchell & Maillard (1993) measured the $^{13}CO$ (1,0) lines toward NGC 2264 IRS and determined an abundance ratio $^{12}CO/^{13}CO = 94 \pm 13$ through comparison with the $^{12}CO$ (2,0) measurements of Black et al. (1990).

The light sources used in the IR absorption spectroscopy are often luminous young stellar objects that are deeply embedded in molecular clouds. The foreground column densities are usually large enough to make the CO (1,0) lines highly saturated. However, Mitchell et al. (1989, 1990) found, in addition to the thick absorbing column of CO at low temperature ($T \approx 15 - 70$ K), a second thin component at very high temperature ($T \approx 100 - 1000$ K) in several cases. The hot molecules evidently probe the molecular environment very close to the embedded star. In one case, AFGL 2591, the infrared and radio observations of molecules have been combined to make a comprehensive picture of the star-forming cloud core (Carr et al. 1995). The CO absorption toward Orion BN also shows interesting complexity (Scoville et al. 1983). Mitchell, Maillard & Hasegawa (1991) have observed multiple velocity components in absorption toward luminous embedded stars, which they interpret in terms of episodic molecular outflows: blue-shifted features in foreground absorption are unambiguous kinematical signatures of outflow.

## 3. Solid CO in the Interstellar Medium

Solid CO in the interstellar medium can be observed by means of an absorption feature at $\lambda 4.67$ $\mu$m (Lacy et al. 1984, Geballe 1986). The study of interstellar ices has been reviewed recently by Whittet (1993). Laboratory investigations have determined the absorption coefficient and have shown how the band shapes are related to the chemical structure of the ice matrix (Sandford et al. 1988, Schmitt, Greenberg, & Grim 1989, Tielens et al. 1991, Gerakines et al. 1995). Solid CO in a polar $H_2O$-ice matrix produces a broad, shallow component, while CO in a non-polar matrix produces a narrow component. The integrated absorption strength of the CO 4.67 $\mu$m feature is measured to be $\int (\sigma/c) d\nu = 1.1 \times 10^{-17}$ cm molecule$^{-1}$ at $T = 14$ K, independent of the dominant ice constituent but slightly dependent on temperature (Gerakines et al. 1995). This value is approximately 70% smaller than in earlier studies (Sandford et al. 1988, Tielens et al. 1991); therefore, published abundances of solid CO may need to be increased accordingly. The corresponding integrated absorption strength of the $v = 1 \leftarrow 0$ band of gaseous CO is $\frac{\pi e^2}{mc^2} f = 1.1 \times 10^{-17}$ cm molecule$^{-1}$.

The signatures of both polar and non-polar ices are observed (e.g., Eiroa & Hodapp 1989, Tielens et al. 1991). Because these two components probably have quite different volatilities, their amounts (in relation to each other and to the amount of $H_2O$ ice) can be used as probes of the photochemical and thermal processing that has occurred in interstellar dust mantles. For example, lines of sight toward field stars behind the Taurus dark cloud show a non-polar component appropriate for a CO-dominated, $H_2O$-deficient mixture at $T = 10$ K; while lines of sight toward embedded stars in the same cloud exhibit a non-polar component consistent with a $CO_2$:CO mixture at $T = 10$ K, suggestive of efficient production of $CO_2$ ice by radiative processing in environments near young stars (Chiar et al. 1995). The relative amounts of gaseous and solid CO in the same cloud indicate the extent to which gas has frozen out onto grain surfaces and thus probe the evolutionary state of the cloud (Léger, Jura, & Omont 1985; Whittet & Duley 1991; Sandford & Allamandola 1993; Willacy & Williams 1993; Caselli et al. 1994). There remain interesting questions about the mechanisms of desorption that prevent all heavy molecules from freezing out in quiescent, dark clouds (e.g., Willacy, Williams, & Duley 1994; Hasegawa & Herbst 1993; Dzegilenko & Herbst 1995).

In regions where CO can be observed in both solid and gaseous forms, the column-averaged fraction of all CO in solid form shows an interesting dependence on environment. Observations in cold, dark clouds like Taurus show a solid fraction $\delta(\mathrm{CO}) = 7 - 39\%$ (e.g., data from Chiar et al. 1995), while measurements toward luminous "protostellar" sources in hot cores show $\delta(\mathrm{CO}) = 1 - 10\%$ (data from Tielens et al. 1991).

## 4. Stellar and Circumstellar Spectroscopy

In stellar studies, the infrared lines of CO can be used to make accurate determinations of the abundances of the isotopes of carbon and oxygen, which are probes of nucleosynthesis and mixing in stars. The shapes of CO line profiles in spectra of young stellar objects yield valuable information about the kinematics of disks around them (e.g., Calvet et al. 1991, Carr et al. 1993). In the environments surrounding young stellar objects, the gas is sometimes so highly disturbed that intense far-infrared rotational line emission (Watson et al. 1985, Boreiko & Betz 1989) and even emission in vibration-rotation lines (Thompson 1985, Geballe & Garden 1990) are excited. Infrared spectroscopy of CO in evolved stars with circumstellar envelopes can reveal radial variations of density and temperature (e.g., Keady et al. 1988). Carbon monoxide is observed in the infrared spectra of novae (cf. Gehrz 1988) and supernovae (e.g., Liu, Dalgarno, & Lepp 1992; Bouchet & Danziger 1993).

Individual giant and supergiant stars can be identified at 2.3 $\mu$m in the Galactic Center region (e.g., Eckart et al. 1995; Sellgren et al. 1990); the sharp CO band edges in the spectra of the late-type stars can be used to determine their radial velocities and hence to map the stellar kinematics of the Galactic Center. In this way the mass distribution can be determined from the inner bulge at 100–300 pc (Rieke & Rieke 1988, Blum et al. 1995) into the central 0.5 pc (Krabbe et al. 1995). Such observations have strengthened the arguments for the presence of a possible black hole of $\approx 3 \times 10^6$ $M_\odot$ in the Center, which have relied in the past on kinematical studies of the gas (Genzel, Hollenbach, & Townes 1994). In external galaxies where late-type giants and supergiants dominate the K-band light, the CO bands in the unresolved spectra can be used both to assess the evolved supergiant segment of a starburst population and to determine stellar velocity dispersions (Shier et al. 1994, Doyon et al. 1994, Goldader et al. 1995). The derived dynamical masses in the centers of luminous starburst galaxies are an important complement to gas masses that have been estimated from the intensity of mm-wave CO emission.

## References


Black, J.H. 1980, in *Interstellar Molecules*, IAU Symp. 87, B.H. Andrew, ed., Dordrecht: Reidel, p. 257

Black, J.H., & van Dishoeck, E.F. 1988, ApJ, 331, 986

Black, J.H., & Willner, S.P. 1984, ApJ, 279, 673

Black, J.H., van Dishoeck, E.F., Willner, S.P., & Woods, R.C. 1990, ApJ, 358, 459

Black, J.H. et al. 1996, in preparation

Blum, R.D., Carr, J.S., Sellgren, K., Terndrup, D.M. 1995, ApJ, 449, 623

Boreiko, R.T. & Betz, A.L. 1989, ApJ, 346, L97

Bouchet, P., & Danziger, I.J. 1993, A&A, 273, 451

Calvet, N., Hartmann, L., & Kenyon, S.J. 1991, ApJ, 383, 752

Carr, J.S., Evans, N.J., Lacy, J.H., & Zhou, S. 1995, ApJ, 450, 667

Carr, J.S., Tokunaga, A.T., Najita, J., Shu, F.H., & Glassgold, A.E. 1993, ApJ, 411, L37

Caselli, P., Hasegawa, T.I., & Herbst, E. 1994, ApJ, 421, 206

Chiar, J.E., Adamson, A.J., Kerr, T.H., & Whittet, D.C.B. 1995, ApJ, 455, 234

Doyon, R., Wells, M., Wright, G.S., Joseph, R.D., Nadeau, D., & James, P.A. 1994, ApJ, 437, L23

Dzegilenko, F., & Herbst, E. 1995, ApJ, 443, L81

Eckart, A., Genzel, R., Hofmann, R., Sams, B.J., & Tacconi-Garman, L.E. 1995, ApJ, 445, L23

Eiroa, C., & Hodapp, K.-W. 1989, A&A, 210, 345

Federman, S.R., Glassgold, A.E., Jenkins, E.B., & Shaya, E.J. 1980, ApJ, 242, 545

Federman, S.R., Cardelli, J.A., Sheffer, Y., Lambert, D.L., & Morton, D.C. 1994, ApJ, 432, L139

Geballe, T.R. 1986, A&A, 162, 248

Geballe, T.R., & Garden, R.P. 1990, ApJ, 365, 602

Geballe, T.R., & Oka, T. 1989, ApJ, 342, 855

Gehrz, R.D. 1988, ARA&A, 26, 377

Genzel, R., Hollenbach, D.J., & Townes, C.H. 1994, Rep. Prog. Phys., 57, 417

Gerakines, P.A., Schutte, W.A., Greenberg, J.M., & van Dishoeck, E.F. 1995, A&A, 296, 810

Goldader, J.D., Joseph, R.D., Doyon, R., & Sanders, D.B. 1995, ApJ, 444, 97

Goldberg, L., Parkinson, W.H., & Reeves, E.M. 1965, ApJ, 141, 1293

Goorvitch, D. 1994, ApJS, 95, 535
Goorvitch, D., & Chackerian, C., Jr., 1994a, ApJS, 91, 483
Goorvitch, D., & Chackerian, C., Jr., 1994b, ApJS, 92, 311
Hall, D.N.B., Kleinmann, S.G., Ridgway, S.T., & Gillett, F.C. 1978, ApJ, 223, L47
Hasegawa, T.I., & Herbst, E. 1993, MNRAS, 263, 589
Hogerheijde, M.R., de Geus, E.J., Spaans, M., van Langevelde, H.J., & van Dishoeck, E.F. 1995, ApJ, 441, L93
Keady, J.J., Hall, D.N.B., & Ridgway, S.T. 1988, ApJ, 326, 832
Kleinmann, S.G., Hall, D.N.B., Ridgway, S.T., & Wright, E.L. 1978, AJ, 83, 373
Knacke, R.F., Geballe, T.R., Noll, K.S., & Tokunaga, A.T. 1985, ApJ, 298, L67
Krabbe, A. et al. 1995, ApJ, 447, L95
Kuiper, G.P. 1962, Comm. Lunar Planet. Lab., 1, 1
Lacy, J.H., Evans, N.J., Achtermann, J.M., Bruce, D.E., Arens, J.F., & Carr, J.S. 1989, ApJ, 342, L43
Lacy, J.H., J.S. Carr, Evans, N.J., Baas, F., Achtermann, J.M., & Arens, J.F. 1991, ApJ, 376, 556
Lacy, J.H., Knacke, R., Geballe, T.R., & Tokunaga, A.T. 1994, ApJ, 428, L69
Lacy, J.H., Baas, F., Allamandola, L.J., Persson, S.E., McGregor, P.J., Lonsdale, C.J., Geballe, T.R., van de Bult, C.E.P. 1984, ApJ, 276, 533
Lambert, D.L., Sheffer, Y., Gilliland, R.L., & Federman, S.R. 1994, ApJ, 420, 756
Léger, A., Jura, M., & Omont, A. 1985, A&A, 144, 147
Levshakov, S.A., Foltz, C.B., Chaffee, F.H., & Black, J.H. 1989, AJ, 98, 2052
Liu, W., Dalgarno, A., & Lepp, S. 1992, ApJ, 396, 679
Lyu, C.-H., Smith, A.M., & Bruhweiler, F.C. 1994, ApJ, 426, 254
Mitchell, G.F., & Maillard, J.-P. 1993, ApJ, 404, L79
Mitchell, G.F., Curry, C., Maillard, J.-P., & Allen, M. 1989, ApJ, 341, 1020
Mitchell, G.F., Maillard, J.-P., Allen, M., Beer, R., & Belcourt, K. 1990, ApJ, 363, 554
Mitchell, G.F., Maillard, J.-P., & Hasegawa, T.I. 1991, ApJ, 371, 342
Morton, D.C. & Noreau, L. 1994, ApJS, 95, 301
Oka, T. 1981, Phil. Trans. Roy. Soc. London A, 303, 543
Rieke, G.H., & Rieke, M.J. 1988, ApJ, 330, L33
Sandford, S.A., & Allamandola, L.J. 1993, ApJ, 417, 815
Sandford, S.A., Allamandola, L.J., Tielens, A.G.G.M., & Valero, J. 1988, ApJ, 329, 498
Schmitt, B., Greenberg, J.M., & Grim, R. 1989, ApJ, 340, L33
Scoville, N.Z., Hall, D.N.B., Kleinmann, S.G., & Ridgway, S.T. 1983, ApJ, 275, 201
Sellgren, K., McGinn, M.T., Becklin, E.E., & Hall, D.N.B. 1990, ApJ, 359, 112
Sheffer, Y., Federman, S.R., Lambert, D.L., & Cardelli, J.A. 1992, ApJ, 397, 482
Shier, L.M., Rieke, M.J., & Rieke, G.H. 1994, ApJ, 433, L9
Sinton, W.M. 1962, Appl. Opt., 1, 105
Smith, A.M., & Stecher, T.P. 1971, ApJ, 164, L43
Smith, P.L., Stark, G., Yoshino, K., & Ito, K. 1994, ApJ, 431, L143
Snow, T.P., et al. 1996, ApJ, submitted
Spitzer, L., Jr., & Jenkins, E.B. 1975, ARA&A, 13, 133
Thompson, R.I. 1985, ApJ, 299, L41
Tielens, A.G.G.M., Tokunaga, A.T., Geballe, T.R., & Baas, F. 1991, ApJ, 381, 181
van Dishoeck, E.F. & Black, J.H. 1988, ApJ, 334, 771
Wannier, P.G., Penzias, A.A., & Jenkins, E.B. 1982, ApJ, 254, 100
Watson, D.M., Genzel, R., Townes, C.H., & Storey, J.W.V. 1985, ApJ, 298, 316
Whittet, D.C.B. 1993, in *Dust and Chemistry in Astronomy*, T.J. Millar & D.A. Williams, eds., Bristol: IOP Publ., p. 9
Whittet, D.C.B., & Duley, W.W. 1991, A&AR, 2, 167
Willacy, K., & Williams, D.A. 1993, MNRAS, 260, 635
Willacy, K., Williams, D.A., & Duley, W.W. 1994, MNRAS, 267, 949
Wilson, R.W., Jefferts, K.B., & Penzias, A.A. 1970, ApJ, 161, L43

# MILLIMETER-WAVE ABSORPTION STUDIES OF THE ONSET OF DARK-CLOUD CHEMISTRY

R. LUCAS
*Institut de Radioastronomie Millimétrique,*
*300 rue de la Piscine, F-38406 Saint Martin d'Hères, France*

AND

H.S. LISZT
*National Radio Astronomy Observatory,*
*520 Edgemont Road, Charlottesville, VA, USA 22903-2475*

We have recently observed mm-wave absorption lines of $^{13}CO$, $HCO^+$, HCN, $C_2H$, HNC, CN in front of distant, compact radio sources (Lucas and Liszt 1993, 1994). Molecules are easily detected, while the extinction is moderate (typically 1 mag.); $HCO^+$ spectra are richer, broader, and more complex than $^{12}CO$ emission observed with single dishes in the same directions (BL Lac, NRAO 150, B0212+735). Molecular abundances are similar in magnitude to those in dark clouds, while the UV radiation field should be dominant. These results suggest that a chemically rich component of the interstellar medium could be present, but not well represented in CO emission surveys. The clouds we have detected appear to be very similar to diffuse clouds such as $\zeta$ Oph, where Liszt and Lucas (1994) have detected, in emission, a comparable column density of $HCO^+$. We briefly summarize here further recent progresses.

To obtain a better characterization of those clouds, we used the Plateau de Bure array to systematically survey $HCO^+$ absorption in front of 30 mm-wave continuum sources (Lucas and Liszt 1996). Observed spectra are shown in Fig. 1.

Nearly every source below $b = 15°$ was detected. The search for $HCO^+$ emission in the same directions, with the NRAO 12m antenna, occasionally yielded some weak lines (0.05 K), even for $\tau > 1$, implying that $T_{EX}$ must be very close to 2.73 K; thus accurate column densities may be deduced from the 3 mm optical depth alone. Detection statistics show that absorption occurs $\sim 30\%$ as often as for 21 cm HI. The optical depth distribution of individual absorption components have very similar slopes for $HCO^+$



*W.B. Latter et al. (eds.), CO: Twenty-Five Years of Millimeter-Wave Spectroscopy,* 95–97.

and 21 cm HI.

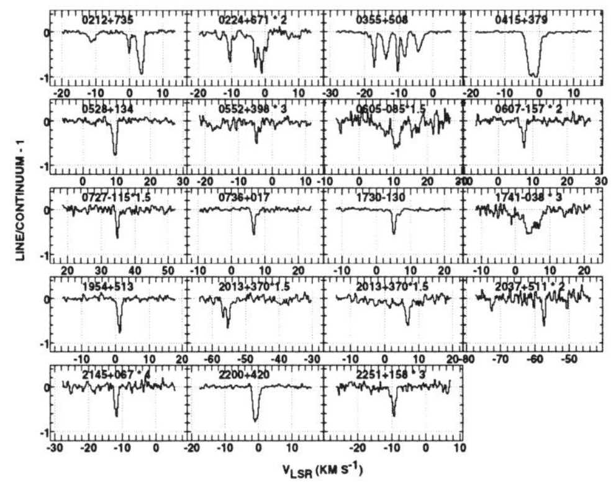


*Figure 1:* $HCO^+$ survey detections. Our sample contained all known sources with $S_{3mm} > 2.5$ Jy for $b > 15°$, and $S_{3mm} > 2.0$ Jy for $b < 15°$ ; we added all sources with CO emission in the survey of Liszt and Wilson (1993). A bandwith of 100 MHz was searched, with a resolution of 140 kHz (336 km $s^{-1}$and 0.47 km $s^{-1}$).

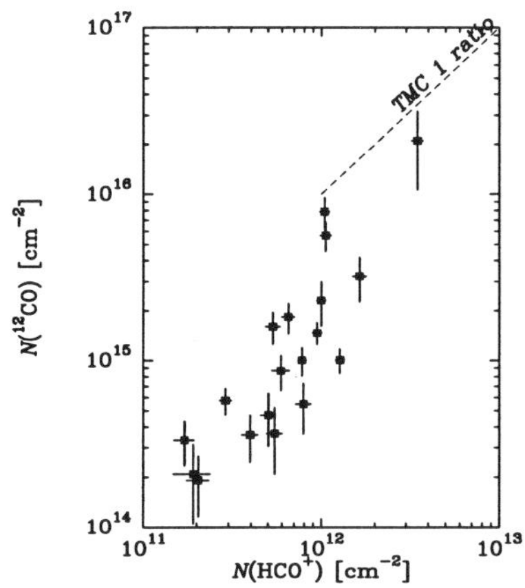


*Figure 2:* $^{12}CO$ versus $HCO^+$ column densities: there is clearly an underabundance of $^{12}CO$ for $N(HCO^+)$ lower than $\sim 10^{12}$ $cm^{-2}$; similarly, the strength of the $^{12}CO$ emission increases abruptly in that range. The same phenomenon is observed in UV absorption studies at $N(H_2) \sim 4\,10^{20}$ $cm^{-2}$. We are seeing the onset of dust- and self-shielding at extinctions below 1 mag, in diffuse clouds and/or outer regions of dark clouds.

We have recently obtained $^{12}CO$ emission and absorption data in some of the surveyed directions (B0212+735, B0224+671, NRAO150, B0528+134, NRAO530, and 3C454.3 (Liszt and Lucas, work in preparation); we use available data for B2013+370 (Lequeux et al. 1993). By direct comparison of emission and absorption, we deduce $T_{EX}$ for each cloud; the results are in the range 3–8 K, indicative of excitation conditions where $n(H_2)T_K \sim 3000 - 10000$ $cm^{-3}$ K. Further constraints from $HCO^+$ emission, which is excited by electrons, imply $T_K > 20 - 40$ K for most sources. At these excitation temperatures, about half of the CO molecules are in the lowest rotational state; the CO column densities can be accurately deduced from the $J = 1 - 0$ optical depth after correcting for the excitation. Fig. 2 shows a plot of $^{12}CO$ versus $HCO^+$ column densities.

The absorption optical depths of $^{12}CO$ are surprisingly low: one would expect larger optical depths on the basis of the observed *emission* intensity ratio between $^{12}CO$ and $^{13}CO$. Since this occurs in nearly every source, one may exclude accidentally low column densities in the direction of the sources. Furthermore, the $^{13}CO$ emission is quite strong while the $C^{18}O$ is very weak wherever we searched for it: the $^{13}CO$ to $C^{18}O$ ratio is much larger than terrestrial. In front of 3C111, this ratio, both in absorption and in emission, is larger than 25. Direct comparison of $^{12}CO$ to $^{13}CO$ *absorption* gives ratios of about 20 in most sources where this ratio may be computed.

Chemical fractionation from exchange with $^{13}C^+$ would increase the $^{13}CO$ abundance for $T_K$ lower than 35 K, while selective photodissociation with $^{12}CO$ being self-shielded has the effect of increasing the $C^{18}O$ abundance, with respect to $^{12}CO$, by $\sim 5-8$. The same effect should also apply to $^{13}CO$. Thus, the high $^{13}CO$ abundances probably result from the chemical fractionation overcoming the selective photodissociation, and *the kinetic temperatures should be low.*

We have measured 18 cm OH absorption with the VLA on some of the sources from the $HCO^+$ survey. Column densities were computed assuming $T_{EX} = (2.73 + 1)$ K, since absorption/emission observations of some of our sources, and similar ones, (surprisingly) lead to very low $T_{EX}$ for 18 cm OH (Dickey et al. 1981). *OH and $HCO^+$ are well correlated*: with an $OH/HCO^+$ ratio of 0.03 to 0.05, which is about the same ratio as in dark clouds. However, simple mechanisms linking OH to $HCO^+$, initiated, e.g., by $C^+ + OH \rightarrow CO^+ + H$, should result in higher values for the same ratio. New chemical models of diffuse clouds involving supersonic turbulence, and its dissipation, succeed to produce $CH^+$ (and $HCO^+$) with endothermic reactions; but they produce also too much OH (Falgarone et al. 1994, Spaans et al. 1994).

Liszt and Lucas (1995) have shown, using 2 mm and 6 cm $H_2CO$ absorption, that $H_2CO$, like CO, turns on at $N(HCO^+) \sim 10^{12}$ cm$^{-2}$. Work in progress shows a similar effect is also seen in HCN, CN, and HNC.

## References

Dickey, J. E., Crovisier, J., Każes, I., 1981: A&A 98, 271
Falgarone, E., Pineau des Forêts, G., Roueff, E. 1995: A&A 300, 870
Lequeux J., Allen R. J., Guilloteau S. 1993: A&A 280, L23
Liszt H.S., Lucas R. 1994: ApJ 431, L131
Liszt H.S., Lucas R. 1995: A&A, 299, 847
Liszt, H. S., Wilson, R. W. 1993: ApJ, 403, 663
Lucas R., Liszt H.S. 1993: A&A 276, L33
Lucas R., Liszt H.S. 1994: A&A 282, L5
Lucas R., Liszt H.S. 1995: A&A 299, 847
Lucas R., Liszt H.S. 1996: A&A 307, 237
Spaans, M., Black, J. H., van Dishoeck, E. F. 1995: preprint

# GALACTIC CARBON MONOXIDE ISOTOPE RATIOS

WILLIAM D. LANGER
*MS 169-506*
*Jet Propulsion Laboratory*
*California Institute of Technology*
*Pasadena, CA 91109*

## 1. Introduction

For twenty-five years, almost since the first detection of CO, isotopic abundance ratios of molecules in interstellar clouds (ISC) have been measured by radio techniques. These observations have been used to: 1) determine Galactic chemical evolution (nucleosynthesis, star formation rate, injection and mixing of gas in the interstellar medium); 2) interpret physical conditions in ISCs; and, 3) test models of interstellar chemistry. The carbon isotopic species are important because 12 and 13 carbon trace primary and secondary nucleosynthesis, respectively, and carbon molecules are pervasive in ISCs. To appreciate the difficulty of these measurements, keep in mind that we are looking for small differences by usual astronomical standards, about a factor of two across the Galaxy and 25 percent locally.

## 2. Observations of Carbon Isotopic Ratios

Across the Galaxy carbon isotopic ratios have been measured in the radio, mainly with CO and $H_2CO$, and locally ($< 1$ kpc) also measured in the optical with $CH^+$ and CN. CO is pervasive and easily observed in emission, however the most abundant isotopes have large optical depths. The earliest efforts relied on $^{13}C^{16}O$ and $^{12}C^{18}O$ but unfortunately even $^{13}C^{16}O$ is optically thick in the line center. Wannier et al. (1976) circumvented this problem by using the line wings where $\tau < 1$. With dramatic improvements in receiver technology, recent measurements have been able to make use of $^{12}C^{18}O$ and the very weak $^{13}C^{18}O$. Formaldehyde is observed in absorption towards strong continuum sources in its $^{12}C$ and $^{13}C$ isotopes. To use these probes one must address line transfer, chemical fractionation, line of sight



W.B. Latter et al. (eds.), CO: Twenty-Five Years of Millimeter-Wave Spectroscopy, 98–100.

origins of the signals, including the uncertain effects of clumpiness on the trapping correction.

The optical absorption studies are generally limited to sources within 1 kpc (towards very bright nearby sources) and require high velocity resolution to separate the isotopic lines. The determination of ratios is complicated by baseline subtraction and line fitting (cf. Vladilo and Centurion 1991). Isotopic enhancement is probably not important for $CH^+$ formation towards these sources because of the high temperatures in the $CH^+$ regions, but may be a factor for CN and CO.

To analyze the carbon isotopic ratio across the Galaxy, Langer & Penzias (1990) studied nine interstellar clouds observed in the rare and doubly rare isotopes of carbon monoxide, $^{12}C^{18}O$ and $^{13}C^{18}O$. These low abundance species minimize radiative trapping, and observations towards dense shielded cores reduce isotopic chemical and photodissociation enhancements. They found a systematic gradient in the $^{12}C/^{13}C$ ratio across the Galaxy from 30 in the inner part at 5 kpc to 70 at 12 kpc, similar to the trend derived from formaldehyde observations (Henkel, Wilson, & Beiging 1982). However, the formaldehyde ratios are generally higher across the disk and show more scatter than the CO measurements (see Figure 1 in Langer 1992). CO and formaldehyde both yield a Galactic Center value about 24.

Because there are many techniques and species available for measuring the ratio in the solar neighborhood, accurate determinations in this region are important for fixing the local value. Langer & Penzias (1990) derived an average value $^{12}C/^{13}C = 57\pm3$ locally, although individual sources ranged from 49 to 67. To improve the local solar neighborhood value they later studied three sources not associated with HII regions: B5, B335, and L134N (L183); and, observed another position in Orion far from the HII region near K-L (Langer & Penzias 1993). In these 4 sources the ratio ranges from 58 to 71 with a weighted average of $61\pm2$. This value is considerably smaller than the formaldehyde result of $80\pm7$, and the terrestrial value of 89. Orion is the most extensively studied source. From $^{13}C^{18}O$ Langer & Penzias found $79\pm7$ in KL and $67\pm3$ a few arcmin away. Using the optically thin line wings of far-IR emission from excited $^{12}C^{16}O$ and $^{13}C^{16}O$, Genzel et al. (1988) derived a ratio of $85\pm20$. In the near-IR bands of CO Scoville et al. (1983) found $96\pm5$, while the ratio way out in the line wings of J=1-0 emission yield $74\pm9$.

The most reliable optical determinations of the carbon isotope ratio probably come from $CH^+$. Those from CN may be affected by fractionation (Langer 1992). Stahl et al. (1989), Crane et al. (1991), and Stahl & Wilson (1992) find a value about $67\pm2$ from $CH^+$. The best values derived from the $CH^+$ absorption measurements are of the nearby cloud Zeta Oph, $71\pm3$ (Stahl & Wilson 1992), $67\pm2$ (Crane et al. 1991), and $67\pm3$ (Centu-

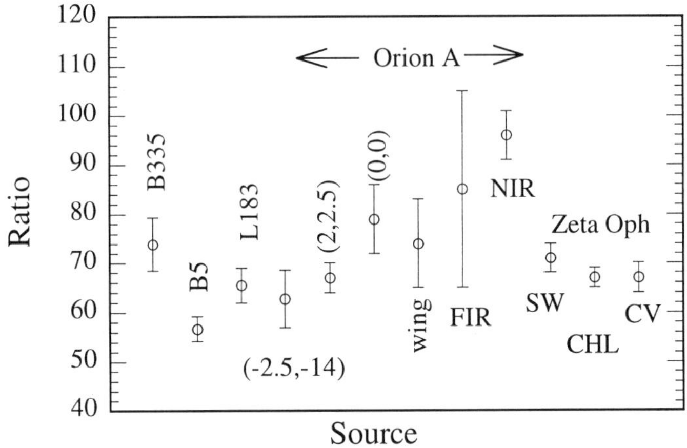


*Figure 1.* Carbon isotopic ratios in the local solar neighborhood using different probes. $^{13}C^{18}O$ is used for B5, B335, L183, Orion A (-2.5,-14), Orion A (0.5,2.5) from Langer & Penzias; $CH^+$ used for Zeta Oph where the letters SW, CHL, and CV refer to Stahl & Wilson, Crane et al., and Centurion & Vladilo, respectively; the other Orion A data uses the Far-IR (FIR), Near-IR (NIR), and CO line wings (wings) as referenced in the text.

rion & Vladilo 1991). These radio and the optical data in the local solar neighborhood are summarized in Figure 1. The carbon ratio ranges from 57 to 96 with a weighted mean value of 67.3±1.5.

This work was performed at the Jet Propulsion Laboratory, California Institute of Technology, under contract to the National Aeronautics and Space Administration.

## 3. References

Centurion, M. & Vladilo, G., 1991, A&A, **251**, 245
Crane, P., Hegyi, D. J., & Lambert, D. L. 1991, ApJ, **378**, 181
Genzel, R., Poglitsch, A & Stacey, G. J. 1988, ApJ Letters, **333**, 59
Henkel , C., Walmsley, M., & Wilson, T. L. 1980, A&A, **109**, 344
Langer, W. D. 1992, IAU Symposium No. 150, Astrochemistry of Cosmic Phenomena, Ed. P. D. Singh, p193 (Kluwer:Dordrecht)
Langer, W. D. & Penzias, A. A. 1990, ApJ, **357**, 477 (LP90)
Langer, W. D. & Penzias, A. A. 1993, ApJ, **408**, 539 (LP93)
Scoville, N., Kleinman, S., Hall, D. & Ridgway, S. 1983, ApJ, **275**, 201
Stahl, O., Wilson, T., Henkel, C., & Appenzeller, I. 1989, A&A., **221**, 321
Stahl, O. and Wilson, T. L. 1992, A&A, **254**, 327.
Wannier, P., Penzias, A., Linke, R., & Wilson, R. 1976, ApJ. **204**, 26

# USING CO ISOTOPES TO PROBE THE ISM

GLENN J. WHITE
*Physics Department, Queen Mary and Westfield College,*
*London University, Mile End Road, London E1 4NS, England*

## 1. Introduction

Observations of CO and its most abundant isotopomers $^{13}CO$, $C^{18}O$ and $C^{17}O$ have been extensively used over the past 25 years to infer the excitation conditions and to study the physical characteristics of the interstellar medium. These measurements have allowed studies of

- Molecular abundances – forming the foundation of ISM chemistry
- Masses of galaxies – probing the star formation history and 'X-factor'
- Cooling of molecular clouds – tracing the ISM thermodynamics
- Energetics of molecular outflows
- Masses of molecular clouds
- Star Formation histories.

The usual method for estimating the gas properties has at its heart the following chain of inferences connecting observational quantities to $H_2$ column densities

$$\int T_a^* dv \overset{\text{LTE}}{\longrightarrow} N_{\text{col}}(^{13}\text{CO}) \overset{X(^{13}\text{CO/CO})}{\longrightarrow} N(\text{CO}) \overset{X(\text{CO/H}_2)}{\longrightarrow} N(\text{H}_2).$$

There are several assumptions involved in this estimate; that the Local Thermodynamic Equilibrium (LTE) approximation can be used to estimate the total $^{13}CO$ column density; that we can estimate the excitation temperatures and opacity; and that we have good knowledge of the inter-isotopomeric abundance ratios of the $^{13}CO/CO$ and $CO/H_2$ and the four dimensional (x, y, z, and velocity) structure of a molecular cloud—which has often been assumed homogeneous and spherical with systematic velocity gradients. In fact, looking back over much of the past 25 years of the history of molecular line astronomy, it is remarkable how few reliable



*W.B. Latter et al. (eds.), CO: Twenty-Five Years of Millimeter-Wave Spectroscopy,* 101–109.

determinations of these important conversion factors have been made, particularly since they have had a profound and wide ranging influence on almost all aspects of the field. Recent observational and theoretical models show that neither these conversion factors nor the excitation conditions remain constant within individual clouds, and that many of the 'authoritative' column densities, cloud masses, and molecular abundances that are spread throughout the literature have systematic inaccuracies.

## 2. Observational Methodology

The column density of a diatomic molecule may be estimated using the emission from a line (such as the $(J+1)\rightarrow J$ transition) of an optically thin(ner, than CO) isotopomer; for the example of the $J = 2$–1 transition;

$$N_{col} = \frac{3.34 \times 10^{14}}{\left[\frac{\nu}{\mathrm{GHz}}\right]\left[\frac{\mu}{\mathrm{Debye}}\right]^2 \left(1 - e^{-\frac{h\nu}{kT_{\mathrm{ex}}}}\right)\left(e^{-\frac{Jh\nu}{2kT_{\mathrm{ex}}}}\right)} \frac{\tau}{1 - e^{-\tau}} \int T_r^* dV \quad \mathrm{cm}^{-2}$$

where $\mathrm{T_{ex}}$ is estimated from the peak CO antenna temperature, $\int T_r{}^* \, dV$ is the integral of the isotopomer's antenna temperature over the line, $J$ is the lower rotational quantum number (1 for the CO $J$= 2–1 transition), $\nu$ is the line frequency, $\mu$ is the dipole moment and the opacity, $\tau$, is estimated using the peak intensity of the CO and $^{13}$CO lines, remembering that measured antenna temperatures are related to kinetic temperature through the relationship

$$T_{\mathrm{peak}} \cong \frac{h\nu}{k} \frac{(1 - \exp(-\tau))}{\left(\exp\left(\frac{h\nu}{kT_{\mathrm{kin}}}\right) - 1\right)}.$$

The equation for column density assumes that the Planck corrected CO peak temperature is the same as the kinetic temperature (which is then assumed to be equal to the excitation temperature and constant for all the CO rotational levels), that the isotopomer is in LTE, and that the escape probability $\beta = \tau \,/\, (1\text{–}e^{-\tau})$ accurately accounts for all optical depth effects. It appears from multi-isotopomeric maps we have made of several sources, that LTE column density estimates are unreliable, and systematically lead to under-estimates of true column densities. Indeed, Monte Carlo calculations of several dark clouds (Avery *et al.* 1987) have already shown that the $J$= 2–1 and 3–2 CO transitions are sub-thermally excited, and that the cloud cores are often hidden by a surrounding lower density envelope. This means that although the most optically thin lines (such as C$^{17}$O) trace emission from the denser regions at the core, the column densities derived using LTE techniques for the more abundant isotopomers are unreliable.

Another LTE technique has been developed to calculate the absorption of a background source by a foreground cloud (Phillips *et al.* 1988; Greaves *et al.* 1995). This models the depth of the self-absorption feature with increasing $J$-transition, allowing the excitation temperature, $T_{ex}$, of a linear molecule to be estimated from observations of two adjacent rotational transitions (1 and 2)

$$\frac{\int T_{r_1} dv}{\int T_{r_2} dv} = \frac{\nu_1}{\nu_2} \frac{J}{J+1} \frac{\left(e^{h\upsilon_1 kT_{\rm ex}} - 1\right)}{\left(1 - e^{-h\upsilon_2 kT_{\rm ex}}\right)} \frac{T_{\rm back_1}}{T_{\rm back_2}},$$

where $\int T_{r1}\ dv$ and $\int T_{r2}\ dv$ are the integrated absorption line intensities for adjacent transitions denoted by (1) ($J \rightarrow J$ -1) and (2) ($J$ +1$\rightarrow J$ ), at frequencies $\nu_1$ and $\nu_2$, against a background source of temperature $T_{back}$, assuming both transitions have the same $T_{ex}$.

## 3. The Physical Situation

A further example of the difficulties faced by the data modeller is variability of the CO/$^{13}$CO/C$^{18}$O ratio—in some dark clouds the $^{13}$CO column density is enhanced by selective photodissociation and ion-molecule exchange reactions (van Dishoeck & Black 1988). The photodissociation rates for some other isotopic species are larger than for $^{13}$CO owing to smaller self-shielding of the less abundant isotopes. As $^{13}$CO is more abundant than C$^{18}$O, it benefits from greater self-shielding. Since the isotopic lines are shifted in frequency from those of $^{12}$CO, mutual shielding will not be significant – detailed modelling of isotopomeric ratios has been presented by several authors including most recently Gierens *et al.* (1992); Köster *et al.* (1994); and Jansen (1995). Further, the $^{13}$CO abundance in cool material may also be enhanced due to the ion molecule exchange reaction

$$^{13}\mathrm{C}^+ + {}^{12}\mathrm{CO} \Leftrightarrow {}^{12}\mathrm{C}^+ + {}^{13}\mathrm{CO} \quad (+36\ \mathrm{K}),$$

since the reaction from left to right is exothermic, the right-to-left sequence is not efficient in the low temperature gas that forms the bulk of material in molecular clouds—leading to a $^{13}$CO overabundance.

## 4. Recent Observational Data

Recent advances in on-the-fly mapping observing techniques and in low-noise millimetre and submillimetre receiver technology now allow the routine acquisition of maps containing tens of thousands of independent data points, contrasting with the situation just a few years ago, when 'typical' maps contained only a few tens of data points. These modern data sets allow statistical techniques to be applied to search for trends which relate the behaviour of the different isotopomers. The few sources which have

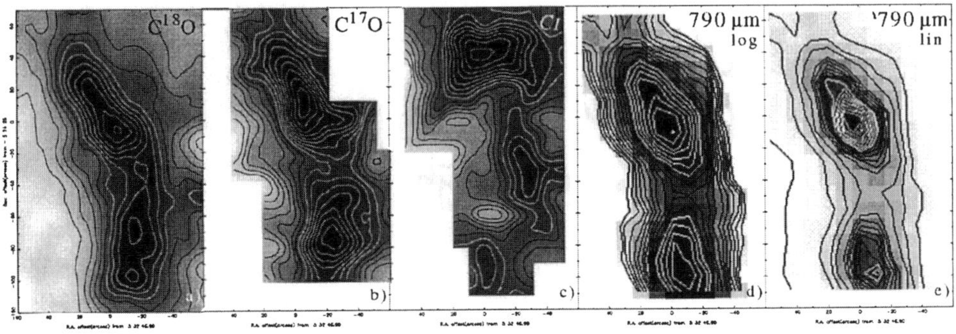


*Figure 1.* CO and isotopes in the core of the Orion Molecular Cloud

been studied in this way show a complexity far beyond that which simple homogeneous models are able to explain.

In a recent study of gas at the centre of the Orion Nebula, White and Sandell (1995) mapped the CO $J$=2–1, 3–2 and 4–3 transitions of CO, and various isotopomers including $^{13}$CO, C$^{18}$O, C$^{17}$O, and 790 $\mu$m continuum emission towards the central region of the Orion molecular cloud (OMC-1 cloud), as shown in Figure 1.

Analysing the data in a consistent way, the inter-isotopomeric abundance ratios were estimated over a wide range of column densities as seen in Figure 2. Under favourable geometries, column densities can be related to line-of-sight extinction values. Although this technique may not always trace the extinction between ultraviolet sources which ionise the molecular cloud, this scenario may hold reasonably well for the Orion molecular cloud, since the dominant ionisation sources for the region—the Trapezium Cluster—lie between the earth and the OMC-1 core. The $^{13}$CO lines towards Orion are optically thick with the $^{13}$CO abundance enhanced in lower column density material because of strong isotopic fractionation. No isotope selective enhancement of the other CO isotopomers was seen; C$^{18}$O may in fact show a slight depletion in more exposed material. The C$^{18}$O and C$^{17}$O lines are optically thin, and correlated with the 790 $\mu$m dust continuum emission. Observations of one of the photoionised products of CO, atomic carbon CI, showed that its emission comes from hot, optically thin gas; the abundance ratios of [CI]/[CO] are typically 0.05–0.3, with the larger ratios towards the northern section of the Orion ridge. The CI abundance ratios are also high along the edge of the Bright Bar which is adjacent to the HII region (and the Trapezium cluster which excites it), but decrease in the dense shielded material behind the Bar.

Several studies have suggested that the ratio of $^{13}$CO to C$^{18}$O can

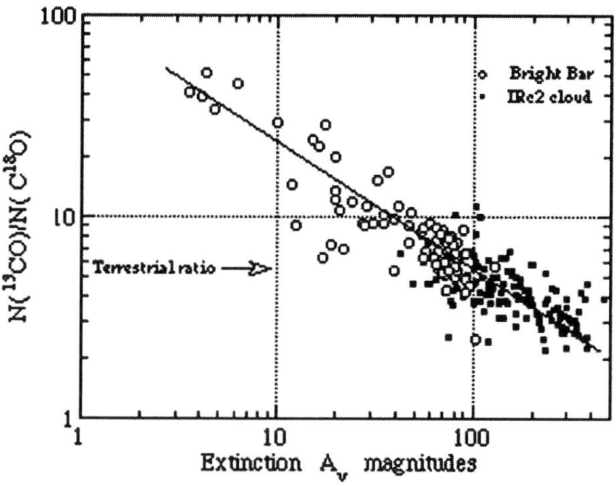


*Figure 2.* Graphs of $N(^{13}CO)/N(C^{18}O)$ against extinction as estimated from the $C^{18}O$ data. In a) the merged data from the OMC1 cloud and the Bright Bar are shown together. The best fit power law shown is for the relationship $N(^{13}CO)/N(C^{18}O) = 80A_v^{-0.6}$ with a correlation coefficient of 0.91. The regions where the fractionation is high are dominated by emission close to the Bright Bar ionisation front, where the extinction to the Trapezium stars is low, providing strong evidence for $^{13}CO$ fractionation, whilst the case for a $^{13}CO$ overabundance in the OMC1 core is less convincing. The bottom axis is given in units of extinction, where this is assumed to be proportional to the $C^{18}O$ column density as shown by Lada *et al.* (1994). This assumption may be realistic for Orion, where we are viewing molecular material face on illuminated by the Trapezium stars, which lie between the Earth and Orion, but questionable for other clouds.

exceed the terrestrial ratio (5.5) because of $C^{18}O$ photodissociation, and chemical fractionation of $^{13}CO$ (van Dishoeck & Black 1988; Turner *et al.* 1992; Turner 1993; Fuente *et al.* 1993; Köster *et al.* 1994). The ratio $N(^{13}CO)/N(C^{18}O)$ has been shown to exceed the terrestrial value close to the surfaces of several dark clouds (Frerking *et al.* 1982; Bachiller & Cernicharo 1986; and Lada *et al.* 1994).

Similar trends in inter-isotopomeric intensities have been observed towards about another half dozen sources including S140 (Minchin, Ward-Thompson, & White 1995 a); NGC1977 (Minchin, White, & Ward-Thompson 1995 b); IC1848, B35, Cepheus B (all in preparation); NGC253 (Israel, White, & Baas 1995); and the Serpens Nebula (which is shown below) (White, Casali, & Eiroa 1995). Data towards the Serpens Nebula are shown in Figure 3. A similar analysis was applied to the Serpens data and is illustrated in Figure 4.

Although the $^{13}CO$ emission from most sources appears to be optically thick (which invalidates the use of $^{13}CO$ emission in the column density equation—unless the opacity is carefully corrected for), there are just a few sources where $^{13}CO$ has been shown to be optically thin; including the SE

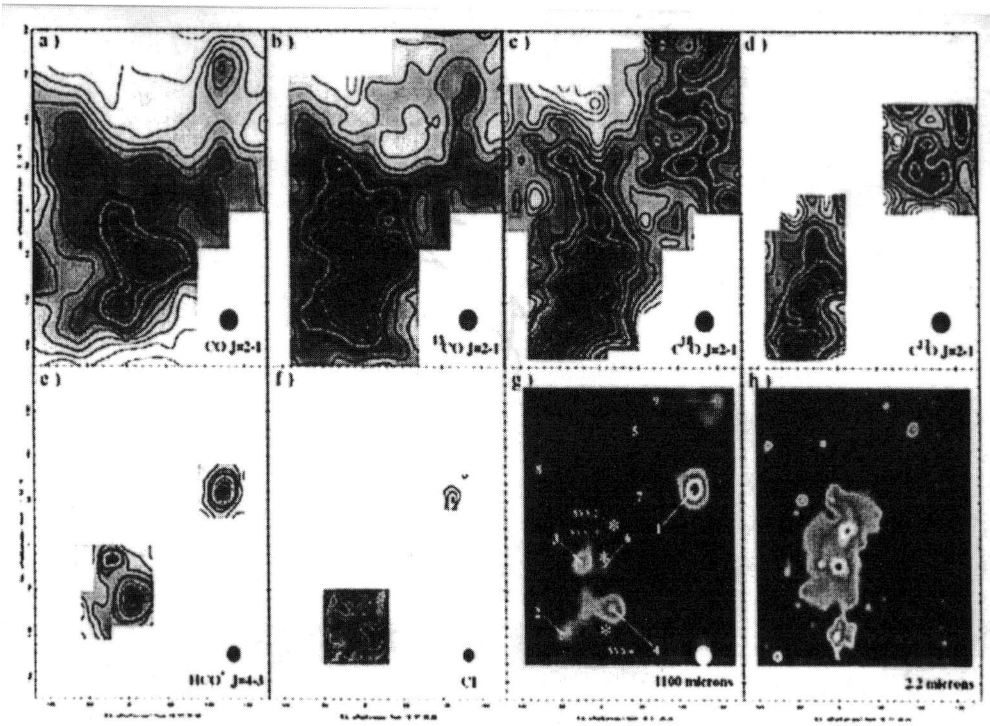


*Figure 3.* CO and continuum data for the Serpens Nebula obtained with the JCMT

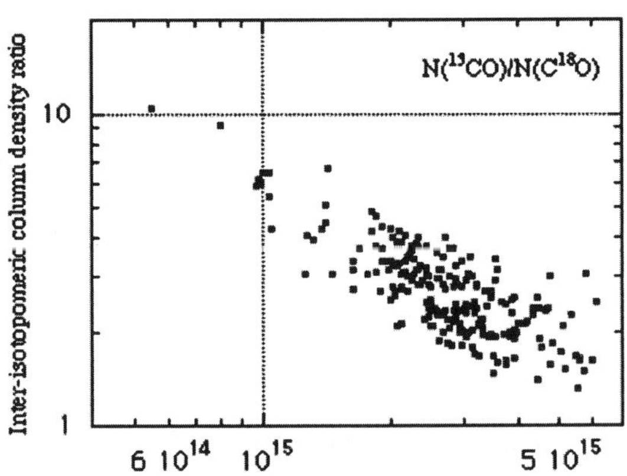


*Figure 4.* Isotopomeric ratios observed towards the Serpens Nebula

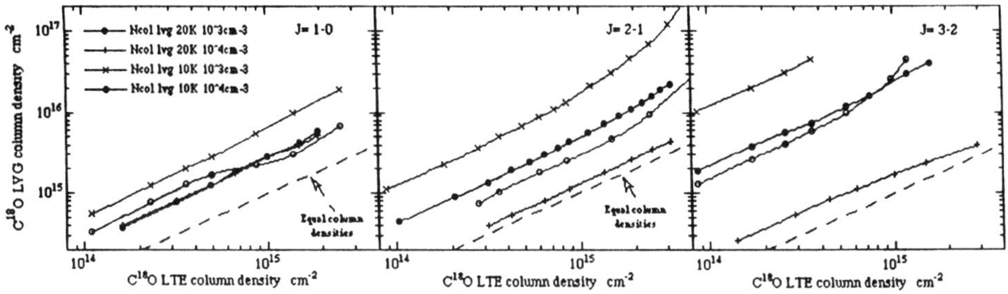


*Figure 5.* Comparison of $C^{18}O$ $J=$ 1–0, 2–1, and 3–2 column densities derived using LVG and LTE techniques. The LTE estimates are consistently lower, by factors of 4–70, of those from LVG analysis.

lobe of the archetypal molecular outflow source L1551 (Phillips *et al.* 1988; Fridlund & White 1989; Fridlund *et al.* 1996) and in some of the molecular clouds at the edge of the blast wave of the IC443 supernova remnant (White 1994).

## 5. Modelling the Problem

To examine the uncertainties which might result from use of the LTE analysis, a series of calculations have been made for simple slab geometries, comparing the column densities which would be derived from a set of data analysed separately with an LTE and LVG analysis. The objective of these models is to demonstrate how the choice of model technique can bias the estimate of a column density for different rotational transitions. LVG calculations were performed for clouds with kinetic temperatures of 10 K and 20 K, to determine the column density that would be estimated for various antenna temperatures (for discussion of the technique see Israel, White, & Baas 1995). The $H_2$ particle density, $n(H_2)$ was set to values of $10^3$ and $10^4$ $cm^{-3}$, and column densities calculated for the $C^{18}O$ antenna temperature range observed towards Serpens ($\sim 0.5 - 5$ K). Column density estimates for the J= 1–0, 2–1, and 3–2 transitions were then compared with values derived using the relationships in the LTE equations and are shown in Figure 5.

The LVG column densities estimated from the $J=$ 2–1 lines are $\sim$ 4–8 times, and $\sim$ 12–70 times greater than the LTE estimates, for kinetic temperatures of 20 and 10 K respectively. The difference between LVG and LTE estimates is greatest at the higher $J-$ transitions and in cooler material. Despite the problems highlighted above, it has been a widespread practice amongst many molecular line astronomers to use the LTE approximation and to unjustifiably assume that the $^{13}$CO emission is optically thin. It is likely that many of the LTE-based estimates of cloud masses, using the higher rotational transitions of C$^{18}$O lines to estimate column densities, have under-estimated the true masses. Mass estimates using the $^{13}$CO lines to estimate column densities are even more unreliable, due to the finite opacity in this isotopomer. However, in the warm ($\sim$ 20 K) moderate density ($\sim 10^4$ cm$^{-3}$) gas in the central region of the Serpens Nebula, the LTE and LVG estimates agree more closely.

These calculations show that column densities estimated by the LTE model differ by perhaps one order of magnitude compared to those inferred using an LVG model for the cases of low-density, cool gas (see also discussion by Stenholm 1985; Richardson *et al.* 1985). In all cases, the column densities are lower than those in the LVG calculation. This result is consistent with the sub-thermal excitation of the CO isotopomers suggested by Monte Carlo modelling (Avery *et al.* 1987), towards dark clouds observed in the $J=$ 2–1 and 3–2 transitions. This modelling suggested that the cores are often surrounded by lower density envelopes, which can modify the line shapes and temperatures due to material in the core. Schilke *et al.* (1994) have subsequently arrived at similar conclusions towards the Taurus dark cloud core TMC-1. The discrepancies between LTE and LVG estimates are more pronounced at the lower densities and temperatures found in dark clouds, than towards warmer clouds (where most of the published C$^{18}$O observations to date have been made). This is consistent with the suggestion of Richardson (1994) that the LTE method is only effective at estimating column densities (at the centre of a line) down to an optical depth of $\sim$ unity.

These simple calculations have shown the uncertainties which can occur from use of the LTE technique. Undoubtedly, the problem is more complex—we know that sources are inhomogeneous, clumpy, UV-permeated, have temperature and density gradients, are shocked or turbulent, and exist in chemically evolving regions. There have been many attempts to provide a framework for a more rigorous treatment of the radiative transfer through more realistic complex regions; for example, recent papers by Castets *et al.* (1990); Gierens *et al.* (1992); Köster *et al.* (1994); and references cited within these papers. Discussion of these is beyond the intent of this paper. There remains an urgent need to develop further radiative transfer code which can corrcctly account for the complex processes we are now beginning to recognise as being present. The best candidate for these would seem to be based around a Monte Carlo approach.

## 6. Summary

- $H_2$ and CO column densities based on $^{13}$CO data are very uncertain, despite having occupied a central role in the development of molecular line astrophysics
- Many of the estimates of column densities and abundances made in the past are subject to systematic uncertainties as a consequence of the use of the LTE analysis
- The excitation temperatures of the various CO isotopes are often poorly determined, and incorrectly assumed
- $^{13}$CO fractionation is common in molecular clouds, and its abundance varies within clouds—this has not been adequately taken account of in estimating cloud masses
- The LTE approximation often inaccurately estimates column density
- Further work on robust Monte Carlo-based radiative transfer models are needed to compliment an explosion of new data on this topic

## References

Avery, L.W., White, G.J., Williams, I.P., & Cronin, N.J., 1987, ApJ, 312, 848
Bachiller, T., & Cernicharo, J. 1986, A&A, 166, 283
Fridlund, M., & White, G.J., 1989, A&A, 223, L13
Fridlund, M., White, G.J., Pilbratt, G., & Tauber, J. 1996, A&A, in prep
Fuente, A.,Martin-Pintado, J.,Cernicharo,J.Bachiller, R. 1993, A&A, 276, 473
Gierens, K., Stutzki, J., & Winnewisser, G. 1992, A&A, 259, 271
Greaves, J.S., Ohishi, M., White, G.J., & Sunada, K, 1995, A&A, 299, 591
Israel, F.P., White, G.J., & Baas,F., 1995, A&A, 302, 343
Jansen, D. 1995, PhD thesis, University of Leiden
Köster, B., Störzer, H, Stutzki, J., & Sternberg, A. 1994, A&A, 284, 545
Minchin, N., White, G.J., & Ward-Thompson, D., 1995a, A&A, 301, 894
Minchin, N., White, G.J., & Ward-Thompson, D., 1995b, A&A, 302, L25
Phillips, J.P., White, G.J., *et al.*, 1988, A&A, 190, 237
Richardson, K.J., White, G.J., Avery, L.W., Lesurf, J.C.G., & Harten, R., 1985, ApJ, 290, 637
Schilke, P., Carlstrom, J.E., Keene, J., & Phillips, T.G. 1993, ApJ, 417, L67
Stenholm, L.G. 1985, A&A, 284, 545
Turner, B.E., Xu, L., & Rickard, B.J. 1992, ApJ, 391, 158
Turner, B.E. 1993, ApJ, 405, 229
van Dishoeck, E., & Black,J. 1988, ApJ 334, 771
Wang, Z., & Scoville, N.Z. 1992, ApJ, 386, 158
White, G.J., Rainey, R., Hayashi, S.S., & Kaifu, N., 1987, A&A, 173, 337
White, G.J. 1994, A&A 283, L25, 1994
White, G.J. Ellison, B., Claude, S., Dent, W.R.F., & Matheson, D.N. 1994, A&A 284, L23
White, G.J. Casali, M., & Eiroa, C., 1995, A&A, 298, 594
White, G.J., & Sandell, G., 1995, A&A, 299, 179

# CLOUDS AND CORES

# PROBING GIANT MOLECULAR CLOUD CORES WITH MILLIMETER AND SUBMILLIMETER OBSERVATIONS OF $C^{18}O$ AND DUST

PAUL F. GOLDSMITH
*NAIC, Cornell University, Ithaca NY 14853*
EDWIN A. BERGIN
*FCRAO, University of Massachusetts, Amherst MA 01003*
AND
D. C. LIS
*Calif. Inst. of Technology, MS 320-47, Pasadena CA 91125*

## 1. Introduction

Giant molecular cloud cores are extremely complex regions. Despite extensive observations starting with the first detection of carbon monoxide in interstellar space, we really do not understand their density structure, nor the velocity distribution of material within them. These parameters are obviously critical to our understanding of the process of massive star formation that occurs in GMC cores, and the mass spectrum of stars that results. The dramatic improvement in observational capability at millimeter and submillimeter wavelengths during the past years now allows the observation of optically thin tracers of material to be carried out over large parts of GMC cores. It is thus increasingly important to understand the capabilities of different tracers, and in this presentation we describe work carried out using both gas and dust as tracers of material in these regions. Each has its advantages and disadvantages; the dust emission is largely free of excitation issues and is certainly optically thin at these wavelengths, but the emissivity of dust is not well known. Molecular tracers suffer from uncertainties due to excitation, and the interstellar chemistry may produce variations in abundance. However, spectral lines do offer important kinematic information.

In this project, we have observed three GMC cores: Orion A, M17, and Cepheus A. Using the FCRAO antenna, we have observed the $J$ = 1–0



*W.B. Latter et al. (eds.), CO: Twenty-Five Years of Millimeter-Wave Spectroscopy*, 113–115.

transition, and with the CSO the $J = 2$–1 and $J = 3$–2 transitions of $C^{18}O$. From the three lowest transitions, the total column density can be determined with high accuracy. Using the CSO we have observed these same sources in the continuum at 1300 and 800 $\mu$m, and for Orion at 450 $\mu$m as well, giving us valuable information about the dust distribution. This summary highlights some conclusions from $C^{18}O$ observations; a more complete analysis of the gas and dust emission is in preparation.

## 2. Observations and Analysis of $C^{18}O$ Emission

$C^{18}O$ emission is a useful tracer of GMC cores because its lines are optically thin, while being strong enough to be readily mapped across large parts of these regions. In addition, the fractional abundance of carbon monoxide in well-shielded regions appears to be relatively constant. The major uncertainty that remains is the degree of molecular excitation and thus obtaining the total column density from the transition(s) observed.

For relatively well-excited ($T_{ex} > h\nu/k$) lines, the integrated intensity of a given line observed directly yields the column density in the upper level of the transition. We define the **correction factor**, **CF**, to be the ratio of the total molecular column density to the column density derived from the set of transitions observed, as determined from a statistical equilibrium calculation. The correction factor when the three lowest transitions of $C^{18}O$ have been observed is shown in the figure. As long as the density is greater than $\sim 10^3$ cm$^{-3}$, CF is essentially independent of both temperature and density - showing a total variation of less than a factor of 1.5:1. This set of the three lowest $C^{18}O$ transitions therefore enables us to accurately determine the column density in GMC cores without having to assume LTE or adopting any particular excitation model.

We have assumed a constant value of CF equal to 1.75, based on the range of gas temperatures previously determined for these GMC cores (Bergin et al. 1994). The peak $C^{18}O$ column densities found are 1.5, 3.2, and $1.9 \times 10^{18}$ cm$^{-2}$ for Orion A, M17, and Cepheus A, respectively. The total masses of the cloud cores are 580, 11500, and 500 $M_\odot$, assuming a $C^{18}O$ fractional abundance of $1.7 \times 10^{-7}$. The radii of the cores are 1.2, 4.9, and $1.3 \times 10^{18}$ cm. Together with the observed FWHM linewidths averaged over the cloud cores, we obtain virial masses (assuming an $r^{-1}$ density distribution) of 350, 3000, and 400 $M_\odot$, respectively. The relatively good agreement between molecular and virial masses suggests that the $C^{18}O$ fractional abundance is not very far from the value that we assumed.

From tho cloud masses and dimensions we derive mean volume densities equal to 3.7, 1.9, and $4.3 \times 10^4$ cm$^{-3}$ for Orion A, M17, and Cepheus A. These densities are less by about 2 orders of magnitude than those derived from multitransition studies of high dipole moment molecules such as $HC_3N$ (Bergin, Snell, & Goldsmith 1996). The linewidths and line center velocities of the $C^{18}O$ and $HC_3N$ are quite similar, despite their very different excitation requirements. This is a significant constraint on models for what appear to be very inhomogeneous GMC cores.

## 3. Observations of the Dust and Comparison With the Gas

The observations of the dust were carried out with the CSO facility bolometer. Based on other studies, the contribution from spectral lines may be appreciable at the positions of embedded sources but in the general cloud volumes this should not be a significant effect. The free-free emission represents only a minor correction. The dust opacity is obtained by assuming that the dust temperature is equal to the kinetic temperature of the dense gas as determined from observations of $CH_3C_2H$ (Bergin et al. 1994). A preliminary reduction of the Orion data suggests that the spectral index of the dust optical depth between 1300 $\mu$m and 450 $\mu$m ($\tau(\nu) \sim \nu^{\beta}$) is $\beta$ = 1.2. In M17 and Ceph A, the dust opacity per unit column density of $C^{18}O$ is essentially independent of molecular column density (or equivalently, of position within the cloud) and is equal for the two GMC cores. In distinction, the positions in Orion exhibit a large scatter, and a mean value approximately an order of magnitude less than that for the other two clouds studied.

## References


Bergin, E.A., Goldsmith, P.F., Snell, R.L., & Ungerechts, H. 1994, *ApJ*, **431**, 674.

Bergin, E.A., Snell, R.L., & Goldsmith, P.F. 1996, *ApJ*, 460, 343.


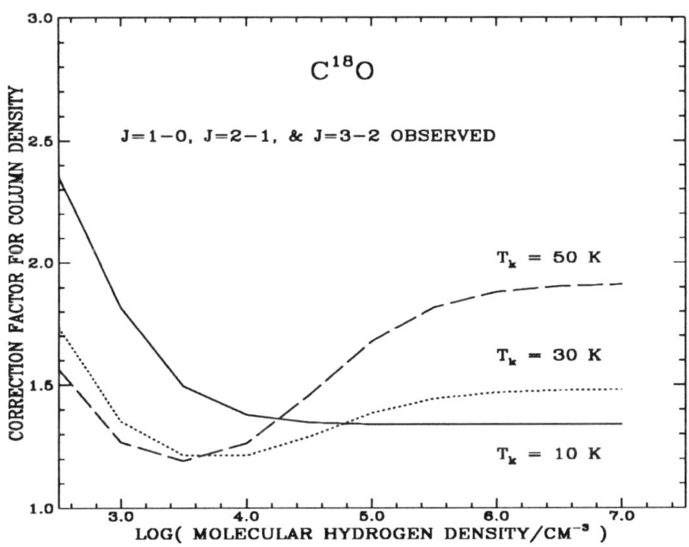

# VELOCITY COHERENCE IN DENSE CORES

A.A. GOODMAN, J.A. BARRANCO AND D.J. WILNER
*Harvard-Smithsonian Center for Astrophysics*
*60 Garden Street, Cambridge, MA 02138*

AND

M.H. HEYER
*FCRAO/UMASS*

**Abstract.** By analyzing the dependence of line width on size in $NH_3$ maps of dense cores, and in $C^{18}O$ and OH maps of the gas around the cores, we find that: (1) on scales larger than the FWHM contour of an $NH_3$ core, line width is correlated with size such that $\Delta v \propto R^a$ where $a \sim 0.4$; and (2) within the cores themselves, the power-law slope of the line width-size relation goes to zero (i.e., line width is constant). We interpret this result as evidence for a breakpoint between "turbulent" self-similar molecular clouds and "velocity coherent cores" which can collapse to form individual stars or small groups of stars.

Low-mass ($\sim$ 1 to 10 $M_\odot$) dense cores appear to be relative islands of calm in a more turbulent sea. These cores, which are the birthplace of many of the stars in our Galaxy, have average densities $\sim 10^4$ cm$^{-3}$, sizes of approximately a few tenths of a parsec, and are characterized by very small internal velocity dispersions ($\sim$ several tenths of a km s$^{-1}$; cf. Benson & Myers 1989; Myers & Benson 1983) . In our recent work (Barranco & Goodman 1996; Goodman et al. 1996), high-sensitivity $C^{18}O$, OH, and $NH_3$ mapping of the cores and their environs have shown that the line width-size power-law scaling relation discovered more than a decade ago by Larson (1981) appears to have a break in slope at a size scale roughly comparable to the full-width-half maximum (FWHM) contour of an $NH_3$ map ($\sim$ 0.1 pc).

On scales larger than an $NH_3$ $(J, K) = (1, 1)$ FWHM emission contour, extensive $C^{18}O$ and OH maps of core environs show a radial dependence of line width on size such that $\Delta v \propto R^a$ where a $\sim$ 0.4. The left panel of Figure



*W.B. Latter et al. (eds.), CO: Twenty-Five Years of Millimeter-Wave Spectroscopy*, 116–118.

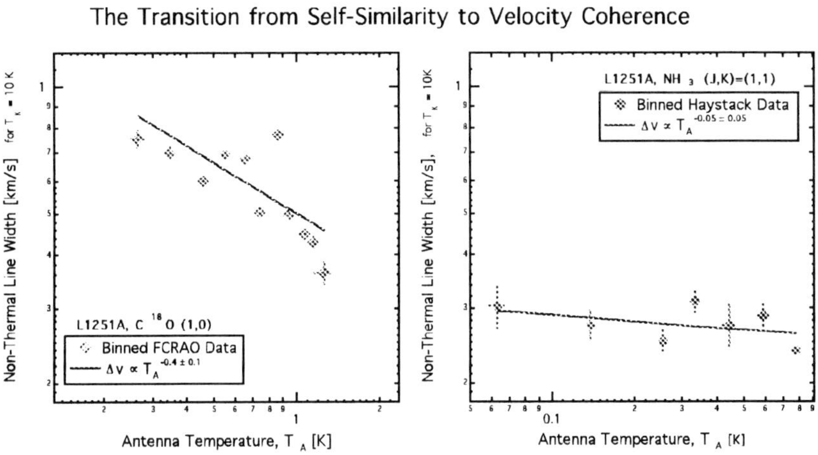


*Figure 1.* Assuming that "size" scales as $1/T_A$, the implied line width–size scaling relation for the L1251A dark cloud environment *(left)* shows a "Larson's Law" dependence of line width on size, whilst observations of an individual dense core within the cloud *(right)* exhibit "velocity coherence" (i. e., virtually no dependence of line width on cloud size).

1 shows one example of this line width-size relation, for the region around the dense core designated L1251A. In practice, the "radial" dependence of line width is derived by approximating $R \propto 1/T_A$ and fitting a straight line to $\log(\Delta v)$ vs. $\log(T_A)$. Barranco & Goodman (1996) describe the fitting procedure in detail and provide empirical justification for the $R \propto 1/T_A$ approximation.

A demonstration of "Larson's Law" (left panel of Figure 1) is interesting on its own, since line width-size relations have rarely been studied in single-tracer maps of individual clouds. (Most studies use multiple lines and/or cloud-to-cloud comparisons (e.g. Fuller & Myers 1992; Larson 1981; Myers & Goodman 1988) . Nonetheless, the most interesting result of our analysis is a "coherence" effect associated with the cores. The $NH_3$ maps show that, within the cores themselves, Larson's Law effectively ends: the line width "inside" the core is constant to a very high degree of significance. This velocity "coherence" effect is illustrated by the nearly constant line width in the L1251A $NH_3$ data shown in the right-hand panel of Figure 1.

Since the measured velocity dispersion, $\sigma$, cannot go below the thermal velocity dispersion, $\sigma_T$, ($\sim 0.07$ km s$^{-1}$ for $NH_3$ at 10 K), it is obvious that the Larson's Law correlation between line width and cloud size must end at some small size (i.e., line width cannot decrease interminably with cloud size). But, the velocity dispersion at which the cores become coherent is always significantly above a purely thermal velocity dispersion. For L1251A, the $\sim$ constant "non-thermal" velocity dispersion is $\sigma_{\mathrm{NT}} = (\sigma^2 - \sigma_{\mathrm{T}}{}^2)^{0.5} = \Delta v_{\mathrm{NT}}/2.35 \sim 0.1$ km s$^{-1}$.

The size scale of the break in the line width-size scaling relationship we find in our molecular line data is just a few times larger than the size scale at which Larson (1995) has recently discovered a break in the power-law slope of the function describing the clustering of newly-formed stars in the ISM. The size scale of the break in the star-clustering power-law has been likened to a Jeans size by Larson. We expect that our result and Larson's are similar, in that we are finding the Jeans mass clumps with roughly constant line width, and he is studying the fruits of those clumps.

In summary, we find that line width decreases with cloud size as Larson's Laws would predict for size scales larger than the FWHM of an $NH_3$ core, but asymptotically approaches a constant value within the cores. We hypothesize that this effect results when the self-similar cascade that characterizes the larger-scale ISM effectively "ends" inside the dense cores, where self-gravity becomes a dominant force. It is possible that the location of the cores corresponds to what once were simply "eddies" or regions of reduced kinetic energy, in the larger-scale turbulent flow of the ISM. In such regions, gravity would gain the edge it needs over kinetic pressure support, in order to initiate a slow collapse.

## References

Barranco, J.A. and Goodman, A.A. 1996, ApJ, in prep.
Benson, P.J. and Myers, P.C. 1989, Ap. J. Suppl., **71**, 89
Fuller, G.A. and Myers, P.C. 1992, ApJ, **384**, 523
Goodman, A.A., Barranco, J.A., Heyer, M.H. and Wilner, D.J. 1996, ApJ, in prep
Larson, R.B. 1981, MNRAS, **194**, 809
Larson, R.B. 1995, MNRAS, **272**, 213
Myers, P.C. and Benson, P.J. 1983, ApJ, **266**, 309
Myers, P.C. and Goodman, A.A. 1988, ApJ, **329**, 392

# STRUCTURE OF MOLECULAR CLOUDS AND TURBULENCE

EDITH FALGARONE
*Radioastronomie Millimétrique, CNRS URA 336,*
*Ecole Normale Supérieure*
*24 rue Lhomond, 75005 Paris, France*

## 1. Introduction

Millimeter wave spectroscopy first revealed the existence, within molecular clouds, of the most opaque cores in which stars form and since then has provided most sophisticated probes of their physical properties. One of them is due to the heterodyne detection itself which allows unmatched spectral resolution in the observations, close to the thermal velocities of molecules at low temperature. Gas motions in molecular clouds are supersonic because the Doppler line broadening is much larger than what would be expected from thermal motions at the low temperatures deduced from line excitation analysis. In 1974, Zuckerman & Evans proposed that these motions were turbulent.

Twenty years later, the nature of the observed supersonic motions is not fully understood. Supersonic turbulence has been discarded for a while because of its large anticipated dissipation rate in radiative shocks, with no comparable injection rate from the known sources of kinetic energy in the medium (explosions of supernovae, HII regions expansion, stellar outflows,...). Recent measurements of the intensity of magnetic field in molecular clouds (see Crutcher et al. 1993) have somewhat released this energetic constraint but the issue of the real nature of the disordered gas motions is still open. Whether these motions are random, turbulent or due to magneto–hydro–dynamics waves is an issue of importance because of their different impact on the physics of the clouds. For instance, as was first proposed by Bonazzola et al. (1987) and further developed in Bonazzola et al. (1992), turbulence with a Kolmogorov spectrum (*i.e.* more power in the large scale motions) is able to stabilize clouds against gravitational instability. This paper reviews the various pros and cons of the turbulent



*W.B. Latter et al. (eds.), CO: Twenty-Five Years of Millimeter-Wave Spectroscopy,* 119–127.

nature of the clouds velocity field and discusses the possible observational signatures of the existence of turbulence.

## 2. Possible Signposts of Turbulence

### 2.1. CHARACTERISTICS SHARED WITH INCOMPRESSIBLE TURBULENCE

The gas motions in interstellar clouds are suprathermal, disordered and exhibit features of incompressible turbulence. The Reynolds numbers are extremely large Re $\sim 10^9$ for scale sizes L $\sim$ 1kpc. Much larger scales might even be involved in the cascade since interstellar media in galaxies open up in intergalactic media. Power–law scaling relations have been found in the velocity field of molecular clouds (Fig. 1). Fig. 1a displays the internal non–thermal velocity dispersion versus the size of structures identified in a large variety of data sets. A structure is a connected volume in the space–velocity cube of CO or $^{13}$CO emission. The physical meaning of such entities is not yet clear. It involves the complexity of radiative transfer of line photons and the non–linear formation and destruction laws of molecules like CO. In Fig. 1b, the size–linewidth relation is derived from azimuthally averaged structure functions of the velocity field (Miesch & Bally, 1994).

The data points of Fig. 1a reveal a large scatter, particularly at small scale. As an ensemble, they are close to a law $\Delta v \propto l^h$ with $h \sim 1/3$ over an impressive range of sizescales. Entities in virial balance between self–gravity and kinetic energy for which the confining pressure is negligible ($h \sim 1/2$) coexist with structures for which the flatter exponent of the size–linewidth scaling law ($h \sim 1/3$) is reminiscent of the Kolmogorov cascade in incompressible turbulence, as was first proposed by Larson (1981). A comparable dispersion in the slopes of this power–law (between 1/2 and 1/3) is visible in Fig. 1b although the method followed by Miesch & Bally (1994) is different. These authors discuss the possible origins of the different slopes and find they agree with theoretical predictions for the velocity power spectrum when compressibility and magnetic field are taken into account.

Another signpost of turbulence is the fractal dimension of the cloud brightness contours (Scalo, 1990; Falgarone, Phillips & Walker 1991; Zimmermann & Stutzki 1992). This property has been brought together with the multifractal geometry of the dissipation field in laboratory turbulence (see the review of Sreenivasan (1991) and references therein) and it was proposed that the cold cloud population in the interstellar medium traces the regions of enhanced dissipation rate of kinetic energy in a turbulent interstellar medium (Falgarone, Phillips & Walker 1991). One should mention, however, that several other interpretations of the cloud fractal geometry have been proposed, such as collisional fragmentation (Nozakura 1990),

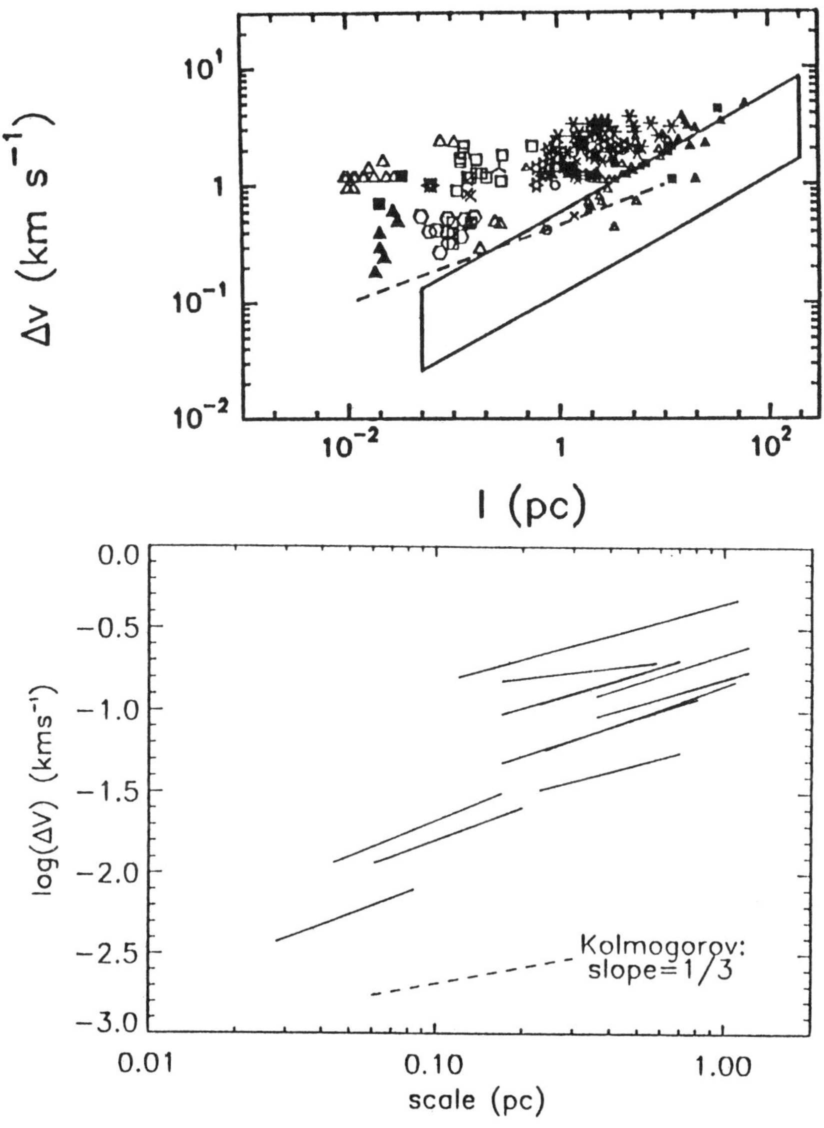


*Figure 1.* Non–thermal velocity dispersion versus size in an ensemble of molecular clouds. In *(a)*, the velocity dispersion is derived from the extent in velocity space of structures isolated by the level of their line emission in the space–velocity data cubes. The references and corresponding symbols are given in Falgarone, Puget & Pérault (1992). The slope (1/3) predicted by Kolmogorov (1941) for incompressible turbulence is shown. The parallelogram encompasses a large number of data points which correspond to self–gravitating entities. In *(b)*, the power–laws are derived from the structure function of the velocity field built on large data sets (Miesch & Bally 1994).

gravitational clustering and virial balance (Pfenniger & Combes 1994).

No characteristic scale has yet been determined. Unresolved structure is still present in the highest resolution maps of non–starforming clouds and several recent observations bring support to the fact that the threshold of the self–similar hierarchy is indeed very small (down to a few 10 AU) in many kinds of cold clouds. The existence of structures even smaller than 10AU and denser than $n_{\mathrm{H}_2} \sim 10^6\,\mathrm{cm}^{-3}$ has been inferred from secular changes in the formaldehyde absorption line profile toward two extragalactic sources whose lines of sight cross ordinary regions of nearby molecular clouds (Marscher, Moore & Bania 1993; Moore & Marscher 1995). Last, early searches for a finite correlation length in the velocity field of molecular clouds have been unsuccessful (see the reviews of Dickman, 1985, and Scalo, 1987). More recent searches have been similarly inconclusive (Pérault, Falgarone & Puget 1986; Kleiner & Dickman 1987; Kitamura et al. 1993). Further specific signatures of turbulence have been searched for in the shapes of line profiles. They are described below.

### 2.2. THE SHAPES OF THE LINE PROFILES

The shapes of line profiles, when observed with high spectral resolution, have long been thought of as powerful probes of the nature of the velocity field in clouds although the discussion of the various regimes relevant to the line formation mechanism in terms of systematic to turbulent velocities ratio in the region of line formation has proved quite intricate (Leung, 1978). In 1976, Baker showed that most of the paradoxical behaviours of the CO line shapes and line ratios between isotopic species could be understood if lines form in a macroturbulent medium. New developments along different but complementary lines of thought have brought some clarity in this complex issue. CO line profiles have been computed in random velocity fields with one correlation length (Albrecht & Kegel, 1987; Kegel, Piehler & Albrecht, 1993) which quantitatively illustrates the major role played by the value of a correlation length compared to the photon mean free path.

Three-dimensional numerical simulations of unforced compressible turbulence with a resolution of $512^3$ grid points (Porter, Pouquet & Woodward 1994) have been used to synthesize line profiles at different epochs of the simulation, *(i)* when strong shocks have just formed, *(ii)* once shock interactions have generated vorticity, thus transferring most of the kinetic energy to non–compressible modes and *(iii)* in the phase of self–similar decay (Falgarone et al. 1994). In the two latest phases, the synthesized profiles are statistically very similar to observed line profiles, with almost Gaussian profiles coexisting with double–peaked and asymmetric profiles.

The same diversity and similar lineshapes, however, have been found by Dubinski, Narayan & Phillips (1995) who computed line profiles from a random Gaussian velocity field with a Kolmogorov–type power spectrum, under the same approximations as above. The only difference between the profiles synthesized in a random velocity field with Kolmogorov spectrum and those synthesized in a turbulent flow lies in the existence of the weakest and broadest linewings, which trace the presence of local structures of high vorticity, as explained in Falgarone et al. (1994). She, Jackson & Orszag (1990) have shown that structures also exist in a random Gaussian velocity field, in particular vorticity filaments are present. In a turbulent field, though, the intensity of these structures is much larger than in a random velocity field, as illustrated in Figure 2. These intense (coherent) structures are thought to be responsible for the property of intermittency of turbulence, at the origin of the non–Gaussian probability distributions of velocity increments, derivatives and vorticity (Vincent & Meneguzzi 1991; Porter, Pouquet & Woodward 1994). The broadest and weakest linewings in molecular clouds possibly trace the rarest (and most intense) peaks of the vorticity distribution as was proposed in Falgarone & Phillips (1990). There is not a sharp cut–off between the non–Gaussian features due to the Kolmogorov power spectrum itself and those due to intermittency. In a turbulent velocity field, the local enhancements of vorticity, for instance, are just more prominent and more coherent than in a random velocity field. It is this property of turbulence which may imprint very specific signatures on the evolution of interstellar clouds because the regions of enhanced vorticity in turbulent flows are closely associated with those of the largest dissipation rate of kinetic energy. The two subsets of space where these quantities are concentrated are highly structured and intermittent (Brachet 1990; Vincent & Meneguzzi 1991) and the viscous dissipation of turbulence, which eventually heats the gas, does so at a very uneven rate in space and in time.

## 3. The Intermittent Dissipation of Turbulent Kinetic Energy

### 3.1. THE GAS TEMPERATURE DISTRIBUTION IN ATOMIC CLOUDS

The local dissipation rate of kinetic energy is $\epsilon_D = \eta/2 \sum_{i,j}(\partial_j v_i + \partial_i v_j)^2$ where $\eta = \rho\nu$ is the dynamical viscosity and $\rho$ the gas mass density. This quantity is best provided by experiments on flows at large Reynolds numbers which allow to estimate the probability distributions of the velocity derivatives down to lengthscales very close to the dissipation length, unlike numerical simulations or observations of the interstellar medium. Flow experiments reported in Gagne (1987) provide the probability distribution functions (PDFs) of velocity increments $\delta v$ over a large range of lags $\delta l$, down to only a few dissipation scales. Estimates of the PDFs of the dissipation rates (averaged over various sizes) can be derived from the distribu-

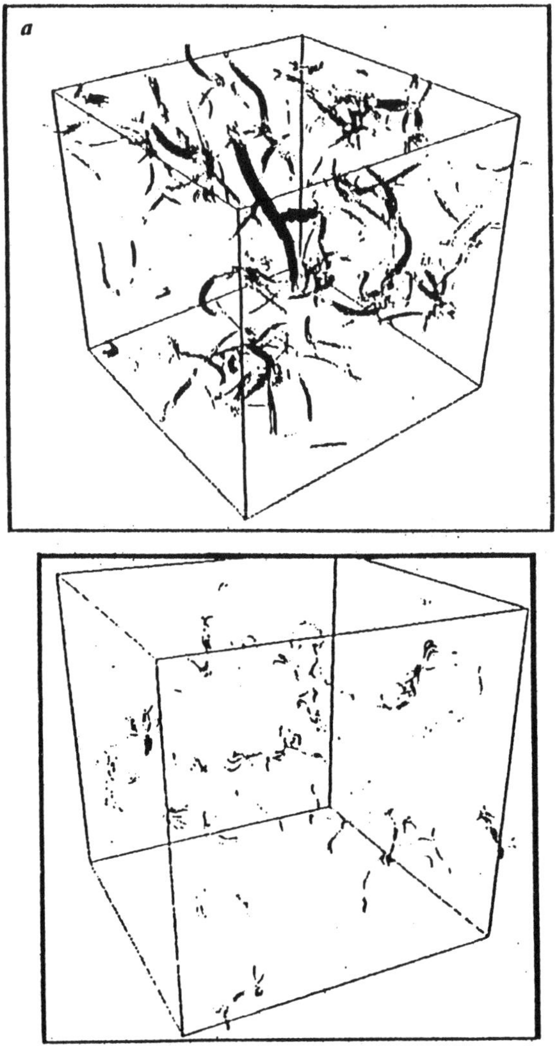


*Figure 2.* Upper panel: vorticity field in Navier–Stokes simulations of incompressible turbulence. Lower panel: vorticity distribution (same threshold) in a random Gaussian velocity field with same Kolmogorov power spectrum (from She, Jackson & Orszag 1990)

tions of velocity increments. These results have been scaled to interstellar turbulence by Falgarone & Puget (1995) under the assumption that the statistical flow properties in turbulence only depend on the kinetic energy transfer rate $(\delta v)^3/\delta l$ and on the value of the dissipation scale, $l_D$ which, in atomic clouds, is estimated to be $l_D \sim 10$AU. The resulting locally averaged dissipation rates are found to be local heating sources of interstellar (atomic) clouds that exceed all the others (due to UV photons, coupling to dust grains, cosmic rays...) by orders of magnitude. Moreover, the intense vortices at the origin of the dissipation bursts have been observed to have lifetimes one to two orders of magnitude larger than their own anticipated turnover timescale $\sim \delta v/\delta l$ (see for example the experiment of Douady, Couder & Brachet 1991) and their persistence allows thermal balance to be reached. The lifetime of an intermittent structure of thickness $\delta l$ ~60 AU and characteristic velocity $\delta v_l \sim 0.3$ km s$^{-1}$ is therefore estimated to be $\sim 10^4$ to $10^5$ yr. This lifetime being larger than the cooling time associated to radiative processes, thermal balance is reached between the heating provided locally by the intermittent large dissipation of kinetic energy and the radiative cooling of the gas. The unexpected result is that gas temperatures locally as large as $T_k = 10^3$ to $10^4$K are found to occur with non negligible probabilities (or filling factors) of $10^{-2}$ to $10^{-3}$ within atomic clouds where the bulk of the gas is at $T_k$ =50K.

### 3.2. IMPLICATIONS FOR THE CHEMICAL REACTIONS

Many chemical reactions with activation barriers of several thousand Kelvin and endothermic reactions can therefore be activated in regions of intermittent large dissipation rate. Time–dependent calculations (Falgarone, Pineau des Forêts & Roueff 1995) show that, as expected, the abundances of molecules like OH, $H_2O$, $CH^+$, $HCO^+$, and therefore CO, in these regions can be orders of magnitude larger than in the surrounding colder gas. A fraction as small as 10% of warm gas at $T_k = 10^3$K on a line of sight across an atomic cloud is sufficient to reproduce the observed quantities of $CH^+$, a fragile molecule, for which no formation mechanism so far could reproduce the large observed abundance. The large abundances of $HCO^+$ and other molecules recently discovered in absorption in clouds poorly shielded from the ambient UV radiation field (Lucas & Liszt 1993, 1994, 1996) require a very efficient mechanism for the formation of these molecules. A plausible scenario is the above mechanism, locally powered by the intermittent viscous dissipation of turbulence in otherwise mostly atomic clouds.

### 3.3. LOCAL CHANGES OF THE DISTRIBUTION OF RELATIVE VELOCITIES BETWEEN SMALL DUST PARTICLES

Intermittency of a turbulent velocity field also affects the relative velocity of two nearby small dust particles. Völk et al. (1980) have shown that turbulence increases the relative velocity of large dust particles but does not affect that of the lightest particles which stay closely coupled to the gas motions. Intermittency introduces timescales for the changes in the topology of the velocity field which are short enough to be comparable to (or smaller than) the friction time of the smallest dust particles so that the distribution of their relative velocities is modified in the regions of strong shear or vorticity (Falgarone & Puget 1995). This process may induce local changes in the size distribution of small particles and may explain the IRAS color variations of dust thermal emission of nearby clouds (Boulanger et al. 1990) or the variations in the shapes of the extinction curves across neighbouring similar clouds (Boulanger et al. 1994). An additional possible confirmation that such processes occur even in diffuse clouds is the correlation between the mid–infrared 12$\mu$m continuum emission of high latitude clouds and their $CH^+$ equivalent width (Boulanger 1994).

## 4. Summary

All the above cannot be taken as a proof that turbulence exists in interstellar clouds but it puts together a set of significant signatures of turbulence. In spite of the complexity of the physics of interstellar clouds, and the many processes which generate non–linear behaviours, it seems that the velocity patterns, the shapes of molecular lines, the line ratio properties, all suggest that a Kolmogorov–type cascade is present with the additional property of intermittency characteristic of turbulence. Unlike what is commonly thought, the fact that interstellar clouds share so many properties with incompressible turbulence is not in contradiction with their compressibility. Recent numerical simulations (Porter et al. 1994) have shown that shock interactions rapidly generate vorticity, increasing the energy in the incompressible modes at the expense of that in compressible modes which builds up a Kolmogorov power spectrum.

The property of intermittency is responsible for the local characteristics of the gas to considerably depart from their average values, as derived for instance from the observations. These large amplitude and non–Gaussian fluctuations induced at small scale in the temperature or the chemical content of the gas may be at the origin of the most profound difficulty met in modelling interstellar clouds and understanding their small–scale evolution. At large scale, however, the gross properties of the cycle of the interstellar medium through its different phases, up to star formation, have been successfully reproduced in numerical simulations which take into account the interplay of self–gravity, turbulence and compressibility of the flows (Vásquez–Semadeni, Passot & Pouquet 1995).

## References

Albrecht, M.A. & Kegel, W.H., 1987, A&A, 176, 317
Baker P.L., 1976, A&A, 50, 327
Bonazzola S., Falgarone E., Heyvaerts J., Pérault M. & Puget J.-L., 1987, A&A, 172, 293

Bonazzola S., Pérault M., Puget J.L., Heyvaerts J., Falgarone E. & Panis J.F., 1992, J. Fluid Mech., 245, 1
Boulanger F., Falgarone E., Puget J.L., & Helou G., 1990, ApJ, 364, 136
Boulanger F., Prévot M.L., & Gry C., 1994, A&A, 294, 956
Boulanger F. 1994 *First Symposium on infrared cirrus and diffuse molecular clouds*, eds. R. Cutri & W. Latter, The Arizona University Press
Brachet, M.E., 1990, C. R. Acad. Sci. Paris, 311, 775
Crutcher R.M., Troland T.H., Goodman A.A., Heiles C., Kazès I. & Myers P.C., 1993, ApJ, 407, 175
Dickman R.L. 1985 in *Protostars and Planets II*, D.C Black and M.S. Matthews eds., The university of Arizona Press
Douady S., Couder Y., & Brachet M.E., 1991, Phys. Rev. Letters, 67, 983
Dubinski J., Narayan R. & Phillips T.G., 1995, ApJ, 448, 226
Falgarone E. & Phillips T.G., 1990, ApJ, 359, 344
Falgarone E., Puget J.-L. & Pérault M., 1992, A&A, 257, 715
Falgarone E., Phillips T.G. & Walker C., 1991, ApJ, 378, 186
Falgarone E., Lis, D.C., Phillips, T.G., Pouquet A., Porter, D.H. & Woodward, P.R., 1994, ApJ, 436, 728
Falgarone, E. & Puget, J.-L., 1995, A&A, 293, 840
Falgarone E., Pineau des Forêts G. & Roueff E., 1995, A&A, 300, 870
Gagne, Y. 1987, Thèse d'Etat, Université de Grenoble
Kegel, W.H., Piehler, G. & Albrecht, M.A., 1993, A&A, 270, 407
Kleiner S.C. & Dickman R.L., 1987, ApJ, 312, 837
Kitamura Y., Sunada Y., Hayashi M. & Hasegawa T., 1993, ApJ, 413, 221
Kolmogorov A.N., 1941, Dokl. Akad. Nauk. SSSR, 26
Larson R.B., 1981, MNRAS, 194, 809
Leung C.M., 1978, ApJ, 225, 427
Lucas R. & Liszt H., 1993, A&A, 276, L33
Lucas R. & Liszt H., 1994, A&A, 282, L5
Lucas R. & Liszt H., 1996, A&A, 307, 237
Marscher A.P., Moore E.M. & Bania T.M., 1993, ApJL, 419, L101
Miesch M.S. & Bally J., 1994, ApJ, 429, 645
Moore E.M. & Marscher A.P., 1995, ApJ, 452, 671
Nozakura T., 1990, MNRAS, 243, 543
Pérault M., Falgarone E. & Puget J.-L., 1986, A&A, 157, 139
Pfenniger D. & Combes F., 1994, A&A, 285, 94
Porter D.H., Pouquet A. & Woodward P.R., 1994, Phys. Fluid, 6, 2133
Scalo, J. M. 1987 in *Interstellar processes*, D.J. Hollenbach & H.A Thronson eds., Reidel Pub.:Dordrecht
Scalo J.M. 1990 in *Physical processes in Fragmentation and Star Formation*, eds R. Capuzzo–Dolcetta et al., Kluwer Acad. Pub.
She Z.-S., Jackson E. & Orszag S.A., 1990, Nature, 344, 226
Sreenivasan K.R., 1991, Ann. Rev. Fluid Mech., 23, 539
Vásquez–Semadeni E., Passot T., & Pouquet A., 1995, ApJ, 441, 702
Vincent, A. & Meneguzzi, M., 1991, J. Fluid Mech., 225, 1
Völk H.J., Jones F.C., Morfill G.E. & Röser S., 1980, A&A, 85, 316
Zimmermann T. & Stutzki J., 1992, Physica A, 191, 79
Zuckerman B. & Evans N.J., 1974, ApJL, 192, L149

# $^{13}$CO (6–5) IN THE ORION BAR: A CRITICAL OBSERVATIONAL TEST FOR PDR MODELS

D. C. LIS, P. SCHILKE, AND J. KEENE
*California Institute of Technology*
*Downs Laboratory of Physics 320–47*
*Pasadena, CA 91125*

## 1. Introduction

The Orion Bar is a limb–brightened edge of the cavity carved out in the Orion molecular cloud by the expanding HII region surrounding the Trapezium cluster. The Bar is one of the best studied Photon–Dominated Regions (PDRs; see Genzel in this volume for a review). Tauber et al. (1995) found a tight correlation between the distributions of the neutral atomic carbon and $^{13}$CO (2–1) emission toward the Bar, which is difficult to explain in terms of classical PDR models. A possible explanation for this correlation was given by the geometric model of Hogerheijde et al. (1995) in which the PDR geometry changes from face–on to edge–on and back with increasing distance from the ionizing stars. The observed maximum of the molecular emission toward the Bar thus corresponds to a molecular column density peak and there are no significant variations in the excitation conditions across the Bar. In the present paper we investigate the uniformity of the excitation conditions across the Bar through observations of the $^{13}$CO (6–5) and $^{12}$CO (6–5) emission. We argue that the $^{13}$CO (6–5) emission provides a critical test for models of this region and PDRs in general.

## 2. Discussion

The $^{13}$CO (6–5) and $^{12}$CO (6–5) emission toward the Orion Bar was observed in 1995 January using the Caltech Submillimeter Observatory on Mauna Kea, Hawaii. The FWHM beam size was $\sim 11''$. All temperatures reported here have been corrected for an efficiency of 60%, intermediate between the main beam efficiency ($\sim 45\%$) and the Moon efficiency ($\sim 70\%$). The (6–5) transitions were supplemented with maps of the (2–1) and (3–2)



*W.B. Latter et al. (eds.), CO: Twenty-Five Years of Millimeter-Wave Spectroscopy,* 128–130.

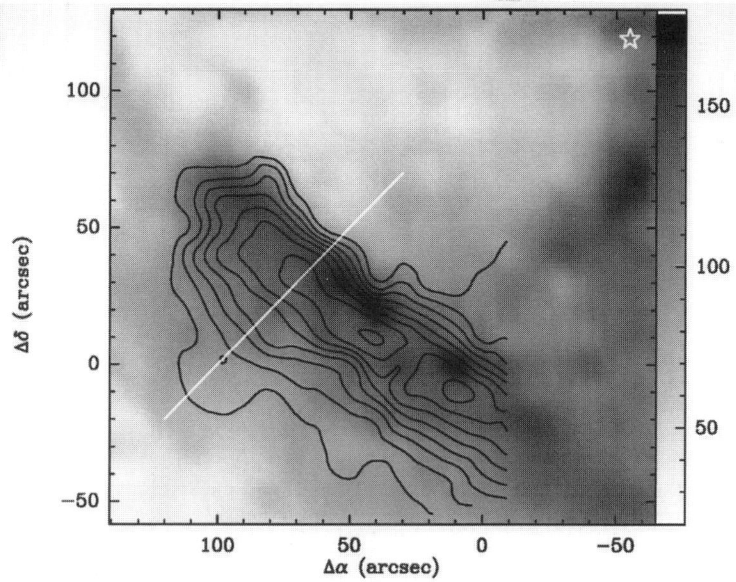


*Figure 1.* Overlay of the $^{12}$CO (6–5) brightness temperature (gray–scale) toward the Orion Bar and the $^{13}$CO (6–5) integrated intensity in the 5–15 km s$^{-1}$ velocity range (contours). Contour levels are 40 to 180 in intervals of 20 K km s$^{-1}$. The star marks the location of $\theta^1$C Ori.

transitions of the two isotopomers. We also observed one position toward the Bar in C$^{18}$O (6–5).

An overlay of the $^{13}$CO (6–5) integrated intensity in the 5–15 km s$^{-1}$ velocity range and the $^{12}$CO (6–5) brightness temperature is shown in Figure 1. The peak $^{12}$CO (6–5) brightness temperature toward the Bar ($\sim$ 175 K) is much higher than that observed in low–$J$ $^{12}$CO transitions in a similar beam ($\sim$ 120 K). This suggests that the $^{12}$CO (6–5) transition becomes optically thick in outer PDR layers with higher kinetic temperatures. The $^{12}$CO (6–5) emission is displaced toward the ionizing stars, relative to the $^{13}$CO (6–5) emission. The peak brightness temperature of the $^{13}$CO (6–5) emission is $\sim$ 75 K; such high intensities of the $^{13}$CO (6–5) emission appear common in Galactic star–forming regions (Graf et al. 1993).

We computed preliminary PDR models for the Bar, assumed to be nearly edge–on, and the face–on "Hole" near the N–W end of the cut shown as the white line in Figure 1, using an improved version of the code of Le Bourlot et al. (1993). Our models can reproduce the observed intensities toward the Bar and the Hole for all CO transitions we observed *with the exception of* $^{13}$CO (6–5). The intensities of $^{13}$CO (6–5) are underestimated by factors of 2 and 3.5 toward the Hole and the Bar, respectively. The models fail because they predict a kinetic temperature of only $\sim$ 30 – 50 K for most of the $^{13}$CO emitting gas, while we derive a lower limit of 120 K from

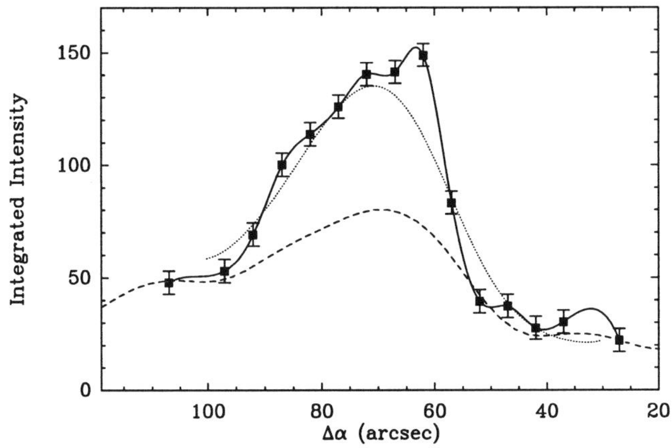


*Figure 2.* The variation of the $^{13}$CO (6–5) and $^{13}$CO (2–1) integrated intensity along the cut perpendicular to the Bar shown in Fig.1 (solid and dashed lines, respectively). Notice the very sharp drop of the $^{13}$CO (6–5) emission on the side facing the ionizing stars. The $^{13}$CO (2–1) intensities (from Tauber et al. 1995) were multiplied by a factor of 1.5. The distribution of $^{13}$CO (6–5) integrated intensity calculated from an image convolved to the 30″ resolution of the $^{13}$CO (2–1) map is shown as dotted line.

the observed $^{13}$CO (6–5) to (2–1) intensity ratio.

The variation of the $^{13}$CO (6–5) and (2–1) integrated intensities across the Bar is shown in Figure 2. The intensity ratio increases from $\sim$ 1.5 toward the Hole to $\geq$ 2.5 toward the Bar. Since the $^{13}$CO emission is at most only moderately optically thick ($\tau \leq 1$, based on the observed $^{13}$CO to C$^{18}$O intensity ratio), this suggests a *significant difference* in the excitation conditions (kinetic temperature, density, or a combination of both) between the two regions.

We conclude that additional heating mechanisms for the interior (like shocks or turbulence) or time–dependent PDR models (e.g., Goldshmidt & Sternberg 1995) may be required to explain the high column density of *warm* $^{13}$CO in the Orion Bar implied by our observations. We believe that this conclusion is true for PDRs in general and that observations of the $^{13}$CO (6–5) emission provide a critical test for PDR models.

Research at the CSO is supported by the NSF under grant AST 93–13929.

## References

Goldshmidt, O., & Sternberg, A., 1995, ApJ, **439**, 256
Graf, U. U., et al. 1993, ApJ, **405**, 249
Hogerheijde, M. R., Jansen, D. J., & van Dishoeck, E. F. 1995, A&A, **294**, 792
Le Bourlot, J., Pineau des Forêts, G., Roueff, E., & Flower, D. R. 1993, A&A, **267**, 233
Tauber, J. A., et al. 1995, A&A, **297**, 567

# THE C/CO RATIO PROBLEM: CHEMICAL EFFECTS OF TURBULENCE

TAOLING XIE
*Laboratory for Millimeter-wave Astronomy*
*Department of Astronomy, University of Maryland*
*College Park, MD 20742; email: tao@astro.umd.edu*

**Abstract.** We discuss the suggestion that the chemistry in a photon-dominated region is coupled to that in the UV-shielded region behind it by turbulent transport processes. In addition to diffusion time-scales, we discuss why MHD waves will likely cause, instead of prohibit, transport.

## 1. On the C/CO Ratio Problem

The study of interstellar chemistry has so far focused mainly on two physical regimes; regions subject to abundant ultraviolet photons and regions that are shielded from UV photons. The former are often referred to as photon-dominated regions (so-called PDRs) where the chemistry is dominated by photon-related reactions, and the latter are often thought to be cosmic ray-dominated regions (hereafter CDRs) where cosmic rays sustain the ion-neutral reaction scheme.

The large observed C/CO ratio ($\sim 0.1$) in molecular clouds has been a major challenge to theoreticians working on either regime (cf. Sorrel 1992). One common opinion in the field is that the clumpiness of a cloud allows an enhanced UV penetration into the cloud interior that will dissociate the molecules and create many small photon-dominated regions on the clump surfaces. One severe difficulty with this picture, however, seems to be the large observed abundances of many large organic molecules. Since large organic molecules do not survive UV photodissociation in PDRs, the observed large organic molecules must be deep inside the clumps, which requires the co-existence of abundant neutral carbon (Chièze et al. 1991). The co-existence of neutral carbon with CO in the same volume is strongly indicated by the similarity of their line profiles (cf. Walker et al. 1993) and



*W.B. Latter et al. (eds.), CO: Twenty-Five Years of Millimeter-Wave Spectroscopy*, 131–133.

the tight correlation between their emission intensities (cf. Tauber et al. 1995).

## 2. On Turbulent Transport and Its Time-scales

What feasible mechanism can bring a lot of neutral carbon deep inside each clump without those UV photons that also destroy large organic molecules? An easy choice seems to be particle transport, which is known to occur universally in random gas motions. Xie, Allen & Langer (1995, hereafter XAL) formulated the idea in the framework of the Prandtl mixing-length theory for turbulent diffusion and found that the transport may give rise to a C/CO ratio in accord with observations.

Transport of particles occurs as a result of molecular diffusion. If there is a wind, convection, or a large-scale random component in gas motions, then the transport due to any of these processes is likely to dominate over molecular diffusion because of the much larger spatial scales and amplitudes over which these processes occur in interstellar clouds. In general, random motions cause fluctuations of a physical quantity in a non-uniform multi-component gas. Defining the fractional abundances $f_i = n_i/n(H_2)$ for species $i$, the transport flux can be written as (cf. Colegrove et al. 1966; XAL), $\phi_i[\mathrm{cm}^{-2}\ \mathrm{s}^{-1}] = n(H_2) < V_t \delta f_i >= -Kn(H_2)\frac{df_i}{dz} = -Kn_i(\frac{1}{n_i}\frac{dn_i}{dz} - \frac{1}{n(H_2)}\frac{dn(H_2)}{dz})$, where $<>$ denotes the time average of a quantity, and the diffusion coefficient $K[\mathrm{cm}^{-2}\ \mathrm{s}^{-1}] = < V_t L >$ with $V_t$ being the velocity for the random motion and $L$ the mixing-length scale.

Now with a transport term, the continuity equation at a cloud position is $\frac{\partial n_i}{\partial t} + \frac{\partial \phi_i}{\partial z} = P_i - L_i$ where $P_i$ and $L_i$ are the chemical production and loss terms, respectively. Defining a chemical time-scale, $\tau_i^{chem} = n_i/(P_i - L_i)$, and a diffusion time-scale, $\tau_i^{diff} = n_i/\frac{\partial \phi_i}{\partial z}$, the continuity equation reduces to a simpler form,

$$\frac{\partial n_i}{\partial t} = n_i(\frac{1}{\tau_i^{chem}} - \frac{1}{\tau_i^{diff}}). \tag{1}$$

Defining a time constant, $\tau = R^2/K$, that is typically large compared to the chemical time-scales, Williams & Hartquist (1991) argued that diffusion will not be important. It is not difficult to see that the real situation is much more complicated than this. First, $\tau = R^2/K$ has essentially no physical meaning in the formulation. Both the chemical and diffusion time-scales in the above continuity equation are position-, time-, and species-dependent. Even when we fix the question to a particular moment, both time-scales still depend in a complicated manner on the detailed physical and chemical conditions at a cloud position. The diffusion time-scale may be simplified to $\tau_i^{diff} = H_i H/K$, where the scale heights $H_i = n_i/\frac{\partial n_i}{\partial z}$ and $H = f_i/\frac{\partial f_i}{\partial z}$, but it would not reduce to $\tau = R^2/K$ without orders of magnitude error. Secondly, diffusion is generally coupled to the highly non-linear chemistry and cannot be treated as a pure diffusion in this case; a small change in some key species due to diffusion between neighboring layers can boost the local generation or destruction of certain species (XAL).

## 3. In the Case of MHD Waves

It is quite possible that the observed "turbulence" in molecular clouds is due to magneto-hydrodynamic waves (MHD waves). A perfectly coherent wave particle carried up will be carried down in a cycle, so there will be no net transport. Would MHD waves be perfectly coherent in molecular clouds, given their irregularities in magnetic fields, densities, and other physical conditions and even their star forming activities? The magnetic Reynolds number is so large that any such waves will be turbulent and random in nature on relevant spatial scales (Myers & Khersonsky 1995), which will naturally lead to net particle transport (Lindzen 1971; Hunten 1975). In fact, even if waves are perfectly coherent, the non-uniform and non-homogeneous chemistry will still unavoidably cause a net transport of certain chemical species, as chemicals carried up could be destroyed and will thus never be carried down, or vice versa. Lindzen (1971) suggested that the diffusion coefficient in the case of interrupted waves can be expressed as $K \sim v\lambda/(\pi x)$, where $v$ is the average wave velocity, $\lambda$ is the averaged wavelength, and $x$ is the number of wave cycles (non-integer) before being interrupted. Notice that diffusive transport along the average magnetic field would be enough to have significant effects on the chemistry. The simple-minded approach outlined by XAL is not sensitive to the nature of the observed "turbulence" in molecular clouds.

TX's research is supported in part by an NSF grant to the Laboratory for Millimeter-wave Astronomy at the University of Maryland.

## References

Chièze, J.P., Pineau des Forêts, & Herbst, E. 1991, *ApJ*, **373**, 110
Colegrove, F.D., Johnson, F.S., & Hanson, W.B. 1966, *J. Geophys. Res.*, **71**, 2227
Hunten, D.M. 1975, in *Atmospheres of Earth and the Planets*, ed. B.M. McCormac (Dordrecht: Reidel), 59
Lindzen, R.S. 1971, in *Mesospheric Models and Related Experiments*, ed. G. Fiocco (Dordrecht: Reidel), 122
Myers, P.C., & Khersonsky, V.K. 1995, *ApJ*, **442**, 186
Sorrell, W.H. 1992, *Comm. Astrophys.*, **16**, 123
Tauber, J.A., Lis, D.C., Keene, J., Schilke,P., & Buttgenbach 1995, *AA*, **297**, 567
Walker, C.K., et al. 1993, *ApJ*, **415**, 672
Williams, D.A. & Hartquist, T.W. 1991, *MNRAS*, **251**, 351
Xie, T., Allen, M., & Langer, W.D. 1995, *ApJ*, **440**, 674 (XAL)

# STAR FORMATION

# TURBULENCE AND COLLAPSE IN STAR-FORMING MOLECULAR CLOUDS

P. C. MYERS
*Harvard-Smithsonian Center for Astrophysics*
*60 Garden Street*
*Cambridge MA 02138 USA*

**Abstract.** We review developments in our knowledge of interstellar turbulence, and star-forming gravitational collapse. The turbulent motions in interstellar line widths are supersonic, by a factor exceeding $\sim 10$ on size scales $\sim 1$ pc, and are progressively smaller on smaller size scales, becoming subsonic by a factor $\sim 10$ for low-mass dense cores of size $\sim 0.1$ pc. Line width–size relations $\Delta v \sim R^{1/2}$ in a single line over a wide range of cloud sizes probably reflect little more than virial balance. Line width–size relations over several lines within a cloud vary from low-mass to massive cores, and probably reflect the structure of the density and magnetic field within the cloud. Searches for infall in star-forming clouds rely on line broadening, line asymmetry, spatial variation of the line centroid, and red-shifted absorption. Early claims of infall based on CO lines, coarse angular and spectral resolution, and distant sources were sensitive only to large-scale homologous collapse. More recent observations are sensitive to "inside-out" collapse over sizes 0.01 pc and velocities 0.05 km $s^{-1}$, and appear consistent with gravitational infall in several well-studied low-mass sources.

## 1. Introduction

Since radio astronomers first detected interstellar spectral lines, they have tried to apply the physical information derived from the lines to astrophysical problems of the interstellar gas. In this paper we describe work on two of the most fundamental of these problems—the turbulent motions in molecular clouds and the collapse motions associated with star formation.



*W.B. Latter et al. (eds.), CO: Twenty-Five Years of Millimeter-Wave Spectroscopy,* 137–147.

## 2. Turbulence

### 2.1. SUPERSONIC LINE WIDTHS

The earliest radio measurements of interstellar lines of H I (Ewen & Purcell 1951), OH (Weinreb et al. 1963), and CO (Jefferts, Penzias & Wilson 1970) showed that their widths are supersonic—they generally exceed the thermal line width by a significant factor. These observations led to the idea that the supersonic motions arise from large-scale gravitational collapse (Goldreich and Kwan 1974). However, such ubiquitous cloud collapse would imply a star formation rate much higher than the few $M_\odot$ yr$^{-1}$ known from stellar observations (Zuckerman & Evans 1974). A more promising explanation is that the supersonic line widths are magnetic in origin, and correspond to the propagation of magnetohydrodynamic waves (Arons & Max 1975). The ionization of a molecular cloud, with log electron fraction of order $-7$ to $-5$, can be maintained by cosmic rays and ultraviolet photons. Even this weak ionization is sufficient to couple the magnetic field and the neutrals together via ion-neutral collisions. If sufficiently excited, the neutral gas will move at a speed close to the wave propagation speed, or the Alfven speed, $v_A = B/(4\pi\rho)^{1/2}$. Here $B$ is the magnetic field strength and $\rho$ is the mass density of neutrals and ions. Typical estimates of $v_A$ are of order 1 km s$^{-1}$, comparable to observed CO line widths.

### 2.2. SUBSONIC LINE WIDTHS

After OH and $H_2O$ were discovered, the 1.3 cm $(J, K)$ = (1,1) rotation-inversion line of $NH_3$ revealed the third radio molecule in the interstellar gas (Cheung et al. 1968). The ability of $NH_3$ lines to reveal properties of quiescent molecular gas was not realized until nearly a decade later, when sufficiently sensitive maser amplifiers were developed. $NH_3$ line observations of "dense cores" —opaque spots of high obscuration in dark clouds on the Palomar optical photographs—showed emission localized to $\sim$ 0.1 pc from gas with temperature 10 K, density $\sim$ $10^4$cm$^{-3}$, and FWHM line width $\sim$ 0.3 km s$^{-1}$. This width is subsonic, i.e., narrower than the thermal width of $H_2$ gas (0.44 km s$^{-1}$; Rydbeck et al. 1977; Ho et al. 1977). These subsonic line widths of $NH_3$ and other heavy molecules, such as $HC_3N$, were soon found to be widespread in the condensations in nearby dark clouds (Churchwell, Winnewisser & Walmsley 1978; Myers, Ho & Benson 1979; Myers & Benson 1983; Myers 1983). One of the smallest nonthermal components is in L1512, where $\sigma_{NT}$ = 0.038 $\pm$ 0.005 km s$^{-1}$, less than one-tenth of the thermal velocity dispersion $\sigma_T$ (Fuller & Myers 1993).

## 2.3. LINE WIDTHS AND MAP SIZES

As dense cores were mapped in different lines, it became apparent that the size of a map varies from line to line, from $\sim$ 0.1 pc for $NH_3$ to $\sim$ 1 pc for CO. This was understood as a consequence of decreasing critical density for excitation $n_{cr}$, since from $NH_3$ to CO $n_{cr}$ declines from $\sim 10^4 cm^{-3}$ to $\sim 100 cm^{-3}$, taking radiative trapping into account. In addition, the width $\Delta v$ of the line generally increases with a measure of its map size R, such as the geometric mean of the maximum and minimum FWHM diameters. This increase can be described as $\Delta v \sim R^p$, with $p$ = 0.3–0.5 (Ho et al. 1977; Martin & Barrett 1978; Myers et al. 1978; Snell 1981). When only a few examples were known, this "line width–size relation" was interpreted in terms of large-scale sytematic motion in TMC-2 in Taurus (Ho et al. 1977), and in terms of radial inward motion which is slower than free-fall, in L1688 in Ophiuchus (Myers et al. 1978).

By the early 1980s nearly 100 clouds within a few kpc of the Sun had been mapped in the 2.7 mm $J$ = 1–0 line of $^{13}CO$. Compiling the information in these $^{13}CO$ maps, and in maps of $NH_3$ and $H_2CO$ lines, Larson (1981) showed that the increase of line width with map size is a distinct trend, not only for the few lines which could be observed in a cloud, but also for the many clouds recently mapped in the $^{13}CO$ line. Larson's linear fit through a log-log plot of $\Delta v$ v. $L$, where $L$ is the largest map dimension, has slope 0.38. Larson also showed that the mean density $n$ in a map tends to decrease with increasing map size as $L^{-1}$, suggesting that the observed clouds have constant column density of a few $\times 10^{21} cm^{-2}$ within scatter of nearly a factor of 10. With similar large scatter, Larson showed that his cloud data are also consistent with "virial equilibrium," where the cloud mass $M$, the line width $\Delta v$, and the map size $R$ follow $\Delta v^2 \sim GM/R$, with $G$ the gravitational constant.

These "Larson's Laws" $\Delta v \sim R^p$, and $n \sim R^{-q}$, with $p \sim 0.4$ and $q \sim 1$, can be understood to follow from two independent relations. One is generally accepted to be virial equilibrium, but the physical basis of the second is still uncertain. Larson (1981) proposed that the second relation is the Kolmogoroff turbulent energy cascade from large to small scales, which for the "inertial range" of size scales, where no dissipation occurs, would yield $\Delta v \sim R^{1/3}$. More recent line width–size studies for large samples tend to find $p$ closer to 0.5 (e.g., Falgarone & Puget 1986; Falgarone, et al. 1992).

The line width–size relation from cloud to cloud in a line is much more distinct than from line to line in a cloud. Two surveys of the Galactic plane in the $^{12}CO$ line showed $\Delta v \sim R^{0.5}$ (Dame et al. 1986; Solomon et al. 1987), and a compilation of various observations from cloud to cloud, including those used by Larson (1981), also showed $\Delta v \sim R^{0.5}$ over three decades in $R$ (Myers & Goodman 1988). In contrast, the line width–size relations from line to line in a cloud shows slopes and intercepts differing according to the cloud, and also to the mass of the star characteristic of the cloud (Fuller & Myers 1992; Myers & Fuller 1993; Caselli & Myers 1995).

### 2.4. LINE WIDTH–SIZE RELATION FROM CLOUD TO CLOUD IN A LINE

The line width–size relation from cloud to cloud in a line, with $\Delta v \sim R^{0.5}$, has a simple explanation based on virial equilibrium and the relatively small variation of the observed column density $N$ compared to the relative variation of the cloud size $R$ (Larson 1981, Myers 1990). To illustrate this point, consider a uniform cloud of radius $R$, mean column density $N$, velocity dispersion $\sigma$, mean molecular mass $m$, and ratio of kinetic and gravitational energy density $\mathbf{K/G}$. Then it follows that

$$\sigma^2 = (2\pi Gm/5)\ (K/G)\ NR \tag{1}$$

Eq. (1) shows that $\sigma \sim R^{1/2}$ if most clouds in the sample are close to virial equilibrium ($\mathbf{K/G} \approx 1$) and if the relative range of $R$ is much larger than the relative range of $N$. This suggests that the line width-size relation from cloud to cloud in a line, with a large range in cloud sizes $R$, is little more than a statement of virial equilibrium. In a survey of clouds at a wide range of distances, $R$ can range over several decades. The relatively small range of $N$, less than one decade, can arise for several reasons, including sample selection, instrumental sensitivity, or, if the magnetic field strength $B$ is close to its equipartition value, a small range in $B$ from cloud to cloud (Myers & Goodman 1988a).

### 2.5. LINE WIDTH–SIZE RELATIONS FROM LINE TO LINE IN A CLOUD

On the other hand, the line width–size relation from line to line in a cloud varies substantially from one cloud to the next, and contains more physical information about the cloud structure than does eq. (1). Observations among "low-mass" cores in Taurus, Ophiuchus, Cygnus, and Cepheus are summarized by Fuller & Myers (1992; FM) and Myers & Fuller (1992, 1993). Observations of both low-mass cores, as in FM, and "massive" cores in Orion A and B are discussed by Caselli & Myers (1995; CM). These observations indicate that low-mass cores have narrower lines, which increase more rapidly with size scale than do massive cores. Thus low-mass cores have a transition from a nonthermally-dominated exterior to a thermally-dominated interior at a radius $\sim 0.1$ pc, while massive cores remain nonthermally dominated for all observed scales. Figure 1 summarizes observations of line width–size relations for low-mass and massive cores.

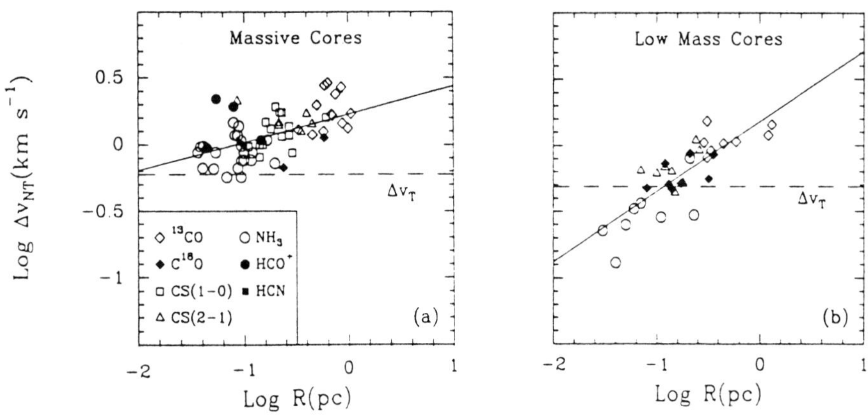


*Figure 1.* Line width–size relations between the nonthermal component of the line width and the map radius for (a) 18 massive cores in Orion, and (b) 8 low-mass cores in Perseus, Taurus, Orion, Ophiuchus, and Cepheus. Dashed lines labelled $\Delta v_{\rm T}$ represent the thermal part of the line width of the molecule of mean mass, assuming temperature 18 K (massive cores), or 10 K (low-mass cores). Massive cores have broader lines and a slower rate of increase of line width with size than do low-mass cores (Caselli & Myers 1995).

"TNT" models of hydrostatic equilibrium, including thermal motions due to uniform kinetic temperature, and nonthermal motions which increase as a power law of radius, match the line width–size relations of low-mass and massive cores, and predict typical column densities of $\sim 3$ and $20 \times 10^{21}\rm cm^{-2}$, in good agreement with observations (CM). In these models the density profile of the typical low-mass core gradually changes from $n \sim r^{-2.0}$ to $n \sim r^{-1.1}$, with equal contributions to the velocity dispersion from thermal and nonthermal motions at $r = 0.1$ pc. In contrast, the typical massive core maintains a density profile $n \sim r^{-1.6}$ over all observed scales.

The internal structure of the CM cores has also been modelled as a self-gravitating, pressure-bounded "pure logotrope" with equation of state

$$P = P_c[1 + A\ \ln(\rho/\rho_c)] \qquad (2)$$

where $P$ and $\rho$ are the kinetic pressure and the mass density, the subscript $c$ denotes values at the center, and $A$ is a constant determined from fitting to the CM data to be 0.18 (McLaughlin & Pudritz 1996). In this model, low-mass cores have flatter density structure than massive cores, which are closer to critical stability against collapse. This model provides a good fit to the line width–size relations and provides a uniform framework in which to model both low-mass and massive cores.

The TNT and logotropic models fit the observed line width–size data well, but do not specify the physical basis of the nonthermal motions. The coupling of the magnetic field to the neutrals seems necessary to maintain the amplitude of the observed nonthermal motions, and estimates of ionization in molecular cloud cores suggest that ion-neutral coupling is sufficient to support MHD waves, even in the densest cores (Myers & Khersonsky 1995). Further, the relatively shallow dependence of the kinetic pressure on radius, $P \sim r^s$, with $0 < s < 1$, from the line width–size relations, combined with the evidence for similarity of magnetic and kinetic energy density in some clouds (Myers & Goodman 1988b), suggest that the magnetic field must also vary very slowly with radius in dense cores.

Magnetic support of molecular clouds, and its relation to the line width–size relations, are the subject of many recent theoretical discussions, including those of Henriksen (1991); McKee et al. (1993); Fatuzzo & Adams (1993); McKee & Zweibel (1995); Zweibel & McKee (1995); Fleck (1996); and Gammie & Ostriker (1996). Gammie & Ostriker (1996) use numerical simulations to investigate a system where nonlinear MHD waves propagate, interact, and decay. They find that the turbulent spectrum of velocities approaches that expected for $\Delta v \sim R^{1/2}$. If the line width–size relations originate in nonlinear MHD turbulence, it will take substantial computational power to simulate these phenomena for a compressible, self-gravitating medium in three dimensions.

## 3. Kinematic Studies of Infall

### 3.1. EARLY ATTEMPTS 1974 – 1992

In the same year that Goldreich and Kwan (1974) proposed the homologous collapse of GMCs, Loren, et al. (1974) observed spatially concentrated CO line broadening toward R CrA , LkH$\alpha$ 198, and Mon R2 which they interpreted as indicating infall. This spatial concentration would soon have another, more convincing explanation, in terms of molecular outflow.

A more specific diagnostic of infall compares optically thick and thin line profiles. In an infalling system with centrally elevated excitation temperature, the peak of the thicker profile is skewed to the blue, with a self-absorption dip for sufficiently large optical depth (Hummer & Rybicki 1968; Lucas 1976, Snell & Loren 1977; Leung & Brown 1977; Leung 1978; Zhou 1992, Walker, Narayanan, & Boss 1994). Model profiles demonstrate this effect, with more pronounced self-reversals for inward-increasing velocity laws (e.g., $v \sim r^{-1/2}$) than for homologous collapse ($v \sim r$). Such skewing was observed in Mon R2 by Loren et al. (1974), and further examples, including self-reversal, were reported in $\rho$ Oph A by Encrenaz, Falgarone, & Lucas (1975); in NGC 1333 by Loren (1976); and in seven other regions, primarily of massive star formation, by Loren et al. (1981) and by Phillips et al. (1981). All of these regions showed self-reversal of their CO line emission, but in about half the cases the brighter peak was skewed to the blue, and in half to the red. The variety and complexity of the CO line profiles was more impressive than the case for infall. Similarly, asymmetrical $^{13}$CO lines in the starless dark clouds TMC2 and L134N were modelled in terms of inward motions (Myers 1980), but further maps of the $^{13}$CO line showed a complex pattern which was difficult to interpret.

In 1977 an important physical model of the collapse process was developed, starting not from uniform gas, but instead from centrally concentrated gas–a structure which would be expected in an isothermal, self-gravitating cloud (Shu 1977). This initial condition was a much more realistic approximation to the recent observations than maps of higher-density tracer lines, such as $NH_3$ and $HC_3N$, which have smaller extent than do maps of lower-density tracer lines, such as $^{13}$CO and $^{12}$CO. The self-gravitating isothermal sphere was also well-known theoretically (Chandrasekhar 1939). In the form of a singular isothermal sphere (SIS)—infinitely extended and infinitely dense at the center—it is simpler to model than the pressure-bounded sphere (Bonnor 1956; Ebert 1955), since the SIS has only one parameter—the kinetic temperature. Furthermore, Shu's similarity solution to the infall problem gave an important physical insight into the possible nature of star-forming collapse.

The mass shells of a collapsing SIS do not all fall in together, as in homologous collapse, but in sequence, from the inside out. Each collapsing shell removes the pressure support for the next, so the instantaneous locus where infall starts is an "expansion wave" moving outward at the sound speed. Inside-out collapse presents a much harder detection problem than does uniform collapse. The collapsing region moves slowly, with the infall speed close to the sound speed (0.19 km s$^{-1}$ for gas at 10 K), and is small, extending outward only to a few hundredths of a pc in order to form a star of 1 $M_\odot$.

The first spectroscopic evidence for infall toward a previously unknown IRAS source was based on CS 5–4 and 2–1 observations toward IRAS 16293−2422 in the L1689 cloud in Ophiuchus (Walker et al. 1986). The asymmetrical CS 5–4 profile, with deep self-absorption, and the more symmetrical CS 2–1, C$^{34}$S 5–4, and C$^{34}$S 2–1 profiles were well-fit by infall models with inward-increasing velocity law $v \sim r^{-1/2}$, the asymptotic form of the inside-out collapse model. However, this infall interpretation was challenged by Menten et al. (1987), who mapped the CS 3–2 emission with resolution (17″) finer than that of Walker et al. (1986; 27″). Menten et al. found that the sense of asymmetry reverses from infall-like to outflow-like in about one-quarter of the map positions. They concluded the absorption and emission come from physically independent regions, and the shift in line asymmetry is due to rotation. Further complications arose as new observations indicated the "source" has two peaks at 6 and 2 cm, separated by $\sim$ 750 AU (Wootten 1989), and has four CO outflow lobes (Mizuno et al. 1990).

Indications of large-scale collapse in regions of massive star formation have been reported, via red-shifted absorption against the continuum emission from H II regions, for W49 in the $HCO^+$ line by Welch et al. (1987), and for G10.6-0.4 in the $NH_3$ line (Keto, Ho, & Haschick 1987; Keto 1990). In these cases the relatively large spatial scales suggest that the infall is not necessarily associated with the formation of a single star, but instead represents gas falling toward a massive cluster of stars.

### 3.2. RECENT ATTEMPTS 1993 – 1996

The most detailed case, with the fewest objections, has been made for infall of the gas near the far infrared source in the isolated globule B335, observed in high-density tracer lines of CS and $H_2CO$ (Zhou et al. 1993). There the effects of rotation and outflow appear much smaller than in IRAS 16293-2422, and the thick and thin profiles maintain their infall-asymmetry relationship over the map. Monte Carlo models of the line profiles, assuming the velocity and density laws of inside-out collapse, indicate an expansion wave radius of 0.03 pc, collapse age of $1 \times 10^5$ yr, mass infall rate of $3 \times 10^{-6} M_\odot yr^{-1}$, and a stellar mass of 0.4 $M_\odot$ (Choi et al. 1995).

Similar evidence for infall has been presented for L1527 in Taurus (Mardones et al. 1994; Zhou et al. 1994; Myers et al. 1995), and several other protostellar candidates are under intensive study. These relatively few sources cannot show the general prevalence of infall in young stellar objects, since in such few cases the line properties could arise from peculiarities in source structure and outflow motions.

Several efforts are underway to overcome the limitations of small-number statistics, of confusion from outflows, and of slow infall speeds. If infall candidates are well-selected to be at similar distances and are truly collapsing, surveys of such objects, made with sufficient resolution and sensistivity, should reveal a statistical excess of sources with infall asymmetry, since outflows should mask this effect in at most half of the cases. Surveys have been undertaken in lines of $H_2CO$ and $C^{18}O$ (Wang et al. 1995), $H_2CO$, CS, $C_3H_2$, and $N_2H^+$ (Mardones et al. 1996), and $HCO^+$ and $H^{13}CO^+$ (Gregerson et al. 1996).

The slow infall speeds deduced from filled-aperture observations are masked by the random motions of the ambient dense core gas, complicating the detection problem. Interferometric observations so far have focussed on the relatively bright lines of CO and its isotopes, which are sensitive to densities $\sim 10^{3-4} cm^{-3}$. With tracer lines such as 1mm $H_2CO$, sensitive to density greater by a factor $> 10$, and with sufficient sensitivity, it should be possible to detect "infall wings" on small scales at high density, just as one detects "outflow wings" on large scales at lower density. At sufficiently short wavelengths, in the submillimeter, the continuum brightness temperature of the protostar should exceed the excitation temperature of the infalling gas. Then submillimeter interferometers should reveal red-shifted absorption against the protostar.

As techniques for observing infall have improved, so too have the tools for diagnosing and modelling infall systems. The velocity centroid of an optically thick line from an infall system becomes bluer as the map position approaches the center of infall, whereas for rotation the centroid becomes bluer on one side of the map and redder on the other (Adelson & Leung 1988; Walker, Narayanan, & Boss 1994). Monte Carlo codes have improved to the point where one need not use the large-velocity-gradient approximation for radiative transfer in situations where the expected gradients are small (Choi et al. 1995). A simplified analytic model of infall determines the characteristic infall speed of the gas forming the observed line via fitting of the line profile, and for two-peak lines via a formula depending only on observables (Myers et al. 1996).

### 3.3. OUTLOOK

This survey of old and new searches for infall raises the question whether the newer claims of infall evidence are also destined for refutation, once more detailed observations are made. To address this point, we note important changes in the way the more recent data have been taken. The recent observations of B335, L1527, and similar sources are made with much finer linear resolution ($\sim 0.01$ pc) and spectral resolution ($\sim 0.05$ km s$^{-1}$) than previous searches, and have been made using lines requiring much higher densities ($> 10^5$cm$^{-3}$ for $H_2CO$ rather than $\sim 10^3$cm$^{-3}$ for $^{13}CO$) for excitation. These changes have become possible largely because of the improvements in resolution and sensitivity of large millimeter-wavelength telescopes. Figure 2 illustrates this point with three versions of the $2_{12} - 1_{01}$ line of $C_3H_2$ at 85 GHz from the inside-out infall candidate protostar in L1527. With fine spectral resolution but coarse angular resolution (top), or with fine angular resolution but coarse spectral resolution (bottom), the two-peak infall signature is absent. Only with the combination of both fine spectral resolution and fine angular resolution is the infall signature visible.

These improvements suggest that the quantitative study of the kinematics of gravitational collapse is now underway, and that in the near future we may expect dramatic improvements in our knowledge of how stars form.

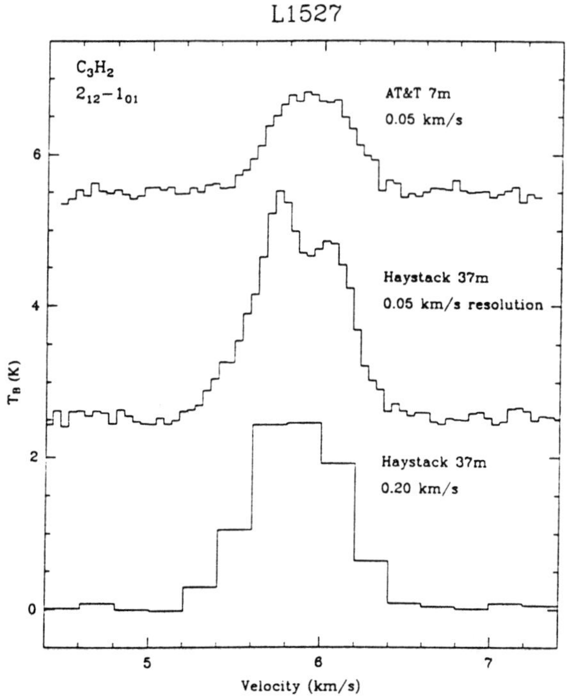


*Figure 2.* Spectra of the 85 GHz $J_{K-K+} = 2_{12} \rightarrow 1_{01}$ line of $C_3H_2$ in the star-forming core L1527, observed with differing angular and spectral resolutions, at the AT&T 7 m telescope by R. W. Wilson, and at the Haystack Observatory 37 m telescope (Myers et al. 1995). Only with the combination of both fine angular resolution (FWHM 25″) and fine spectral resolution (0.05 km s$^{-1}$) is the two-peak signature of infall motions evident.

## References

Adelson, L. M., & Leung, C. M. 1988, MNRAS, 235, 349.
Arons, J., & Max, C. E. 1975, ApJ, 196, L77.
Bonnor, W. B. 1956, MNRAS, 116, 351.
Caselli, P., & Myers, P. C. 1995, ApJ, 446, 665.
Chandrasekhar, S. 1939, Stellar Structure (Chicago: Univ. of Chicago Press)
Cheung, A. C., et al. 1968, Phys. Rev. Lett. 21, 1701.
Choi, M., Evans, N.J., Gregersen, E. M., & Wang, Y. 1995, ApJ, 448, 742.
Churchwell, E., Winnewisser, G., & Walmsley, C. M. 1978, A&A, 67, 139.
Dame, T. M., Elmegreen, B. G., Cohen, R. S., & Thaddeus, P. 1986, ApJ, 305, 892.
Ebert, R. 1955, ZAp, 37, 217.
Encrenaz, P., Falgarone, E., & Lucas, R. 1975, A&A, 44, 73.
Ewen, H. I., & Purcell, E. M. 1951, Nature, 168, 356.
Falgarone, E., & Puget, J. 1986, A&A 162, 235.
Falgarone, E., Puget, J., & Pérault , M. 1992, A&A, 257, 715.
Fatuzzo, M., & Adams, F. C. 1993, ApJ, 412, 146.
Fleck, R. C. 1996, ApJ, 458, 739.
Fuller, G. A., & Myers, P. C. 1992, ApJ, 384, 523.
Fuller, G. A., & Myers, P. C. 1993, ApJ, 418, 273.
Gammie, C. F., & Ostriker, E. C. 1996, ApJ, in press.
Goldreich, P., & Kwan, J. 1974, ApJ, 189, 441.

Gregerson, E., Evans. N. J., Zhou, S., & Choi, M. 1996, in preparation.
Henriksen , R. N. 1991, ApJ, 377, 500.
Ho, P. T., Martin, R. N., Myers, P. C., & Barrett, A. H. 1977, ApJ, 215, L29.
Hummer, D. G., & Rybicki , G. B. 1968, ApJ, 153, L107.
Jefferts, K. B., Penzias, A. A., & Wilson, R. W. 1970, ApJ, 161, L87.
Keto, E. R. 1990, ApJ, 355, 190.
Keto, E. R., Ho, P. T., & Haschick, A. D. 1987, ApJ, 318, 712.
Larson, R. B. 1981, MNRAS, 194, 809.
Leung, C. M., & Brown, R. B. 1977, ApJ, 214, L73.
Leung, C. M. 1978, ApJ, 225, 427.
Loren, R. B. 1976, ApJ, 209, 466.
Loren, R. B., Peters, W. L., & Vanden Bout, P. A. 1974, ApJ, 194, L103.
Loren, R. B., Plambeck, R. L., Davis, J. H., & Snell, R. L. 1981, ApJ, 245, 495.
Lucas, R. 1976, A&A, 46, 473.
Mardones, D. M., Myers, P. C., Caselli, P., & Tafalla, M. 1994, in Clouds, Cores, and Low Mass Stars, eds. D. Clemens & R. Barvainis (San Francisco: ASP), p. 192.
Mardones, D. M., Bachiller, R., Myers, P. C., & Tafalla, M. 1996, in preparation.
Martin, R. N., & Barrett, A. H. 1978, ApJS, 36, 1.
McKee, C. F., Zweibel, E. G., Goodman, A. A., & Heiles, C.1993, in Protostars & Planets III, eds. E. H. Levy & J. I. Lunine (Tucson: U. of Arizona Press), 327.
McKee, C. F., & Zweibel, E. G. 1995, ApJ, 440, 686.
McLaughlin, D. E., & Pudritz, R. E. 1996, ApJ, in press.
Menten, K. M., Serabyn, E., Güsten, R., & Wilson, T. L. 1987, A&A, 177, L57.
Mizuno, A., Fukui, Y., Iwata, T., Nozawa, S., & Takano, T. 1990, ApJ, 356, 184.
Myers, P. C. 1980, ApJ, 242, 1013.
Myers, P. C. 1983, ApJ, 270, 105.
Myers, P. C. 1991, in Fragmentation of Molecular Clouds and Star Formation, eds. E. Falgarone, F.Boulanger & G. Duvert (Dordrecht: Kluwer), p. 221.
Myers, P. C., et al. 1978, ApJ, 220,864.
Myers, P. C., et al. 1995, ApJ, 449, L65.
Myers, P. C., et al. 1996, ApJ, submitted.
Myers, P. C., & Benson, P. J. 1983, ApJ, 266, 309.
Myers, P. C., & Fuller, G. A. 1992, ApJ, 396, 631.
Myers, P. C., & Fuller, G. A. 1993, ApJ, 402, 635.
Myers, P. C., & Goodman, A. A. 1988a, ApJ, 326, L27.
Myers, P. C., & Goodman, A. A. 1988b, ApJ, 329, 392.
Myers, P. C., & Khersonsky, V. K. 1995, ApJ, 442, 186.
Phillips, T. G., et al. 1981, ApJ, 245, 512.
Rydbeck, et al. 1977, ApJ, 215, L35.
Shu, F. H. 1977, ApJ, 214, 488.
Snell, R. L. 1981, ApJS, 45, 121.
Snell, R. L., & Loren, R. B. 1977, ApJ, 211, 122.
Solomon, P. M., Rivolo, A. R., Barrett, J., & Yahil, A. 1987, ApJ, 319, 730.
Walker, C. K., et al. 1986, ApJ, 309, L47.
Walker, C. K., Narayanan, G., & Boss, A. P. 1994, ApJ, 431, 767.
Wang, Y., Evans, N. J., Zhou, S., & Clemens, D. P. 1995, ApJ 454, 217.
Weinreb, S., Barrett, A. H., Meeks, M. L., & Henry, J. C. 1963, Nature, 200, 829.
Welch, W. J., et al. 1987, Science 238, 1550.
Wootten, A. 1989, ApJ, 337, 858.
Zhou, S. 1992, ApJ, 394, 204.
Zhou, S., Evans, N. J., Kompe, C., & Walmsley, C. M. 1993, ApJ, 404, 232.
Zhou, S., Evans, N. J., Wang, Y., Peng, R., & Lo, K. Y. 1994, ApJ, 433, 131.
Zuckerman, B. E., & Evans, N. J. 1974, ApJ, 192, L149.
Zweibel, E. G., & McKee, C. F. 1995, ApJ, 439, 779.

# STAR FORMATION IN THE GEM OB1 MOLECULAR CLOUD COMPLEX

JOHN M. CARPENTER

*Institute for Astronomy, University of Hawaii*

AND

RONALD L. SNELL AND F. PETER SCHLOERB

*Five College Radio Astronomy Observatory, University of Massachusetts*

**Abstract.** We have undertaken a study of the global star formation activity in the Gem OB1 molecular cloud complex using a combination of molecular line, near-infrared, and far-infrared data. The most striking features in the molecular line images are the numerous arc and ring structures which we believe to have been formed by the interactions of expanding HII regions and wind blown bubbles with the ambient molecular gas. The most massive cores in this complex are found within these arcs of gas and most show signs of star formation. We suggest that these cores have been formed by gravitational instabilities in the swept up shells of molecular gas, and that star formation proceeds rapidly after the massive dense cores are formed.

## 1. Introduction

We have conducted a detailed study of the molecular gas and star formation activity in the Gem OB1 molecular cloud complex. The primary goal of this study is to determine the large scale properties and structure of the molecular gas, and to determine how this structure is related to both the past and current star formation activity. The salient results of this analysis are presented here. A more detailed description of the conclusions from this work can be found in Carpenter, Snell, & Schloerb (1995a,b).



*W.B. Latter et al. (eds.), CO: Twenty-Five Years of Millimeter-Wave Spectroscopy,* 148–150.

## 2. Large Scale Structure of the Gem OB1 Complex

The Gem OB1 complex was mapped in $^{12}$CO(J=1–0) and $^{13}$CO(J=1–0) using the focal plane receiver QUARRY at FCRAO. The beam size of the FCRAO telescope at these transitions is 45″ and 47″ for $^{12}$CO and $^{13}$CO respectively. In Figure 1, we present a grey scale image of the observed peak $^{12}$CO antenna temperature. The map covers over 32 deg$^2$ of the sky, which corresponds to a physical scale of 177 pc x 221 pc for the adopted distance of 2.0 kpc. The $^{13}$CO map covers a comparable area. The most striking features in the Gem OB1 cloud complex are the series of arc and ring shaped structures found on spatial scales from $\sim$ 1 pc to $>$ 35 pc. We have concluded that these structures have resulted from the interactions of relatively massive stars with the ambient molecular material based upon:

- The morphology of the structures.
- The association of some of these structures with optical HII regions.
- In some instances, the kinematics of the arcs and rings.
- Comparison of the sizes, masses, and kinetic energies of the arcs and rings with models of expanding HII regions and wind blown bubbles.

## 3. Dense Cores in the Gem OB1 Complex

We searched for dense cores within the Gem OB1 complex by mapping regions with extended $^{13}$CO emission in CS(J=2–1). From this CS survey we identified 11 dense cores with masses ranging from 96 $M_\odot$ to 2200 $M_\odot$. Most of these cores are contained within the arc and ring shaped structures found in the $^{12}$CO images. We examined the star formation content within these dense cores using the IRAS point source catalog and near–infrared imaging surveys. At least 8, and possibly 10, of these cores are associated with IRAS point sources, and many of these cores already contain a substantial cluster of stars. The high fraction of massive cores with star formation suggests that stars form rather quickly after the formation of a massive core. Assuming that we are not observing the last episode of star formation in the Gem OB1 complex, this implies that dense cores are continuously formed throughout the lifetime of a molecular cloud complex.

## 4. Synthesis

Based on our molecular line observations we propose the following overall picture for the evolution of the Gem OB1 molecular cloud complex. While we cannot say how the previous generation of stars in this region formed, it is apparent that they have left their imprints within the molecular gas as evidenced by the prevalent "arcs" and "rings" of molecular material. These structures are the preferential location of the more massive cores

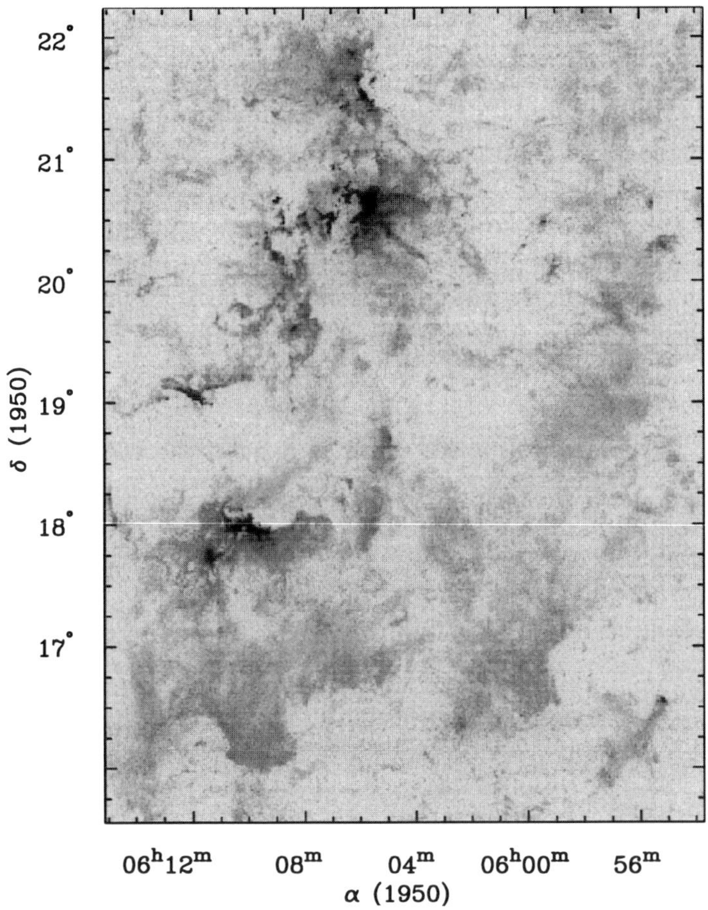


*Figure 1.* Map of the observed peak $^{12}$CO(J=1–0) antenna temperature.

(mass > 100 $M_\odot$) within the Gem OB1 complex. Thus we suggest that the current population of massive dense cores in the Gem OB1 complex have formed primarily in swept up shells of molecular gas. Since most of the cores already contain substantial signatures of star formation activity, we further suggest that the onset of star formation in *massive* dense cores occurs rather quickly after the core is formed.

**Acknowledgments:** J.M.C. acknowledges support from the James Clerk Maxwell Telescope Fellowship.

## References

Carpenter, J. M., Snell, R. L., & Schloerb, F. P. 1995a, ApJ, 445, 246
Carpenter, J. M., Snell, R. L., & Schloerb, F. P. 1995b, ApJ, 450, 201

# PROTOSTELLAR AND PROTOPLANETARY DISKS

ANNEILA I. SARGENT
*California Institute of Technology*
*Department of Astronomy 105-24*
*Pasadena CA 91125*

## 1. Introduction

It is barely a decade since the first molecular line images of disks around embedded and pre-main sequence stars were published (Beckwith *et al.*, 1986; Mundy *et al.*, 1986; Sargent & Beckwith 1987). In the intervening years there has been steady progress in detecting and analyzing these disks, which are an expected by-product of the star formation process (*cf.* Shu *et al.*, 1987; Shu *et al.*, 1993), and there is now a veritable flood of new and intriguing results as millimeter-wave instrumentation improves (*cf.* Sargent & Welch 1993).

Observations across the wavelength spectrum provide indirect evidence on size scales from 0.01 to 1000 AU that disks are associated with very young stars (*cf.* Beckwith 1994, and references therein; Sargent 1995, and references therein). Indeed, infrared and millimeter-wavelength continuum emission from many solar-mass pre-main sequence objects can be reproduced by models of star-disk systems (Adams *et al.*, 1987), furnishing a powerful tool for characterizing disks. Masses appear to range from 0.001 to 2 $M_{\odot}$; radii are of order 100 AU; and at 1 AU radius, temperatures are less than a few $\times$ 100 K (*cf.* Adams *et al.*, 1990; Beckwith *et al.*, 1990; Beckwith & Sargent 1993a). These properties are quite consistent with those of the pre-solar nebula so that studies of young disks may inform our understanding of how planetary systems form and evolve.

Detailed disk investigations require high spatial resolution since 100 AU is equivalent to $0.7''$ at the 140 pc distance of the nearest star-forming clouds. Aperture synthesis imaging of millimeter-wave line and continuum emission can not only achieve the desired resolution to sample the physical and chemical properties of the cool, outer disks, but also discriminates against the smooth, large-scale emission expected from their parent molec-



*W.B. Latter et al. (eds.), CO: Twenty-Five Years of Millimeter-Wave Spectroscopy,* 151–158.

ular clouds. The hotter, inner disk regions can be probed at near infrared wavelengths through, for example, observations and modelling of the 2 $\mu$m CO bandhead emission (*cf.* Chandler *et al.*, 1995; Carr 1995, and references therein). In view of the Symposium theme, my review will focus mainly on gas in disks. The dust component is discussed in a number of other reviews (*e.g.* Beckwith 1994; Kenyon 1995; Sargent 1995). There are many open questions, particularly with regard to circumstellar grain properties (*cf.* Pollack *et al.*, 1994) and the nature of the envelopes that may surround many star/disk systems (*cf.* Calvet *et al.*, 1994; Natta 1994; Chandler *et al.*, 1995).

## 2. Outstanding Questions

Although much has been learned about proto-stellar and proto-planetary disks, many questions remain. Among the most important problems:

- It is not clear how the fraction of young stars exhibiting disks varies with such factors as age and environment. Since these may be planetary systems in the making, disk frequency is important for theories of cosmogony.
- Most stars form in binaries or multiple groups (Simon *et al.*, 1992; Leinert *et al.*, 1993; Ghez *et al.*, 1993; Reipurth & Zinnecker 1994), but the effects of companions on circumstellar or circumbinary disks is not well understood (*cf.* Mathieu 1994).
- It is not known how disks evolve, although this has important implications for star and planet formation. Morphological evolution has been inferred from the fact that the infrared emission from some systems can be reproduced by invoking the presence of disks with cleared, central holes (Skrutskie *et al.*, 1990; Beckwith *et al.*, 1990; Marsh & Mahoney 1992; Strom *et al.*, 1993; Osterloh & Beckwith 1995). However, there are alternative interpretations (Boss & Yorke 1993).
- Kinematic evolution of disks remains an enigma. Gravitational infall is expected for true protostars, but supporting observational evidence is limited (*cf.* Zhou *et al.*, 1993; Chandler & Sargent 1993; Zhou & Evans 1994). Images of the disks around pre-main sequence systems have been interpreted in terms of infall, rotation, and outflow motions (*e.g.* Sargent & Beckwith 1991; Koerner *et al.*, 1993; Hayashi *et al.*, 1993; Guilloteau & Dutrey 1994; Cabrit *et al.*, 1995; Koerner & Sargent 1995), but no evolutionary pattern has been identified.
- Disk mass estimates based on dust continuum fluxes are frequently much larger than those derived from molecular line maps (*e.g.* Guilloteau & Dutrey 1994; Koerner & Sargent 1995; Dutrey *et al.*, 1995). The cause of the discrepancy has yet to be established. It could result from the fact that disk emission in the low-lying rotational transitions of CO and $^{13}$CO is

often optically thick (Beckwith & Sargent 1993b) or because the molecular species used to trace the presence of molecular hydrogen is depleted (*cf.* Blake *et al.*, 1992; Dutrey *et al.*, 1994, 1995).

• Dispersal times for the dust and gas components of disks are tied to the mass determinations and are therefore difficult to establish. Nevertheless, these time-scales are critical for proto-planetary disks since they strongly constrain theories of how giant planets form (*cf.* Lissauer *et al.*, 1994; Podosek & Cassen 1994).

## 3. Recent Pertinent Observations

### 3.1. ENVIRONMENT

Far-infrared and mm-wave continuum excesses indicate that a substantial fraction, perhaps 50%, of young, low-mass stars in the nearby Taurus, Ophiuchus and Chamaeleon clouds are surrounded by disks (Strom *et al.*, 1989; Beckwith *et al.*, 1990; Henning & Thamm 1994; André & Montmerle 1994; Mannings & Emerson 1994; Osterloh & Beckwith 1995). Near-infrared excesses also suggest a very high fraction of associated disks for the members of young embedded clusters (Ladd *et al.*, 1993; Lada & Lada 1995).

Recently, arcsec-size, externally-illuminated nebulae have been detected in the Trapezium cluster in Orion (O'Dell *et al.*, 1993; O'Dell & Wen 1994). Many are associated with young stars of age $\sim 10^6$ years (Prosser *et al.*, 1994; McCaughrean & Stauffer 1994). It seems likely that these are star/disk systems where the neutral circumstellar material is being ablated by radiation from nearby OB stars, (Bally *et al.*, 1995). This interpretation is strengthened by the observation that some disk-like structures, each associated with a young star, are seen silhouetted against the nebula. As in the quiescent dark clouds, inferred disk properties are consistent with those of the pre-solar nebula (McCaughrean & O'Dell 1996). The fraction of stars that might support pre-planetary disks does not appear to be diminished even in a dense cluster environment like the Trapezium.

### 3.2. MULTIPLICITY

Very few binary systems have been mapped to date but the results are intriguing. The IRAS 16293-2422 disk was one of the earliest imaged (Mundy *et al.*, 1986). In subsequent higher resolution centimeter- and millimeter-wavelength maps, two continuum sources, separated by about 800 AU, were discovered within a rotating, circumbinary disk (Wootten 1989; Mundy *et al.*, 1992; Walker *et al.*, 1993). The detection of vibrationally excited $C^{32}S$, possibly from self-gravitating instabilities in an accretion disk (Walker *et al.*, 1994), suggests that at least one element of the binary is a star/disk system.

Many pre-main sequence stars have recently proven to be binaries. The prototype of the class, T Tau, is no exception. The optically-visible T TauN and its infrared companion, T TauS, are separated by $\approx$ 100 AU. Evidence for a circumbinary disk was presented by Weintraub *et al.*, (1989) but very high resolution, 267 GHz continuum images imply that only T TauN is surrounded by an accretion disk (van Langevelde *et al.*, 1996).

The separations of the two sources in IRAS 16293-2422 and T Tau are important. On the whole, mm-wave continuum flux, presumably from a circumbinary disk, appears to be suppressed when binary elements are separated by less than about 100 AU (*cf.* Beckwith & Sargent 1993a; Osterloh & Beckwith 1995; Jensen *et al.* 1996a). Interferometer maps of the triple system UZ Tau at 1 mm by Jensen *et al.* (1996b) nicely illustrate this point. Strong 1 mm continumm and CO(2$\rightarrow$1) emission, consistent with a rotating disk, are associated with UZ TauE; only weak continuum and no line emission are detected from the companion at 530 AU, UZ TauW, which is itself a close binary with element separation 50 AU. Circumbinary disks are in fact likely to be truncated at an inner radius defined by the binary orbit, with concomitant reduction in disk surface area and infrared/millimeter flux (Artymowicz & Lubow 1994). Indeed, high resolution, $^{13}$CO(1$\rightarrow$0) and continuum images of the binary GG Tau can be fit by a model circumbinary disk of radius 800 AU with a cleared inner cavity of radius 180 AU (Dutrey *et al.*, 1994). The subject of how disks evolve in a binary environment needs to be pursued.

### 3.3. DEPLETION

Dutrey *et al.* (1994) point out that the disk mass derived from their $^{13}$CO measurements of GG Tau is much lower than estimated from the dust continuum fluxes (*e.g.* Beckwith & Sargent 1991). Similar discrepancies (up to factors of 100) are noted in other T Tauri stars (Koerner *et al.*, 1993; Guilloteau & Dutrey 1995; Koerner & Sargent 1995; Dutrey *et al.*, 1995). Masses based on continuum flux are probably correct to within a factor of 3 (*cf.* Pollack *et al.*, 1994; Agladze *et al.*, 1995) but factors such as beam dilution, optical depth effects (Beckwith & Sargent 1993b), and depletion of some molecular species (Aikawa *et al.*, 1995) could result in erroneously low masses of molecular hydrogen (*cf.* Dutrey *et al.*, 1995). For example, a multi-transition study of CS suggests it is much less abundant in the HL Tau disk than in interstellar clouds (Blake *et al.*, 1992). Not surprisingly, very few disks are detected around visible pre-main sequence stars in interferometric CS surveys (Ohashi *et al.*, 1991; Ohashi *et al.*, 1994;

Ohashi & Hayashi 1995).

### 3.4. KINEMATIC EVOLUTION

The spectral line shapes that result from radially infalling and rotating gas in disks are disconcertingly similar. It is sometimes possible to distinguish between these kinematic extremes using aperture synthesis molecular line maps (*e.g.* Koerner & Sargent 1995), but uncertain disk inclination or azimuthal structure can cause confusion. Thus, in early $^{13}CO(1\rightarrow0)$ aperture synthesis maps of HL Tau, it appeared that the circumstellar gas was centrifugally supported out to radii of 1000 AU (Sargent & Beckwith 1987; 1991). More recent position-velocity ($p$-$v$) diagrams indicate a relatively small, rotating inner disk surrounded by a substantial radially-infalling envelope (Hayashi et al. 1993), although an entrained outflow interpretation has also been proposed (Cabrit *et al.*, 1995).

Unambiguous interpretation of molecular line emission from disks can be effected by comparing synthetic maps based on model disks with the observations (*e.g.* Koerner *et al.*, 1993; Dutrey *et al.*, 1994; Koerner, this volume). For this purpose, imaging is best carried out in the CO or $^{13}CO$ $(2\rightarrow1)$ lines at $\lambda = 1.3$ mm, where the Planck function and the larger statistical weight of the higher rotational level combine to ensure stronger emission. In this way, Koerner *et al.* (1993) showed that a 0.1 $M_\odot$ disk with radius $\sim$ 1000 AU is gravitationally bound to the pre-main sequence star GM Aur; there is little evidence for infalling material. A similar analysis for DM Tau demonstrates that rotation *and* infall are probably occurring in this disk (Spangler *et al.*, 1995; but see also Saito *et al.*, this volume)

Reliable knowledge about the velocity fields in a sample of proto-stellar and proto-planetary disks could help determine how initially infall-dominated motions gradually evolve to a rotationally-dominated state where planets may form. Table 1 is a compilation of relevant properties for apparently single stars. Stellar ages and masses, as well as disk masses, are from Beckwith *et al.* (1990), except for HL Tau, where the age is that proposed by Stapelfeldt *et al.* (1995) based on their discovery that this star is still heavily embedded. Inside the centrifugal radius, $R_C$, all material is rotating; $R_C$ and the outer radius, $R_D$, are derived from mm-array maps and synthetic images (Koerner 1994; Koerner & Sargent 1995; Spangler *et al.*, 1995; Sargent & Koerner 1996, in preparation). The region of radial infall is defined by $(R_D - R_C)$.

The values of $R_C$ in Table 1 are still tentative. However, it is noteworthy that for HL Tau the rotating core radius derived from molecular line analysis, 80 AU, is very close to the disk radius, 70 AU, obtained from sub-arcsecond images at 2.6 mm (Mundy *et al.*, 1995), and through sin-

TABLE 1. Disk Infall Regions ($R_D - R_C$)

| Star | $M_*$ $M_\odot$ | Age yrs | $M_D$ $M_\odot$ | $R_C$ AU | $R_D$ AU |
|---|---|---|---|---|---|
| HL Tau | 1.1 | $\leq 10^5$ | 0.1 | $\sim 80^*$ | 1000 |
| RY Tau | 1.25 | $2 \times 10^5$ | 0.04 | $\sim 150$ | 200 |
| DO Tau | 1.6 | $6 \times 10^5$ | 0.02 | 350 | 500 |
| DL Tau | 0.25 | $1 \times 10^6$ | 0.08 | 200 | 200 |
| GM Aur | 0.72 | $2 \times 10^6$ | 0.08 | 1000 | 1000 |
| DM Tau | 0.62 | $3 \times 10^6$ | 0.03 | 160 | 500 |
| Elias 3-1 | 1.03 | $\leq 10^7$ | 0.01 | $\sim 400$ | 600 |

gle baseline interferometry at 850 $\mu$m (Lay *et al.*, 1994). In addition, disk masses derived from rotation curves agree quite well with $M_D$. There is no obvious trend in ($R_D - R_C$) with stellar age, but the sample is very small. As might be expected, ($R_D - R_C$) is dramatically larger for the extremely young HL Tau.

## 4. Future Steps

The observations summarized above should be regarded as the tip of the iceberg. With the advent of infrared arrays, upgrades to Hubble Space Telescope, and the launch of the Infrared Space Observatory (ISO), great strides will be made in understanding disk frequency as a function of environment. Increasing sensitivity with telescopes such as Keck and Gemini will help determine which young stars are in multiple systems and what are the properties of each star, as well as providing samples for molecular line disk studies. The increased resolution and sensitivity expected from the expanding millimeter-wave arrays, from the SAO sub-millimeter array, and from the planned national mm-wave array will allow more rapid surveys of disks in a wide variety of molecular species so that mass problems, and related questions regarding dust and gas dispersal times, can be settled. Detailed investigations of disk morphology and velocity fields will also be more straightforward, increasing the viability of disk evolution studies. For example, the size of the HL Tau dust disk was established with a relatively modest gain in resolution (Mundy *et al.*, 1995), and similar measurements for other disks should soon become routine.

Acknowledgements: I am grateful to Drs. V. Mannings and D. Koerner for critical readings of this manuscript.

## References


Adams, F.C., Emerson, J.P. & Fuller, G.A. (1990) *Astrophys. J.*, **357**, pp. 606–620.
Adams, F.C., Lada, C.J. & Shu, F.H. (1987) *Astrophys. J.*, **312**, pp. 788–806.
Agladze, N.I., Sievers, A.J., Jones, S.A., Burlitch, J.M. & Beckwith, S.V.W. (1995) *Nature*, **372**, pp. 243–245.
Aikawa, Y., Miyama, S.M., Nakano, T., & Umebayashi, T. (1995) *Astrophys. J.*, in press.
André, P. & Montmerle, T. (1994) *Astrophys. J.*, **420**, pp. 837–862.
Artymowicz, P., & Lubow, S. H. (1994) *Astrophys. J.*, **421**, pp. 651–667.
Bally, J., Devine D., & Sutherland, R. (1995) *Rev. Mex. Astron. Astrofisica, Serie de Conferencias*, **1**, pp. 19–27.
Beckwith, S.V.W. (1994) in W. Duschl, J. Frank, F. Meyer, E. Meyer-Hofmeister and W. Tscharnuter (eds.), *Theory of Accretion Disks -2*, (Dordrecht:Kluwer) pp. 1–18.
Beckwith, S.V.W., & Sargent, A.I. (1991) *Astrophys. J.*, **381**, pp. 250–258.
Beckwith, S.V.W., & Sargent, A.I. (1993a), in E.H. Levy and J.I. Lunine (eds.), *Protostars and Planets III*, (Tucson:University of Arizona Press), pp. 521–541.
Beckwith, S.V.W. & Sargent, A.I. (1993b), *Astrophys. J.*, **402**, pp. 280–291.
Beckwith, S.V.W., Sargent, A.I., Chini, R.S. & Güsten, R. (1990) *Astron. J.*, **99**, pp. 924–945.
Beckwith, S., Sargent, A.I., Scoville, N.Z., Masson, C.R., Zuckerman, B., *et al.*, (1986), *Astrophys. J.*, **309**, pp. 755–761.
Blake, G.A., van Dishoeck, E.F. & Sargent, A.I. (1992), *Astrophys. J.*, **391**, pp. L99–L103.
Boss, A.P., & Yorke, H.W. (1993), *Astrophys. J.*, **411**, pp. L99–L102.
Cabrit, S., Guilloteau, S., André, P., Bertout, C., Montmerle, T. & Schuster, K. 1995, *Astron. Astrophys.*, in press.
Calvet, N., Hartmann, L., Kenyon, S.J., & Whitney, B.A. (1994) *Astrophys. J.*, **434**, pp. 330–340.
Carr, J.S. (1995) *Astrophys. Space Sci.*, **224**, pp. 25–28.
Chandler, C.J., Carlstrom, J. E., & Scoville, N. Z. (1995) *Astrophys. J.*, **446**, pp. 793–800.
Chandler, C.J., & Sargent, A.I. (1993) *Astrophys. J.*, **414**, pp. L29–L32.
Chandler, C.J., Koerner, D.W., Sargent, A.I., & Wood, D.O.S. (1995) *Astrophys. J.*, **449**, pp. L139–L142.
Dutrey, A., *et al.*, (1995) *Astron. Astrophys.*, in press.
Dutrey, A., Guilloteau, S. & Simon, M. (1994) *Astron. Astrophys.*, **286**, pp. 149–159.
Ghez, A.M., Neugebauer, G. & Matthews, K. (1993) *Astron. J.*, **106**, pp. 2005–2023.
Guilloteau, S., & Dutrey, A. (1994) *Astron. Astrophys.*, **291**, pp. L23–L26.
Hayashi, M., Ohashi, N., Miyama, S. M. (1993), *Astrophys. J.*, **418**, pp. L71–L74.
Henning, T. & Thamm, E. (1994), *Astrophys. & Space Science*, **212**, pp. 215–220.
Jensen, E.L.N.,Mathieu, R.D., & Fuller, G.A. (1996a), *Astrophys. J.*, in press.
Jensen, E.L.N., Koerner, D.W., & Mathieu, R.D. (1996b), in preparation.
Kenyon, S.J. (1995) *Astrophys. & Space Sci.*, **223**, pp. 3–17.
Koerner, D.W. (1994) Ph.D. Dissertation, California Institute of Technology.
Koerner, D.W., & Sargent, A.I. (1995) *Astron. J.*, **109**, pp. 2138–2145.
Koerner, D.W., Sargent, A.I. & Beckwith, S.V.W. (1993), *Icarus*, **106**, pp. 2–10.
Lada, E.A., & Lada, C.J. (1995) *Astron. J.*, **109**, pp. 1682-1696.
Ladd, E.F., Lada, E.A. & Myers, P.C. (1993) *Astrophys. J.*, **410**, pp. 168–178.
Lay, O.P., Carlstrom, J., Hills, R.J. & Phillips, T.G. (1994) *Astrophys. J.*, **434**, pp. L75–L78.
Leinert, C., Zinnecker, H., Weitzel, N., Christou, J., Ridgway, S.T. et al. (1993), *Astron. Astrophys.*, **278**, pp. 129–149.
Lissauer, J.J., Pollack, J.B., Wetherill, G.W., & Stevenson, D. J. 1994, in D. P. Cruickshank & M. S. Matthews (eds.), *Neptune and Triton*, (Tucson:University of Arizona

Press), in press.
Mannings, V. & Emerson, J.P. (1994) *Mon. Not. R. Astron. Soc.*, **267**, pp. 361–378.
Marsh, K.A., & Mahoney, M.J. (1992) *Astrophys. J.*, **395**, pp. L115–L118.
Mathieu, R.D. (1994) *Ann. Rev. Astron. Astrophys.*, bf 32, pp. 465–530.
McCaughrean, M.J., & O'Dell, C.R. (1996) *Astron. J.*, in press.
McCaughrean, M.J. & Stauffer, J.R. (1994) *Astron. J.*, **108**, pp. 1382–1397.
Mundy, L.G., Looney, L.W., Erickson, W., Grossman, A., Welch, W.J., *et al.*, (1995), *Astrophys. J.*, in press.
Mundy, L.G., Wilking, B.A. & Myers, S.T. (1986) *Astrophys. J.*, **311**, pp. L75–L79.
Mundy, L.G., Wootten, H.A., Wilking, B.A., Blake, G.A. & Sargent, A.I. (1992) *Astrophys. J.*, **385**, pp. 306–313.
Natta, A. (1993) *Astrophys. J.*, **412**, pp. 761–770.
O'Dell, C.R., Wen, Z. & Hu, X. (1993) *Astrophys. J.*, **410**, pp. 696–700.
O'Dell, C.R. & Wen, Z. (1994) *Astrophys. J.*, **436**, pp. 194–202.
Ohashi, N., & Hayashi, M. (1995) in H.J. Staude & S.V.W. Beckwith (eds.), *Disks and Outflows around Young Stars*, (Berlin:Springer Verlag), in press.
Ohashi, N., Kawabe, R., Hayashi, M. & Ishiguro, M. (1991) *Astron. J.*, **102**, pp. 2054–2065.
Ohashi, N., Kawabe, R., Ishiguro, M. & Hayashi, M. (1994) *Astrophys. & Space Science*, **212**, pp. 239–250.
Osterloh, M. & Beckwith, S.V.W. (1995) *Astrophys. J.*, **439**, pp. 288–302.
Podosek, F.A., & Cassen, P. (1994) *Meteoritics*, **29**, pp. 6–25.
Pollack, J.B., *et al.*, (1994) *Astrophys. J.*, **421**, pp. 615–639.
Prosser, C.F., *et al.*, (1994) *Astrophys. J.*, **421**, pp. 517–541.
Reipurth, B., & Zinnecker, H. (1993) *Astron. Astrophys.*, **278**, pp. 81–108.
Sargent, A.I. (1995) in H.J. Staude & S.V.W. Beckwith (eds.), *Disks and Outflows around Young Stars*, (Berlin:Springer Verlag), in press.
Sargent, A.I. & Beckwith, S. (1987) *Astrophys. J.*, **323**, pp. 294–305.
Sargent, A.I. & Beckwith, S.V.W. (1991) *Astrophys. J.*, **382**, pp. L31–L35.
Sargent, A.I., & Welch, W. J. (1993) *Ann. Rev. Astron. Astrophys.*, **31**, pp. 297–343.
Spangler, C.S., Sargent, A.I., & Koerner, D.W. (1995), in preparation.
Stapelfeldt, K.R., *et al.*, (1995) *Astrophys. J.*, **448**, pp. 888–893.
Shu, F.H., Adams, F.C. & Lizano, S. (1987) *Ann. Rev. Astron. Astrophys.* , **25**, pp. 23–81.
Shu, F.H., Najita, J., Galli, D. & Ostriker, E. (1993) in E.H. Levy & J.I. Lunine (eds.), *Protostars and Planets III*, (Tucson:University of Arizona Press), pp. 3–45.
Simon, M., Chen, W.P., Howell, R.R., Benson, J.A. & Slowik, D. (1992) *Astrophys. J.*, **384**, pp. 212–219.
Skrutskie, M.F., *et al.*, (1990) *Astron. J.*, **99**, pp. 1187–1195.
Strom, K.M., Strom, S.E., Edwards, S., Cabrit, S., & Skrutskie, M.F. (1989) *Astron. J.*, **97**, pp 1451–1470.
Strom, S.E., Edwards, S. & Skrutskie, M.F. (1993) in E.H. Levy and J.I. Lunine (eds.), *Protostars and Planets III*, (Tucson:University of Arizona Press), pp. 837–866.
van Langevelde, H.J., van Dishoeck, E.F., & Blake, G.A. (1996), in preparation.
Walker, C., Carlstrom, J.E., & Bieging, J.H. (1993) *Astrophys. J.*, **402** pp. 655–666.
Walker, C., Maloney, P.R., & Serabyn, E. (1994) *Astrophys. J.*, **437**, pp. L127–130.
Weintraub, D.A., Masson, C.R., Zuckerman, B. (1989) *Astrophys. J.*, **344**, pp. 915–924.
Wootten, H.A. (1989) *Astrophys. J.*, **337**, pp. 858–864.
Zhou, S., Evans, N.J., Kompe, C., Walmsley, C.M. (1993) *Astrophys. J.*, **404**, pp. 232–246.
Zhou, S., & Evans, N.J. (1994) in D. Clemens & R. Barvainis (eds.) *Clouds, Cores and Low Mass Stars: The Fourth Haystack Observatory Conference*, pp. 183–191.

# 350 AU SCALE CIRCUMSTELLAR ROTATING GASEOUS DISK AROUND DM TAURI

MASAO SAITO AND RYOHEI KAWABE

*Nobeyama Radio Observatory, Nobeyama, Minamimaki, Minamisaku Nagano 384-13, Japan*

SHOKEN M. MIYAMA

*National Astronomical Observatory 2-11-1, Oosawa, Mitaka, Tokyo 181, Japan*

TOSHIHIRO HANDA

*Institute of Astronomy, Faculty of Science, University of Tokyo 2-11-1, Oosawa, Mitaka, Tokyo 181, Japan*

AND

YOSHIMI KITAMURA

*Institute of Space and Astronautical Science 3-1-1, Yoshinodai, Sagamihara, Kanagawa 229, Japan*

## 1. Introduction

We undertook a survey of the CO emission from protoplanetary gas disks around T Tauri stars using the Nobeyama 45m telescope. The purpose of these observations was to study the structure and evolution of the gas disks and study the process of planet formation. Through the survey, CO emission was found toward DM Tau (Handa et al. 1995). The following observations of DM Tau were made using the Nobeyama Millimeter Array (NMA) to investigate the distribution and kinematics of molecular gas associated with DM Tau. Here, we present the results for DM Tau. DM Tau is a single classical T Tauri star in the Taurus region(140 pc). The stellar mass and age are estimated to be 0.48 $M_\odot$ and 1 x $10^6$ yr from the H-R diagram.



*W.B. Latter et al. (eds.), CO: Twenty-Five Years of Millimeter-Wave Spectroscopy,* 159–161.

## 2. Observations and Results

### 2.1. NOBEYAMA 45M TELESCOPE

Figure 1 shows the profile maps of $^{12}$CO, $^{13}$CO, and C$^{18}$O obtained toward DM Tau with the Nobeyama 45m telescope. It shows clearly two components; one narrow component at $V_{LSR}$=9 km s$^{-1}$, and the broader one at 6 km s$^{-1}$ that is consistent with the systemic velocity of the DM Tau-obtained optical observation. The 6 km s$^{-1}$ component is seen just toward DM Tau, though the 9 km s$^{-1}$ component is extended. These facts suggest that the 6 km s$^{-1}$ component is physically associated with the star. The 6 km s$^{-1}$ feature in $^{13}$CO emission shows a double symmetric peak suggesting a rotating disk (Handa et al. 1995).

### 2.2. NOBEYAMA MILLIMETER ARRAY (NMA)

Figure 2a shows a map of the $^{12}$CO intensity integrated over the velocity range 4.4 to 6.8 km s$^{-1}$ obtained with the NMA. The map clearly shows that the compact disk-like structure is associated with the star. The radius of the disk is estimated to be 300 - 350 AU by the separation of the peaks of the blue- and red-shifted emission in Figure 2b (Saito et al. 1995).

## 3. Discussion

The observed CO kinematics are interpreted as indicating the presence of the rotating disk, based mainly on three observational results. First, the $^{13}$CO spectrum obtained by the 45 m telescope has a clear symmetric double-peaked profile, as expected from the Keplerian rotating disk. Second, there exists the velocity gradient along the major axis of the disk. Finally, the velocity width of the $^{12}$CO emission is much smaller than typically observed for outflow sources and the $^{12}$CO emission is positioned close to the star.

The line-of-sight rotational velocity is 0.6 km s$^{-1}$ from the $^{12}$CO intensity weighted $V_{LSR}$. As there exists the velocity structure, the disk inclination is not face-on, and is estimated to be roughly 40° (the inclination is 0° for face-on). The obtained parameters are roughly consistent with the line-of-sight velocity of Keplerian rotation at r = 350 AU, 0.70 km s$^{-1}$ which is estimated assuming i = 40°.

We estimated that the $H_2$ mass of the disk is 0.019 $M_\odot$ from the flux density at 2.6 mm, 23 mJy. Although the disk around DM Tau is larger than the solar system, the disk may be a protoplanetary disk from the viewpoint that the mass of the disk is comparable to the minimum mass of the solar nebula ($\sim$ 0.01 $M_\odot$) and the kinematics of the disk is consistent with Keplerian rotation.

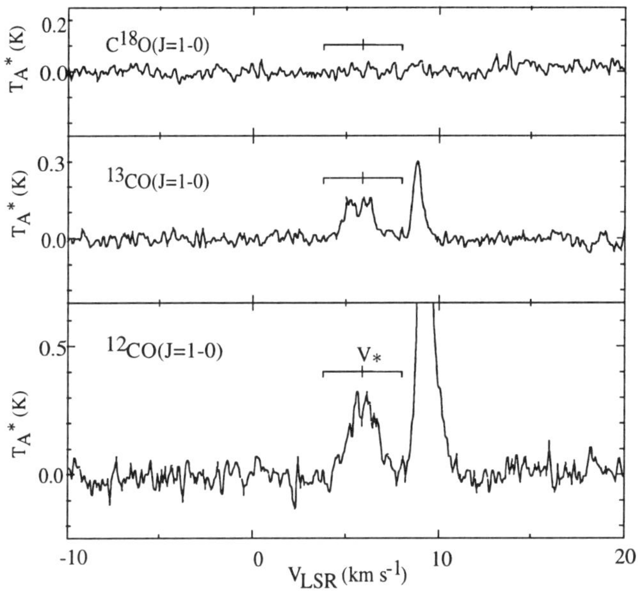


*Figure 1.* $^{12}$CO, $^{13}$CO, and C$^{18}$O profile maps of DM Tau obtained with the Nobeyama 45m Telescope.

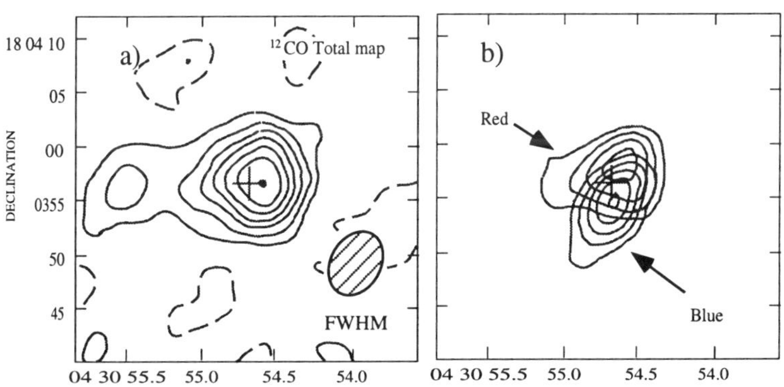


*Figure 2.* a) Map of $^{12}$CO emission around DM Tau integrated over a velocity range of 4.0 - 7.3 km s$^{-1}$ with NMA. b) Map of superposition blue- (4.0 - 5.6 km s$^{-1}$) and redshifted (5.6 - 7.3 km s$^{-1}$) $^{12}$CO components around DM Tau.

## References

Handa, T. et al. 1995, ApJ, in press

Saito, M., Kawabe, R., Ishiguro, M., Miyama, S. M., Hayashi, M., Handa, T., Kitamura, Y., and Omodaka, T. 1995, ApJ, in press

# KINEMATICS OF DISKS AROUND T TAURI STARS

DAVID W. KOERNER
*Jet Propulsion Laboratory,*
*California Institute of Technology*
*Mail Stop 169-506*
*4800 Oak Grove Dr.*
*Pasadena, CA 91109*

## 1. Introduction

The presence of circumstellar disks around T Tauri stars (TTs) is strongly supported by narrow- and broad-band spectroscopic evidence at many wavelengths (cf. Sargent 1995). Spatially resolved observations of the associated velocity field are needed to establish whether or not the circumstellar material is centrifugally supported and available to planet formation. Past studies relied on averaged presentations of the data such as rotation curves or Position-Velocity diagrams and often led to ambiguous or conflicting interpretations. Synthetic spectral line maps generated by a kinematic model provide a more precise identification of the circumstellar velocity field, as illustrated by modeling of rotating disks around the TTs GM Aurigae (Koerner *et al.*, 1993) and the T-Tauri multiple system GG Tauri (Dutrey *et al.*, 1994). To adequately study younger stars, a radial velocity component must be added to such a model and parameterized in a physically appropriate way.

## 2. Synthetic Spectral Line Maps

Synthetic maps of a disk inclined 45° illustrate the expected contrast for Keplerian rotation and free fall in Fig. 1a and 1b, respectively. Patterns are similar at velocities which differ by a factor of $\sqrt{2}$, with infall patterns at right angles to their rotational counterparts. Since the principal difference lies in the orientation of patterns with respect to the disk, the disk orientation must be well identified to distinguish between kinematic states. To potentially aid kinematic analysis of disk-like structures with rotation *and*



*W.B. Latter et al. (eds.), CO: Twenty-Five Years of Millimeter-Wave Spectroscopy,* 162–164.

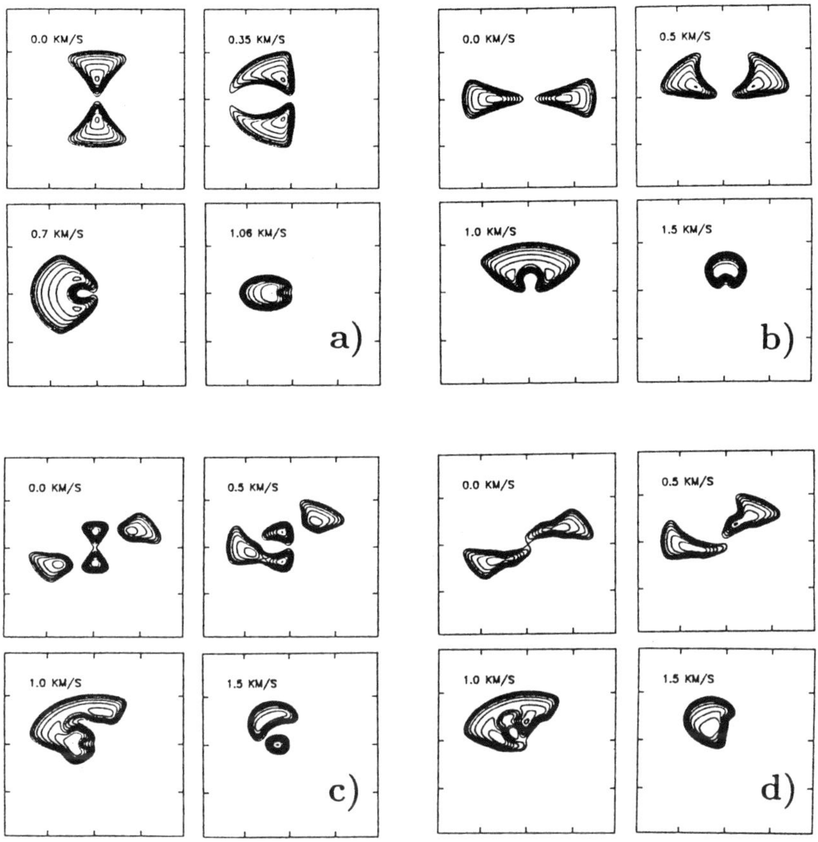


*Figure 1.* Synthetic Spectral Line Maps from a disk of gas with velocity structures given by: (a) Keplerian rotation, (b) free fall, (c) combined disk/envelope, and (d) an inviscid outer region. Maps are at velocities which correspond to the center, inner slope, peak, and outer slope of one half of the corresponding double-peaked spectral line.

infall, synthetic maps from two hybrid models are displayed in Fig. 1c and 1d. Fig. 1c illustrates a flat infall envelope surrounding a region of centrifugal support; a pattern discontinuity occurs at the radius where the infalling gas meets a shock boundary of the accretion disk. In Fig. 1d, the velocity field of the outer disk is that derived by Stahler *et al.*, (1994) for an inviscid disk formed during collapse of a uniformly rotating molecular cloud. The radial momentum of infalling gas is largely conserved in the outer disk and gradually decreases to zero with approach to a critical radius; no break occurs in the corresponding patterns.

## 3. Kinematic Evolution of T-Tauri Disks

The need to consider both radial and tangential velocity components in kinematic analyses of T-Tauri disks is reinforced by detection of a strong contribution from infalling gas to small-scale CO emission from HL Tauri (Hayashi *et al.*, 1993; Sargent & Koerner, 1995), although recent HST observations indicate it is embedded (Stapelfeldt *et al.*, 1995). Images of circumstellar gas around the TTs RY Tau, DL Tau, DO Tau, and AS209 reveal smaller disks (R $\sim$ 100 AU) which appear to be centrifugally supported (Koerner & Sargent, 1995) with some infall around the youngest object (Koerner, 1994). The older TTs, GM Aur, exhibits Keplerian rotation in a somewhat larger disk (Koerner *et al.*, 1993). Consideration of this small sample suggests that young low-mass stars undergo the following stages in the transition from an embedded to the T-Tauri phase: **I.** "Flat spectrum" sources are surrounded by large (1000 AU) disks dominated by an infalling velocity component with an inner circular Keplerian region. The outer disk is plausibly represented by Fig. 1d; $R_C$ expands rapidly in the late stages of infall, since accreting material arrives with progessively greater angular momentum ($R_C \propto t^3$), but this newly accreted material is not necessarily on *circular* Keplerian orbits. **II.** Young T-Tauri stars have smaller disks dominated by Keplerian rotation, but are still surrounded by a remnant of dynamically accreting gas. **III.** Dynamic infall ceases altogether, but viscous transport of angular momentum drives continued expansion of the disk. Although the stages suggested here are generally accepted in some form, the classification of late embedded objects and TTs within this scheme remains uncertain until detailed kinematic modeling of high-resolution observations can be carried out for a larger sample.

## References

Dutrey, A., Guilloteau, S. and Simon, M. (1994) Images of the GG Tauri Rotating Ring, *Astron. & Astrophys.*, **Vol. no. 286**, pp. 149–159

Hayashi, M., Ohashi, N. and Miyama, S.M. (1993) A Dynamically Accreting Gas Disk Around HL Tauri, *Astrophys. J. Lett.*, **Vol. no. 418**, pp. L71–L74

Koerner, D.W., Sargent, A.I. and Beckwith, S.V.W. (1993) A Rotating Gaseous Disk around the T Tauri Star GM Aurigae, *Icarus*, **Vol. no. 106**, pp. 2–10

Koerner, D.W. and Sargent, A.I. (1995) Imaging the Small-Scale Circumstellar Gas around T Tauri Stars, *Astron. J.*, **Vol. no. 109**, pp. 2138–2145

Koerner, D.W. (1994) The Kinematics of Circumstellar Disks around T Tauri Stars, *Ph. D. Dissertation*, California Institute of Technology

Sargent, A.I., and Koerner, D.W. (1995), in preparation

Sargent, A.I. (1995), Disk Observations, in *Disks and Outflows around Young Stars* eds., S.V.W. Beckwith, A Natta, and J. Staude (Berlin: Springer-Verlag), in press

Stahler, S.W., Korycansky, D.G., Brothers, M.J. and Touma, J. (1994) The Early Evolution of Protostellar Disks, *Astrophys. J.*, **Vol. no. 431**, pp. 341–358

Stapelfeldt, K.R., *et al.*, (1995) WFPC2 Imaging of the Circumstellar Nebulosity of HL Tauri, *Astrophys. J.*, in press

# CO OUTFLOWS FROM YOUNG STARS

BO REIPURTH
*European Southern Observatory*
*Casilla 19001, Santiago 19, CHILE*

AND

RAFAEL BACHILLER
*Centro Astronómico de Yebes*
*Apartado 148, 19080 Guadalajara, SPAIN*

## 1. General Properties of CO Outflows

Following the detection of extended CO wings in Orion A by Kwan & Scoville (1976) and Zuckerman et al. (1976) came the discovery in 1980 of the first bipolar CO outflow from a low luminosity source, L1551 IRS 5 (Snell, Loren, Plambeck 1980). It was soon realized that bipolar outflows are common in star forming regions, as documented in a highly influential paper by Bally & Lada (1983), and a series of papers by Edwards & Snell (1982, 1983, 1984). Many of the presently best studied molecular outflows were already identified in these early papers, and it was established that there is an association between the optically emitting shocks seen as Herbig-Haro objects and the large-scale moving CO gas. Further details of the early years of molecular outflows are discussed by Lada (1985).

Molecular outflows are commonly found in regions of active star formation, and it is probable that all newborn stars must pass through a phase of prodigious mass loss. Indeed, there is evidence that mass loss may begin already in the protostellar phase, as soon as a stellar core has started to build up. CO outflows are interpreted as the effect of this mass loss on the ambient molecular material, which is gradually being swept up.

Most CO outflows are bipolar, although a few unipolar or amorphous flows are known. Collimation factors typically range from 2 to 5, but recent higher resolution observations have uncovered several much more collimated flows (e.g., Richer et al. 1992). It is also becoming increasingly evident that the lobes of CO outflows are not uniform, but show a clumpy structure



*W.B. Latter et al. (eds.), CO: Twenty-Five Years of Millimeter-Wave Spectroscopy,* 165–174.

(e.g., Bachiller et al. 1994), which suggests that the driving sources may undergo episodic eruptions.

The sizes of CO outflows range from a few tenths of a parsec to about 5 parsec, with masses from less than a solar mass to more than 100 $M_\odot$; there is some tendency for the more massive flows to be driven by higher luminosity sources (Bally & Lada 1983). Such massive flows exclude the possibility that they originate from the embedded stars. Rather, these outflows must consist of swept-up ambient material. Velocities are for most of the outflowing material only a few $km\,s^{-1}$, but very high velocities of more than 100 $km\,s^{-1}$ have been found in highly collimated "molecular jets" (see Sect. 3).

## 2. CO Outflows and Herbig-Haro Flows

An essential question is: How are CO flows driven? The interaction between the driving agent and the ambient molecular material can be either adiabatic or radiative. In the first case the flow is energy driven by the expansion of a high-pressure bubble. In the second case it is momentum driven, with momentum being transferred via entrainment (Dyson 1984).

It is becoming increasingly clear that CO flows, at least from low-mass sources, are momentum driven, and are accelerated by Herbig-Haro jets. Already in the early work of Edwards & Snell (1983,1984), it was realized that the two phenomena often co-exist. However, the low estimates of momentum rates of HH flows by Snell et al. (1985) and Mundt et al. (1987) appeared to suggest that they were not capable of producing the momentum observed in CO flows. In a more sophisticated analysis, Raga (1991) has shown that the two flow phenomena have the same mass and momentum rates, and that a "unified" flow model is viable. A number of theoretical models have explored the details of how the momentum transfer occurs. It appears that we are right now seeing a paradigm shift from the belief that CO flows and HH flows are two aspects of the same underlying process (winds), to the view that CO flows are the direct results of HH jets ramming through ambient molecular material.

HH jets are generally highly collimated, with collimation factors of up to 100. They are very narrow, with angular widths of less than 1 arcsec, and physical widths of a few hundred AU. They almost always emanate from deeply embedded low-luminosity sources, and have space velocities of several hundred $km\,s^{-1}$. They consist of chains of small knots, which are interpreted as internal shocks caused by variations in the flow due to variability of the underlying source (Raga et al. 1990). Where the jet rams into the ambient medium a working surface is formed, which consists of two shocks, a bow shock where the ambient material is shocked and accelerated, and a jet shock, or Mach disk, where the jet material is decelerated. It is common to find multiple working surfaces along the axis of an HH flow, testifying to recurrent eruptions of the underlying source, probably FU Orionis accretion disk events (Reipurth 1989). The fact that the working surfaces normally have very high space velocities indicates that the jets are considerably more dense than their surroundings. For a recent review of the properties of HH jets, see, e.g., Reipurth & Heathcote (1993).

HH jets can transfer momentum to their surroundings in two different ways. In the *hard* interaction, gas is accelerated at the working surface at the head of the jet. Seen in a frame at rest with the working surface, material enters through both the bow shock and through the jet shock and is subsequently ejected sideways. In the *soft* interaction, lateral entrainment occurs along the boundary between the body of the jet and its environment. While both mechanisms are probably active, the principal momentum transfer is likely to occur at the working surface at the tip of the jet.

A fine example of a Herbig-Haro jet is HH 46/47, seen in Fig.1 in a combination of Hubble Space Telescope images taken through H$\alpha$ and [SII] filters (Heathcote et al. 1996). The jet flows at several hundred $\mathrm{km\,s^{-1}}$ away from an infrared source (in the lower right corner) embedded in an isolated Bok globule. A large reflection nebula outlines a conical cavity in the globule, through which the jet is seen to escape. The jet itself consists of a long thin body of numerous knots and an extended working surface. The knots are meandering away from the source, and inspection of the individual interference filter images show that at each knot where the body of the jet has a change of direction, the knot, which is [SII] bright, drives an H$\alpha$ strong shock into the ambient medium. Kinematically, this interaction causes the outer part of the jet to move slower relative to the inner part, as seen in the Fabry-Perot images of Hartigan et al. (1993). The wiggling of the jet is possibly due to precession of the driving source. As a result, the jet interacts with a larger volume of its environment than if it was straight.

The jet is presently entering the working surface at an angle, which accounts for the asymmetry of the working surface. The material ejected sideways from the working surface interacts with the ambient material that is streaming by, and a contact discontinuity with a turbulent mixing layer is formed. Theoretical models suggest that this hot layer expands and fills the cavity behind the working surface, forming a turbulent wake which may be observable as a molecular outflow (Raga & Cabrit 1993). Alternatively, the swept-up shell may itself be interpreted as the molecular outflow (Masson & Chernin 1993). The HST image of HH 46/47 shows a faint emission veil streaming backwards from the bow shock, and this is most likely direct observational evidence of the abovementioned interaction zone. In other words, *we here directly see a molecular outflow in the making.*

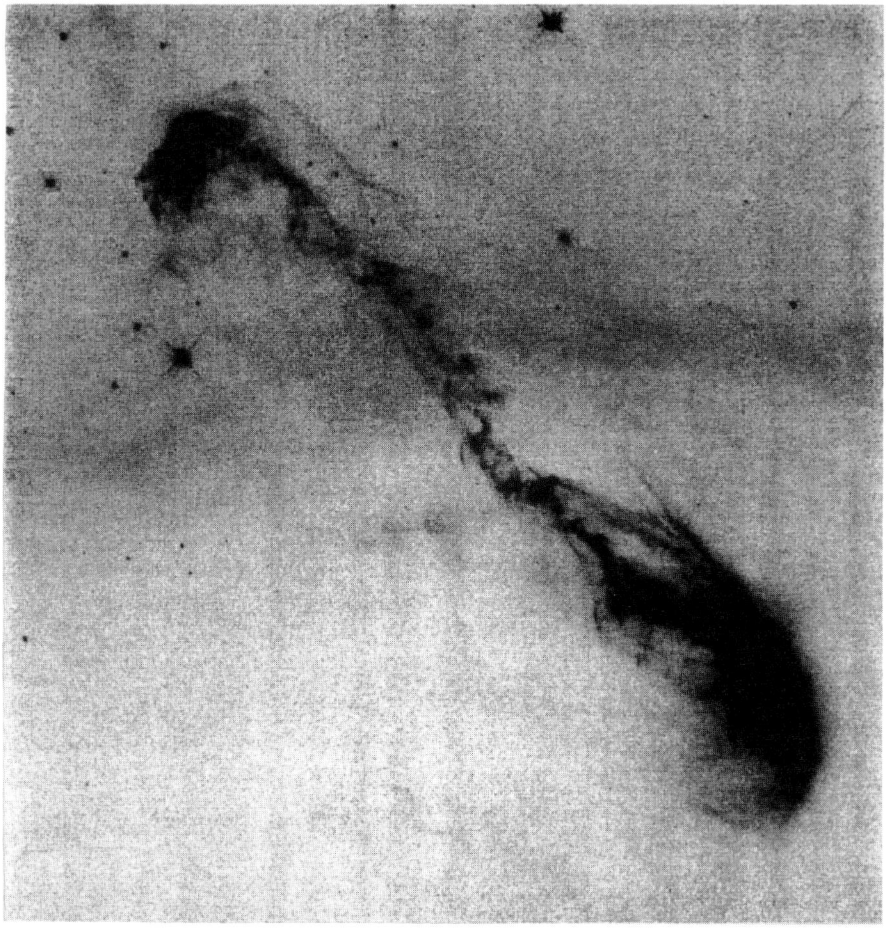

*Figure 1.* A Hubble Space Telescope image of the HH 46/47 jet complex, made by combining images taken through H$\alpha$ and [SII] filters. From Heathcote et al. (1996).

The HH 46/47 jet is indeed observed to drive a CO outflow, albeit a small one (Chernin & Masson 1991, Olberg et al. 1992). Particularly the blue lobe, which coincides with the visible HH jet, is weak, simply because there is little molecular material left to sweep up. Massive luminous OB stars near the Bok globule have blown away the material which would normally obscure the jet. The HH 46/47 jet is thus unusual in the sense that we have the opportunity to directly watch the HH jet, which in most other cases are still deeply shrouded, and only detectable in near-infrared $H_2$ emission, if at all.

The best studied association between an HH jet and a CO flow is HH 111, a large highly collimated jet that drives a major molecular outflow

(Reipurth & Olberg 1991). The CO flow is also very collimated, and contains a number of "bullets" providing evidence of multiple ejection events in the source (Cernicharo & Reipurth 1996). IRAM interferometer observations at Plateau de Bure reveal that the base of the HH jet is surrounded by a tube of fast CO gas providing direct observational evidence for CO entrainment by an HH jet (Cernicharo et al. 1996).

## 3. Jet-driven CO Outflows: The Case of L 1448

The discovery of a new class of highly-collimated molecular outflows, which are similar in many respects to the optical HH jets, is providing much support for the new unified picture of molecular outflows and HH jets (e.g., Bachiller & Gómez-González 1992, Bachiller 1996).

The fast bipolar outflow in L 1448 is one of the best examples of highly collimated CO outflows (Bachiller et al. 1990). This is because the outflow is nearby (300 pc), and its axis is highly inclined (by about 70°) with respect to the line of sight. The CO profiles observed toward the outflow display a rich structure: in addition to the classical (smooth) CO line wings, there are "extremely-high-velocity" (EHV) peaks at radial velocities up to $\pm 70$ km s$^{-1}$. The EHV CO emission is located on the main outflow axis in the form of a narrow molecular jet which consists of a chain of very well defined clumps moving with space velocities in excess of 200 km s$^{-1}$. These high-velocity molecular "bullets" have typical sizes of $\leq 15''$ (0.02 pc), masses of a few $10^{-4}\, M_{\odot}$, and exhibit strong SiO emission (Bachiller et al. 1991b, Guilloteau et al. 1992, Dutrey et al. 1996). High-velocity water maser spots have been observed to share the outflow motion in the proximity of the central young object (Chernin 1995). The central molecular jet is surrounded by a kind of thin $H_2$ envelope or cocoon which emits strong 2.12 $\mu$m lines (e.g. Bally et al. 1993; Davis et al. 1994), and that seems to be a near-IR counterpart of the classical HH flows observed in the visible. The $H_2$ emission is thought to arise in the mixing layer where the jet is entraining ambient molecular material.

The central jet in L 1448 is also surrounded by a more extended envelope of slow-moving CO, a "standard-high-velocity" (SHV) flow. Recent observations carried out at the IRAM interferometer on Plateau de Bure have revealed that this SHV CO emission arises in limb-brightened cavities with a biconical morphology (Bachiller et al. 1995a). The kinematical structure of the cavities is well accounted for with a model in which the CO moves on a narrow biconical shell with the velocity field parallel to the cone walls. Some near-IR continuum emission is observed to arise from the blueshifted part of the cavity (Bally et al. 1993). Very likely, this emission is produced by light escaping from the central object (L 1448-mm) which is scattered at the walls of the cavity through which the blueshifted flow emerges. Indeed we should expect another nebulosity to be formed at the southeast of L 1448-mm, but due to the inclination of the bipolar outflow, the receding lobe is more embedded within the ambient cloud, and the nebula formed at the redshifted CO lobe remains hidden by a higher intervening column density of dust. In summary, the configuration observed at the base of the L 1448 outflow suggests that we are witnessing the formation of a cometary nebula. As the outflow disperses more ambient material, the nebula will become optically visible giving a similar configuration to, for instance, Hubble's nebula illuminated by R Mon.

All the existing observations strongly suggest that the bipolar molecular outflow in L 1448 is being driven by the central fast jet which lies at the outflow axis. These observations, and those of similar objects (Bachiller & Gómez-Gónzalez 1992) have motivated a number of jet-driven models for molecular outflows (e.g. Richer et al. 1992; Raga & Cabrit 1993; Masson & Chernin 1993; Stahler 1994; Chernin et al. 1994a). Such models have allowed to make predictions about, e.g., the kinematics of the molecular gas (Stahler 1994), shapes of the flow lobes (Raga & Cabrit 1993), and other attributes of the molecular gas which can be observed now with the available instrumentation.

In all these models, a highly collimated jet injects mass and momentum into the surrounding ambient molecular clouds, but the assumptions differ from model to model. For instance, Stahler (1993) proposes that the standard CO outflow corresponds to a "turbulent envelope" in which the jet and entrained ambient material are mixed. However, one of the main problems faced by this kind of jet-driven models to explain the morphology of the L 1448 outflow (and most other highly-collimated outflows) is that the cavity observed at the base of the outflow is well developed, with typical opening angles of 30–45°. In other words, the ambient molecular gas is observed to be entrained over long transverse distances. In contrast, the mixing turbulent layer produced by the propagation of the highly supersonic jet is expected to be very narrow (Cantó & Raga 1991). This discrepancy could, in principle, be solved by allowing the direction of the jet to wander (Masson & Chernin 1993). Indeed, if a narrow jet wanders around a given direction, it will excavate a much larger area than in the case of a stationary jet. However, the existing observational evidence for precessing molecular jets (see, for instance, Gueth et al. 1996 for the case of L 1157) indicate that, when present at all, the precession happens in much narrower cones than those observed at the base of highly-collimated outflows.

Probably the most promising jet-driven models of bipolar molecular outflows are those in which the jet entrains ambient molecular material through the formation and propagation of large "bow shocks" which sweep up the ambient molecular gas (see Sect.2). In this kind of model, the jet is able to disturb the ambient gas at relatively large distances from the jet axis, which could explain the low collimation of most observed CO outflows. Moreover, such models are able to explain the formation of large cavities, and predict the geometrical form of the cavities by assuming that the characteristic size of the bow shock increases slowly with distance (see the simulations by Raga & Cabrit 1993). This occurs, for example, if the density of the surrounding ambient gas falls as a power law of index $\beta$ in the range –4 to 0, but a similar behavior is obtained if the mass-loss rate or the ejection velocity from the working surface increases with distance from the source.

In the case of L 1448, strong support for a model in which successive bow shocks are responsible for the formation of the standard CO outflow, and the cometary nebula, comes from a comparison of the $H_2$ images of the jet (Bally et al. 1993) with the CO images of the cavity (Fig. 2). The walls of the cavity seen at blueshifted velocities are complementary to the bow shock traced by the $H_2$ emission, i.e., the arc structure delineated by the innermost blueshifted $H_2$ bow shock seems to "close" the conical CO cavity. This morphology clearly favors the jet-driven models based on entrainment of ambient material through the propagation of large bow shocks. A similar situation, with large bow shocks and biconical cavities, is also clearly observed in the L 1157 molecular outflow (Gueth et al. 1996).

However, not all the details are well accounted for by current models of gas entrainment by bow shocks. In particular, the shapes of the cavities observed in L 1157 are very difficult to reproduce. Also, as pointed out by Lada & Fich (1996), the velocity field over the length of the bow shock is very badly understood at present. Finally, detailed calculations produce outflow lobes which are narrower than the observed ones (e.g. Raga et al. 1993).

The source driving the L 1448 outflow, L 1448-mm, is an extremely cold object which has been detected at mm, cm, and FIR wavelengths (Bachiller et al. 1991a; Curiel et al. 1990); its spectral flux distribution points to an extremely young object, a "Class 0" source (André et al. 1993). It appears that most of the highly-collimated molecular outflows similar to L 1448 (e.g., IRAS 03282, NGC 2264G, L 1157, L 1527, VLA 1623, NGC 1333/IRAS 4, L 483) are driven by this kind of Class 0 source. Such sources are believed to be the youngest (proto-)stellar objects known at present, and have the peculiarity of being extremely efficient in transferring momentum to their associated molecular outflows. This extraordinary outflow activity, as well as other properties of Class 0 sources, is still far from being well understood.

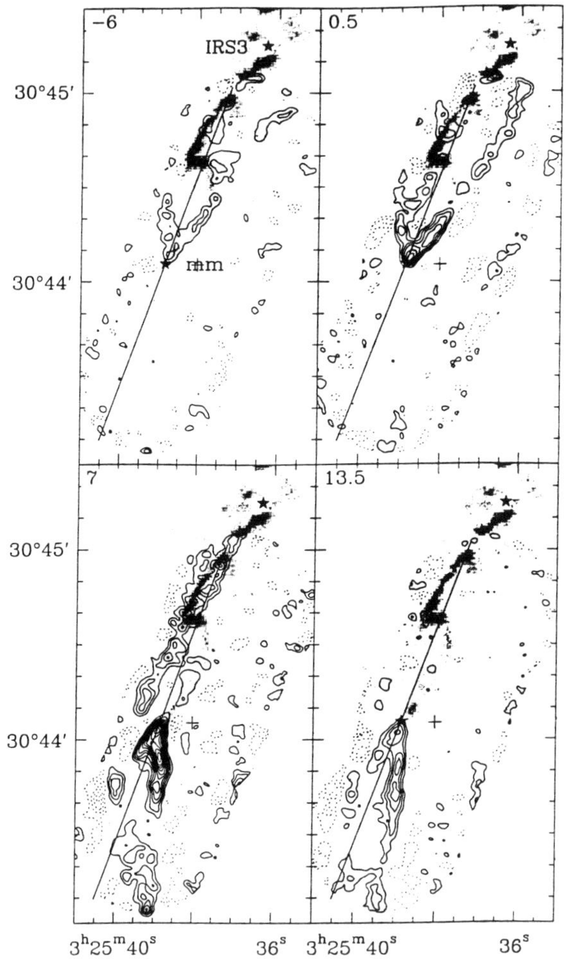


*Figure 2.* Superposition of a grey-scaled $H_2$ image of the L 1448 jet (from Bally et al. 1993) and the CO images of the cavity observed with the IRAM interferometer (from Bachiller et al. 1995a).

## 4. The Role of Chemistry: L 1157

The shock waves generated by the high Mach-number jets from young stars propagate through the ambient molecular cloud and alter the physical conditions and the chemical composition of the gas. In particular, shock-processing of the dust grains leads to an injection of certain atoms and molecules into the gas phase. Particularly spectacular is the case of the refractory SiO molecule which is found to be enhanced by several orders of magnitude at the shocked regions of high-velocity outflows (Bachiller et al. 1991b; Martín-Pintado et al. 1992; McMullin et al. 1994). Sulfur-bearing molecules are also known to be enhanced (e.g., Chernin et al. 1994b).

The blueshifted lobe of the bipolar outflow in L 1157 is one of the best targets to study the region in which an outflow is interacting with the ambient medium. This lobe presents manifest signs of shock-excitation in line emission from CO (Umemoto et al. 1992), SiO (Mikami et al. 1992), and $H_2$ (Hodapp 1994). This is an ideal laboratory to study the chemical alterations induced by the shocks. For instance, ammonia ($NH_3$) and methanol ($CH_3OH$) are two molecules which were known to be abundant in interstellar ice mantles, and were thus expected to be desorbed from grains. Recent observations of L 1157 in lines of these two molecules have confirmed that both species have enhanced abundances in the regions shocked by the L 1157 outflow (Bachiller et al. 1993, 1995b). Methanol is enhanced by a factor of $\sim$ 400 toward the blueshifted lobe of L 1157, whereas ammonia is enhanced by at least one order of magnitude. Measuring the abundance of $CH_3OH$ is crucial, since this molecule is thought to dominate the chemical processes where it is abundant, as in the Orion compact ridge. The (3,3) line of $NH_3$ near $\lambda$ 1.3 cm is particularly intense toward the L 1157 outflow lobes, and it becomes possible to image the flow with the VLA (which provides a resolution of $\sim$5″ in its D configuration). Such VLA images reveal that the high-velocity ammonia arises from a series of bow shocks along the outflow axis (Tafalla & Bachiller 1995).

A description of the complex shock-chemistry found around bipolar outflows clearly requires high-resolution multiline observations of a number of different molecular species. Recent molecular line surveys of cold low-mass stars (NGC1333/IRAS4: Blake et al. 1995; IRAS16293: Blake et al. 1994, van Dishoeck et al. 1995) are proving to be very useful in examining the relative importance of different processes such as sputtering from the dust grains, depletion onto the mantles, and gas-phase reactions.

Finally, we stress that the detection of molecular lines from the shocked regions is interesting not only because of the chemical studies, but also because it opens the possibility of estimating the physical conditions of the gas. The case of ammonia is particularly interesting, since this is the best interstellar thermometer. In L 1157, the shocked gas is found to be heated to more than 60–80 K, i.e., at least a factor of 5 over the ambient quiescent gas. Furthermore, multiline studies of methanol can provide estimates of the gas volume density; in the L 1157 shocked region, Bachiller et al. (1995b) found densities in excess of $10^6 cm^{-3}$.

## References

André, P., Ward-Thompson, D., Barsony, M.: 1993, ApJ 406, 122
Bachiller, R.: 1996, Ann. Rev. Astron. Astrophys., in press
Bachiller, R., Gómez-González, J.: 1992, A & A Reviews 3, 257
Bachiller, R. et al.: 1990, A&A 231, 174

Bachiller, R., André, P., Cabrit, S.: 1991a, A&A 241, L43
Bachiller, R., Martín-Pintado, J., Fuente, A.: 1991b, A&A 243, L21
Bachiller, R., Martín-Pintado, J., Fuente, A.: 1993, ApJ 417, L45
Bachiller, R., Tafalla, M., Cernicharo, J.: 1994, ApJ 425, L93
Bachiller, R., Guilloteau, S., Dutrey, A., Planesas, P., Martín-Pintado, J.: 1995a, A&A, 299, 857
Bachiller, R., Liechti, S., Walmsley, C.M., Colomer, F.: 1995b, A&A 295, L51
Bally, J., Lada, C.J.: 1983, ApJ 265, 824
Bally, J., Lada, C.J., Lane, A.P., 1993, ApJ 418, 322
Blake, G.A. et al.: 1994, ApJ 428, 680
Blake, G.A. et al.: 1995, ApJ 441, 689
Cantó, J., Raga, A.: 1991, ApJ 372, 646
Cernicharo, J., Reipurth, B.: 1996, in preparation
Cernicharo, J., Neri, R., Reipurth, B., Bachiller, R.: 1996, in preparation
Chernin, L.M.:1995, ApJ 440, L97
Chernin, L.M., Masson, C.R.: 1991, ApJ 382, L93
Chernin, L.M., Masson, C.R., Gouveia Dal Pino, E.M., Benz, W.: 1994a, ApJ 426, 204
Chernin, L.M., Masson, C.R., Fuller, G.: 1994b, ApJ 436, 741
Curiel, S., Raymond, J.C., Rodriguez, L.F., Cantó, J., Moran, J.M.: 1990, ApJ 365, L85
Davis, C.J. et al.: 1994, MNRAS 266, 933
Dutrey, A., Guilloteau, S., Bachiller, R.: 1996, in preparation
Dyson, J.E.: 1984, Astrophys. Space Sci. 106, 181
Edwards, S., Snell, R.L.: 1982, ApJ 261, 151
Edwards, S., Snell, R.L.: 1983, ApJ 270, 605
Edwards, S., Snell, R.L.: 1984, ApJ 281, 237
Gueth, F., Guilloteau, S., Bachiller, R.: 1996, A&A, in press
Guilloteau, S., Bachiller, R., Lucas, R., Fuente, A.: 1992, A&A 265, L49
Hartigan, P., Morse, J.A., Heathcote, S., Cecil, G.: 1993, ApJ 414, L121
Heathcote, S. et al.: 1996, in preparation
Hodapp, K.W.: 1994, ApJSS 94, 615
Kwan, J., Scoville, N.: 1976, ApJ 210, L39
Lada, C.J.: 1985, Ann.Rev.Astron.Astrophys. 23, 267
Lada, C.J., Fich, M.: 1996, ApJ, 459, 638
Martín-Pintado, J., Bachiller, R., Fuente, A.: 1992, A & A, 254, 315
Masson, C.R., Chernin, L.M.: 1993, ApJ 414, 230
McMullin, J.P., Mundy, L.G., Blake, G.A.: 1994, ApJ 437, 305
Mikami, H., Umemoto, T., Yamamoto, S., Saito, S.: 1992, ApJ 392, L87
Mundt, R., Brugel, E.W., Bührke, T.: 1987, ApJ 319, 275
Olberg, M., Reipurth, B., Booth, R.: 1992, A&A 259, 252
Raga, A.C.: 1991, AJ 101, 1472
Raga, A.C., Cantó, J., Binette, L., Calvet, N.: 1990, ApJ 364, 601
Raga, A.C., Cabrit, S.: 1993, A&A 278, 267
Raga, A.C. et al. 1993: A&A 276, 539
Reipurth, B.: 1989, Nature 340, 42
Reipurth, B., Olberg, M.: 1991, A&A 246, 535
Reipurth, B., Heathcote, S.: 1993, in *Astrophysical Jets*, Cambridge Univ. Press p.35
Richer, J.S., Hills, R., Padman, R.: 1992, MNRAS 254, 525
Snell, R.L., Loren, R.B., Plambeck, R.L.: 1980, ApJ 239, L17
Snell, R.L., Bally, J., Strom, S.E., Strom, K.M.: 1985, ApJ 290, 587
Stahler, S.W.: 1994, ApJ 422, 616
Tafalla, M., Bachiller, R.: 1995, ApJ 443, L37
Terebey, S.: 1991, Mem.S.A.It. 62, 823
Umemoto, T. et al.: 1992, ApJ 392, L83
van Dishoeck, E.F., Blake, G.A., Jansen, D.J., Groesbeck, T.D.: 1995, ApJ, 447, 760
Zuckerman, B., Kuiper, T.B.H., Kuiper, E.N.R.: 1976, ApJ 209, L137

# CO $J$=6–5 OBSERVATIONS OF PROTOSTELLAR OUTFLOWS

LAWRENCE M. CHERNIN
*Dept. of Astronomy*
*U. of California, Berkeley CA 94720, USA*
COLIN R. MASSON
*Center for Astrophysics*
*60 Garden St. Cambridge, MA 02138 USA*
AND
JOCELYN KEENE
*California Institute of Technology*
*320-47, Pasadena, CA 91125, USA*

## 1. Introduction

Although the CO $J$=1–0 transition is the most frequently used tracer of molecular outflows from protostars, CO $J$=6–5 is a better probe of warm, dense gas in outflows, since the upper state energy of the $J$=6 level is 110 K above ground in comparison to 5.5 K for $J$=1. In this short contribution we compare CO $J$=6–5 to lower $J$ observations of the NGC 2071 outflow, and concentrate the discussion on the outflow structure, overall density and temperature, and the extremely high velocity (EHV) CO features. We present data here for the NGC 2071 outflow, but in an associated journal article we will also present data on several other outflows. The CO $J$=6–5 data were obtained in December 1994 at the Caltech Submillimeter Observatory (CSO) on Mauna Kea, Hawaii.

## 2. General Properties of the Outflow

In Figure 1 we show channel maps of the NGC 2071 outflow(s) in both the CO 6–5 and CO 3–2 lines. Generally, the maps of the two transitions look quite similar, except that the 6–5 observations show more detailed structure since the beam size is half that of the 3-2 observations ($11''$ vs $22''$). The average physical conditions in the outflow implied from the line ratio



*W.B. Latter et al. (eds.), CO: Twenty-Five Years of Millimeter-Wave Spectroscopy*, 175–177.

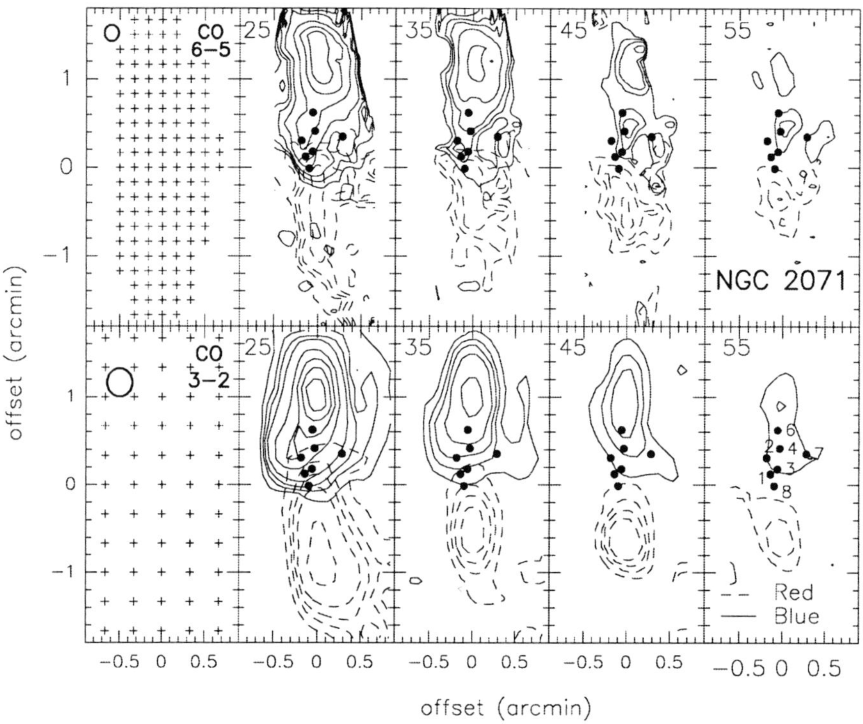


*Figure 1.* Channel Maps of the CO 6-5 and CO 3-2 emission from the NGC 2071 Outflow. The channel velocity relative to the cloud is shown in the top right corner. The maps have been rotated to lie along PA=36°. The black dots in the center represent the IR sources within the dense core (Walther et al. 1993).

are temperatures of 30–60 K and densities of $10^3$-$10^4$ $cm^{-3}$. The temperatures are significantly warmer than the ambient cloud, which is 10–15 K, as derived from the optically thick cloud emission profile.

The 25-45 km $s^{-1}$ J=6–5 channel maps show a conical shape for the opening of the blue shifted outflow lobe originating near IRS 8 (–0.1, 0). Similar conical lobe shapes have been seen in recent interferometer observations of other outflows (cf. Bachiller et al. 1995; Masson, Chernin & Carlstrom, in preparation). The conical shape is most likely produced by the combination of a density gradient in the core, and impulses provided by the jet shocks. We also identify a second, parallel outflow from the IRS 7 source at (0.3, 0.3). Interferometer observations of NGC 2071 are planned to investigate these features in more detail.

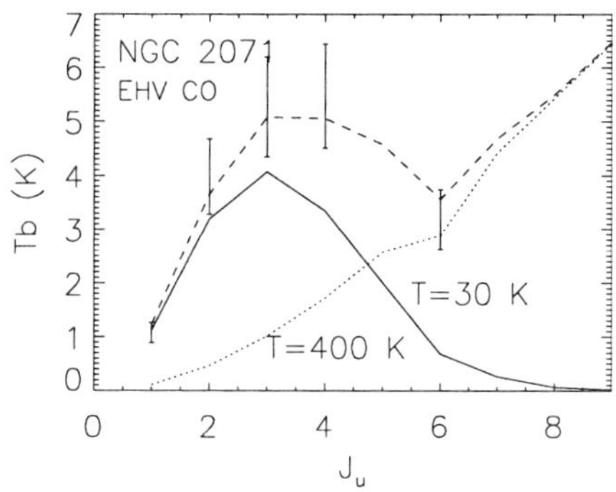


*Figure 2.* Brightness temperatures of CO transitions at the position and velocity range of the EHV CO feature (cf. Chernin & Welch 1995).

## 3. Extremely High Velocity CO Features

It has been argued that the EHV features directly trace where the jet is accelerating ambient gas (cf. Chernin & Masson 1992). The EHV CO velocities range from 40–60 km $s^{-1}$, and therefore the CO is probably re-formed after a dissociative shock (Hollenbach & McKee 1989; Neufeld & Dalgarno 1989a,b). If the pre-shock density is greater than about $10^4$ $cm^{-3}$, the shock models predict a molecule re-formation plateau in the post-shock region which is held at a temperature of a few hundred degrees by the formation heating of $H_2$.

Figure 2 shows the brightness temperatures derived from BIMA (1–0) and CSO observations, and the expected optically thin emission from collisionally excited 30 and 400 degree gas. From our data in Figure 2, and given the error bars on the individual measurements, it is not clear whether we detect a $\approx$ 400 K plateau, but this will be more tightly constrained with $J$=7–6 and $J$=8–7 observations, which will soon be possible with further improvements in submillimeter receivers.

The CSO is supported by the National Science Foundation under grant AST 93-13929.

## References

Bachiller, R. et al. 1995, A&A, 299, 857
Chernin, L. M. & Masson, C. R. 1992, ApJ, 396, L35
Chernin, L. M. & Welch, W. J. 1995, ApJ, 440, L21
Hollenbach, D. & McKee, C. F. 1989, ApJ, 342, 306
Neufeld, D. & Dalgarno, A. 1989a, ApJ, 340, 869
Neufeld, D. & Dalgarno, A. 1989b, ApJ, 344, 251
Walther, D. et al. 1993, ApJ, 418, 310

# DENSE AND COLD GAS IN THE CHAMELEON I CLOUD: CO AND IRAS OBSERVATIONS

L. HAIKALA, M. TORISEVA, K. MATTILA, J. HARJU, AND T. LILJESTRÖM
*University of Helsinki Observatory*
*P.O.Box 14*
*Fin-00014 University of Helsinki*
*Finland*

**Abstract.** The southern low mass star forming cloud Chameleon I has been observed in CO $J = 1 \rightarrow 0$ using the 15m Swedish-ESO Submillimeter Telescope. The cloud has been fully mapped in $C^{18}O$ and is found to be higly fragmented. Two cloud cores were detected south of the known star-formation sites in the cloud. The total mass of the gas in Chameleon I, as observed in $C^{18}O$ is 240 $M_\odot$ and is fragmented into nine separate clumps the masses of which range from 12 $M_\odot$ to 70 $M_\odot$. The fragments associated with newly born stars are warmer and/or less dense than the quiescent fragments lying farther away from the star formation centers. Good correlation of the cold dust IR emission (as calculated from IRAS data) with the observed $C^{18}O$ emission was found. The regions of active star formation have been further observed in the $^{12}CO$ and $^{13}CO$ $J = 1 \rightarrow 0$ transitions. Numerous molecular outflows were detected. Some of these are associated with known young stellar objects and some do not have an obvious central source.

## 1. Introduction

Molecular line observations play a major role when the structure of a star-forming cloud is studied. The basic structure of molecular clouds is best studied in nearby high latitude dark clouds and globules. Because of the lack of large millimeter telescopes in the southern hemisphere before SEST, the necessary basic information regarding the distribution of dense molecular matter in many of the southern nearby molecular clouds does not yet exist. $C^{18}O$ $J = 1 \rightarrow 0$ is the most convenient molecule when mapping the large



*W.B. Latter et al. (eds.), CO: Twenty-Five Years of Millimeter-Wave Spectroscopy*, 178–180.

scale structure of dense molecular clouds. Practice has shown (eg. Myers and collaborators) that this CO isotope is a good tracer of dense cloud cores. Because of the low optical depth of the $C^{18}O$ emission (radiative trapping is not important), the effective critical density of this isotopic species is an order of magnitude higher than that of the more abundant $^{12}CO$. On the other hand the critical density is still lower than that of, e.g., CS. Besides within the cores, the $C^{18}O$ molecule can also be used to map the less dense regions where no emission from the CS molecule is expected.

The Chameleon I dark cloud is an active site of recent star formation with some 75 associated pre-main-sequence stars. For a review see, e.g., Schwartz (1991). The cloud is nearby and lies at a high galactic latitude ($b = -15°$). Therefore, it is ideally suited for detailed molecular line study of a star-forming cloud.

The SEST telescope has been used to map large areas of this cloud in the CO $J = 1 \rightarrow 0$ transition. A first report of the observations was given by Mattila *et al* (1989) where also the discovery of two major outflows was discussed. A similar study of the CrA star-forming site has been conducted at SEST (Harju *et al* 1993)

## 2. Observations

The SEST telescope has been used during several observing runs since 1988 to map Chameleon I in $C^{18}O$ and partly in the more abundant CO isotopes. The $C^{18}O$ mapping (altogether some 1600 positions) was done in a 1min grid and is therefore slightly undersampled. The known star-forming sites around the three bright reflection nebulae in the cloud and the infrared nebula in the south were further observed in the main CO isotope. The emission in the latter isotope is always strongly self absorbed and can be used only to locate the possible molecular outflows. The observed $^{13}CO$ lines are often flat topped and show very little structure. This is an indication of a large optical depth for this isotope. The optical depth of the $C^{18}O$ isotope is mostly below 1 and is higher only in the centers of the star-forming cores around HD 97300 in the north, Ced 110 in the centre and HD 97048 in the south.

Strong 100 micron emission is seen in IRAS data in the direction of the cloud. Much of this emission comes from warm (24K), IR cirrus-like material and the distribution of this emission does not trace the molecular cloud accurately. The emission from cold dust at 60 microns is small and the contribution of the cirrus-like emission at 100 microns can be approximated by multiplying the IRAS 60 micron emission by 5. This corresponds roughly to the 100 to 60 micron colour ratio from 24K dust. Subtracting this emission from the 100 micron IRAS map, only the part of the emission coming from cold dust remains. The correlation of this "cold dust emission" and the $C^{18}O$ emission in Chameleon I is excellent, although, due to the course resolution of the IRAS data, much more details are seen in the CO map.

## 3. Results

In $C^{18}O$ , nine separate (spatially or in velocity) cores can be separated. The $C^{18}O$ line widths, corrected for line opacity, range from 0.5 to 0.9 km $s^{-1}$ and velocities between 3.8 and 4.9 km $s^{-1}$. The maximum column densities in the cores are between 1.5 $10^{22}cm^{-2}$ and 2.5 $10^{22}cm^{-2}$. The properties of the clumps have been calculated assuming a distance of 140pc to the cloud.

The northernmost core has already been discussed in Mattila *et al* (1989). The position velocity diagram of this region indicates a possible rotating toroid in the center of which is the center of a molecular outflow. The mass of this structure is $\approx$20 $M_\odot$.

The reflection nebula Ced 110 lies near the centre of the most massive core in the region. This 0.5 by 0.2 pc and 70 $M_\odot$ core is elongated in the NE-SW direction. Projected on this core, near its center, is a smaller (20 $M_\odot$) 0.15 pc diameter core. The two cores are easily separated in velocity.

The active star-forming location around HD 97048 lies at the eastern rim of a 20 $M_\odot$ core. Several CO outflow lobes were detected. Due to the large number of young stars, it is difficult to identify the central sources but at least the stars HD 97048 (or Sz 22), Sz 19 and T33 seem to be associated with moderate molecular outflows. The infrared nebula, located SE from HD 94078 is associated with a small size, not very massive core and possibly with a bipolar outflow.

Two cloud cores, which are elongated in the same direction as the core associated with Ced 110, lie to SW from HD 94078. The masses of these cores are 60 and 20 $M_\odot$. No star formation is known in their direction and the $C^{18}O$ emission correlates here well with the IRAS "cold dust emission".

## References

Harju, J., Haikala, L.K., Mattila, K., Mauersberger, R., Booth, R.S., Nordth, H.L. (1993), A&A **278**, 569

Mattila, K., Liljeström, T. and Toriseva, T. (1989), in ESO Workshop on *Low Mass Star Formation and Pre Main Sequence Objects*, Proceedings No. 33, ed. B. Reipurth (ESO: Garching bei München), p. 153

Schwartz, R. D., 1991 in Low Mass Star Formation in Southern Molecular Clouds, ESO Scientific report No. 11, ed. B. Reipurth (ESO: Garching bei München), p. 93

# SURVEYS WITH HETERODYNE FOCAL PLANE ARRAYS AT MM WAVELENGTHS: QUARRY AT FCRAO

HANS UNGERECHTS
*IRAM*
*Avenida Divina Pastora 7-NC, E-18012 Granada, Spain*

## 1. Introduction

In early 1991 the focal plane array receiver QUARRY (Erickson *et al.*, 1992) for the 3 mm wavelength band went into operation on the FCRAO 14 m telescope. QUARRY uses $3 \times 5$ Schottky diode mixers with beams spaced $100'' \times 50''$ on the sky, while the beamwidth is $45''$ at 115 GHz. Maps sampled every full or half beamwidth are obtained by stepping the telescope to fill in the intermediate points. Backends used from the start were filterbanks with 32 channels of 250 kHz and 1 MHz for each receiver; and accordingly the first observations with QUARRY concentrated on galactic giant molecular clouds. More recently an autocorrelator (Predmore, 1994) has become available allowing a much wider range of targets including cold clouds requiring high spectral resolution and galactic surveys which need wide bandwidths.

The most extensive projects – or "surveys" – with QUARRY are large scale maps of GMC complexes using CO and other tracers, described in §2, and multiline studies of cloud cores with many molecular species, summarized in §3. A complete review of all projects with QUARRY is beyond the scope of this article; on the other hand details of several surveys are reported elsewhere in these proceedings (Gem OB 1: Carpenter, 1996; Second Galactic Quadrant: Heyer, 1996; and see the list of references).

## 2. Largescale Surveys of Clouds

A distinctive feature of CO surveys of GMCs with QUARRY is the large ratio of size covered to resolution or sampling. For example, the CO maps of Gem OB1, extend over about 180 pc × 220 pc sampled every 0.5 pc (Carpenter *et al.*, 1995a). The results are images of molecular clouds that are rich



*W.B. Latter et al. (eds.), CO: Twenty-Five Years of Millimeter-Wave Spectroscopy,* 181–188.

TABLE 1. Largescale Surveys with QUARRY

| Region | Molecule | Reference |
|---|---|---|
| Gem OB1 | $^{12}$CO and $^{13}$CO, [CS] | J. Carpenter *et al.* (1995a, 1995b, 1996) |
| "Ceph Bubble" & IC 1396 | $^{12}$CO and $^{13}$CO | N. Patel *et al.* (1994, 1995, 1996) |
| "Maddalena's" G216-2.5 | $^{12}$CO and $^{13}$CO [CS, C$^{18}$O, HCO$^{+}$] | Y. Lee *et al.* (1994) |
| Mon R2 | $^{12}$CO | T. Xie *et al.* (1993, 1994, 1996) |
| Sh 140, Sh 155, Sh 235 | $^{12}$CO and $^{13}$CO | M. Heyer *et al.* (1996) |
| Galactic Center | $^{12}$CO and HCN | J. Jackson *et al.* (1996) |
| Second Galactic Quadrant | $^{12}$CO | M. Heyer (1996) |

Note: lines in brackets [ ] were observed in restricted regions selected from the main survey.

in detail and reveal a wide range of processes at work shaping the clouds. Before QUARRY, some similar images of GMCs had been obtained; for example, $^{13}$CO measurements of regions such as Orion A obtained with the Bell Labs Telescope (Bally *et al.*, 1987), and single-beam measurements of the $^{12}$CO emission toward Orion A (Heyer *et al.*, 1992). We now have more and larger surveys with such characteristics, and for most $^{12}$CO and $^{13}$CO were mapped. An overview is given in Table 1, which gives full references; to save space in the following text these references are not always repeated when reporting results from one of these surveys.

For most of the clouds the $^{12}$CO and $^{13}$CO images look at first glance very similar. In $^{12}$CO there are extended regions of low intensity emission and the strongest lines are found toward a relatively small fraction of lines of sight. The strongest $^{13}$CO lines are generally seen where the main isotope is strong; the contrast to the low intensity extended emission is generally higher in $^{13}$CO, and the area of detected emission is smaller in $^{13}$CO. The latter is in part due to the lower signal-to-noise ratio of most $^{13}$CO images, but there is also a substantial variation in the $^{12}$CO to $^{13}$CO intensity ratio (Carpenter *et al.*, 1995a).

Most of the cloud mass is in regions of low column density and temperatures of 10 K or less (e.g., Carpenter *et al.*, 1995a, Lee *et al.*, 1994). High column densities and larger temperatures are generally associated with H II-regions or star-forming activity. Densities averaged over a large complex like Gem OB1 are as low as 1 molecule cm$^{-3}$, only a few times the interstellar average, but more typically on the order of 100 cm$^{-3}$ for the extended emission of individual clouds and above $10^4\,\mathrm{cm}^{-3}$ in cloud cores.

Column densities can be compared to the total luminosity of the $^{12}$CO emission to derive estimates of the $X$-ratio relating $^{12}$CO luminosity and cloud mass: e.g., for Gem OB1 and G216−2.5 a typical value is $X = 1.0\ 10^{20}\,\mathrm{cm}^{-2}/(\mathrm{K\,km\,s}^{-1})$ with a spread of a factor of about 1.5. Many clouds with and without star formation appear approximately virialized when the mass from LTE column densities is related to the velocity dispersion seen in the line width, the major exception is, of course, Maddalena's cloud, G216−2.5, which has no signs of star formation but very broad lines (Maddalena and Thaddeus, 1985, Lee *et al.*, 1994).

It is not clear how typical these results are for the majority of clouds in the Galaxy: the clouds mapped with QUARRY were selected to be in the local arm or in the outer Galaxy, and the properties of the GMCs within the solar circle may be different.

The structure of the regions of strong CO emission can in some cases directly be linked to physical causes. Arcs or rings, sometimes only partial or broken into globules, are found around H II-regions (Gem OB1: Carpenter *et al.*, 1995a; IC1396: Patel *et al.*, 1995). In Mon R2 the $^{12}$CO images show a shell swept up around the blue lobe of the bipolar outflow (Xie *et al.*, 1993). Other arcs and shells are found without an evident cause: e.g., in G216−2.5 an embedded shell of diameter $\approx 20\,\mathrm{pc}$ with a full three dimensional kinematic signature (Lee *et al.*, 1994). In both G216−2.5 and Mon R2, the large-scale structures of the clouds show the kinematics of a hemispherical expanding shell (Lee *et al.*, 1994; Xie and Goldsmith, 1994). Some morphological arcs and shells lack clear kinematic evidence for expansion. The globules around IC1396 appear on an expanding ring rather than a three dimensional shell, possibly because the H II-region formed in a molecular cloud that had previously been compressed with the formation of a much larger shell, the "Cepheus-Bubble", around which many more molecular clouds are found (Patel *et al.* 1995, Patel 1996).

Cores of enhanced density are indicated by strong $^{13}$CO lines and high column densities. Some CO surveys have been followed by observations of the CS $J = 2 \rightarrow 1$ line in regions of high column density. In Gem OB1 eleven CS cores with masses 100 to $2000 M_\odot$ are located in the CO arcs and filaments (Carpenter *et al.*, 1995b). In G215−2.5 only a few CS cores of lower mass, $\approx 10 M_\odot$, are found which resemble cloud cores in cold dark clouds (Lee *et al.*, 1994).

To trace the relative amount of high density gas in clouds near the Galactic Center, Jackson *et al.* (1996) have surveyed CO and HCN and compared the results with active galactic nuclei. Around the Galactic Center the density peaks correlate with the CO luminosity peaks, and starburst galactic nuclei have a larger mass of dense gas, larger average density, and larger regions of high density.

The location of young stellar objects in the molecular clouds, in particular YSO candidates identified from the IRAS point source catalog, has been studied for Gem OB1, IC 1396, and Mon R2; there are no known YSOs associated with G215−2.5. In Gem OB1 the more luminous sources tend to be associated with more massive CS cores (Carpenter *et al.*, 1995b). In Mon R2 the IRAS sources are concentrated in the bright emission ridges (Xie, 1996); in IC 1396 they mostly fall onto the expanding ring of dense molecular gas (Patel *et al.*, 1995). On the other hand, sources of lower luminosity, as well as some bright sources, are more widely distributed and not necessarily associated with massive dense cores. A survey in the J, H, K, and L bands around Sh 247 in the Gem OB1 complex revealed four star clusters, all in massive CS cores (Carpenter *et al.*, 1995b).

A part of the CO survey of Mon R2 was used by Jarrett *et al.* (1994) for a comparison with the magnetic field morphology determined from polarimetric observations of stars: the magnetic field appears aligned with one of the prominent ridges or filaments in the CO map, coincident with the axis of the Mon R2 bipolar outflow, and on a smaller scale distorted or bent toward the long axis of the main cloud core.

In summary, the overall picture presented by the CO images of the large cloud complexes appears consistent with scenarios in which H II-regions (Elmegreen and Lada 1977) or stellar outflows and winds (Norman and Silk 1980), caused by one generation of stars, sweep up shells of enhanced density. Dense cloud cores form within these density enhancements, most of which soon form stars, which disperse the cores and subsequently cause new shells. G215−2.5 could be in a period between episodes of star formation (Lee *et al.* 1994).

## 3. Multiline "Chemistry" Surveys of Cloud Cores

In an attempt to study the relation between physical conditions and chemical abundance variations, QUARRY has been used to obtain fully sampled maps of many molecules and transitions in selected cloud cores. The early surveys with the filterbanks were of GMC cloud cores: the Orion A ridge (Ungerechts *et al.*, 1996), M17SW, and Cepheus A (Bergin *et al.*, 1996c); recently the autocorrelator has been used for the cold cloud cores TMC 1 and L183, a.k.a. L134N (Pratap, 1996) (compare Table 2). A similar study using single beam receivers on the IRAM 30 m telescope has been done of the NGC 7023 photodissociation region (Fuente *et al.*, 1993).

The molecules in the QUARRY surveys of GMC cores are CO, $^{13}CO$, $C^{18}O$, CS, $C^{34}S$, $HCO^+$, $H^{13}CO^+$, $N_2H^+$, CN, $C_2H$, HCN, $H^{13}CN$, HNC, $HN^{13}C$, SO ,$^{34}SO$, $SO_2$, $CH_3OH$, $C_3H_2$, $HC_3N$, and $CH_3CCH$; except for

TABLE 2. QUARRY Chemistry Surveys of Cloud Cores

| Source | Size of maps | Reference |
|---|---|---|
| GMC Cores: | | |
| Orion A | $5' \times 12'$ | (Ungerechts *et al.*, 1996) |
| M17SW | $4' \times 5'$ | (Bergin *et al.*, 1996c) |
| Cepheus A | $4' \times 5'$ | (Bergin *et al.*, 1996c) |
| Cold cloud cores: | | |
| TMC 1 | $4' \times 15'$ | (Pratap, 1996) |
| L183 (L134N) | $8' \times 10'$ | (Pratap, 1996) |

$C_3H_2$ they are being mapped in the cold cores as well. For $N_2H^+$, CN, and $C_2H$ hyperfine components are resolved, for $CH_3OH$, SO, $CH_3CCH$, and $HC_3N$ several rotational transitions were mapped, which are sensitive to different excitation conditions. Finally, the $HC_3N$ observations were complemented with observation of additional transitions at other observatories.

The strongest differences between maps of various species are found along the Orion A ridge. While all distributions are similar to first order – all show the ridge and the molecular bar – on a closer look they can be grouped into a range of morphological classes (Figure 1): i) distributions that are very strongly peaked toward KL and to a lesser degree toward Orion-1.5′S: CO, $CH_3OH$, SO, $SO_2$, $HC_3N$, HCN, and $H^{13}CN$; ii) those that very clearly show the ridge with moderate peaks at KL and Orion-1.5′S: CS, $C^{34}S$, and $^{13}CO$; iii) maps that show the ridge but no predominant peak: especially CN, but also $C^{18}O$, and $C_3H_2$; iv) distributions that are strongest $\approx 4'$ north of KL around the radical-ion-peak "RIP" or show an overall north-south gradient: $N_2H^+$, $H^{13}CO^+$, $C_2H$, and $HN^{13}C$. Additionally some species have particularly low intensities toward the molecular bar around offsets $(1.7', -2.9')$: $HC_3N$, $N_2H^+$, and $C^{34}S$. The distinction between the categories above is somewhat open, – HNC and $CH_3CCH$ have characteristics of ii) and iv) – and if one examines all the details in the maps, each one shows some idiosyncrasy making it distinctly individual compared to all others.

Variations in temperatures can be estimated from the peak intensity and hence excitation temperature of the CO line for the outer layers of the cloud cores and from the relative strength of the $K$ components of $CH_3CCH$ for the denser cloud interior. All along the Orion ridge as well as for most positions in M17SW the interior temperature is lower, and in Orion both temperatures decrease with distance from KL (Bergin *et al.*, 1994).

The two $HC_3N$ transitions mapped with QUARRY, $J = 10 \rightarrow 9$ and $12 \rightarrow 11$, when combined with maps of one lower, $4 \rightarrow 3$, and one higher

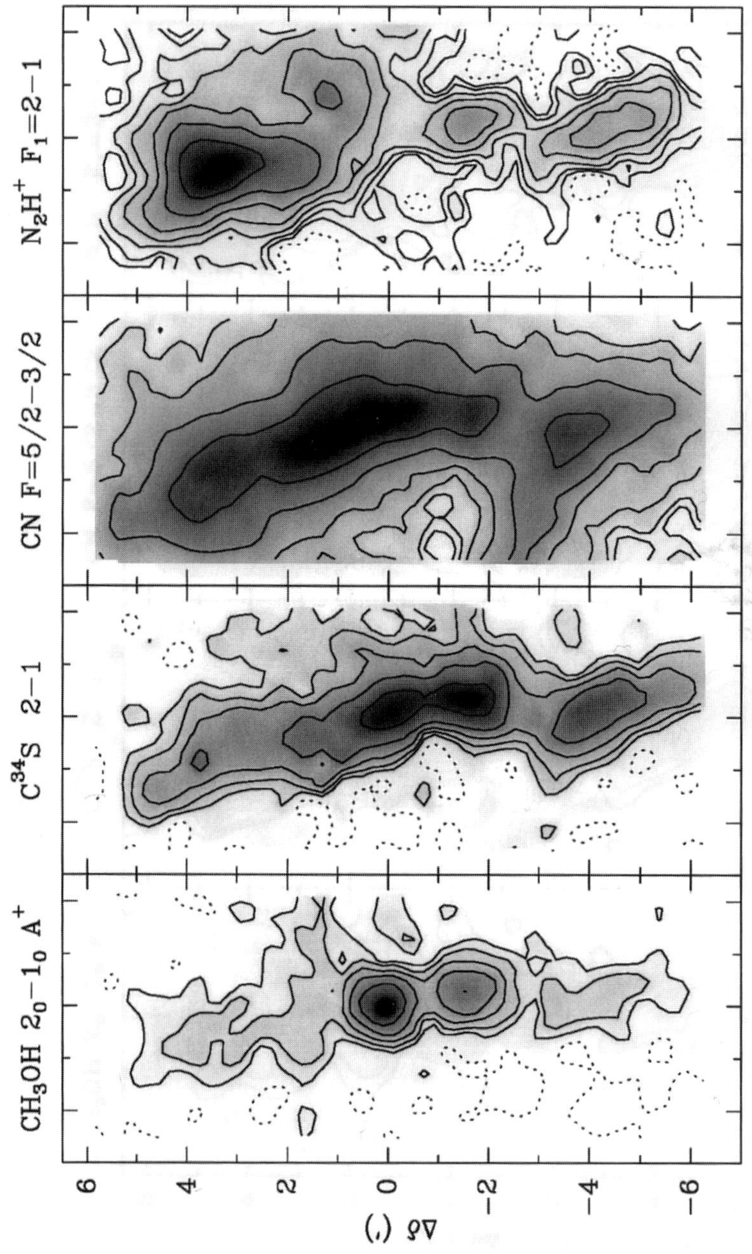


*Figure 1.* Maps of intensity integrated over velocity for selected molecules along the Orion A molecular ridge. These maps are representative of the different classes of distributions: i) $CH_3OH$ $J_K = 2_0 \to 1_0A^+$: strongly peaked toward Orion-KL (0,0) and 1.5′S; ii) $C^{34}S$ $J = 2 \to 1$: clearly outlining the ridge with moderate peaks at KL and 1.5′S; iii) CN $F = \frac{5}{2} \to \frac{3}{2}$: showing ridge and bar with no prominent peaks; iv) $N_2H^+F_1 = 2 \to 1$: strongest around the RIP (radical-ion-peak) about 4′N of KL.

line, $16 \rightarrow 15$, provide an estimate of density. In all GMC cores studied, the derived densities are high, around $10^6\,cm^{-3}$, with no systematic variation with position (Bergin *et al.*, 1996a). The kinematic evidence is consistent with the assumption that $HC_3N$ and $CH_3CCH$ trace conditions in the interior of the cores, from which the emission of most optically thin species in the survey arises. It is, therefore, reasonable to use the derived temperatures and densities with the observed intensities of optically thin lines to estimate molecular abundances relative to $C^{18}O$ (Bergin, 1996b).

The strongest abundance variations are the well-known enhancements toward KL and to a lesser degree toward Orion-1.5′S; this is the case for $HC_3N$, SO, $CH_3OH$, and HCN. A smaller, but distinct abundance variation corresponds to the NS-gradient or difference between the RIP and the cloud core 4′S of KL. These two cores have very similar densities as well as temperatures, but $N_2H^+$, $H^{13}CO^+$, $C_2H$, and to a lesser degree $CH_3CCH$ and $HN^{13}C$ are more abundant around the RIP (Bergin, 1996b). Finally the relative weakness of some species toward the molecular bar translates into lower abundances than on the main ridge.

The strong enhancements around KL have, of course, for some time been explained by chemical processes directly linked to the massive star formation in this area (Blake *et al.*, 1987). The observed abundances in the more quiescent Orion ridge and the two other GMC cores are rather high when compared to models of the gas phase chemistry (Bergin *et al.*, 1995). Under the observed physical conditions, both a high C/O ratio and "early" times in the chemical evolution are necessary to match the observations (Bergin, 1996b). The observed secondary abundance variations – the NS contrast and the weakness of some species in the molecular bar – are possibly related to variations in the clumping and volume filling factor and in the UV radiation field.

## 4. Outlook

A $2 \times 2$ array with SIS receivers for the 3 mm band has been used on the Nobeyama 45 m telescope (Ukita and Tsuboi, 1994) and is to be replaced by a larger array. An 8-element SIS array for 1.3 mm will soon be available at the NRAO 12 m telescope (B. Latter and P. Jewell, priv. comm.). The next generation of focal plane arrays at 3 mm may directly employ HEMT amplifiers, offering the prospect of much lower noise temperatures than those of QUARRY as well as more elements. Especially when combined with faster observing modes like on-the-fly sampling, these systems may provide another order-of-magnitude gain in the speed of spectral-line observations at mm wavelengths.

The total area of CO emission in the composite galactic survey with the 1.2 m Columbia-Chile "Mini" telescopes (Dame *et al.*, 1987) is about 10 times larger than that of the second quadrant survey currently underway with QUARRY (Heyer, 1996). The next generation of focal plane arrays should enable us to map most of the galactic molecular gas with current telescopes of the 15 m size class with full sampling and angular resolution better than one arc minute.

**Acknowledgments.** I would like to thank everyone who helped with the GMC core "chemistry" survey or provided preprints, figures, or information: Ted Bergin, John Carpenter, Paul Goldsmith, Mark Heyer, Bill Irvine, Phil Jewell, Bill Latter, Amy Lovell, Phil Maloney, Doug McGonagle, Tim Paglione, Nimesh Patel, Preethi Pratap, Pete Schloerb, Ron Snell, Tao Xie, the FCRAO staff, and the QUARRY team.

## References

Bally, J., Langer, W.D., Stark, A.A., and Wilson, R.W. (1987) *Ap.J.L.* **312**, L45
Bergin, E.A., Goldsmith, P.F., Snell, R.L., and Ungerechts, H. (1994) *Ap.J.* **431**, 674
Bergin, E.A., Langer, W.D., and Goldsmith, P.F. (1995) *Ap.J.* **441**, 222
Bergin, E.A., Snell, R.L., and Goldsmith, P.F., (1996a) *Ap.J.* **460**, 343
Bergin, E.A. (1996b) Poster Presentation at *this IAU symposium*
Bergin, E.A. *et al.* (1996c) in preparation for *Ap.J.*
Blake, G.A., Sutton, E.C., Masson, C.R., and Phillips, T.G. (1987) *Ap.J.* **315**, 621
Carpenter, J.M., Snell, R.L., and Schloerb, F.P. (1995a) *Ap.J.* **445**, 246
Carpenter, J.M., Snell, R.L., and Schloerb, F.P. (1995b) *Ap.J.* **450**, 201
Carpenter, J.M. (1996) Contributed Presentation at *this IAU symposium*
Dame, T.M. *et al.* (1987) *Ap.J.* **322**, 706
Elmegreen, B.G., and Lada, C.J. (1977) *Ap.J.* **214**, 725
Erickson, N.R. *et al.* (1992) *IEEE Trans. on Microwave Theory and Techniques* **40**, 1
Fuente, A., Martín-Pintado, J., Cernicharo, J., and Bachiller, R. (1993) *A.&A.* **276**, 473
Heyer, M.H., Morgan, J., Schloerb, F.P., Snell, R.L., and Goldsmith, P.F. (1992) *Ap.J.L.* **395**, L99
Heyer, M.H. (1996) Contributed Presentation at *this IAU symposium*
Heyer, M.H., Carpenter, J.M., and Ladd, E.F. (1996) *Ap.J.* **463**, 630
Jackson, J.M., Heyer, M.H., Paglione, T.A.D., & Bolatto, A.D. (1996) *Ap.J.L.* **456**, L91
Jarrett, T.H., Novak, G., Xie, T., and Goldsmith, P.F. (1994) *Ap.J.* **430**, 743
Lee, Y., Snell, R.L., and Dickman, R.L. (1994) *Ap.J.* **432**, 167
Maddalena, R.J., and Thaddeus, P. (1985) *Ap.J.* **294**, 231
Norman, C., and Silk, J. (1980) *Ap.J.* **238**, 158
Patel, N.A., Heyer, M.H., Goldsmith, P.F., Snell, R.L., Hezel, T., and Pratap, P. (1994) in *Clouds, Cores, and Low Mass Stars*, eds. D.P. Clemens and R. Barvainis, p. 81
Patel, N.A., Goldsmith, P.F., Snell, R.L., Hezel, T., and Xie, T. (1995) *Ap.J.* **447**, 721
Patel, N.A. (1996) Poster Presentation at *this IAU symposium*
Pratap, P. (1996) Poster Presentation at *this IAU symposium*
Predmore, C.R. (1994) in *Workshop on Multi-feed Systems for Radio Telescopes*, Astronomical Society of the Pacific Conference Series **75**, p. 259
Ukita, N., and Tsuboi, M. (1994) *Proceedings of the IEEE* **82**, 725
Ungerechts, H. *et al.* (1996) in preparation for *Ap.J.*
Xie, T., Goldsmith, P.F., and Patel, N. (1993) *Ap.J.L.* **419**, L33
Xie, T., and Goldsmith, P.F. (1994) *Ap.J.* **430**, 252
Xie, T. (1996) Poster Presentation at *this IAU symposium*

# INSTRUMENTATION

# SINGLE DISHES: THE POTENTIAL OF PRESENT-DAY BIG DISHES AND THE PROMISE OF THE LMT

A.I. HARRIS
*Department of Physics & Astronomy and FCRAO*
*619 Lederle Graduate Research Center*
*University of Massachusetts*
*Amherst, MA 01003, USA*

## 1. Introduction

Single dish antennas are familiar facilities at all millimeter wave astronomical observatories, whether as stand-alone instruments or parts of interferometers. Baars (1983) and an issue of the *Proceedings* of the IEEE (Imriale & Thorburn 1994) contain summaries of many aspects of modern radio astronomical single dish antennas. Filled-aperture instruments are only one possible form for antennas, and many other types are important at radio astronomy facilities. At millimeter waves, however, the only other principal configuration is the aperture synthesis interferometer, the subject of Jack Welch's talk (Welch 1995) and the excellent book by Thompson, Moran, and Swenson (1986). Some analysis of the relative strengths and limitations of these instruments is an essential preliminary to properly speculating on their potential best uses. The critical areas for comparison between the two types of instruments are sensitivity, imaging properties, and spatial resolution.

## 2. Sensitivity and Imaging

### 2.1. POINT SOURCE SENSITIVITY

The simplest sensitivity comparison is to find the time that both instruments need to detect the same point source. Solving for the same $S/N$ ratio for the same source flux, the integration time ratio is

$$\frac{\tau_I}{\tau_S} = \left( \frac{1}{K} \frac{T_{\mathrm{rec},S}}{T_{\mathrm{rec},I}} \frac{\eta_{A,S}}{\eta_{A,I}} \frac{A_S}{A_I \sqrt{n(n-1)}} \right)^2 \frac{\Delta\nu_S}{\Delta\nu_I}, \tag{1}$$



*W.B. Latter et al. (eds.), CO: Twenty-Five Years of Millimeter-Wave Spectroscopy*, 191–198.

for $n$ identical elements each with area $A_I$ in the interferometer, a single dish with area $A_S$, aperture efficiencies $\eta$, receiver temperatures $T_{\rm rec}$, $\Delta\nu$ detection bandwidths, $K$ an efficiency term about $\sqrt{2}$ to 2 for conventional observing modes, and $S$ and $I$ subscripts denoting terms corresponding to the single-dish and interferometer. Since the interferometer's total area is $\sqrt{n(n-1)} \sim n$ times the area of an individual element, the required time scales approximately as the square of the total area and linearly with bandwidth. In the case of spectral line observations, the line's width defines the bandwidth $\Delta\nu$. Expanding the bandwidth can bring a large increase in $S/N$ for observations of sources with wideband continua, however.

The rising spectrum of dust and falling spectrum of free-free emission cross in the millimeter band; dust continuum tends to dominate emission at wavelengths shorter than a few millimeters, with emission from hot plasmas dominating at longer wavelengths. In the case of dust, Figure 1 shows that a wavelength near 1 mm is optimum for survey work. For the 50-m telescope (Fig. 1*a*), the peak signal-to-noise ratio lies at about 260 GHz (1.1 mm). Increasing dust emissivity (dust flux scales as $\nu^2 B_\nu$ in Fig. 1) and a more accurate surface shift the peak to about 860 GHz for a precision telescope in excellent weather (Fig. 1*b*). The higher frequency atmospheric windows close faster than the lower frequency windows, however, and the peak lies near 260 GHz for all conditions but those with very low water vapor.

Direct detectors can be substantially more sensitive than heterodyne receivers for wideband signals. Figure 1*c* shows the relative $S/N$ for a coherent detector. Figure 1*d* gives the ratio of $S/N$ for the incoherent to coherent detectors. Near 260 GHz, where the two systems' sensitivities peak, the incoherent system has a factor 7 better signal to noise than the heterodyne system. The heterodyne system's signal-to-noise ratio will always be a factor of about two or more lower than the incoherent system's.

Single dishes can easily accommodate direct detectors, but interferometers require a rather cumbersome optical beam combination to use incoherent detectors. Especially at submillimeter and longer wavelengths, coherent techniques will dominate interferometry, and this will limit instantaneous bandwidths to a few GHz; single dish instruments will support bandwidths that cover entire atmospheric windows.

### 2.2. EXTENDED SOURCE SENSITIVITY

One approach for sensitivity comparisons for extended sources is to find the time ratio necessary to fully map the area contained by the primary beam of a smaller interferometer antenna with a larger single dish. Comparing

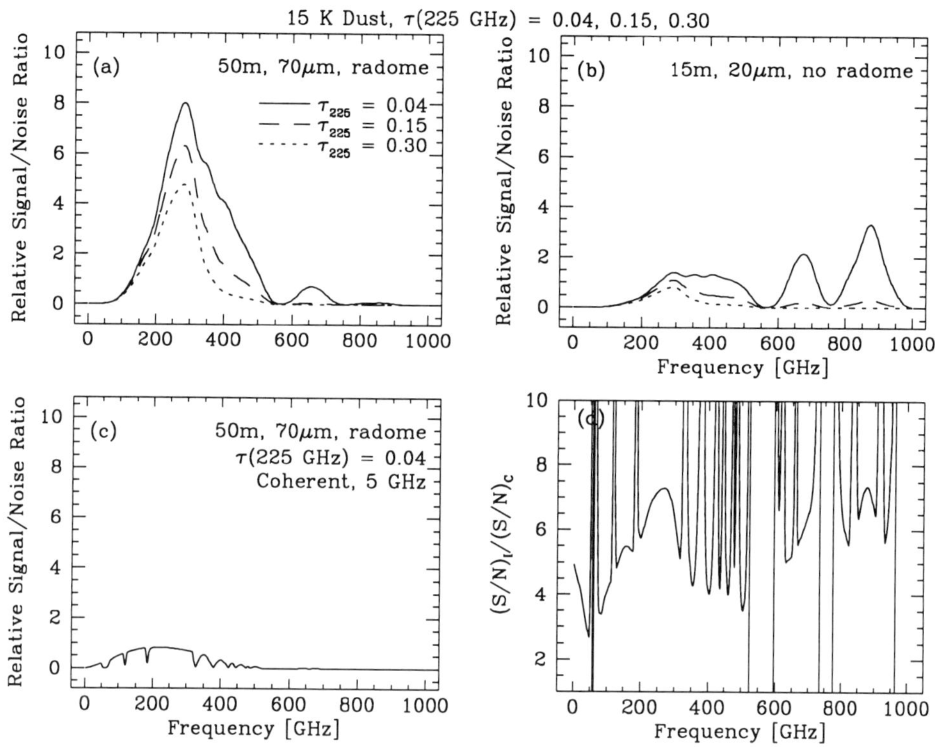


*Figure 1.* $S/N$ ratios for 15 K dust from a point source, observed with an incoherent detector on a *(a)* 50-m diameter telescope with a 70 $\mu$m rms surface, in a radome, and *(b)* a 15-m telescope with a 20 $\mu$m surface and no radome. The three curves in each panel are for different atmospheric optical depths corresponding to about 1, 4, and 8 mm of precipitable water along the line of sight. Panel *(c)* is the $S/N$ ratio for the 50m telescope but with a coherent detector with a bandwidth of 5 GHz. Panel *(d)* is the ratio of incoherent and coherent detection $S/N$ ratios with the 50-m telescope. The signal-to-noise scales are arbitrary but consistent from one figure to another. The incoherent detector is background-limited and single mode with a 50 GHz FWHM Gaussian bandpass; the heterodyne receiver has a 5 GHz rectangular bandwidth and noise temperature $3h\nu/k$. The radome has 20% transmissive loss, scattering 80% of this power onto the sky and 20% onto ambient.

solid angles on the sky, there are

$$N = \frac{\Omega_P}{\Omega_S} = \frac{\lambda^2 A_S}{\lambda^2 A_I} = \left(\frac{D_S}{D_I}\right)^2 \tag{2}$$

single dish beam solid angles $\Omega_S$ within the interferometer's primary beam solid angle $\Omega_P$. The primary beam response modulates the signal by a factor of two from center to half-width edge; the average factor of 2 $\ln(2) \sim \sqrt{2}$ for a Gaussian beam is a representative correction. The time ratio for the same $S/N$ ratio for a given average signal level is

$$\frac{\tau_I}{\tau_S} = \left( \frac{2\,\ln(2)}{K} \frac{T_{\mathrm{rec},S}}{T_{\mathrm{rec},I}} \frac{\eta_{B,S}}{\eta_{B,I}} \frac{A_S}{\sqrt{n(n-1)}A_I} \right)^2 \frac{\Delta\nu_S}{\Delta\nu_I} \frac{M_S}{4\,M_I}. \tag{3}$$

Multiple receivers or polarizations are accommodated by factors of $M_S$ for an $M_S$ element focal plane array on the single dish and $M_I$ at each interferometer antenna. The difference in sampling the power and the field accounts for the factor of four. This comparison contains the implicit constraint, through Eq. 2, that the number of interferometer antennas $n$ must be within a distance equal to the diameter of the single dish to weigh in with full efficiency.

This time ratio again scales as the square of the total collecting areas and linearly with bandwidth. Eq. 3 confirms an intuitive feeling that instruments with comparable numbers of equal receivers, and comparable areas contained within a given diameter, will have very similar sensitivities: differences will derive more from switching strategies, illumination, packing, and other details that are not fundamental to either type of instrument. The main practical difference is that interferometers can be spread over large areas to produce much higher spatial resolution maps than single dishes can obtain. The price for an interferometer's higher resolution is lower sensitivity to extended emission and more complex imaging characteristics.

### 2.3. IMAGE FORMATION AND SPATIAL FREQUENCY RESPONSE

The differences in imaging characteristics are easiest to see in the spatial frequency plane ($(u,v)$ plane), which closely approximates the Fourier transform of the image plane. A single dish equipped with a power detector has a spatial frequency response given by $g \star g$, where $g$ is the field "grading," or distribution, across the aperture and the star denotes autocorrelation. (Bracewell & Roberts (1954) contains the definitive discussion of this topic). Radiation temperatures across a map are the convolution of the true brightness distribution of the source with the autocorrelation function of the field grading (the power beam distribution):

$$T_{\mathrm{map}} = T_{\mathrm{source}} \,*\, (g \star g). \tag{4}$$

Figure 2*a* is the radial spatial frequency response for a circular antenna. This instrument has a centrally peaked frequency response that is most sensitive to large spatial structures (components of structures with low spatial frequencies). Although it is hard to see at the plot's scale, the edge of the dish forces a hard cutoff for spatial frequencies greater than $D/\lambda$, where $D$ is the antenna diameter. This cutoff is the origin of the sampling theorem's $\lambda/2D$ minimum angular spacing requirement.

Careful design can produce blocked, axis-symmetric apertures with performance that approaches the clean spatial frequency response of unblocked designs. For instance, Lawrence et al. (1994) report a series of measurements on their 5.5 m antenna, which is designed for continuum measurements of the cosmic background. They adopted the secondary support scheme developed for the BIMA antennas to achieve a reduction of differential elevation-dependent pickup from 4.3 mK to 0.14 mK.

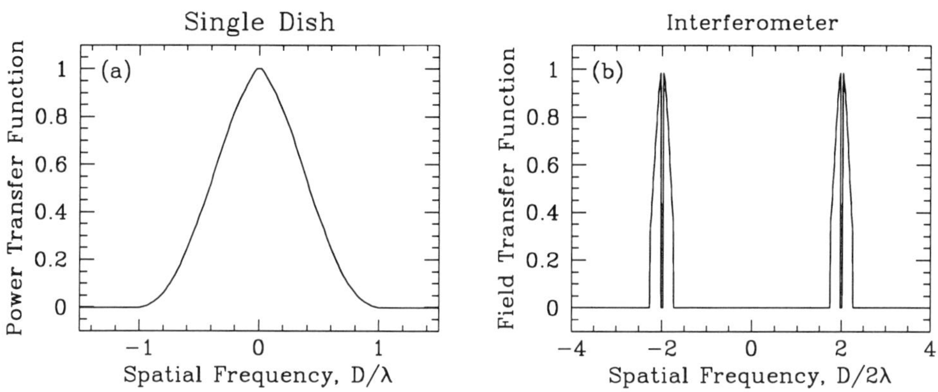


*Figure 2.* Difference in spatial frequency response for single dishes and interferometers with same field grading $g$: a 10 dB power edge taper and 10% radial central blockage.

Interferometers retain field amplitude and phase information rather than power intensity alone, and produce maps that are the convolution of the true brightness distribution of the source with the field grading $g$ itself,

$$T_{\text{map}} = T_{\text{source}} * g. \tag{5}$$

The effective grading, and therefore the spatial frequency responses, depend on the locations of the antennas in the array. At any instant the interferometer is sensitive to points in the $(u,v)$ plane corresponding to the antenna separation (Fig. 2$b$), at one spatial frequency and one angle on the sky. Unless the antennas can overlap, this distance is necessarily greater than the antenna diameters. The best known effect of the lack of short projected spacings is the central "hole" in interferometers' spatial frequency responses. Interferometers consequently produce maps with complicated spatial and spectral distortion when the sources have spatial extent comparable to the primary beam or larger. Wilner and Welch (1994) give a clear analysis of this effect for Gaussian and disk sources. Such spatial frequency filtering is sometimes useful, for instance in removing extended emission to emphasize compact structure. Although far from a fundamental issue, this can also add to the complexity of interpretation.

Mapping an extended source with a single dish samples larger spatial frequencies to fill in detail near the center of the $(u,v)$ plane. The spatial frequency filter function $g \star g$ does not severely attenuate below $\sim D/2\lambda$, but there is no information at all past $D/\lambda$. Combination of single-dish and interferometer data profits from large single dishes whose edges are as far out in the $(u,v)$ plane as possible, with good response at the interferometer's shortest spacings.

## 3. Reflector and Phase Front Tolerances

The chief size—and therefore spatial resolution—limit on single dishes is the requirement of high mechanical precision needed to maintain phase coherence across the antenna. The strength of materials gives a semi-fundamental limit to the size of conventional steerable telescope structures of about 680 m (von Hoerner 1969), but flexure rather than strength is the practical limit on large antennas. Scaling small, stiff structures to much larger size becomes impractical: the deflection of one dimensional structures, such as struts, scales as length cubed, and two dimensional structures scale as radius squared. These considerations limit steerable cm-wave antennas to diameters of order 100 m and mm-wave antennas to about half that.

Perhaps the most germane measure of precision is the on-axis gain, or coupling to a point source. For a single mode, diffraction limited system random errors in the surface reduce the gain by a factor $\exp(-4\pi\delta/\lambda)^2$, where $\delta$ is the rms deviation of the surface from a perfect parabola and $\lambda$ is the wavelength (Ruze 1966). It is interesting to note that the usual requirement, $\lambda/20$, corresponds to an efficiency of only 67%. The remainder of the power scatters into a wide angle "error beam" with size $\lambda/l_c$, where $l_c$ is the correlation length characteristic of the distance between surface errors. An incoherent system does not necessarily suffer as large a degradation in gain as do the phase-coherent systems because it can accept multiple modes and collect some power from the error beam.

### 3.1. COMPENSATION FOR NON-IDEAL EFFECTS

It is rare to find a modern single dish antenna that does not surpass its surface accuracy design goals. An important element of this has been developing methods to accurately measure, and then set, the surface. Baars (1983) reviews many of these methods. Opto-mechanical methods are accurate to a few hundred microns and are essential for initial surface adjustment. Purely optical methods have been developed to bring the surface to its final accuracy (e.g., Morris et al. 1988; Rogers et al. 1993; Serabyn et al. 1991).

Once set to high accuracy, the surface must compensate for a variety of mechanical and thermal effects that conspire to distort the surface. Compensation falls into four categories: none at all, passive open-loop, active open-loop, and active closed-loop compensation. A very familiar form of compensation is the "homologous" mechanical design, a passive open-loop design in which the surface deforms with changing elevation but maintains a parabolic form (von Hoerner 1967). Telescopes with homologously deforming surfaces need an active open-loop control of the secondary mirror's position to compensate for focal length and optical axis shifts. Pointing models and systems that appropriately deform secondary mirrors to compensate for large-scale distortions in the primary mirror (e.g., Ingalls et al. 1994) are other active open-loop compensation schemes.

Clever structures and construction can passively eliminate many problems for small antennas, but the trend is to increase the number of active control systems as telescopes increase in size. A relatively new and promising use of active closed-loop compensation is to precisely control to the primary surface shape. This technique has been proven by the Keck telescope at optical and infrared wavelengths (Mast & Nelson 1985). Cortés-Medellín & Goldsmith (1994) have investigated the optical properties of a segmented mirror that operates in the diffraction-limited mode to prove its feasibility and investigate allowable errors at millimeter wavelengths. The 100 m Green Bank Telescope will adjust its surface periodically to maintain accuracy, and the planned 50 m Large Millimeter Telescope (Schloerb et al. 1995) is designed to have a surface with an "update" rate of a few Hertz.

## 4. Conclusion

Sensitivity and imaging considerations indicate that single dish antennas with focal plane arrays will remain the most efficient way to map sources with resolution to $\sim \lambda/D_{\rm dish}$, while interferometers will be capable of imaging at much higher spatial resolution. Total area and bandwidth will determine which instrument is more sensitive for detecting point sources. Especially at wavelengths near and shorter than 1 mm, single dishes' wide-band optics will permit very sensitive continuum detections with broad-band direct detectors. Very large aperture single dish instruments will rely heavily on active measurement and control systems to keep the surface sufficiently precisely shaped. Both types of instruments borrow heavily from each other's successful technologies (e.g. antenna design, correlation techniques, focal plane arrays, on-the-fly mapping, and seeing correction schemes) and this healthy exchange will doubtless continue.

## 5. References

Baars, J.W.M. 1983, "Technology of Large Radio Telescopes for Millimeter and Submillimeter Wavelengths," *Infrared and Millimeter Waves* **9**, 241-281, Academic Press, New York

Bracewell, R.N. & Roberts, J.A. 1954, "Aerial Smoothing in Radio Astronomy," *Aust. J. Phys.* **7**, 615-640

Cortés-Medellín, G. & Goldsmith, P.F. 1994, "Analysis of Active Surface Reflector Antenna for a Large Millimeter Wave Radio Telescope," *IEEE Trans. Antennas Propagat.* **42**, 176-183.

Imriale, W.A. & Thorburn, M, eds. 1994, "Special Issue on Radio Telescopes," *Proc. IEEE* **82**, 633-811..

Ingalls, R.P., et al. 1994, "Upgrading the Haystack Radio Telescope for Operation at 115 GHz," *Proc. IEEE* **82**, 742-755.

Lawrence, C.R., Herbig, T., & Readhead, A.C.S. 1994, "Reduction of Ground Spillover in the Owens Valley 5.5-m Telescope," *Proc. IEEE* **82**, 763-767.

Mast, T.S. & Nelson, J.E. 1985, "Fabrication of the Keck Ten Meter Telescope Primary Mirror," *Proc. SPIE* **542**, 48-59

Morris, D., Baars, J.W.M., Hein, H., Steppe, H., Thum, C., & Wohlleben, R. 1988, "Radio-holographic Reflector Measurement of the 30-m millimeter radio Telescope at 22 GHz with a Cosmic Signal Source," *Astron. Astrophys.* **203**, 399-406.

Rogers, A.E.E., Barvainis, R., Charpentier, P.J., & Corey, B.E. 1993, "Corrections for the Effects of a Radome on Antenna Surface Measurements Made by Microwave Holography," *IEEE Trans. Antennas Propagat.* **41**, 77-84.

Schloerb et al. 1995, this *Proceedings*

Serabyn, E., Phillips, T.G., & Masson, C.R. 1991, "Surface Figure Measurements of Radio Telescopes with a Shearing Interferometer," *Appl. Optics* **30**, 1127-1241.

Thompson, A.R., Moran, J.M., & Swenson 1986 (TMS), *Interferometery and Synthesis is Radio Astronomy,* pp. 155-168, Wiley, New York

Ruze, J. 1966, "Antenna Tolerance Theory – A Review," *Proc. IEEE* **54**, 633-640.

von Hoerner, S. 1967, "Design of Large Steerable Antennas," *Astronom. J.* **72**, 35-47.

Welch, W.J. 1995, this *Proceedings*

Wilner, D. & Welch, W.J. 1994, "Aperture Synthesis Observations of S140," *Astrophys. J.* **427**, 910-913.

# CHARACTERISTICS OF THE MILLIMETER ARRAYS

WM. J. WELCH
*Radio Astronomy Laboratory*
*University of California*
*Berkeley, CA 94720*

**Abstract.** The characteristics of millimeter wavelength arrays are discussed with a view to the general usefulness of the arrays for a wide variety of astronomical programs. Their most important capability is high angular resolution. In addition, their accurate pointing allows detailed comparison of images made at different wavelengths, and permits accurate astrometry. The existing and planned arrays have collecting areas of several hundred square meters and therefore sensitivities comparable to even the largest single reflector antennas. Because of their relative immunity to emission from the atmosphere, they are the instruments of choice for detection of weak signals from point sources. The current and planned systems have flexible correlation spectrometers with high resolution and wide spectral coverage. Their ability to accurately spatially locate emission from new spectral features is important in the identification of new molecular species. The improvement in millimeter arrays has enabled mapping with dynamic ranges of one hundred or more and good enough fidelity to permit quantitative comparison of features on different maps. Finally, large ratios of baseline length to antenna diameter yield maps with large ratios of map field size to map resolution.

## 1. Introduction

The development of interferometers for millimeter wavelengths began in the early 1970's after it was found that strong millimeter wavelength emission could be detected from molecules in the dust clouds in the interstellar medium, in the atmospheres of the planets, and in the envelopes of evolved stars. The high angular resolution that an interferometer could provide promised to give a glimpse of such important processes as the formation of



*W.B. Latter et al. (eds.), CO: Twenty-Five Years of Millimeter-Wave Spectroscopy,* 199–206.

stars in the interiors of the dense clouds. At the present time, there are five such instruments operating at millimeter or sub-millimeter wavelengths, two more that will be in operation within a few years, and another large instrument that is well along in the design and is expected to be funded for construction soon.

Single reflector telescopes have long been thought to be the more general purpose astronomical instruments. However, the millimeter arrays show that, in addition to having high angular resolution, they have many other features that are desirable for a wide range of astronomical problems. In the following paragraphs we discuss these characteristics and call attention to the capabilities of the existing and planned arrays.

## 2. Array Characteristics

It is well known that for high angular resolution studies arrays are necessary. A number of other, mostly related, features make them the instruments of choice for a wide range of investigations. They have inherently more accuracy in pointing, excellent sensitivity, high spectral resolution and spectral flexibility, high dynamic range in map brightness sensitivity,and large spatial dynamic range. Table 1 lists a number of the parameters of the existing (or nearly existing) systems. These parameters determine the capabilities of the instruments in these various categories.

### 2.1. ANGULAR RESOLUTION

The resolution achieved by an array after an earth rotation observation that adequately fills in the *uv* plane is approximately $0.7\lambda/s_{\max}$, where $\lambda$ is the wavelength and $s_{\max}$ is the maximum baseline. Table 1 shows that the existing arrays achieve resolutions of the order one to a few arc-seconds in their operating wavelength ranges. For example, Ohashi et al. (1991), using the Nobeyama Millimeter Array, observed six embedded young stellar objects and five classical TTauri stars with resolutions of $2.8''$ to $8.8''$ in both the $\lambda 3$ mm continuum and in CS(2-1). The TTauri stars showed continuum (probably disks) and no CS, whereas the embedded stars showed, on the whole, no continuum but CS that was displaced from the IR positions, possibly core gas from which the young stars had formed.

A number of studies made with the Owens valley Millimeter Array (e.g, Sargent and Beckwith, 1987) in CO(1–0) at a resolution of $2''$ have shown flattened disk-like structures with radii of order 1000 AU surrounding T Tauri stars. Lay et al. (1994), using the CSO/JCMT sub-millimeter interferometer, found the 0.8 mm continuum emission from HL Tau to be extended, resembling a tilted disk of about 65 AU radius, and Mundy et al. (in press), with the BIMA Array, also resolved the HL Tau continuum at $\lambda 2.7$ mm finding a similar structure. Interferometric resolutions of a few arc-seconds or better were essential to all of these studies.

### 2.2. ACCURATE ANGULAR REGISTRATION

Arrays enjoy a substantial advantage over single reflector telescopes in pointing accuracy. The IRAM 30 m telescope has an angular resolution approaching that of the arrays when it is operated at its shortest wavelength of about one millimeter. However, pointing of this telescope, like other single reflectors, to better than one or two arc-seconds is very difficult. The arrays, on the other hand, routinely point to a fraction of an arc second. One reason is mechanical. The array elements are fastened to the ground, a much stiffer structure than the relatively floppy backup frame of a single reflector. The other reason is that the array is relatively immune to the refractive effects of the atmosphere. The array elements are fastened to the ground, and, to the extent that the atmosphere is a plane layer, the ray paths through the atmosphere are the same to every element independent of source elevation.

Maps, such as those of the T Tauri stars discussed above, have typical pointing accuracies of one arc-second or better, which is helpful for comparison with images made at other wavelengths. With care, pointing to 0.1 arc-second or better is possible. For example, Wright et al. (1990) and Baudry et al. (1995) have measured the positions of SiO masers in a number of Red Giant Stars to accuracies of about $0.1 - 0.2''$. Table 2 shows the results of Baudry et al. (1995) for 13 stars. The agreement with the optical astrometry of the stars is quite good.

### 2.3. SENSITIVITY

The instantaneous sensitivity of an array is proportional to its collecting area, and, if the extent of the observed source fits within the primary beam of the array antennas so that only a single pointing of the array antennas is needed, the brightness sensitivity in the synthesized map is also just proportional to the total collecting area. This number is given for the existing arrays in column 4 of Table 1. For the bigger arrays, it is quite comparable to the collecting areas of the biggest single millimeter dishes. Because the cost of individual millimeter receivers is now modest, it is less expensive to build up the collecting area in an array of antennas rather than in a single large reflector. The excellent sensitivity of arrays is illustrated by the detection of CO (1-0) emission from the QSO 3c48 at $z = 0.3695$ with the earlier three element OVRO array shown in Figure 1 (Scoville et al., 1993) and the map of IRAS 10214+4724 with the IRAM array (Radford et al., 1993).

If the source size is larger than the primary beam of the interferometer elements so that multiple pointings must be used, then the dependence of array brightness sensitivity on antenna diameter is linear rather than quadratic. This is shown for the existing arrays in column 5 of Table 1. Whereas the point source sensitivities vary over a factor of three among the arrays, it is roughly the same for all the arrays in the multiple pointing case. The bigger antenna systems are more suited for the small sources and the smaller antenna systems are better used for the more extended sources. The planned MMA is a compromise between these two sensitivities.

### 2.4. SPECTRAL RESOLUTION AND FLEXIBILITY

The spectrometer for an array, which is also the correlator for $n(n-1)/2$ baseline pairs, is a much bigger electronic component than the corresponding unit for a single reflector telescope. However, its cost relative to that of the antennas is small for almost any sized array, and the units that have been built for all the existing arrays are comparable to the spectrometers of the single reflector antennas. Typically, they have best resolutions of a few kHz, maximum bandwidths of several hundred, up to 1000, MHz, and the ability to observe several spectral lines at the same time in multiple windows. The arrays enjoy a perhaps surprising advantage for interstellar chemistry studies apart from their flexible spectral correlators. From an array observation one gets maps as well as several spectra. The molecular chemistry differs among clouds(e.g., Snyder et al., 1992), and if a new species is being identified by a number of emission spectra, it is important to be certain that all the candidate lines arise from the same cloud.

### 2.5. DYNAMIC RANGE IN BRIGHTNESS

The dynamic range in brightness in array maps depends on such factors as stability of the array electronics, pointing, completeness of the uv coverage, and, most importantly, the basic sensitivity of the system. As the stability of the millimeter systems has improved and the number of antennas has increased, the existing arrays have been producing maps of ever greater dynamic range and fidelity, gradually approaching the quality of maps made with the centimeter wave arrays, such as the VLA. An example is shown in Figure 2, a continuum map of the HII region complex W49 at 87 GHz done with the BIMA array (Wilner et al., in preparation). In addition to array data, this map contains single antenna data taken with the IRAM 30 m antenna, so that it is complete over all scales up to the interferometer primary beam of $120''$. It is a single pointing map. The peak flux is about 100 times the noise level, giving a dynamic range of 100. Comparison with a VLA map at 6 cm suggests an accuracy of the map features of about 10%, for a fidelity of 10. Since this is the array with the least collecting area in Table 1, we can expect even better results with the arrays with greater collecting area.

### 2.6. SPATIAL DYNAMIC RANGE

Spatial Dynamic Range is used here to mean the ratio between the size of the field and the synthesized beam. For a single pointing of the array antennas, this ratio will be close to the ratio of the longest baseline used to the array antenna diameter. This parameter is shown in the last column of Table 1. For multiple pointing observations this dynamic range can, of course, be increased. Figure 2 is also a good example of a field observed with a large dynamic range. The primary beam is 120″ and the synthesized beam is about 0.6″ for a ratio of 200. An inspection of Table 1 shows the complementarity between the various arrays. Part of the large spatial dynamic range of the BIMA array comes at the expense of having a smaller array antenna with corresponding lower collecting area. The MMA has clearly been planned to study complex fields and is a compromise between point source sensitivity and good imaging capability.

## 3. Future Developments

The future will bring further developments with the existing arrays and the construction of new arrays which are being planned.

### 3.1. MILLIMETER WAVELENGTH SEEING

A number of studies have shown that the interferometer phase fluctuations caused by atmospheric turbulence increase roughly monotonically with baseline for kilometer scale baselines with $\sigma_\phi \sim s^{0.7}$ (Sramek, 1990; Wright and Welch, 1990; Olmi and Downes, 1992). Since a maximum array spacing of a given length yields a map with a resolution inversely proportional to wavelength, the above power law implies that, on the average, the attainable resolution increases with wavelength. The resolution is poorer at millimeter wavelengths. Olmi and Downes (1992) find a median seeing of about 2″ at one millimeter wavelength. Imaging of better than one arcsecond will only be possible in exceptional weather.

At the present time, several groups are developing calibration schemes based on the idea that the atmospheric water vapor that produces the phase fluctuations also should produce fluctuating emission at millimeter wavelengths, especially near the water vapor absorption wavelengths. Thus, a measurement of the fluctuation in the brightness of the atmosphere will permit the correction of the phase fluctuation (cf Welch, 1994). It appears that one or all of the schemes will probably work. The correction will not only allow observations at sub-arc-seconds much of the time, but it will also permit elimination of the atmospheric de-correlation that is present in observations most of the time at shorter spacings. More accurate imaging should be the result.

### 3.2. NEW INSTRUMENTS

In addition to the MMA of the NRAO included in Table 1, there are two more arrays being planned. One is a large array for both millimeter and sub-millimeter wavelengths proposed by the NRO. The second is an array with an order of magnitude more collecting area than the MMA. These will allow the milliarc-second resolution that will resolve molecular clouds in the most distant galaxies and that may be necessary to fully understand the formation of stars.

## Acknowledgements

Support for this study was provided by NSF grant NSF-FD93-20238.

## References

Baudry, A., Lucas, R., and Guilloteau, S. 1995, *A&A*, **293**, 594.
Lay, O. P., Carlstrom, J. E., Hills, R. J., and Phillips, T. G. 1994, *ApJ*, **434**, L75.
Ohashi, N., Kawabe, R., Hayashi, M., and Ishiguro, M. 1991, *A. J.*, **102**, 2054.
Olmi, L., and Downes, D. 1992, *A&A*, **262**, 634.
Radford, S. J. E., Brown, R. L., and Vanden Bout, P. A. 1993, *A&A*, **271**, L21.
Sargent, A. I., and Beckwith, S. V. W. 1987, *ApJ*, **323**,294.
Sargent, A. I. and Welch, W. J. 1993, Ann. Rev. *A&A*, **31**, 297.
Scoville, N. Z., Padin, S., Sanders, S., Soifer, B. T., and Yun, M. S. 1993, *ApJ*, **415**, L75.
Snyder, L. E., Hollis, J. M., Ziurys, L. M. and Kuan, Y. J. 1992, in *IAU Symposium 150: The Astrochemistry of Cosmic Phenomena*, ed. P. D. Singh (Dordrecht: Kluwer), 187.
Sramek, R. A. 1990, in *IAU/URSI Symposium on Radio Astronomical Seeing*, eds. J. E. Baldwin and Wang Shouguan, 21.
Welch, W. J. 1994, in *IAU/URSI Colloquium 140: Astronomy with Millimeter and Sub-Millimeter Interferometers*, eds. M. Ishiguro and W. J. Welch
Wright, M. C. H., Carlstrom, J. E., Plambeck, R. L., and Welch, W. J. 1990, *A. J.*, **99(4)**, 1299.
Wright, M. C. H., and Welch, W. J. 1990, *IAU/URSI Symposium on Radio Astronomical Seeing*, eds. J. E. Baldwin and Wang Shouguan, 71.

TABLE 1. Properties of the Millimeter Arrays

| Array | $n$ | $D$ [m] | $n\frac{\pi}{4}D^2$ [m$^2$] | $nD$ [m] | Wavelength [mm] | Baseline EW/NS [m] | $s_{\max}/D$ |
|---|---|---|---|---|---|---|---|
| IRAM | 5 | 15 | 883 | 75 | $\lambda 1,3$ | 430/240 | 33 |
| BIMA | 10 | 6 | 283 | 60 | $\lambda 1,3$ | 1000/1000 | 200 |
| CSO/JCMT | 2 | 10/15 | 212 | — | $\lambda 0.3-1.0$ | 165 | 14 |
| NRO | 6 | 10 | 471 | 60 | $\lambda 1,2,3$ | 560/520 | 60 |
| OVRO | 6 | 10.4 | 509 | 63 | $\lambda 1,3$ | 200/220 | 29 |
| AT | 5 | 15 | 883 | 75 | $\geq$2.6 | 3000(EW) | 136 |
| SMA | 8 | 6 | 226 | 48 | $\lambda 0.3-1.3$ | 800/800 | 133 |
| MMA | 40 | 8 | 2000 | 320 | $\lambda 0.9-9.0$ | 3000(Circ) | 375 |

NOTE: IRAM is Institute de Radio Astronomie Millimetrique; BIMA is Berkeley-Illinois-Maryland-Association; CSO is Caltech Sub-millimeter Observatory; JCMT is James Clerk Maxwell Telescope; NRO is Nobeyama Radio Observatory; OVRO is Owens Valley Radio Observatory; AT is Australia Telescope; SMA is Sub-millimeter Array (Harvard/SAO); MMA is Millimeter Array(NRAO). (Sargent and Welch, 1993)

TABLE 2. Positions of SiO Maser Stars (from Baudry et al., 1995)

| Star | Mean epoch | $\alpha$ (2000.0) [h m s] | $\delta$ (2000.0) [° ′ ″] | $V_{\rm LSR}$ [km s$^{-1}$] | | Estimated errors $\sigma_\alpha$[″] | $\sigma_\delta$[″] |
|---|---|---|---|---|---|---|---|
| $o$ Ceti | 1992.04 | 02 19 20.797 | * | 44, | 46.5 | 0.20 | * |
| S Per | 1991.33 | 02 22 51.713 | 58 35 11.50 | −40, | −38 | 0.20 | 0.20 |
| TX Cam | 1992.04 | 05 00 51.127 | 56 10 53.98 | 6, | 9.5 | 0.15 | 0.15 |
| Ori IRc2 | 1992.05 | 05 35 14.498 | −05 22 30.70 | 14.5, | 18 | 0.15 | 0.20 |
| U Ori | 1992.04 | 05 55 49.171 | 20 10 30.34 | −42, | −39 | 0.15 | 0.25 |
| R Cnc | 1991.32 | 08 16 33.826 | 11 43 34.60 | 11, | 17 | 0.10 | 0.15 |
| R Leo | 1991.33 | 09 47 33.490 | 11 25 43.99 | −5, | −1 | 0.10 | 0.15 |
| RX Boo | 1992.03 | 14 24 11.609 | 25 42 13.73 | −1, | 8 | 0.10 | 0.20 |
| RU Hcr | 1992.05 | 16 10 14.519 | 25 04 14.41 | −13, | −6 | 0.15 | 0.15 |
| VX Sgr | 1991.32 | 18 08 04.062 | −22 13 27.30 | −2, | 11 | 0.20 | 0.25 |
| NML Cyg | 1992.05 | 20 46 25.532 | 40 06 59.35 | −10, | −4 | 0.15 | 0.15 |
| $\mu$ Cep | 1991.33 | 21 43 30.446 | 58 46 48.40 | 24, | 29 | 0.20 | 0.15 |
| R Peg | 1992.03 | 23 06 39.157 | 10 32 36.07 | 22, | 26 | 0.15 | 0.20 |
| R Cas | 1991.33 | 23 58 24.781 | 51 23 19.10* | 21, | 28 | 0.20 | * |

* The declination uncertainty was not reliably estimated.

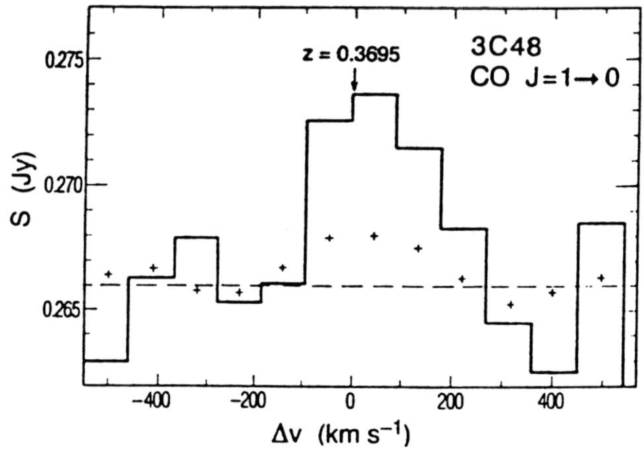


*Figure 1.* Spectrum of the CO emission from the radio-loud quasar 3C48 observed with the OVRO Millimeter Interferometer.

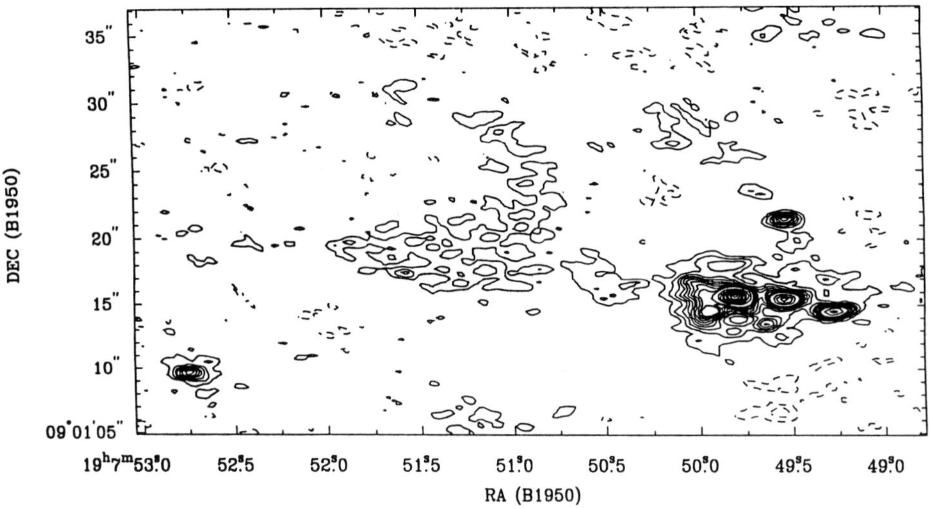


*Figure 2.* Map of the giant HII region W49A made with the BIMA array (Wilner et al, in preparation) at 87 GHz. Resolution is 0.6″, and total flux from the IRAM 30 m is included. The peak is 100 times the RMS noise level.

# INCREASING THE YIELD OF OUR TELESCOPES

DARREL T. EMERSON
*National Radio Astronomy Observatory*
*949 N. Cherry Avenue, Campus Bldg. 65*
*Tucson, Arizona 85721-0655*

## 1. Introduction

Obviously, we all wish to make the best possible use of our telescopes. Mostly, this means striving for the best possible efficiency of observing, and observing simultaneously over as wide a range of parameter space as possible. Some observing techniques are more efficient than others, but there also some pitfalls to avoid in attempts to improve observing efficiency. The emphasis here is on mm-wave observing, but many of the principles are equally applicable to other parts of the radio spectrum. The "Yield" in the title means "useful scientific output," which is much more difficult to quantify than simply bytes of data with a given signal-to-noise ratio.

For the purposes of this presentation, I distinguish between two different observing philosophies:

(1) Observations that use special-purpose instrumentation, or even a special-purpose telescope, and which are truly breaking new ground. If successful, such experiments may have enormous scientific yield, although the telescope efficiency in terms of acquiring bytes of data might be very poor.

(2) "Bread and Butter" observations. These may use an established telescope and instrumentation, and might include large scale surveys, or deep integrations. There is a well-defined goal, for which a considerable quantity of data may be required. The same telescope and instrumentation are used to support many different projects by many independent observers. Efficiency in data acquisition becomes very important; in some sense scientific yield may be proportional to the quantity of reliable data obtained, with some type of non-linear normalization for signal-to-noise ratio.

There are serious dangers in attempting to optimize the efficiency of a telescope. As yet undiscovered and unexpected phenomena might be filtered out and hidden from the observer – pulsars might have escaped detection



*W.B. Latter et al. (eds.), CO: Twenty-Five Years of Millimeter-Wave Spectroscopy*, 207–214.

for many years if today's software interference rejection algorithms had been available in the 1960s. In the name of efficiency, versatility may be lost; an instrument can be finely tuned for efficiency much more easily if it is only ever required to support a single, specialized observing mode. A loss of versatility may discourage observers from trying new, innovative techniques; this risks reducing the potential scientific output of a telescope much more than the possible increase in mean data rate resulting from improved efficiency.

Considerations of efficiency of data acquisition may have little place in mode (1) above. However, it is usually worth investing considerable effort optimizing the efficiency of observations in category (2), provided that the potential pitfalls are recognized.

## 2. Increasing Observing Efficiency

One philosophy is to instrument the telescope so as to observe over the biggest range of parameter space possible, with the highest possible sensitivity. There are potential pitfalls; some instrumentation upgrades help selected types of science, possibly even hindering other scientific studies. An example might be a multiple-beam system; it is often necessary, although still undesirable, to arrive at a compromise in sensitivity of individual elements of a multiple-beam system. It may simply not be feasible to lavish the care and attention on 10 or 100 individual feeds and receivers that could have been devoted to a single receiver system. Such a system may be extraordinarily efficient at mapping large areas, but may be significantly worse at point-source studies – such as searches for new lines in discrete objects – than the original single-beam system. Compromises like this often have to be taken. Some of the factors to be considered are:

1. Improve system noise. All other things being equal, the rate of acquiring data is proportional to $T_{\rm sys}^{-2}$ where $T_{\rm sys}$ is the system noise. This parameter affects **ALL** observing modes equally. Attempts to expand parameter space should not compromise $T_{\rm sys}$. Note that, with modern receivers, the **receiver** noise may only contribute a relatively minor part of $T_{\rm sys}$.

2. The receivers should respond to both orthogonal polarizations. This gives a factor of 2 in observing speed for all observations, discrete source or extended mapping, which are sensitivity limited. A side effect is that polarization studies may become more feasible, but the main incentive for a dual polarization **system** is to improve the sensitivity. This is equivalent to reducing the system noise $T_{\rm sys}$ by $\sqrt{2}$, which may be a much more dramatic improvement than "merely" reducing the **receiver** noise by the same factor.

3. Frequency space: multiple frequency bands. Increasing the instantaneous bandwidth of course helps continuum observations, but the ability

to observe several different spectral lines simultaneously may increase the scientific value of observations. The simultaneous observation may significantly improve the relative calibration of the lines. However, there may not **always** be multiple lines either accessible or of interest in a given source, and even if there are, they are unlikely to be matched in their requirements for signal-to-noise and hence integration time. On average, expanding this parameter space by a factor of $N$ will increase the useful telescope output by a factor significantly lower than $N$.

4. Spatial diversity: multi-beam systems. For $M$ feeds and receiver systems, the gain in mapping speed approaches a factor $M$. Note that, given a choice between $M$ dual-polarization systems and $2 \times M$ single-polarization feeds, receivers and backends, the dual-polarization system will almost always be more productive. It is important that $T_{\text{sys}}$ of individual feeds not be compromised; a degradation of as much as 30% in $(T_{\text{sys}}^{-1})$ will require a doubling of the number of feeds, receivers and backends just to break even on mapping observations. Observations of discrete sources would then take twice as long as they did before the upgrade. Nevertheless, the results of a good multiple-beam system can be outstanding; several spectacular examples of data taken with the FCRAO 15-beam QUARRY system (Erickson 1995) have been presented during this symposium.

## 3. Observing Techniques

The choice of observing strategy is usually a compromise between minimizing any systematic artifacts in a spectrum and minimizing the random noise. From the radiometer equation the uncertainty $dT$ on an observation, using bandwidth $B$ with integration time $t$ is given by $KT_{\text{sys}}/\sqrt{Bt}$, where $K$ is a constant depending on the observing technique.

**Position Switching.** Position switching is perhaps the most common spectral-line observing technique at mm-wavelengths. For the usual case where the telescope is pointed away from the source to blank sky for half of the available time, $K = 2$. Taking account of overhead, in practice $K > 2$. One can do somewhat better by sharing one off-source measurement for several on-source observations. The optimum use of time, neglecting telescope overhead, is when, for $M$ on-source measurements per off-source integration, the telescope spends $\sqrt{M}$ as long integrating on each reference position. The effective value of $K$ then becomes $K = 1 + 1/\sqrt{M}$. As $M$ becomes very large, $K$ tends to 1, the best possible case. At mm-waves, with typical receiver and atmospheric instabilities, a given off-source measurement may only remain valid for about one minute; beyond that, baselines in the calibrated (on-off) spectrum begin to deteriorate. In practice this severely limits the number of on-source to off-source measurements.

**Frequency Switching.** Frequency switching is a very efficient observing mode, where it can be used. The spectrum being studied remains within the bandwidth of the telescope's spectrometer in both phases of the switching cycle, but is switched by an offset of a few MHz and differenced. Provided that the resulting 2 replicas of the spectrum, one now negative and one positive, are sufficiently separated, then one half of the spectrum may be inverted, shifted, and averaged with the other half. $K$ becomes $\sqrt{2}$, an improvement in observing speed of a factor of at least 2, compared with position switching. The disadvantages are (1) the requirement that the total extent of the line be narrower than the frequency switching offset and (2) notoriously bad baselines. The low level standing waves between different parts of the structure, which are always present in radio telescopes and which give rise typically to sinusoidal modulation of the spectral baseline, are not in general cancelled. These disadvantages can be overcome: some algorithm such as that used for restoration of dual-beam observations of extended sources (e.g., Emerson et al. 1979) could be applied in the spectral domain. Modifications to the antenna optics, such as modulation of the path lengths responsible for creating standing wave patterns (e.g., Payne & Ulich 1978) can substantially improve the spectral baselines.

**On-the-fly Observing.** This is a rapid-scanning technique for mapping extended features, either in continuum or in spectral line. It provides a remarkably efficient alternative to the more conventional point-and-step mapping mode and is applicable either to single-beam or to multi-beam systems. The telescope beam scans rapidly and continuously back and forth across the source, recording data all the time. Some of the advantages are:

(1) Telescope overhead is nearly eliminated. Because the true coordinates, rather than demanded coordinates, are recorded, it is possible to continue recording useful data even when the telescope is accelerating or decelerating at the start and finish of individual rows of a map. Telescope tracking tolerances are relaxed, permitting very fast telescope scanning.

(2) Since one coverage of a field can be completed very quickly, the data are much more consistent; in general, there is not time for weather to change dramatically or for receivers to drift significantly during one coverage. Sensitivity is achieved by averaging multiple coverages. Because each coverage is completed relatively quickly, the pointing within a single map coverage is likely to be much more consistent; if necessary, relative pointing between successive coverages can be corrected. Sometimes, a small part of one map coverage may have some defect—e.g., one bad mapping row, caused by a change in atmospheric conditions. When $N$ independent coverages are averaged, the small number of systematic errors is reduced by $(1/N)$ rather than by the $(1/\sqrt{N})$ that applies to Gaussian noise.

(3) If $M$ on-source measurements can share the same off-source refer-

ence observation, then there is a potential gain in sensitivity, for a given usage of telescope time, of $1 + (1/\sqrt{M})$. With the on-the-fly rapid scanning observing mode, there may easily be hundreds of independent on-source measurements per off-source integration – i.e., $M > 100$. The factor $K$ in the radiometer equation, $1 + (1/\sqrt{M})$, tends to $K = 1$. There is an improvement in sensitivity of nearly a factor of 2 compared with conventional position-switched observations, or a factor of nearly 4 in mapping speed.

(4) Although with a multi-beam system some slight compromise is sometimes necessary in the performance of individual receivers, the OTF technique does not require such a compromise; there is no adverse impact on point source observations.

Compared to conventional point-by-point mapping, the gain in observing efficiency can be considerable. The multiple on-per-off aspect (3) gains a speed factor approaching 4. The reduced overhead (1) is a factor of perhaps 1.3. Increased quality and better consistency of data, and greater immunity to systematic drifts, are difficult to quantity, but are certainly significant. Compared with conventional position-switched mapping, an effective overall gain approaching an order of magnitude in observing speed is feasible.

OTF observing has been a standard technique for many years for continuum mapping (e.g., Haslam 1974). However, it is only relatively recently that effective on-the-fly data acquisition with thousand-channel spectrometers has become feasible.

## 4. Example OTF Observations

Probably the most critical aspect of OTF observing is obtaining correct registration of data samples with the instantaneous telescope coordinates. Observations across the limb of the moon provide a very stringent test for potential problems. Figure 1a shows contours, plotted at intervals of 10% of the peak brightness, of part of the eastern limb of the moon. For these data, there was an unwanted 50 ms offset in timing between data acquisition and telescope encoder readout. Since alternate scans are made with respectively increasing and decreasing azimuth, a constant timing offset between data and encoder sampling corresponds to a position error which changes sign in adjacent scans. The scanning speed was 1 degree/minute, so 50 ms corresponds to an error of $\pm 3''$, which is readily seen on the moon edge, even with the $65''$ beam. The ripples in the edge contours of Figure 1a reflect the coordinate offsets between odd and even-numbered rows of the map. Figure 1b shows the same data corrected for this 50 ms timing offset. The relative registration of the gridded data is now within $\sim 1''$.

Figure 2 shows a complete continuum map of the moon at 90 GHz, made with the 12 m telescope in total power mode, using a scanning speed

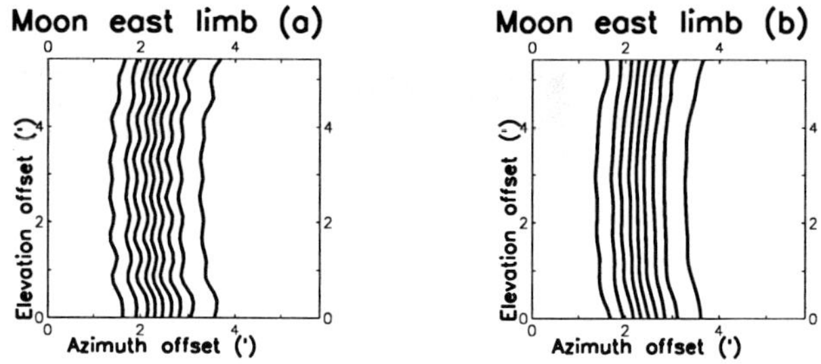


*Figure 1.* The Moon limb, *a (left)* with and *b (right)* without data timing errors.

of 1 degree per minute. The telescope beam is 65″. The fully sampled 45′ by 45′ field, took only one hour to observe. No baseline corrections of any kind have been applied. The observations started at 10:37 UTC on 1 April 1994, and are shown in an az-el coordinate system, centered on the moon. This shows how relatively large areas of sky can be mapped successfully quite quickly, and are immune from the effects of tracking errors.

Figure 3 shows the data from Figure 2, but processed to enhance the contrast; a model of the large-scale moon emission has been subtracted from the data, which have then been edge enhanced. The many crater-like features visible are all real, and remained visible on successive moon maps taken one and two hours later. The line of the terminator can also be discerned, as a vertical change in contrast just to the right of the field center. The small features visible in this image differ from their surroundings by a peak temperature difference of typically $\sim 2$ K.

Figure 4 shows some individual spectral line maps of an interesting molecular cloud CME-1 (Emerson et al. 1996) in the Pleiades, observed at 230 GHz with the 12 m telescope. This area of 30′ by 30′ was observed in OTF mode during a few hours of mediocre weather conditions. After gridding, there are more than 50,000 fully sampled independent spectra in this data set. This demonstrates how substantial quantities of high quality and scientifically useful data can be obtained quite rapidly.

## 5. Conclusions

Expanding the range of parameter space sampled by a given radio telescope can increase dramatically the available useful scientific data. However, it is most important not to compromise the $T_{\rm sys}$ of the overall system in any attempt to increase telescope throughput. It may also be important that flexibility in observing, or in configuring the telescope, is not lost.

The correct choice of observing technique has an important impact on the scientific output of a telescope. The power of modern computer hardware, and the ability to handle large quantities of raw data at reasonable cost, permit techniques such as rapid scanning on-the-fly observing, to be

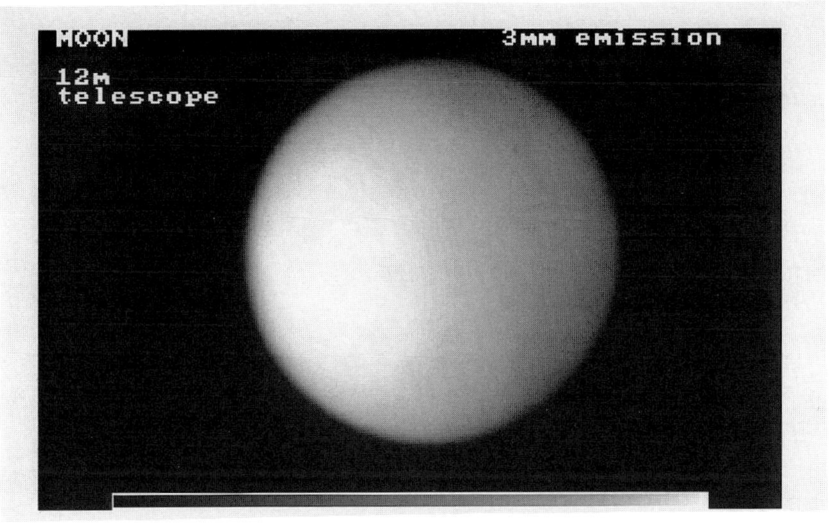


*Figure 2.* The Moon at 90 GHz, total power.

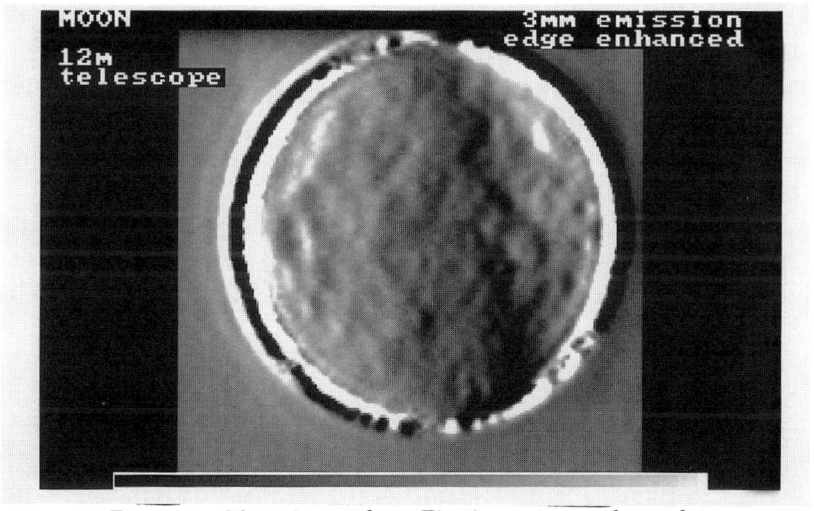


*Figure 3.* Moon image from Fig. 2, contrast enhanced.

exploited relatively easily even with thousand-channel spectrometers. This can lead to an almost unexpectedly high increase in useful scientific output from a telescope. The limiting factor today is often more in data analysis software capability than in the performance of the hardware.

The very first radio astronomical observations (Jansky, 1932) were made in a true on-the-fly mode. The 1932 observations were perhaps for the first time analysed in a way easily interpreted by astronomers some 46 years later (Sullivan, 1978). There is an important lesson to be learned here.

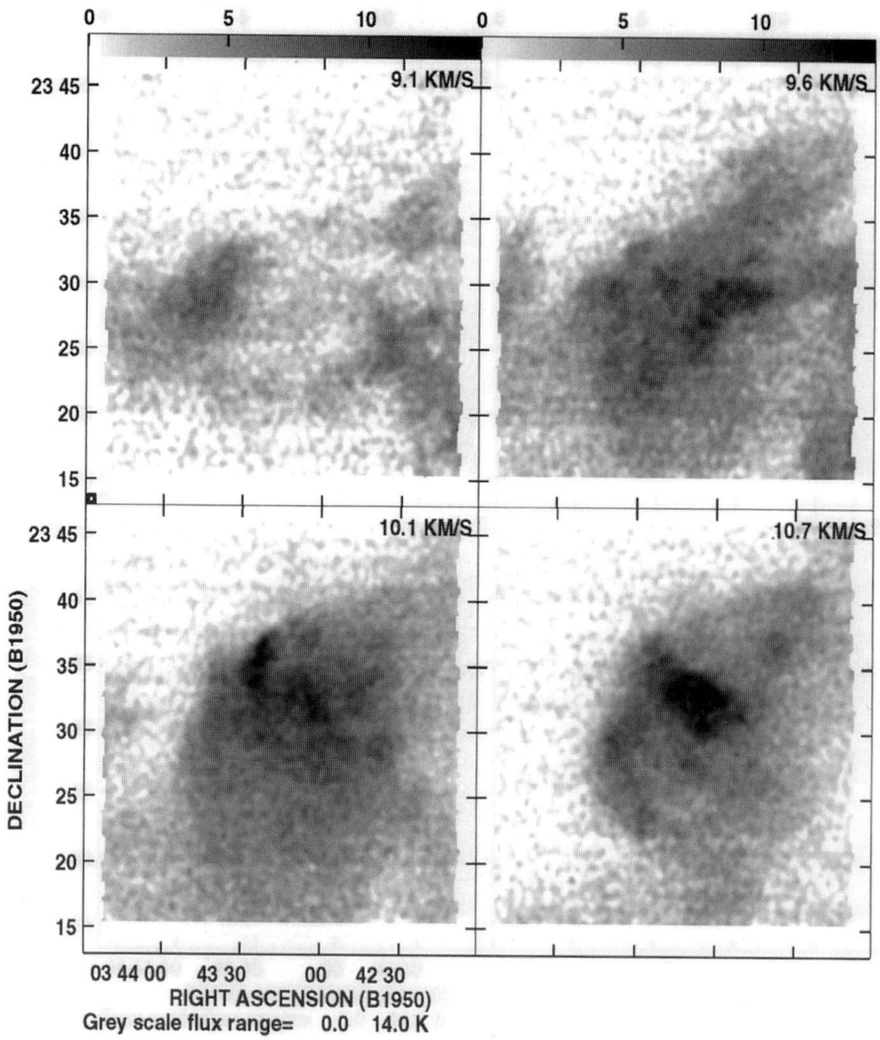


*Figure 4.* CME-1 CO J=2–1, 230 GHz, greyscale image of 4 channel maps.

## References

Emerson, C.M., Mangum, J.G., & Emerson, D.T. (1996), in preparation.

Emerson, D.T., Klein, U., & Haslam C.G.T., 1979, A & A, 76, 92.

Erickson, N.R., 1995, in *Multi-feed Systems for Radio Telescopes*, A.S.P Conference series Vol. 75, Eds. D.T.Emerson & J.M. Payne p. 149.

Haslam, C.G.T., 1974, A. & A.S., 15, 333.

Jansky, K.G. 1932. "Directional Studies of Atmospherics at High Frequencies," Proc. Inst. Radio Eng., 19, 1920-32.

Payne, J.M., Ulich, B.L., 1978. "Path Length Modulator": NRAO internal report.

Sullivan, W.T., III 1978, "A new look at Karl Jansky's original data," Sky & Tel., 56, 101-5.

# SUBMILLIMETER WAVELENGTH FOURIER TRANSFORM SPECTROSCOPY OF ASTRONOMICAL SOURCES

E. SERABYN AND E.W. WEISSTEIN
*California Institute of Technology*
*Pasadena, CA 91125*

At high, dry observatory sites, the transmission of the atmosphere at near-millimeter and submillimeter wavelengths provides access to frequency "windows" which encompass roughly half a terahertz of bandwidth. Broad-band spectroscopic techniques, such as Fourier transform spectroscopy, provide an effective means of exploring these vast tracts quickly, being especially useful for studies of wideband source spectra, and broad-line sources. As is outlined here, the former is applicable to studies of the interstellar medium, while the latter is particularly useful for planetary spectroscopy.

In the tropospheres of the giant planets, pressure broadened absorption lines with widths of several GHz can be expected. Figure 1 shows a spectrum of Jupiter in the highest frequency submillimeter atmospheric window, measured with a Fourier transform spectrometer (FTS) at the Caltech Submillimeter Observatory (CSO). Evident immediately is a strong and wide ($\approx$ 10 GHz FWHM) absorption due to the 3-2 band of $PH_3$, a molecule very useful in tracking vertical transport in the upper troposphere.

As regards the interstellar medium, it is perhaps surprising that twenty five years after the detection of CO in the interstellar medium, wideband millimeter/submillimeter emission line spectra of even well-studied sources have yet to be measured. With an FTS, such wideband measurements are straightforward, and Figure 2 presents the near-millimeter and submillimeter spectrum of the OMC-1 cloud core measured with 0.2 GHz resolution. Five optically thick transitions of CO are evident in the spectrum, and, as Figure 3 shows, a total of over 180 spectral features have been detected. Interestingly, the CO molecule does not dominate the integrated subterahertz line flux emerging from the cloud core; rather, the shock enhanced S-bearing $SO_2$ and SO molecules account, respectively, for the largest and next-largest energy fluxes, with CO in third place.

This work is supported by NSF grant AST-9313929.



*W.B. Latter et al. (eds.), CO: Twenty-Five Years of Millimeter-Wave Spectroscopy,* 215–217.

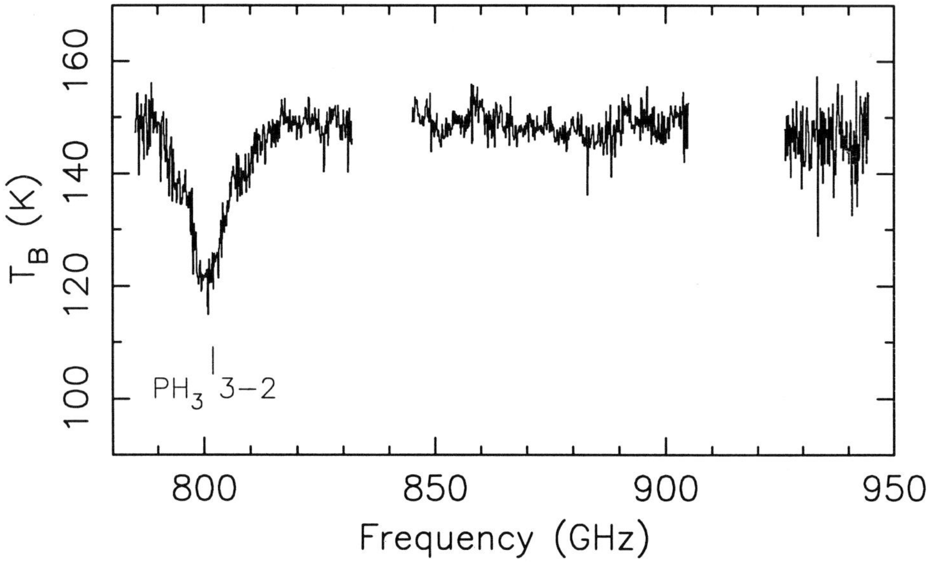


*Figure 1.* The high frequency spectrum of Jupiter, measured at a spectral resolution of 0.2 GHz with a Fourier Transform Spectrometer at the CSO.

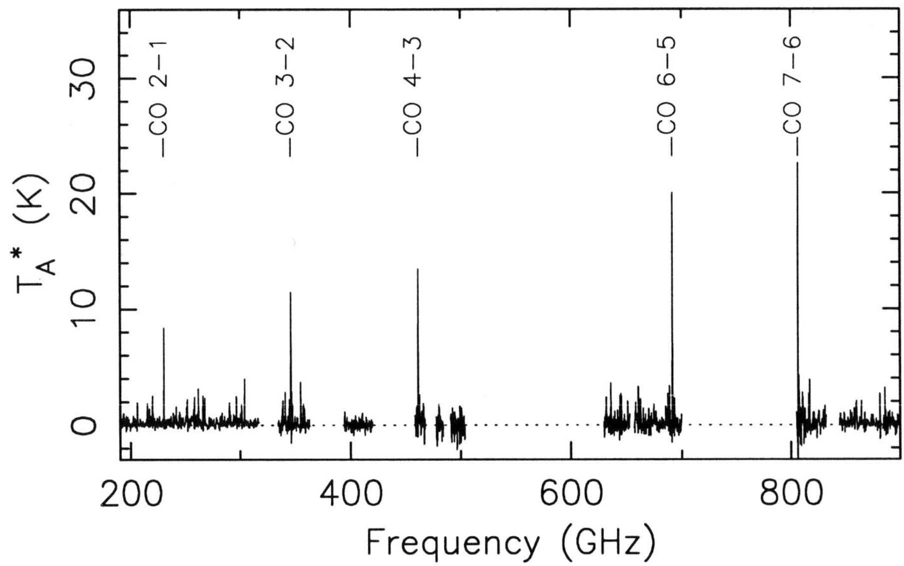


*Figure 2.* The broadband spectrum of OMC-1 measured with the same FTS.

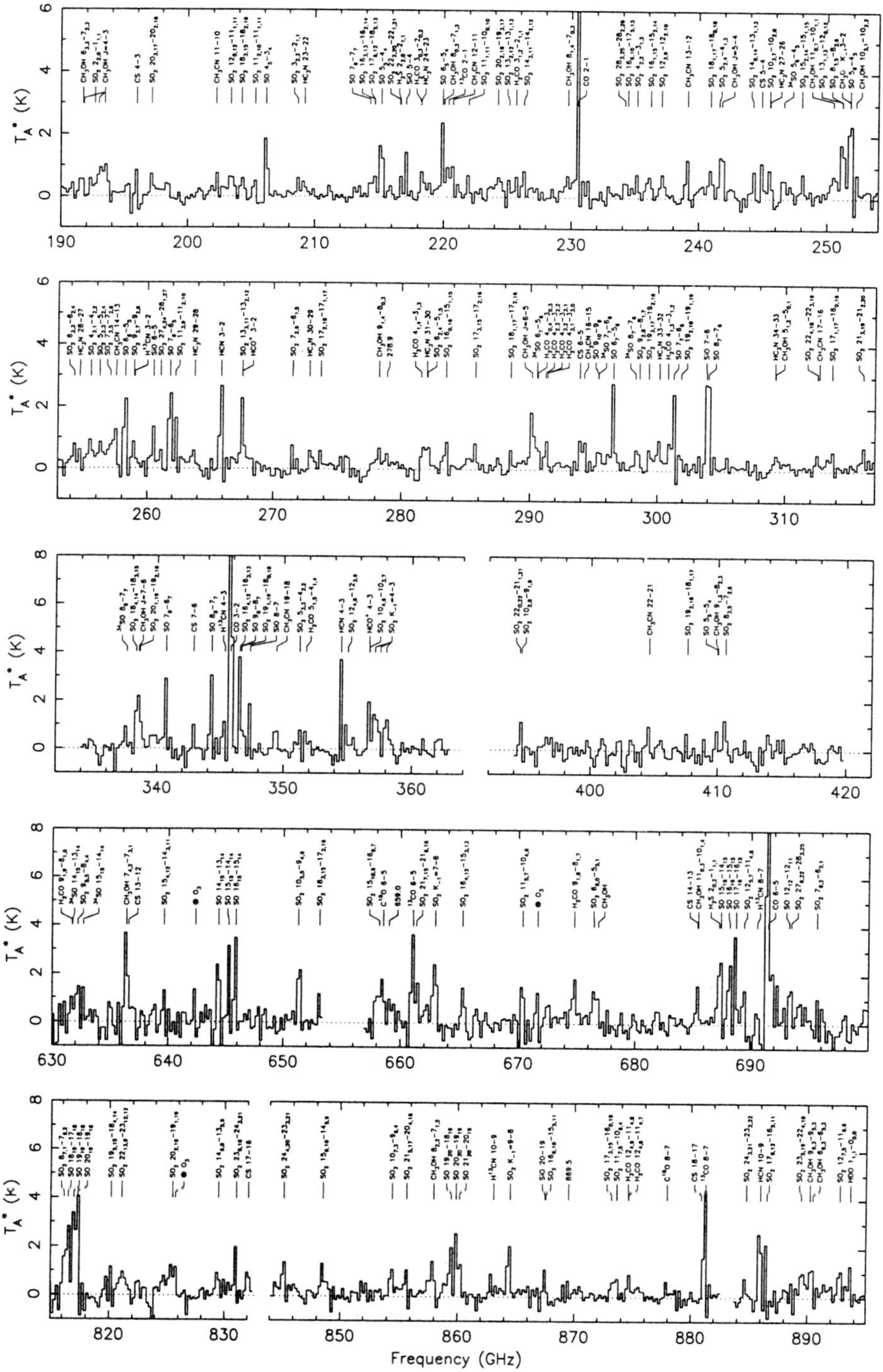

*Figure 3.* An expanded version of the Orion spectrum of Fig. 2.

# THE SUBMILLIMETER TELESCOPE OBSERVATORY

ROBERT N. MARTIN
*SMTO/Steward Observatory*
*University of Arizona*
*Tucson, AZ 85721*

## 1. Introduction

The Steward Observatory of the University of Arizona and the Max-Planck-Institut für Radioastronomie (Germany) have collaborated on the construction and operation of a dedicated submillimeter telescope facility. The Submillimeter Telescope Observatory (SMTO) has been formed to operate the facility for the two partner institutions. The facility is located at an altitude of 3200 meters on Mt. Graham in southeast Arizona. The 10 m diameter telescope is an alt-azimuth mounted f/13.8 Cassegrain homology telescope with two Nasmyth foci. The telescope has diffraction limited performance at a wavelength of 300 microns. During the last year, the telescope reflector surface has been adjusted to 20 microns rms and initial astronomical observations have begun.

## 2. Telescope and Enclosure

The telescope is a 10 m diameter on-axis Cassegrain design with a paraboloid primary and a hyperboloid secondary mirror. We have chosen a rather fast primary mirror (f/0.35) and a small 690 mm diameter secondary. This reduces the telescope axial dimension and minimizes the size (and cost) of the enclosure which is required to protect the telescope from weather damage.

An important feature of the telescope is the construction of the primary and the secondary reflectors out of aluminum core, carbon fiber reinforced plastic (CFRP) face sheet sandwich panels. In addition, the reflector backup structure and the secondary support are made of CFRP structural elements. This design route was chosen because of the low thermal coefficient of expansion and the high strength-to-weight ratio of CFRP. The



*W.B. Latter et al. (eds.), CO: Twenty-Five Years of Millimeter-Wave Spectroscopy*, 218–220.

result is a telescope which meets the required precision for both day and night operation.

The SMT enclosure is designed to protect the telescope from precipitation and high winds which could damage the reflector surface. We have designed a corotating enclosure which, additionally, provides (a) control rooms, (b) convenient access to the telescope at several levels for maintenance, and (c) large instrument rooms with easy access to the telescope foci.

Further details of the SMT design can be found in Martin and Baars (1990).

## 3. Construction Schedule

The recent telescope and enclosure construction is summarized below.

| | | | |
|---|---|---|---|
| 1990 | - | Summer: | site work commences, telescope fabricated |
| 1991 | - | Summer: | start concrete work on site |
| 1992 | - | October: | start assembly of telescope mount on site |
| 1993 | - | August: | enclosure and telescope assembly completed |
| 1993 | - | September: | formal dedication of telescope and enclosure |
| 1994 | - | February: | first light at 1.2 mm wavelength |
| 1994 | - | June: | first holography session to set reflector surface |
| 1994 | - | September: | second holography session |
| 1994 | - | December: | first light at 0.3 mm wavelength |
| 1995 | - | March: | third holography session; reflector rms=20 $\mu$m |
| 1995 | - | April: | preliminary visitor observing tests |
| 1996 | - | January: | regular visitor observing begins |

## 4. Reflector Surface Setting

Considerable time has been spent in the last year to set the reflector surface and to tune up the telescope drive servo system. The surface has been set by using the holographic method with the 38 GHz radio beacon of LES-8 (Mayer et al. 1994). The holographic receiver frontend was borrowed from NRAO Tucson. A special digital logging and data handling backend was developed and built for this experiment by the SMTO.

After several iterations of measurement and surface setting, the reflector is now set with an rms accuracy of 10 microns for ring 1, 15 microns for ring 2, and 25 microns for ring 3. The overall average is 20 microns if weighted by surface area. We believe that this is now the most accurate radio telescope of its size class. The current setting accuracy has been limited by a small instrumental crosstalk in the electronics. We have improved the electronics since the last holography session and expect to pursue further holographic measurements to improve the surface setting. We expect that the instrumentation will now allow measurement accuracy well beyond the surface setting goal of 15 micron or better rms.

A small number of high frequency observations have been made from April 1995 through April 1996. These measurements confirm the reflector surface quality. In addition, daytime measurements confirm the excellent telescope stability even with significant solar heating to the structure.

## References

Martin, R. N., & Baars, J. W. M. 1990, "The Submillimeter Telescope (SMT) Project," *Proc of SPIE*, **1235**, 503-514

Mayer, C. H., Emerson, D. T., & Davis, J. H. (1994) *Proc. IEEE*, **82**, 756-762

# THE LARGE MILLIMETER-WAVE TELESCOPE

F. PETER SCHLOERB
*Five College Radio Astronomy Observatory*
*University of Massachusetts*
*Amherst MA 01003*

## 1. Introduction

The Large Millimeter-wave Telescope (LMT) is a joint project of the United States and Mexico. The project goal is to construct and operate a 50m diameter millimeter-wave antenna. The antenna, shown schematically in Figure 1, will be built on a mountain site in Mexico and will have the highest gain and the highest overall sensitivity of any millimeter-wave antenna ever constructed. The total construction cost is estimated to be approximately 46 million dollars, with completion of the telescope in the year 2000. Initial instrumentation for the antenna will emphasize focal plane arrays for spectroscopy and continuum observations in the 1.3mm wavelength region. Key scientific projects to be carried out with the LMT include studies of: (1) molecular ISM in high redshift systems; (2) spiral structure in external galaxies; (3) chemistry and structure of molecular clouds and star forming cores; and (4) chemistry and physical properties of primitive objects in the solar system.

## 2. LMT Technical Goals and Approach

The technical goals of the LMT are ambitious. The 50m diameter of LMT provides a total collecting area of $\sim$ 2000 m$^2$ which must have sufficient precision to operate from 4mm wavelength to less than 1mm wavelength. Moreover, the large diameter implies a small beam which must be pointed to high precision. For the important 230 GHz frequency window, the requirement for $\lambda/20$ optics leads to an overall error budget of $\leq$70 microns rms. At this frequency, the half power beam width of the antenna is 6 arcsec, so the desirable rms pointing accuracy is about 0.6 arcsec, corresponding to one-tenth of the HPBW.



*W.B. Latter et al. (eds.), CO: Twenty-Five Years of Millimeter-Wave Spectroscopy,* 221–223.

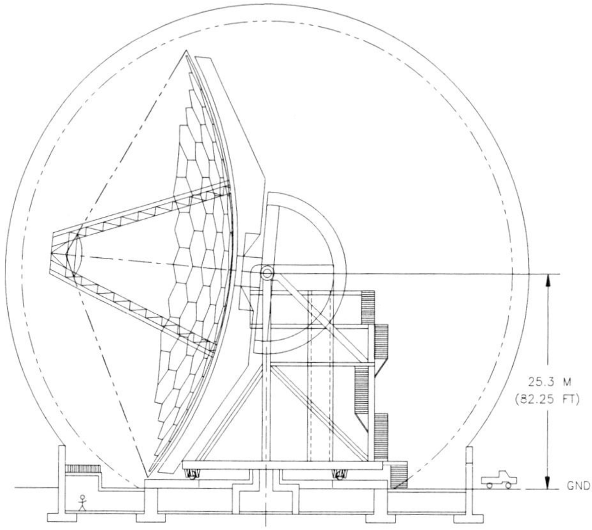


*Figure 1.* Schematic Drawing of the Large Millimeter Telescope.

The LMT will achieve its technical goals through the use of active control systems to maintain the precise optical alignments that are required. This approach has several advantages over a more traditional approach where the accuracy of the structure is obtained through the use of a stiff structure and precision manufacturing. However, for this program, the principal advantage is that it enables construction of a large precise structure at an affordable cost.

The active systems currently planned for the LMT include: (1) an active primary surface following the approach used for the Keck 10m optical telescope; (2) a laser metrology system to sense the alignment of the major optical elements; (3) an active secondary for fine pointing and tracking adjustments; and (4) a wavefront sensing system for correcting pointing shifts due to atmospheric propagation effects. In the following sections, we describe our basic approach for the active surface and the antenna pointing.

## 3. LMT Active Surface System

The LMT will be a "fully active" antenna with a surface that is designed to be controlled at relatively high bandwidth ($\sim$ 1-10 Hz). The basic approach follows the scheme developed for the successful Keck optical telescope and divides the primary surface into 126 hexagonal segments. Each segment consists of a set of precision surface panels mounted on a stiff "subframe". The segment dimension is approximately 5m.

Figure 2 presents a conceptual block diagram of the active surface sys-

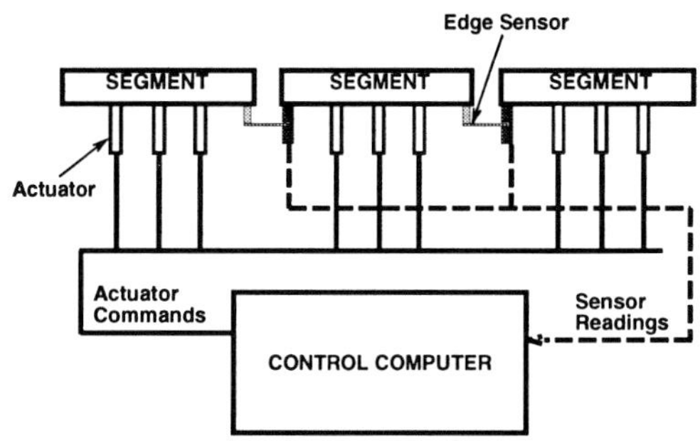


*Figure 2.* Conceptual Block Diagram of LMT Active Surface

tem. Each segment is attached to the backstructure by 3 actuators which permit 3 degrees of freedom per segment to be controlled. These degrees of freedom correspond to tilt of the segment and translation perpendicular to the antenna surface. There are a total of 378 actuators — or 378 controlled degrees of freedom — in the structure. The positions of the segments are sensed by measuring the relative positions of the *edges* of the segments. With two sensors at each segment boundary, there will be 684 total sensor measurements to provide the information required to adjust the 378 actuators. The current error budget for the antenna optics including the active surface system is 69 microns rms.

## 4. LMT Pointing Control System

Sub-arcsecond pointing of a structure the size of the LMT is clearly a major challenge. The antenna must be pointed blind, and preliminary structural studies have indicated that "non-repeatable" pointing errors of approximately 2-3 arcsec are likely. Thus, an approach to dealing with these sources of error is required.

The LMT approach to pointing involves: (1) the active surface system itself, which removes many pointing difficulties associated with surface deformation; and (2) enclosure of the telescope within a radome to protect it from the wind. In addition to these factors, a study by Kaman Aerospace has proposed a system to measure the alignment and spacing of the major optical elements in order to correct the telescope pointing. The budget for mechanical pointing errors with the Kaman system is 0.84 arcsec. Thus, although this approach comes close to the desired level of 0.6 arcsec rms, the problem will require continued study. Moreover, an important point is that the atmospheric "seeing" errors themselves are likely to be comparable contributors to the overall pointing errors, and ideas for actively sensing these errors during observations are now being evaluated.

# FUTURE CO OBSERVATIONS WITH THE JCMT

IAN ROBSON
*Joint Astronomy Centre*
*660 N.A'ohoku Place, University Park, Hilo, HI 96720*

## 1. Introduction

The JCMT is a multi-national 15m diameter telescope located on Mauna Kea and is equipped with a suite of four facility heterodyne and continuum receivers located at Cassegrain and Nasmyth foci for observations over the 2.0mm (150GHz) to 0.35mm (860GHz) atmospheric windows. The receivers are all on-line and remotely selectable by a tertiary mirror. The telescope pointing accuracy is better than 1.5$''$ in each axis and the surface accuracy is better than 25$\mu$m. The JCMT is operated in two shifts, each of eight hours per night and 24 hour operator coverage is provided for extended daytime observing. Facility astronomers are available for user support. A primary astronomical goal for the JCMT is to be able to cover all CO transitions (including isotopic species) observable from Mauna Kea. There is a healthy programme of facility developments to the tune of \$800k a year and this paper discusses those developments which will have an impact on CO observing with the JCMT over the next decade.

## 2. New Instruments

There are two new heterodyne instruments about to be delivered to the JCMT. In 1996 we expect to receive the new 345GHz facility instrument (RxB3) from HIA (Ottawa). This will provide dual polarization capability from 330-360 GHz and has an IF bandwidth in excess of 1 GHz. It will have a cooled single-sideband filter and a receiver temperature <100K at 345GHz using SRON/University of Groningen (RUG) Nb junctions. Laboratory tests have already achieved reproducible values of noise temperatures <80K. It will come equipped for remote operation, including tuning.

MRAO are producing a dual polarization 'wideband' receiver (RxW) to work in the 440-500GHz and 660-690GHz bands. The goals are to have



*W.B. Latter et al. (eds.), CO: Twenty-Five Years of Millimeter-Wave Spectroscopy*, 224–226.

receiver temperatures <200 K at 460GHz and 500K at 670GHz. Laboratory tests have already achieved noise temperatures of <60K at 460GHz using SRON/RUG junctions. The IF bandwidth will also be around 1.5GHz and like RxB3 this receiver will also be fully equipped for remote operation. It is provided with a closed cycle cooler.

## 3. The Upgrade Programme

As part of the facility developments we have introduced a programme to continually upgrade our receivers in order to remain competitive. In pursuit of this, R&D on SIS junction fabrication has been supported at SRON/RUG and this has resulted in the provision of excellent junctions for all the new facility receivers. R&D activity on wide-band and tunerless mixer development is also supported at MRAO and HIA.

We also intend to undertake a fast upgrade for our current 230GHz receiver by utilizing the existing 345GHz receiver (when it is replaced by RxB3) which will be fitted with a new 230GHz mixer using an NRAO designed block and a UVa SIS junction. This will provide single pixel, single polarization coverage from 210 to 280GHz with a target of 50K DSB across the entire frequency range and with a 1.5GHz bandwidth giving an instantaneous velocity coverage of around 1900 km/sec. It will be equipped for fully remote operation and the upgrade will start at HIA in early 1996.

Our current 460GHz receiver will be upgraded to an 800-900 GHz receiver (RxE) following its replacement by RxW. This will be another fast track upgrade and will use either SRON/RUG Nb junctions or PbBi junctions from the University of Kent. It will be a single pixel, single polarization receiver fitted with a cooled single-sideband filter and a 1.5GHz bandwidth IF. It will not be remotely operated or remotely tuned and is being upgraded by the builders at the Rutherford Appleton Laboratory in the UK.

## 4. The 345 GHz Array

The 345 GHz window is capable of providing an abundance of riches of scientific excellence, being available for most of the time on Mauna Kea and being well matched to the high surface accuracy of the JCMT. A particularly exciting development is the provision of a 345GHz focal plane array and associated backend.

There are a number of goals in the design of the array: it should never become background limited; it will be provided with a cooled image sideband rejection filter; the mixers will be 80K or less and in order to remain competitive against single channel devices it must have very low optical losses. The design calls for a 32 pixel array, operated in a dual polarization mode of two $4 \times 4$ array cameras, each with a field of view of $70 \times 70''$. The receiver will be mounted on one of the Nasmyth platforms adjacent to the new correlator and hence dramatically simplify the IF system. At present it is not determined whether two B-band cameras will be provided or, alternatively, a single camera at B-band and, by having the second camera at C or D- band, thereby introducing a dual frequency mapping capability. The cameras are mounted in individual cryostats and are therefore interchangeable as scientific needs dictate.

The correlator is a 32-bank spectrometer serving both cameras. It will have an IF centred at 6GHz and will operate in three modes. These are: extragalactic with 1750 km/s velocity coverage at 5.5 km/s resolution; star formation with 880 km/s at 1.4 km/s and high resolution with 220 km/s at 0.2 km/s. These figures are for a sixteen-bank spectrometer (the day 1 provision) being fed by a single camera. The next sixteen banks will provide higher spectral resolution by almost a factor of two.

The spectrometer will be designed around ultra-fast (4 Gsamples/s) ADCs with instantaneous 2GHz bandwidth sampling. There is a goal for cross-correlation capability. The chip design is being undertaken with Brian Von Herzen and is based on a development of the CSO MkI chip. The use of field programmable gate arrays forms a novel aspect of the chip development.

## 5. Conclusion

The JCMT looks towards an exciting future. As well as the heterodyne developments, there is the eagerly awaited and soon-to-be-installed facility array bolometer (SCUBA) which will provide spectacular sensitivity continuum point-source and mapping capability. The facility developments are also looking towards improving the surface accuracy, perhaps to $15\mu$m, to make high frequency observations even more efficient. The move towards sub-arcsecond imaging provided by links with the SMA promises to be a very exciting prospect, while flexible scheduling should enable the high frequency science to be delivered and the full potential of the facility to be maximised.

## 6. Acknowledgements

The James Clerk Maxwell Telescope is operated by the Particle Physics and Astronomy Research Council on behalf of the National Research Council of Canada (NRC), the Netherlands Organisation for Scientific Research (NWO) and the University of Hawaii.

# ODIN: A SWEDISH SUBMILLIMETER WAVE SPECTROSCOPY SATELLITE FOR ASTRONOMY AND AERONOMY

ÅKE HJALMARSON
*Onsala Space Observatory, S-439 92 Onsala, SWEDEN*

**Abstract.** The salient features of ODIN—a mm/submm wave satellite for astronomy and aeronomy—are presented. Differences with respect to SWAS are emphasized. The ODIN project is a joint venture between the space agencies and scientific communities in Canada, Finland, France, and Sweden ("majority shareholder"). The project cost is estimated to fall below 50 M$ and we aim at launch in October 1997.

## 1. Prologue

Since this presentation is scheduled for such a late hour, I have decided to begin by telling you a bedtime story, which is true and, hopefully, will have a happy ending. Once upon a time—I think it was five years ago—there gathered in the remote snowy kingdom of Sweden a rather wild tribe of astronomers and aeronomers, who had a strong common desire to conquer (not their neighbours but) space by means of a very sensitive submillimeter wave satellite. Their desire was to launch a machine containing an SIS mixer supported by a 4K cooler from the European Space Agency. This would require the space qualification of such a system and would be very useful for ESA's cornerstone FIRST. But confronted with "those evil ghosts and witches"—normally found in bedtime stories—here, in terms of the reality provided by the Swedish Space Company and National Space Board, the aforementioned tribe very soon realized that their desire would be much too expensive in terms of weight and power, and hence money. Subsequently, the Swedish tribe arrived at a more realistic desire (to be described below) and has joined forces with tribes of similar origin in Canada, Finland, and France. We now struggle together towards a happy ending (or rather beginning!!), i.e., a satellite launch in late 1997.



*W.B. Latter et al. (eds.), CO: Twenty-Five Years of Millimeter-Wave Spectroscopy, 227–229.*

## 2. Main Scientific Drivers

ODIN will work in unexplored bands of the electromagnetic spectrum around 0.5 and 2.5 mm, where ground-based observations are hindered by atmospheric absorbtion due to $H_2O$ and $O_2$. Observations of the very same molecules are the main astronomy drivers of the mission, aiming at an improved understanding of interstellar oxygen chemistry and of the cooling processes governing molecular cloud evolution towards star formation. The major aeronomy driver is ozone chemistry. A limb scanning mode will be used for studies of the distributions of $O_3$, ClO, CO, $H_2O$, $O_2$, and other molecules as well as aerosols in the terrestrial atmosphere. Here complementary observations will be performed at ultraviolet, optical and near infrared wavelengths.

## 3. Payload Overview

Also, in this more conservative approach, we still try to emphasize sensitivity. Active cooling is employed (although only to 80K) and the antenna size has been "maximized" (but still to fit a Pegasus XL launcher). To gain at least an order of magnitude in $O_2$ abundance sensitivity for extended (nearby) cold molecular clouds, compared to what is achievable with the submm line, we have managed to include a low noise 119 GHz HEMT amplifier channel. The main ODIN payload features are:

- A 1.1 meter offset Gregorian antenna (surface rms below 10 $\mu$m)
  - Beam size: $2'$ at 0.5 mm, $9'$ at 2.5 mm
  - Resolution: 2 km at 0.5 mm, in aeronomy mode
- Tunable Schottky mixers (tuning range $\pm$1.5%), centered at
  - 572 GHz ($O_3$, ClO, CO, $NH_3$, $H_2S$)
  - 555 & 549 GHz ($H_2O$, $H_2{}^{18}O$, $^{13}CO$, $C^{18}O$)
  - 495 GHz ($H_2O$, HDO, ClO, $O_3$, $O_2$, CI, CS)
- HEMT amplifier / Schottky mixer at 119 GHz (no tuning, just for $O_2$)
- Sterling cycle cooling to 80K, to reduce receiver noise
- Spectrometer capacity:
  - 2 high/low resolution hybrid autocorrelators (150kHz / 1 MHz)
  - Broad band AOS
- Simultaneous operation of $\geq$2 receivers using $\geq$2 spectrometers
- Grating spectrometers with 200 – 800 nm and 1270 nm bands (for aeronomy)

## 4. Differences between ODIN and SWAS

Since ODIN receivers will cover all lines to be observed by SWAS (NASA's Submillimeter Wave Astronomy Satellite), i.e., $H_2O$, $H_2{}^{18}O$, $O_2$, CI, and $^{13}CO$, we will here emphasize important differences:

- ODIN is a joint Astronomy / Aeronomy mission
- ODIN's submm sensitivity is $\geq$3 times that of SWAS (antennas, cooling)
- ODIN carries *tunable* submm mixers—hence allowing spectral scans
- ODIN carries a 119 GHz HEMT channel, which will lead to an order of magnitude higher $O_2$ abundance sensitivity (for cold, extended regions)
- ODIN's launch is scheduled *two years later* than that of SWAS, hence we hope for very useful input information. In fact, collaboration between the SWAS and ODIN astronomy teams already exists in terms of a joint southern hemisphere molecular cloud survey at SEST—the Swedish ESO Submillimetre Telescope.

## 5. Astronomy Research

Most observing time will be spent on studies of interstellar $H_2O$ (557 GHz), $H_2{}^{18}O$ (548 GHz), and $O_2$ (119 & 488 GHz)—missing links in our understanding of oxygen chemistry and cooling of molecular clouds. We foresee CO (5–4) and $^{13}CO$ (5–4) surveys to be valuable complements, especially in accounting for the PDR component of the molecular clouds. Density probing using the $NH_3$ ground state rotational transition (572 GHz) may be useful for the interpretation of the $H_2O$ data.

We also plan to perform spectral scans—of bands not accessible from ground—towards a few selected molecular cloud cores and circumstellar envelopes (this is almost an "obligation", since we have tunable receivers). Some studies of circumstellar envelopes around late type stars, of atmospheric chemistry of planets, and of $H_2O$ outgassing from comets are also being considered. Observations of $H_2O$ and $O_2$ in some relatively nearby galaxies will be attempted. Here comparison of the interstellar medium in the Large Magellanic Cloud with the conditions in the Milky Way may be especially interesting.

*Acknowledgements* The substantial funding from the Swedish National Space Board and from the Swedish Natural Science Research Council—as well as the moral support provided by the ODIN community in terms of my "appointment" to act as Astronomy Mission Scientist—is very much appreciated.

# EUROPEAN PLANS FOR A MILLIMETRE ARRAY

R. S. BOOTH

*Onsala Space Observatory*
*Chalmers University of Technology*
*S-439 92 ONSALA, SWEDEN*

**Abstract.** Since their very first working discussions on a next-generation millimetre telescope, European astronomers have based their ideas on a large array with a collecting area of about 10,000 square metres and a resolution of 0.1 arcsec at a wavelength of 3 mm. They have argued that the array should be built on a high, dry site in Northern Chile, strategically placed to observe the rich harvest of millimetre sources in the central parts of the Galaxy and, at the same time, benefitting from the infrastructure and astronomical base established by ESO. The collecting area of the proposed array is comparable to that of the VLA, and around an order of magnitude greater than is available today. Scientifically, such an instrument will provide a quantum leap in present day millimetre astronomy, and it will open up new directions in astrophysics. It will enable astronomers to exploit the full richness of the millimetre spectral band through the study of all types of millimetre radiation—not only the panoply of molecular lines and dust continuum from the cool gas, but the line and continuum from ionised gas and synchrotron emission from relativistic particles. The starting concept is an array with about $50 \times 15$ m dishes (or $100 \times 10$ m dishes) with 50 microns global precision, to be sited above 2000 m altitude on a site allowing baselines up to 10 km. In order to refine this concept, to review possible sites and to investigate the technology required, the Institute de Radio Astronomie Millimtrique (IRAM), the European Southern Observatory (ESO), Onsala Space Observatory (OSO) and the Netherlands Foundation for Research in Astronomy (NFRA) have recently agreed to pool their resources in a joint European study which has just begun.



*W.B. Latter et al. (eds.), CO: Twenty-Five Years of Millimeter-Wave Spectroscopy,* 231–238.

## 1. Introduction

By way of introduction, it is interesting to take a look at the current European instrumentation for observations of CO.

The Swedish 20 m diameter, radome-enclosed telescope at Onsala was completed in the mid 70s and has recently been resurfaced and generally upgraded to give an aperture efficiency of nearly 50% at 115 GHz. In collaboration with ESO, Sweden also operates the 15 m diameter Swedish-ESO Submillimetre Telescope, SEST, on the ESO site of La Silla in the Chilean Andes. However, the major millimetre facilities in Europe were built and are operated by the French-German-Spanish institute, IRAM. They consist of a 30 m diameter antenna located on Pico Veleta in Spain and a four, soon to be 5 × 15 m element interferometer on Plateau de Bure in the French Alps. (The design of SEST is based on that of the IRAM interferometer telescopes). Finally, we have heard already in this meeting about the 15m James Clerk Maxwell submillimetre telescope on Mauna Kea in Hawaii, operated by the UK, in collaboration with the Netherlands and Canada and the Heinrich Hertz 10 m submillimetre antenna, a collaboration between the Max-Planck Foundation and the Steward Observatory, here in Arizona.

The abundance of results from these and other millimetre telescopes, many of which are presented in the pages of these proceedings, amply justifies the construction of a much more powerful, next generation instrument. The detection of CO and continuum emission from dust in the distant galaxy IRAS 10214+4724 ($z$=2.3) and in the Cloverleaf quasar, H1413+117 ($z$=2.5), show the potential of mm line astronomy as an investigative tool in cosmology and make large collecting area and high resolution imperatives for a new instrument in order that many more high red shift objects may be studied in detail, with sufficient spatial resolution.

The recent mapping of the star IRC+10216 in rare molecules like MgNC and NaCl illustrates the power of mm interferometry in delineating the chemistry of circumstellar envelopes. New studies of the star GG Tauri in $^{13}$CO show the ability of mm interferometry to derive the kinematics of disks around pre-Main Sequence stars. Mosaic maps of galaxies like M51 show the potential of mm interferometry to map giant molecular cloud complexes in nearby galaxies.

In fact, nearly all objects studied in mm astronomy contain weak fine structure, unresolved with current mm telescopes. To resolve this structure, we need to use longer baselines, and because of this we need more sensitivity. This can only be obtained with an interferometric array with a large collecting area.

Arguments for a large millimetre array may be found in papers by Booth (1994), Ishiguro et al (1994), Downes (1994), Booth and Downes (1995) and, in a earlier context, in the MMA Proposal of NRAO (1990). On behalf of the recently formed European array study consortium, Downes, with input from other European scientists, has compiled a strong scientific case for the array discussed here. This article draws freely on that document entitled *A Next-Generation Millimetre Interferometer for Radio Astronomy.*

## 2. The Case for a Next-Generation Millimetre Array

### 2.1. ORIGINS

The millimetre spectral band is especially well matched to the study of the origins of the fundamental building bricks of the Universe - the galaxies, the stars and the planets. This is because it is the domain of radiation processes from cold matter. However, while we are tantalizingly close to important revelations in these exciting areas, we need at least an order of magnitude increase in sensitivity over today's instruments to make a serious impact.

As an example, take the case of the high z galaxies. For the first time radio astronomers can study thermal dust emission and line emission from cool neutral, molecular and atomic matter at epochs close to the formation of galaxies - their first star bursts and their first production of heavy elements. Because all these millimetre signals arise from non-exotic, low temperature emission processes, they should provide reliable estimates of densities, temperatures and masses of this material. However, in this meeting we have learned that the high-z galaxies may only be observed because their emission is amplified by gravitational lens focusing. A millimetre array with a $\times 10$ increase in sensitivity will enable us to study the main population of such galaxies without the help of foreground lensing objects.

### 2.2. COLLECTING AREA AND RESOLUTION

The largest European instruments today are the IRAM 30 m antenna, which has a collecting area of 700 m$^2$, and the IRAM interferometer, which will ultimately have $6 \times 15$ m antennas and a collecting area of about 1000 m$^2$. Based on the lessons of cm/dm radio astronomy, we know that major scientific improvements will come from an order of magnitude increase in sensitivity (the $70-100$ m telescopes v. the earlier 25 m dishes, or the VLA v. the Westerbork interferometer) and since today's millimetre receivers are within a small factor of the quantum limit, this increase will only be achieved by a significant increase in collecting area. Hence our goal of an instrument with a collecting area of $7000-10000$ m$^2$. Such a collecting area may be achieved with $40 - 60 \times 15$ m antennas.

Again, based on today's observations, we know that a resolution of about 1 arcsec is barely sufficient to resolve the high-z galaxies or the nearest protostellar discs. Thus, in resolution, an order of magnitude improvement is also desirable, which leads us to the conclusion that the proposed instrument should have a resolution of 0.1 arcsec at a wavelength of 2.6 mm, or maximum baselines of 10 km.

Observations of quasars will undoubtedly require VLBI baselines and the proposed array, phased as a single dish, will be a powerful VLBI element, improving the sensitivity of today's millimetre VLBI baselines to Chile by more than a factor of 7.

### 2.3. SCIENTIFIC MOTIVATION

Scientifically, a next generation instrument should provide a quantum jump in present-day millimetre astronomy, opening up new directions in astrophysics. This meeting has given us no small insight into the goals for the future but, for completion, the scientific drivers for the array are outlined below.

*Cosmological studies* of a large sample of distant galaxies emitting continuum radiation from dust and red-shifted ($z \geq 2$) lines of CO and other species. Even if the high z galaxies detected today are lensed, as suggested by Radford in these proceedings, the increased sensitivity of the array should be sufficient for a detailed study to be performed on a significant population, giving new insights into the cosmological evolution of galaxies and the production of the first heavy elements.

*Observations of intermediate red-shift galaxies and quasars* in emission and absorption lines, as well as continuum radiation (dust and synchrotron). As pointed out by Wiklind and Combes (1994), it is easier to detect small quantities of gas in absorption than emission making absorption line measurements against background continuum sources a powerful means of studying the chemistry of these intermediate red shift objects. In some cases it might be possible to detect absorption against the central continuum component of a galaxy by gas in one of its spiral arms or possibly even in the dusty torus surrounding its central black hole.

*Sensitive, detailed mapping of nearer galaxies,* their discs and especially their centres, including on the one hand the study of star bursts in interacting systems and on the other, the study of low metallicity systems like the Magellanic clouds.

*Protostars* - the resolution and dynamical studies of protostellar accretion discs and the origins of their bipolar outflows.

*High resolution studies of proto-planetary discs* around young stellar objects, in both line and continuum emission.

*Stars* - sensitive studies of the emission from all types of stars but especially the chemistry and dynamics of the cool envelopes surrounding evolved stars, giving a greater insight into the return of material by these objects into the interstellar medium.

*Solar system* - studies of line and continuum emission from all types of solar system objects from the giant planets and their satellites to comets and asteroids.

## 3. Why the Southern Hemisphere?

The case for the southern hemisphere rests on both scientific reasoning and site considerations. A large fraction of interesting millimetre sources are in the southern sky, and there are a number of good, flat, millimetre-quality sites at high altitude. In addition, through ESO we have a long history of optical data on southern objects which will be enhanced by the construction of the Very Large Telescope, VLT.

### 3.1. SCIENCE

There are compelling scientific reasons for building an array in the southern hemisphere. Although high red shift galaxies may be studied equally well from northern or southern sites, most other objects of interest for mm astronomy are more numerous in the south. These include the Galactic centre itself and the entire central part of the Milky Way, which contains most of the Galactic star forming regions, proto-planetary discs, circumstellar envelopes, planetary nebulae, and compact HII regions. Ironically, many of these objects are already well-known mm sources and their observation with the currently available, northern hemisphere mm interferometers is hampered by atmospheric problems and poor baseline coverage, due to the low elevation of these sources in the northern sky.

At far southerly declinations we also have the Magellanic Clouds, of special interest because of their low metallicity and clues to chemical evolution, and because of their proximity to our Galaxy. The southern sky also contains some spectacular galaxies that are prototypes for their classes. Finally, one of the nearest active galactic nuclei, Centaurus A, at Dec. −42 deg., is best observed from the southern hemisphere.

As mentioned above, there is a wealth of supplementary astronomical information to complement mm studies, as there have been major optical observatories in the southern hemisphere for a long time. Through the optical telescopes in Chile, Australia, and South Africa, and the radio telescopes in Australia, including the recent Australia Telescope, and with the Swedish-ESO Submillimetre Telescope (SEST), there are already very significant astronomical resources in the southern hemisphere to guide and to complement studies with a large millimetre interferometer.

### 3.2. HIGH QUALITY SITES

Some of the world's best millimetre sites lie in the Atacama Desert on the coastal side of the Andes in Northern Chile where there are many large, flat tracts of land at altitudes even as high as 5000 m. Several apparently excellent accessible sites have been identified recently by teams from Japan, Sweden and the USA, with the help of astronomers from the University of Chile, and 22 potential locations were found by Raffin & Kusunoki (1992) in a search for sub-mm interferometer sites. These sites are extremely dry, with precipitable water vapour around 1 to 1.5 mm for large fractions of the time.

Extensive water vapour measurements have been made only on Cerro Paranal (altitude 2650 m), the site for the ESO VLT. These data show that the precipitable water vapour content is below 1.5 mm for at least 30% of the April-November period, and below 3 mm for 60% of this period (Sarazin 1990; Martin 1991). Paranal is only 15 km from the ocean; more recent site testing by US and Japanese groups, again with help from the Chilean astronomers and SEST staff, is beginning to show just how excellent are the higher inland sites. The following papers by Brown and Ishiguro describing site testing at around 5000 m and 4000 m, respectively, will amply demonstrate this point.

The expanse and flatness and low water vapour content of these high altitude sites all suggest that Northern Chile is an ideal location for a large millimetre array.

## 4. The European Study

Four European institutes, representing much of European millimetre wave astronomy, have recently joined forces to study the proposed array. They are the Institute de Radio Astronomie Millimtrique, IRAM; the European Organisation for Astronomical Research in the Southern Hemisphere, ESO; the Onsala Space Observatory; and the Netherlands Foundation for Research in Astronomy, NFRA.

The basic concept is a large millimetre array with a collecting area of 10000 $m^2$, for operation over the frequency band 40 GHz to 350 GHz. It will be located on a high site, above 2500 m altitude, probably in Northern Chile, and will have baselines up to 10 km, giving a clean synthesized beam of 0.1 arcsec at a wavelength of 3 mm. The array will consist of about 50 $\times$ 15 m antennas, or alternatively, depending on site conditions (wind, etc.), 100 $\times$ 10 m antennas.

The detailed study for the array will address several important design aspects. In general, to minimize the need for interventions on a spread-out, remote desert site, a major design goal should be reliability of all components. To keep the cost to a minimum, components should be as simple as possible and capable of easy replication. In addition to site testing, particular sub-studies are foreseen as outlined below.

### 4.1. ANTENNAS

The antennas should be simple and robust, and well-designed to maintain their mm-quality surface (total precision 50 microns r.m.s.) in reasonably windy conditions (up to 15 $ms^{-1}$ as a goal). Since we need to have antennas which can survive in the outdoors, without astrodomes, particular attention must be paid to their pointing accuracy and thermally stable structures must be designed. For a sufficiently large array, it might be possible to have most of the antennas on fixed stations, avoiding the need for rails and transporters.

### 4.2. RECEIVERS

For ease of maintenance, receivers should be modular and of robust construction. They should ideally be tunerless systems with broad instantaneous bandwidth, i.e., they should be designed to avoid complex tuning procedures. Closed-cycle cryogenic systems must be designed to be maintenance-free for long periods. The need for 4 K systems should be avoided by using HEMT technology where feasible.

### 4.3. INTERMEDIATE FREQUENCY SYSTEM

Broad band IF systems are required for sensitivity reasons and early digitization of the IF signal should be considered. Fibre optic transport of the signals (IF and local oscillator) should be studied.

### 4.4. CORRELATOR

The correlator is a major item in the study. The new fast correlator chips should be evaluated and manufacturers identified; a comparison with more units using slower chips should be carried out. The question of separate line and continuum correlators should be addressed. The need for a special chip development and development of fast samplers must also be considered.

### 4.5. ATMOSPHERIC PHASE COMPENSATION

A crucial aspect of the study will be the development of a method to correct the interferometer phases for atmospheric fluctuations. Several schemes have been suggested, but the most promising results have been recorded at IRAM, as reported by Welch at this meeting. The method used is that originally put forward by Wright and Welch (1990) and is to monitor the variation in sky brightness temperature in the direction of observation, preferably with a total power channel in parallel with the receiver used for observations. The brightness temperature variation is proportional to the water vapour fluctuation in that antenna beam and can be used to correct the phase. The method requires high receiver stability (1 part in $10^5$), which is not usually achieved in today's millimetre wave receivers. This, and other phase compensation techniques will be studied using the IRAM interferometer.

## 5. International Collaboration

A large array will be expensive but not excessively so, compared with current optical telescope projects or small scientific satellites. There may even be significant savings through mass production of receivers and antennas. However, in this meeting we will hear about three array projects, all of which favour a Chilean site, and whose combined cost is more than one billion dollars. This is a high cost to astronomy and it must make sense to try to collaborate internationally to fund and build an array which is acceptable to all parties. The Millimetre Array Working Group set up by URSI and now co-sponsored by IAU is an appropriate forum to discuss the compromises and advantages of a collaborative project.

## References

Booth, R. S., 1994, in Astronomy with Millimetre and Submillimetre Wave Interferometry, ed. Ishiguro & Welch. ASP Conf. Series, p.413

Booth, R. S. & Downes, D. 1995 in The Physics and Chemistry of Molecular Clouds, Heidelberg: Springer, in press

Downes, D., 1994, in Frontiers of Space and Ground-Based Astronomy, ed. Wamsteker, Longair & Kondo, Kluwer, Dordrecht, p.133

Ishiguro, M., et al. 1994, in Astronomy with Millimetre and Submillimetre Wave Interferometry, ed. M. Ishiguro & W. J. Welch. ASP Conf. Series, p.405

Martin, R. N. 1990. SMT Technical Memo. UA-90-2 (U. Arizona, Tucson) NRAO, 1990, Proposal for a Millimetre Array. (NRAO, Charlottesville, Va.)

Raffin, P., Kusunoki, A. 1992, SAO Submillimetre Array Tech. Memo, No 59. (Smithsonian Astrophysical Observatory)

Sarazin, M. 1990. ESO VLT Site selection Working Group, Final Rep. No. 62.

Wiklind, T. and Combes, F., 1994 Astron. Astrophys., 286, L9.

Wright, M. C. H. and Welch, W. J. 1990, in: Radio Astronomical Seeing, eds. J. E. Baldwin & Wang Shouguan, Pergamon, Oxford, p.71.

# LMSA: JAPANESE PLANS FOR A LARGE MILLIMETER AND SUBMILLIMETER ARRAY

MASATO ISHIGURO
*Nobeyama Radio Observatory(NRO)*
*Nobeyama, Minamimaki, Minamisaku, Nagano 384-13, Japan*

## 1. Introduction

The LMSA was originally planned as a Large Millimeter Array (LMA) where an array of thirty 10 m telescopes would be realized by adding 25 more telescopes to the Nobeyama Millimeter Array (NMA). However, because the recent discoveries of rotating protoplanetary disks around young stars and CO gas from very distant galaxies, sub-arcsec imaging with the highest achievable sensitivity at millimeter and submillimeter wavelengths is now considered very important as a capability of the proposed array (Ishiguro et al. 1994).

Sub-arcsec imaging is essential, for example, in order to resolve the protoplanetary disks in nearby star forming regions such as the Taurus region where $0.1''$ at a distance of 140 pc corresponds to 14 AU, a scale comparable to our solar system. To achieve an angular resolution of $0.1''$ at 1 mm, an array of 2 km maximum baseline located at a very high and dry site is required. Since submillimeter observations at the Nobeyama site are very difficult, we decided to find a site for the LMSA outside Japan. Mauna Kea and northern Chile are being considered as potential sites for the LMSA.

In the current design concept, the LMSA will consist of 50 telescopes of 8–10 m diameter, the total collecting area of which is equivalent to a single dish of 70 m diameter. The LMSA will cover observing frequencies from 80 GHz through the 800 GHz band. Since 1994, site testing has been conducted extensively in northern Chile using a 220 GHz tipping radiometer and an 11 GHz site-testing interferometer.

Scientific directions, the planned system, and site testing activities in Northern Chile are reported in this paper.



*W.B. Latter et al. (eds.), CO: Twenty-Five Years of Millimeter-Wave Spectroscopy,* 239–246.

## 2. Scientific Program for the LMSA

The LMSA will be used to study all types of radio emission at millimeter and submillimeter wavelenghts, especially molecular line emission from low temperature gas and dust continuum emission. Observations at submillimeter wavelengths provide higher sensitivity to thermal dust emission than at millimeter wavelengths (by a factor of $10^2$–$10^3$). Important astronomical problems, such as the formation and evolution of stars and of protoplanetary systems, could be studied by making use of this advantage. Recently, rotating protoplanetary gas disks around young stars were revealed with the NMA (Kawabe et al. 1992; Saito et al. 1995). However, the resolution (2-3$''$) was insufficient to resolve the structure. Sub-arcsecond imaging is essential in order to resolve protoplanetary disks in nearby star-forming regions such as the Taurus region. With multi-line imaging of protoplanetary disks in high-excitation molecular lines, it will be possible to measure the radial variations of the physical conditions. Continuum observations at multi-frequency bands will provide information on the growth of dust toward planetesimals. Although detection of individual planets will be very difficult, high resolution imaging of face-on protoplanetary disks might show radial gaps due to mass contraction into planets. Polarization observations with the LMSA at 300 GHz–500 GHz will show the magnetic field structure in protoplanetary disks, providing very important information on the role of magnetic fields in the evolution of disks.

The formation and evolution of galaxies and QSOs, in terms of star formation and AGN activities, are also very important topics for the LMSA. The power of spectral lines is approximately proportional to $\nu^3$. Strong dust emission at high $z$ is redshifted to millimeter and submillimeter regions, the strength of which is proportional to $\nu^{2-4}$. Accordingly, observations at higher frequencies are much more productive in detecting this emission. Recent millimeter-wave observations of CO from IRAS F10214+4724 using single dishes and interferometers indicate that the huge amounts of molecular gas are associated with this high-$z$ galaxy, which has an extremely high infrared luminosity (Brown and Vanden Bout 1992; Kawabe et al. 1992; Tsuboi and Nakai 1995). Searching for high-$z$ "proto-quasars" might also be possible by observations of highly redshifted emission of a strong [C II] 158 $\mu$m line (Loeb 1993), which is an important coolant in photon-dominated regions in galaxies. The [C I] 610 $\mu$m line traces photo-dissociation regions in the interstellar medium and can be observed together with high level transitions of the CO molecule (CO[J = 7–6] or CO[J = 4–3]) at submillimeter wavelengths. In these kinds of observations of distant objects, high sensitivity and sub-arcsecond resolution will play very important roles.

Although it is very difficult to summarize all the scientific targets in this paper, the LMSA will open new directions in studying interstellar chemistry, molecular clouds in our galaxy and in nearby galaxies, disks, inflow/outflow around young stellar objects, mass loss from evolved stars, circumstellar shells, etc.

## 3. Description of the Planned System

The scientific requirements described previously define the following capabilities of the LMSA:

1. Wide frequency coverage at millimeter and submillimeter wavelengths (80–800 GHz);
2. Sub-arcsecond imaging;
3. Very high sensitivity at millimeter and submillimeter wavelengths;
4. Quick imaging with high dynamic range.

To achieve an angular resolution of $0.1''$ at $\lambda$ 1 mm, the maximum baseline must be longer than 2 km. In a preliminary design of the array, fifty 10 m antennas are connected to achieve a total collecting area of 4000 $m^2$, equivalent to that of a 70 m diameter single dish. With this collecting area and $T_{SYS} = 300$ K, 100 $\mu$Jy ($5\sigma$) sensitivity will be achieved at 350 GHz. An example of the proposed array configurations is shown in Figure 1, where four different configurations are overlayed. The array sizes under consideration are 100 m, 300 m, 700 m, and 2 km. All antennas should be transportable to realize each of these configurations. The 100m configuration will be used for high-sensitivity surveying of distant galaxies and wide field mosaicking observations. The array configuration has a good instantaneous $(u, v)$ coverage. This will allow us to collect data sets in periods of good radio seeing, so that we can obtain high quality and high resolution images using an ensemble of good snap-shots.

The antennas must be designed for efficient operation at frequencies as high as 850 GHz, requiring a surface accuracy $\sim$15 $\mu$m rms and a pointing accuracy of $\sim 1.0''$. A preliminary design of the antenna structure is shown in Figure 2. The mechanical structure is designed based on conventional Nasmyth optics in order to have enough space for installing a large receiver system. The quadrupod is supported directly from the central hub in order to minimize the gravitational deformation and the radio blockage. The main reflector of the new NMA 10 m antenna has similar design whose mechanical and thermal performances are under test at Nobeyama. Another important aspect of antenna design is mass production and construction at a very high site. The antenna structures should be assembled at lower altitude and brought to the site in sections.

Receivers in the atmospheric windows between 80(40) GHz and 850 GHz are planned for each antenna (see Table 1). HEMT amplifiers will be

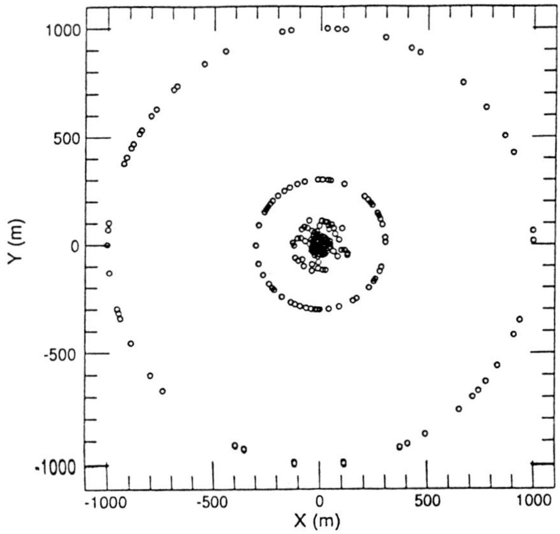


*Figure 1.* An example of the proposed array configurations. Four different configurations for 100 m, 300 m, 700 m, and 2 km maximum baselines are overlayed. In actual design, the north-south size will be elongated to improve the $(u,v)$ coverage over a wide range of declinations.

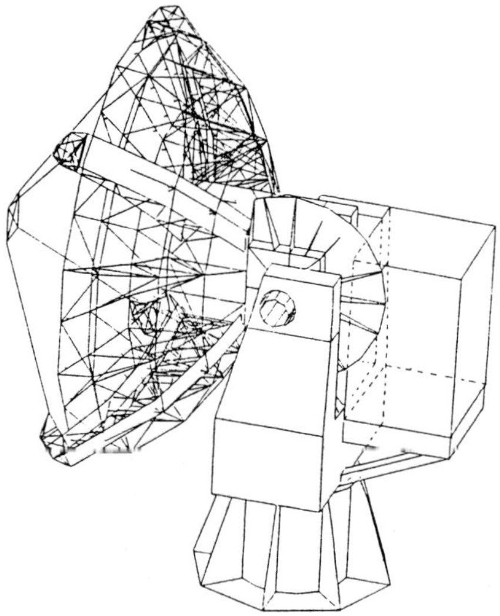

*Figure 2.* A preliminary design of the mechanical structure of the LMSA antenna based on conventional Nasmyth optics.

TABLE 1. Summary of Instrumental Parameters

| | | |
|---|---|---|
| *Array* | | |
| | Number of Antennas | 50 |
| | Total Collecting Area | $\simeq$4000 $m^2$ |
| | Maximum Baseline | 2 km |
| | Max. Angular Resolution | $0.1''$ ($\lambda = 1$ mm) |
| *Antennas* | | |
| | Diameter | 8 m - 10 m |
| | Surface Accuracy | $\sim$ 15 $\mu$m rms |
| | Pointing Accuracy | $\sim 1.0''$ rms |
| *Frequency bands* | | |
| | | (35 - 50 GHz) |
| | | 80 - 140 GHz |
| | | 130 - 200 GHz |
| | | 200 - 310 GHz |
| | | 330 - 420 GHz |
| | | 390 - 500 GHz |
| | | 630 - 710 GHz |
| | | 800 - 900 GHz |
| *Correlator* | FX-type | 1225 cross-correlations |
| | | 50 auto-correlations |
| | Bandwidth | > 4 GHz |
| | Frequency Channel | > 10000 ch / baseline |
| | Max. Freq. Resolution | 40 kHz |
| | | (0.1 km/s at 115 GHz) |

used for 40 GHz and 100 GHz and SIS mixers will be used for the higher frequency bands. In each band a wide tuning range is desirable so that one can observe anywhere within the entire frequency band. A new type of SIS junction called "PCTJ" (Parallel Connected Twin Junction) has been developed at Nobeyama. Experiment with a PCTJ SIS mixer at 100 GHz showed that a very low noise performance was obtained over a wide frequency range without any mechanical tuning elements (Noguchi et al. 1995). This type of SIS mixer is being tested at 500 GHz at Nobeyama. These receivers will have an instantaneous bandwidth of >4 GHz. Dual frequency or dual polarization modes will be included for limited frequency bands.

Phase stabilized optical fibers will be used for transmission of reference signals for phase-locking and IF return signal (Ishiguro, 1994). A wide-band (>4 GHz) FX-type correlator system will be used which can offer flexible operations. A 4 GHz A/D converter is undergoing a test which was developed for the Ultrawide Band Correlator (1 GHz bandwidth) in NMA (Morita, 1994). For spectral-line observations, it can function as several separate spectrometers, all of which may be tuned to different central frequencies from the same or different receivers. The auto-correlation mode is important when using the array as 50 independent single dishes. The instrumental parameters of the LMSA are summarized in Table 1.

One of the key technologies needed to achieve 0.1″ resolution is active phase correction. New techniques of phase correction are proposed and under test at various millimeter interferometers. We are currently investigating differential radiometry and paired antenna methods. In the former method, the difference between atmospheric emission measured by two independent radiometers located at each telescope is used to correct atmospheric phase errors (Welch, 1994). In the latter method, continuous phase correction is made possible by using two closely located antennas as an element of the interferometer, one of which is pointed to a calibrator source continuously.

## 4. The Potential LMSA Sites and Site Testing Activities

The criteria for the potential LMSA site are: 1) atmospheric transparency at millimeter and submillimeter wavelenghts, 2) radio seeing to achieve sub-arcsec resolution, 3) low wind velocity for good pointing accuracy, 4) a flat and wide ($\sim$ 3 km $\times$ 3 km) area to locate the array, 5) accessibility to the site, 6) infrastructure, etc. Under these criteria, the potential LMSA site should be a very dry site at very high altitude.

Mauna Kea (Hawaii) and the Atacama desert in northern Chile are considered as the potential sites for the LMSA in the northern and southern hemisphere, respectively. As site testing has already been conducted on Mauna Kea by other groups (Masson, 1994), we decided to focus our site testing activities in northern Chile. Northern Chile has one of the driest deserts on Earth, with a mean annual precipitation is less than 10 mm. There are very flat areas at high altitude suitable for locating the LMSA.

In 1992 and 1993, we visited about twenty of the possible sites in Northern Chile. These sites are distributed $-22.5°$ to $-25.5°$ in latitude and 2500 m to 5000 m in altitude. Four meteorological stations were set up on selected sites in these regions (see Figure 3). In 1995, we installed a 220 GHz

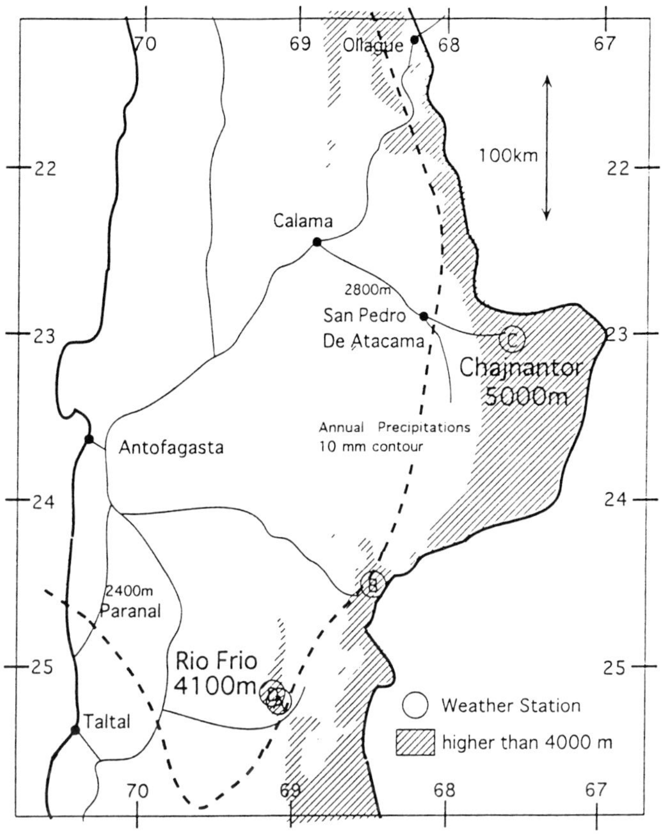


*Figure 3.* Map of northern Chile. The hatched area shows regions higher than 4000 m.

radiometer and an 11 GHz radio seeing monitor at Rio Frio (4100 m) (one of the potential LMSA sites) and started measurements. NRAO installed similar equipment at Chajnantor (5000 m) for MMA site testing. The site testing activities in northern Chile have been conducted in collaboration with SEST/ESO, the University of Chile, and NRAO.

So far, site testing has been based on opacity measurements where atmospheric phase stability has been evaluated indirectly. Recently, it has proven very useful to assess site quality with direct measurements of the atmospheric phase fluctuations by using a small interferometer to receive a signal from a geostationary satellite (Ishiguro et al. 1990). The advantage is that it is possible to monitor the phase fluctuations continuously at a constant elevation angle. This kind of interferometer is called a Radio Seeing Monitor(RSM). The RSM at Rio Frio consists of a pair of 1.8 m antennas, separated by about 300 m, which receive a beacon signal at 11.2 GHz from an INTELSAT satellite. From a zero-baseline test, the RSM has proved to have an accuracy of 0.3° rms, corresponding to 20 $\mu$m rms in pathlength. Thus, although the measuring frequency is relatively low, we can measure the radio seeing at millimeter and submillimeter wavelenghts.

We will choose the LMSA site after collecting data for at least one year and comparing them with those for other sites including Mauna Kea.

## References


Brown, R.L., and Vanden Bout, P.A. (1992) *Ap. J. (Letters)*, **397**, L19

Ishiguro, M., Kanzawa, T.,and Kasuga, T. (1990) *URSI/IAU Symp. on Radio Astronomical Seeing*, eds. J. E. Baldwin and Wang Shouguan, IAP, Beijing, 60

Ishiguro, M., Kawabe, R., Nakai, N., Morita, K.-I., Okumura, S. K., and Ohashi, N. (1994) The Large Millimeter Array, *Astronomy with Millimeter and Submillimeter Wave Interferometry*, eds. M. Ishiguro and Wm.J. Welch, ASP Conference Series, 405

Ishiguro, M. (1994) *IAU Symp. 158, Very High Angular Resolution Imaging*, eds. J. G. Robertson and W. J. Tango, Kluwer Academic Pub., The Netherlands, 235

Kawabe, R., Ishiguro, M., Omodaka, T. Kitamura, Y., and Miyama, S.M. (1992) *Ap. J. (Letters)*, **404**, L63

Kawabe, R., Sakamoto, K., Ishizuki, S., and Ishiguro, M. (1992) *Ap. J. (Letters)*, **397**, L23

Loeb, A. (1993) *Ap. J. (Letters)*, **404**, L37

Masson, C. R. (1994) *Astronomy with Millimeter and Submillimeter Wave Interferometry*, eds. M. Ishiguro and Wm.J. Welch, ASP Conference Series, 87

Morita, K.-I. (1994) *Astronomy with Millimeter and Submillimeter Wave Interferometry*, eds. M. Ishiguro and Wm.J. Welch, ASP Conference Series, 18

Noguchi, T., Shi, S. -C., and Inatani, J. (1995) *IEEE Trans. Appl. Superconductivity*, **vol. 5**, 2228

Saito, M., Kawabe, R., Ishiguro, Miyama, S.M., Hayashi, M., Handa, T., Kitamura, Y.M., and Omodaka, T. (1995) *Ap. J.*, **453**, 384

Tsuboi, M. and Nakai, N. (1994), *Publ. Astron. Soc. Japan*, **46**, L179

Welch, Wm. J. (1994) *Astronomy with Millimeter and Submillimeter Wave Interferometry*, eds. M. Ishiguro and Wm.J. Welch, ASP Conference Series, 1

# NEW INSTRUMENTS, NEW SCIENCE: FUTURE OPPORTUNITIES

ROBERT L. BROWN
*National Radio Astronomy Observatory, Charlottesville, VA*

**Abstract.** The new millimeter and submillimeter wavelength telescopes will provide a detailed view of galaxy formation, the formation of stars and planetary systems, and the chemistry of the envelopes of evolved stars. The keys to the future are larger and more precise telescopes, quantum limited electronics, and superb observatory sites.

## 1. Overview

Millimeter and submillimeter astronomy has developed rapidly over the past 25 years in large measure because the observing technology developed so rapidly over the same time frame. In the past two decades the noise temperature of the receivers used at the CO($J$=1→0) line has decreased from roughly 1000 times the photon noise limit, $h\nu/k$, to a value only a few times this limit, an improvement of more than two orders of magnitude. The sizes of our telescopes have increased from 11 m to 30 m, an increase in angular resolution of a factor of 3 and an increase in collecting area of a factor of 7. Finally, the application of aperture synthesis interferometry at millimeter and submillimeter wavelengths with interferometer baselines of several hundred meters has increased the angular resolution to the point that now images may be made at a resolution equivalent to that obtained at optical and IR wavelengths.

The progression of increasing technical capabilities has given us scientific insight unexpected with the observing capabilities of previous years. We now know that molecular emission can be studied not only in Galactic molecular clouds but also throughout the disks of nearby galaxies and, indeed, in luminous galaxies at cosmological distances. We know that the molecular chemistry in the expelled envelopes of evolving AGB stars is



*W.B. Latter et al. (eds.), CO: Twenty-Five Years of Millimeter-Wave Spectroscopy*, 247–253.

determined by both the star itself and the nature of the interstellar radiation field in which the star is immersed. We know that dusty protostellar disks are ubiquitous, that their dust emission is sufficiently bright to be detectable at millimeter wavelengths, and that once so identified, the disk kinematics can be revealed by molecular spectroscopic observations of sufficiently high angular resolution. But whereas we are aware of what can be learned at millimeter and submillimeter wavelengths from our existing observing technology, we are also aware that in all the cases mentioned above we are presently observing "tip-of-the-iceberg" phenomena: we see only the most luminous cosmological galaxies, only the nearest AGB stars, only the bright, nearby protostars. We cannot put these observations in the larger astronomical context; we are missing the insight they have the potential to provide owing to the limitations of our observing facilities.

The future millimeter and submillimeter telescopes will allow us to overcome these limitations. They will provide us with sufficient increases in sensitivity and angular resolution and with access to enough of the frequency spectrum that our observations will be comparable or superior to the complementary data we will obtain with the new generation of infrared telescopes. In so doing we will achieve the suite of capabilities necessary for the understanding of the physics of cosmic sources of thermal emission. Much of the future advance at millimeter and submillimeter wavelengths will come from larger, high precision telescopes equipped with quantum-limited receivers and located on very dry high elevation sites.

## 2. New Instruments

Now in construction, or well along in the planning stage, are two large millimeter-wavelength single dish telescopes, the NRAO Green Bank Telescope (GBT) and the UMass/Mexico Large Millimeter Telescope (LMT) and four millimeter and submillimeter interferometer arrays; the Smithsonian Submillimeter Array (SMA), the international Millimeter Array (MMA) project led by the NRAO, the Japanese Large Millimeter and Submillimeter Array (LMSA), and the European Large Southern Array (LSA). As presently conceived, these instruments have somewhat comparable capabilities blending different emphases on millimeter-wavelength sensitivity, angular resolution, and submillimeter performance. All will play important roles in defining the future of millimeter-wave astronomy as described below.

## 3. New Science

### 3.1. THE DISTANT UNIVERSE

The discovery that emission from gaseous CO and CI can be detected in galaxies at redshifts greater than two opens the door for investigations of galaxy evolution in the earliest stages wherein a gravitational accumulation of matter in the early universe (a protogalaxy) evolves from being principally gaseous to becoming principally stars. To follow this evolution one needs to observe the gas in emission so that one can map out its spatial distribution and kinematics with the intention of comparing these images to images of light from the earliest stars as detected by the new 8 meter class optical/IR telescopes. Millimeter/submillimeter observations are the only way to image the gas content of protogalaxies, and this will be done very effectively with the new telescopes.

The large single dishes, the GBT and the LMT, will play an important role in detecting the redshifted CO emission from cosmologically distant galaxies. As an example of their effectiveness, let us consider the redshifted CO($J$=3$\rightarrow$2) emission from the $z = 2.3$ galaxy IRAS F10214+4724. Scoville et al. (1995) show that the peak flux density of this line observed at 105 GHz is approximately 15 mJy. The expected performance of the GBT at 100 GHz will be less than half that at the longer wavelengths and the site is limiting, but under good weather conditions with a 3 mm HFET amplifier and a spectral resolution of 25 km s$^{-1}$, the GBT sensitivity should be 5 mJy in one minute of integration (two polarizations). This means that the peak of the F10214 CO(3$\rightarrow$2) line will be detected at 3$\sigma$ in a minute. In an hour of integration we get to 25$\sigma$, and in 10 hours, 75$\sigma$. This suggests that in a long, 3-10 hour, integration "normal" galaxies with 1/10 the CO luminosity of F10214+4724 are detectable with the GBT, at redshifts greater than one. We can expect similar performance from the LMT; although the aperture is smaller, the site is better and the efficiency will be higher.

The large synthesis arrays will image the distribution and kinematics of this gas. The LSA with its planned 10000 m$^2$ of collecting area will be particularly effective for imaging redshifted CO at 3 mm wavelength. We can expect the observing times needed to be comparable with those above for the GBT, namely a few minutes for lines as bright as those from F10214 and a few hours for galaxies with 1/10 the molecular mass of this ultraluminous object. We return to this point below.

The images of redshifted CO obtained with the MMA, LMSA, and LSA are important for our understanding of galaxy formation, but searching for forming galaxies with the CO line observations is not an effective approach. Instead, dusty galaxies at all redshifts can be most effectively discovered through observations of their redshifted thermal dust continuum emission.

TABLE 1. Cumulative Number of IR Galaxies

| Flux Density (mJy) | Frequency 230 GHz | 345 GHz |
|---|---|---|
| 1.0 | $1.5 \times 10^7$ | $7.5 \times 10^7$ |
| 0.1 | $3.8 \times 10^8$ | $1.6 \times 10^9$ |
| 0.01 | $5.0 \times 10^9$ | $1.5 \times 10^{10}$ |

Here the idea is to make use of the "inverted" shape of the Rayleigh-Jeans spectrum which, when redshifted, causes the steep "K-correction" to compensate for the $1/r^2$ distance effect as galaxies are observed at increasing redshift. Detailed calculations (e.g., Blain and Longair 1993) indicate that this compensation is so effective that if a given galaxy is detectable at millimeter/submillimeter wavelengths at $z \approx 0.1$, then that same galaxy will be detectable at $z > 5$; the observed flux density will be essentially unchanged.

How many galaxies are likely to be detectable? The IRAS survey allows us to make such an estimate since we can construct a far-IR luminosity function, refer this to models of stellar formation and stellar evolution, and then carry the calculation out to high redshift (earlier epochs). Such a detailed calculation has been made by Franceschini et al. (1991). Figure 1 shows the number of galaxies per steradian as a function of flux density for various wavelengths between 1300 and 25 microns. Let's just consider two of these wavelengths, 1300 and 800 microns. Integrating over the whole sky, Table 1 gives the number of IR galaxies we expect as a function of flux density.

Over the whole sky there are more than 15 million such galaxies detectable at a level of 1 mJy or greater at 230 GHz and there are more than 75 million detectable at the same level at 345 GHz.

The continuum sensitivity goals for the MMA for a minute of integration at 230 and 345 GHz are 0.1 and 0.2 mJy beam$^{-1}$, respectively. With this sensitivity, the thermal dust continuum emission from 15 million galaxies is detectable at 230 GHz by the MMA at a signal-to-noise ratio of 10 or greater with one minute of observation each. At 345 GHz, 75 million galaxies are similarly detectable at a signal-to-noise ratio greater than five. The sensitivity of the other planned arrays is likely to be similar. This is a very rich scientific field indeed.

Finally, for the MMA and LMSA, which emphasize good submillimeter capability, a galaxy such as the Milky Way provides very luminous submillimeter and far-IR lines that will be redshifted into the observing bands

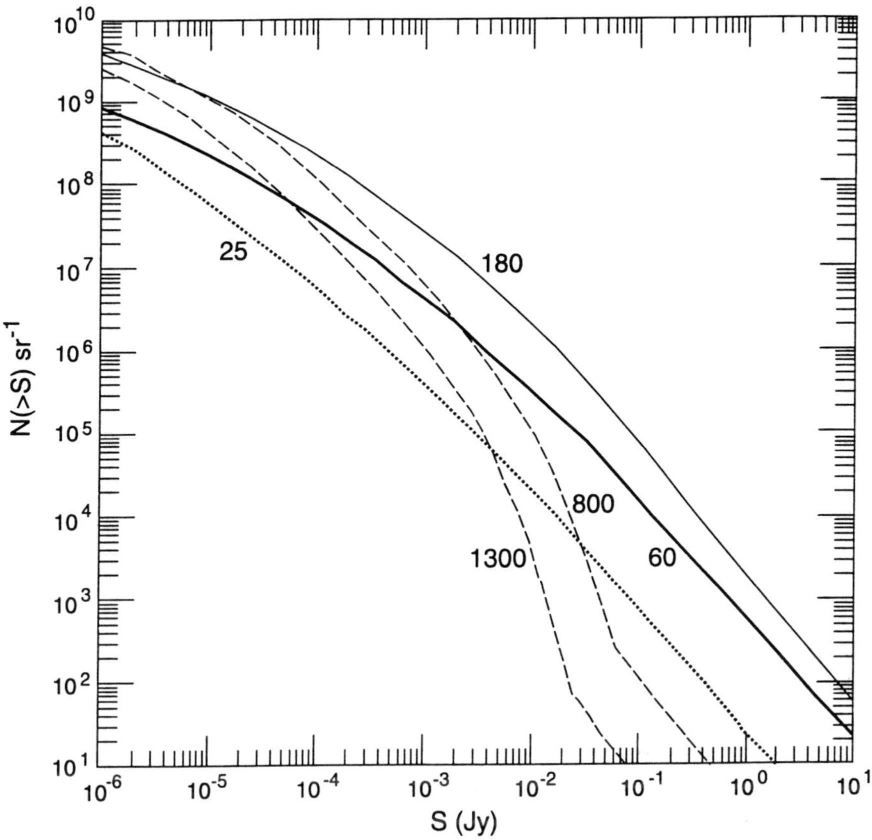


*Figure 1.* Integrated counts of galaxies at the wavelengths (microns) shown. (From Franchesini et al. 1991)

and will provide us with a tool in addition to CO for the study of cosmologically distant galaxies. Table 2 gives the luminosity of important cooling lines from the Milky Way as seen by COBE (Wright et al. 1991). The line flux density $S_L$ in Jy, that we would observe from a galaxy identical to the Milky Way at redshift $z$ from a line of luminosity $L$ is

$$S = S_L \ \Delta v = \frac{0.01(L/10^6 \ L_\odot)}{\nu_{\rm rest}/(1+z)} H_0^2 q_0^4 Q^{-2}$$

where $Q = q_0 z + (q_0 - 1)[(1 + 2q_0 z)^{1/2} - 1]$. For the Milky Way, the lines of greatest observed flux density are the NII and CII lines that will be red-

TABLE 2. Spectral Luminosity of the Milky Way

| Species | (GHz) | $L/L_\odot$ |
|---|---|---|
| CO(1→0) | 115 | $< 5 \times 10^4$ |
| CO(2→1) | 230 | $7.9 \times 10^4$ |
| CO(3→2) | 345 | $1.3 \times 10^5$ |
| CO(4→3) | 461 | $1.3 \times 10^5$ |
| CI | 493 | $2.0 \times 10^5$ |
| CO(5→4) | 577 | $1.0 \times 10^5$ |
| CI | 810 | $3.2 \times 10^5$ |
| NII | 1460 | $5.0 \times 10^6$ |
| CII | 1900 | $5.0 \times 10^7$ |
| OI | 2060 | $6.3 \times 10^5$ |
| NII | 2460 | $7.9 \times 10^6$ |

shifted at $z > 1$ to the short submillimeter bands. The MMA and LMSA will emphasize these submillimeter lines in the study of cosmological galaxies. The LSA will employ its greatest collecting area at millimeter wavelengths to study the fainter, low order CO lines in these same galaxies. Taken together, the instruments provide astronomers with the needed tools for study of the early universe.

### 3.2. THE GALAXY NEARBY

In the Milky Way itself, the great sensitivity of all the new instruments will permit studies to be made of the ensemble of AGB stars throughout the disk of the galaxy. The important question here has to do with chemical variations as a function of galactocentric radius, which will tell us of the pace of star formation in the galaxy over the past ten billion years. Additionally, the sub-arcsecond angular resolution of the MMA, LMSA, and LSA will reveal for us the physics of the mass loss process, what physical and chemical processes occur in the expelled stellar envelope, how dust forms and where it forms in the outflow, and how much mass is ejected into the galaxy that will serve as fuel for the next generation of star formation. These studies are presently impossible because the sub-arcsecond resolution is not available at any wavelength at which the stellar envelopes are transparent and because the spectral sensitivity of existing instruments is inadequate. The new arrays will pioneer this research field.

The sub-arcsecond angular resolution of the new arrays will also give us the first clear image of the way in which planets form around young solar-like stars. A resolution of $0.1''$ at the distance of active sites of star formation near the Sun such as in the Taurus, Ophiuchius, or Coalsack molecular clouds correspond to a linear extent of less than 10 A.U. On this scale we will explore the radial density distribution of protoplanetary disks and seek to confirm observationally the speculation that forming planets create azimuthal gaps in the circumstellar disk. We can uniquely study planets in formation because we observe at wavelengths long enough that the disk is transparent but short enough that the thermal emission from the disk is bright. The chemistry of protoplanetary disks, perhaps even the pre-biotic chemistry, will be a rich and previously unexplorable field of research because the proper angular resolution at the proper wavelength was lacking. The new instruments will change all this.

Finally, and closest to home, the new imaging instruments will allow us to see features on objects in the solar system with a clarity that exceeds that of the HST or the new 8 m class optical telescopes operating at their diffraction limit. The resulting spectroscopic images will reveal for us the winds on Mars, Venus, and the giant planets, the chemical composition of comets, and the out-gassing of volcanoes on Io.

The new millimeter and submillimeter wavelength telescopes will make the meeting "Fifty Years of Millimeter Wave Spectroscopy" in 2020 very different scientifically from the meeting today. I look forward to the excitement of the next 25 years.

## References

Blain, A. W. and Longair, M. S. 1993, MNRAS, 264, 509

Francheschini, A., Toffolatti, L., Mazzei, P., Danese, L., and De Zotti, G. 1991, A&A Suppl., 89, 285.

Scoville, N. Z., Min, Y. S., Brown, R. L., and Vanden Bout, P. A. 1995, ApJ Lett., 449, L1098.

Wright, E. L. 1991, ApJ, 381, 200

# GALAXIES AND GALACTIC NUCLEI

# COMPARATIVE STUDIES OF OTHER GALAXIES

A. ECKART
*Max-Planck-Institut für Extraterrestrische Physik (MPE)*
*Giessenbachstrasse, D 85740 Garching bei München, Germany*

**Abstract.** Knowledge of the distribution and physical properties of the molecular gas content of external galaxies is crucial for understanding their dynamics and, in combination with data obtained across the electromagnetic spectrum, the role of star formation in their evolution. Due to its high abundance and line strength, CO plays a key role in these investigations. This paper will concentrate on recent results obtained during the past five years. Previous work will be briefly summarized. This article covers multi-line studies of ultraluminous infrared galaxies at cosmological distances like F10214+4724 ($z$=2.29) or the Cloverleaf quasar ($z$=2.56), nearby starburst and active galaxies such as IC 342, M82, and Centaurus A, and nearby spiral galaxies like M51.

## 1. Introduction

Since the first detection of $^{12}$CO J=1-0 line emission in an extragalactic system 20 years ago (Rickard et al. 1975; Solomon and de Zafra 1975), CO has played a key role in the investigation of the dense molecular interstellar medium in galaxies. During recent years these studies have been extended towards redshifts as high as $z$=2.56 (Cloverleaf). The results of previous years have been summarized in Young and Scoville (1991) and Henkel, Baan, and Mauersberger (1991). Earlier reviews are by Morris and Rickard (1982); Soifer et al. (1987); Telesco (1988); and Cohen(1989).

The efficiency with which the molecular gas mass $M(H_2)$ in galaxies is transformed into massive stars that are responsible for the far-infrared luminosity $L_{IR}$ can be measured by the $L_{IR}/M(H_2)$ ratio (Young and Scoville 1991; Downes, Solomon, and Radford 1993; Sanders et al. 1988) and is reflected in a correlation between $L_{IR}$ and $M(H_2)$ that has now been extended towards luminosities in excess of $10^{12}L_\odot$ (e.g. Sanders et al. 1988;



*W.B. Latter et al. (eds.), CO: Twenty-Five Years of Millimeter-Wave Spectroscopy*, 257–264.

Brown and Vanden Bout 1992; Barvainis et al. 1994). The validity of this correlation out to large distances can also be taken as an argument that the FIR-luminosity is due to thermal emission rather than non-thermal synchrotron emission (Alloin et al. 1992). Multi-level isotopic line emission can be used in radiative transfer calculations to determine the physical state (density, temperature, and abundances) of the interstellar medium. $^{13}$CO line emission has now been detected towards a large number of galaxies (e.g., Sage and Isbell 1991; Casoli, Dupraz, and Combes 1992) including the nearby QSO I Zw 1 (Eckart et al. 1994). Besides, the detection of high redshifted CO molecular line emission is also becoming statistically more important for cosmology (see analysis of the FCRAO survey in Verter 1991, 1992). Dickey and Kazès (1992) and Schöniger and Sofue (1994, with a sample of 32 galaxies) have shown that a Tully-Fisher relation for the CO line can be established. Although the line shapes of HI and CO are sometimes very different, CO emission evidently extends beyond the peak of the rotation curves in most spiral galaxies. Therefore, one finds a perfect relation between the linewidths in CO and HI. The linewidths are correlated with the absolute magnitude, so they can be used as a distance indicator.

## 2. Ultraluminous Infrared Galaxies

Progress in receiver technology and the operation of large millimeter wavelength single dish telescopes and sensitive interferometers now also allow study of the molecular gas phase in ultraluminous galaxies at cosmological distances. In these objects like the galaxy F10214+4724 at $z$=2.29 (P.Solomon in this edition; Brown and Vanden Bout 1992) or the Cloverleaf quasar at $z$=2.56 (S.Radford and R.Barvainis in this edition; Barvainis et al. 1994), the molecular gas accounts for a substantial fraction of their total dynamical mass. More detailed investigations of these objects may improve our understanding of galaxies at early stages of their evolution. F10214+4724 and the Cloverleaf are the two most luminous high redshift objects in which intense CO emission with an implied molecular hydrogen mass of about $10^{11}$M$_{\odot}$has been discovered. Yamada et al. (1995) report the possible detection of molecular line emission from a quasar at $z$=2.88 which implies a molecular gas mass of $\sim 10^{12}$M$_{\odot}$. For F10214+4724, multi-line observations (Brown and Vanden Bout 1992; Solomon et al. 1992; Sakamoto et al. 1992; Radford, Brown, and Vanden Bout 1993) show the bulk of the molecular gas is warm and dense with $n(H_2) \approx 5 \times 10^3 cm^{-3}$ if $T_{kin} = 50$ K. Of particular interest is the fact that the Cloverleaf is a well-known gravitational lens. The question then arises whether the implied large molecular gas masses are significantly overestimated due to amplification by the foreground lens. This may also be the case for F10214+4724.

## 3. Nearby Starburst and Active Galaxies

With a large number of sensitive interferometer arrays with angular resolutions of the order of 2″, it is now possible to obtain detailed information on the distribution and dynamics of the nuclear molecular gas. Current studies indicate that the structures of nearby starburst galaxies are dominated by molecular rings and bars that in most cases can be interpreted as part of the inner spiral structure of the corresponding galaxies. These structures are in agreement with the hypothesis of infalling gas streaming along a nuclear stellar bar. As in the Seyfert 2 galaxy NGC 1068 (see also contributions by L.J.Tacconi and T.T.Helfer in this edition), these measurements give direct evidence for nuclear molecular gas concentrations and gas feeding in active nuclei as postulated in 'unified models'. Isotopic multiline studies of different molecules indicate for the nuclei rather dense and warm molecular gas with $n = 10^3 - 10^5$ cm$^{-3}$ and $T_{\rm kin} \sim 30$ K. Well studied examples are IC 342, a gas-rich spiral showing a mild nuclear starburst, M82, the archetypical starburst galaxy, and Centaurus A (see also the contribution by W.Wild in this edition), the closest radio galaxy.

**IC342:** Single dish and interferometer multi-line studies (Xie, Young, and Schloerb 1994; Eckart et al. 1990a; Güsten et al. 1993; Wright et al. 1993; Wall et al. 1993; Irwin and Avery 1992; Turner and Hurt 1992) show that the properties of the $\sim 10^8$ $M_\odot$ molecular gas within the central kpc vary as a function of position. In interferometer maps the bar of molecular gas is resolved into two spiral arms that meet in a 100 pc diameter ring. The most recent, highest spatial resolution maps of IC342 (Wright et al. 1993; Turner and Hurt 1992) reveal peaks in the line emission probably corresponding to giant molecular cloud complexes (GMCs) 20-25 pc in size with molecular gas masses of $\sim 10^6$ $M_\odot$. These observations are complemented by subarcsecond imaging spectroscopy in the near-infrared, e.g., with the MPE 3D spectrometer (Böker et al. 1996). This now allows full modeling of the deeply embedded stars and molecular gas in IC342 and other galaxies.

**M82:** Morphology and molecular excitation has been studied extensively in this archetypical starburst galaxy. Recent multi-line and gas dynamical studies are by White et al. (1994); Brouillet and Schilke (1993); Wild et al. (1992); Sofue et al. (1992); Tilanus et al. (1991); Phillips and Mampaso (1989). Although the $^{12}$CO(2–1)/$^{12}$CO(1–0) ratio is generally measured to have values ranging between 1.3 and 2, indicative of a contribution from warm optically thin gas, Wild et al. (1992) find that the strength of the CO J=6–5 emission in comparison with the lower-J lines, as well as the ratios of isotopic lines, indicate that the lower rotational transitions of $^{12}$CO are optically thick. The bulk molecular gas temperature is $\sim$ 45 K, with $n \sim 10^4$ cm$^{-3}$ and $^{peak}N(H_2) \sim 5 \times 10^{22}$ cm$^{-2}$ in a 13″ beam. The centrally

peaked $C^{18}O$ J=2-1 data reveal about $10^8 M_\odot$ of molecular gas within the 400 pc diameter double lobed ring structure seen in $^{12}CO$.
**Cen A:** SEST (Eckart et al. 1990b; Israel et al. 1990; Rydbeck et al. 1993) and CSO (Quillen et al. 1992; Phillips et al. 1987) observations of Centaurus A show that, similar to the ionized gas as seen in H$\alpha$, $2\times10^8 M_\odot$ of molecular gas are distributed in a thin warped disk in a prolate potential. In addition to this, Israel et al. (1990) as well as Rydbeck et al. (1993) find evidence for an edge-on circumnuclear ring with a radius of $\sim$100 pc and rotational velocities of 220 km/s$^{-1}$. This ring is perpendicular to the radio jet and implies an enclosed mass of $1.4\times10^9$ $M_\odot$.

## 4. Nearby Spiral Galaxies

Due to their proximity, nearby spirals can be studied with the highest spatial resolution. Individual GMCs and the distribution and dynamics of the molecular gas in the disks can be investigated. Detailed knowledge of the physical properties of these systems represents an important reference point to which extreme properties of ultraluminous and more distant active galaxies have to be compared. M31 and M33 are covered in the contribution by P. Thaddeus in this edition (see also the review article by Young and Scoville 1991). For M51, single dish and interferometer maps (Guélin et al. 1989; Rand 1993; Adler et al. 1992; Garcia-Burillo, Combes, and Gerin 1993; Garcia-Burillo, Guélin, and Cernicharo 1993) now account for all the line flux and consistently find wider spiral arms (about 1.3 kpc or less) than seen previously (Vogel et al. 1988, Rand and Kulkarni 1990). The average integrated intensity arm-to-interarm contrast is of the order of 5:1 with peak values 4 times higher in the inner arms. A detailed picture of the velocity field in M51 shows streaming motions along the arms and turbulent motion close to the center (Garcia-Burillo, Guélin, and Cernicharo 1993). The structure of the CO, HI, and H$\alpha$ line emission across the arms reflects the influence of a density wave system and the variation of massive star formation (Knapen et al. 1992; Adler et al. 1992). 3D Monte-Carlo simulations (Garcia-Burillo, Combes, and Gerin 1993) indicate that the dynamics as well as the arm-to-interarm contrast can probably be understood as a consequence of an encounter with NGC 5195.

## 5. Molecular Excitation

Under LTE conditions we expect $^{12}CO(2\text{-}1)/^{12}CO(1\text{-}0)$ line temperature ratios near 1.0 for dense (the critical density to populate the J=2 level is $2\times10^4$ cm$^3$), warm (T>20 K), and optically thick gas, and 4 for dense, warm, and optically thin gas. A ratio <1 can indicate low temperatures for optically thick lines. Deviations from the LTE values can also occur when the density is too low to thermally populate the rotational levels, as in many envelopes of clouds in our Galaxy. Several surveys to probe the molecular excitation in nuclei of galaxies have recently been compiled (e.g., Braine and Combes 1992, 1993; Wall et al. 1993; Sage and Isbell 1991; see also the contributions of S. Aalto-Bergman and T. Hasegawa in this edition). They find the $^{12}CO(2\text{-}1)/^{12}CO(1\text{-}0)$ line ratio is of the order of 0.9 and indicates molecular line emission from dense warm molecular gas. Bright $^{12}CO$ line emission from nuclei of nearby galaxies (NGC 253, IC 342, M 82, NGC1068, and NGC 6946) in the submillimeter J=6-5 (Harris et al. 1991) and J=4-3 (Güsten et al. 1993) transition convincingly demonstrates the presence of dense warm molecular material in the nuclei of galaxies with star formation activity. In nearby galaxies where the disk component can be resolved or in distant objects where the beam filling is dominated by the disk, the $^{12}CO(2\text{-}1)/^{12}CO(1\text{-}0)$ ratio is of the order of 0.5 indicative of cold and/or subthermally excited molecular gas (e.g., for M31, Allen and Lequeux 1994).

The molecular gas is probably excited by a combination of mechanisms. The gas in dense molecular clouds may be excited by an enhanced cosmic ray ionization rate (e.g., Goldsmith & Langer 1978; de Jong et al., 1980) that may be two orders of magnitude above the standard Galactic value of a few times $10^{-17}$ $s^{-1}$ in M82 (Schilke et al. 1993). Such an enhancement might be due to the high star formation activity and can probably be measured directly, if the non-thermal radio flux density is proportional to the cosmic ray density. A further efficient heating mechanism is the intense UV radiation field expected from an increased number of OB stars in a starburst or in embedded active nuclei. Photoelectric heating and far-UV pumping of $H_2$ will heat surfaces of molecular clouds up to a few hundred degrees (Tielens & Hollenbach 1985; Sternberg and Dalgarno 1989). This may result in strong far-infrared cooling lines (Crawford et al. 1985) and optically thin emission from hot CO on the photodissociated cloud surfaces. For the nuclei of galaxies, turbulent heating can also be important, as is probably the case in the Galactic Center (Zhao, Goss, and Ho 1995).

## 6. Absorption Spectroscopic Studies

Absorption studies against bright nonthermal nuclear continuum sources are of great value since they provide a "pencil beam" measurement of the characteristics of the molecular clouds responsible for the absorption features and give a deeper insight into the properties of the interstellar medium of the corresponding galaxies. Up to now there are only two sources for which millimeter absorption spectroscopic studies have been published: Centaurus A (Phillips et al. 1987; Israel et al. 1990; Eckart et al. 1990c) and the BL Lac object PKS 1413+135 $z = 0.2467$ (Wiklind and Combes 1994). Barvainis and Antonucci (1994) searched for CO absorption towards Cygnus A without success. This may either indicate the absence of a molecular torus or may be explained by extraordinarily large rotation temperatures due to the presence of the strong nonthermal nucleus (Maloney et al. 1994). For Centaurus A, absorption lines of $^{12}CO$, $^{13}CO$, $HCO^+$, HCN, CN, and CS (Eckart et al. 1990c; Israel et al. 1990) show a number of features with line widths of a few km $s^{-1}$. They can be attributed to individual molecular clouds with properties similar to dense molecular clouds in our Galaxy. CO absorption line ratios, as well as the hyperfine line ratios for CN and HCN, indicate the deepest absorption at the systemic velocity of Centaurus A originates in quiescent molecular clouds that are cold, dense, and clumped and happen to lie along the line of sight to the central compact continuum source.

## 7. Open Questions

In the future, multi-line studies of ultra luminous galaxies at high redshift are required in order to determine the mass and physical properties of their molecular gas. Since the molecular mass apparently constitutes a large fraction of the dynamical mass, such a study is essential to understanding galaxy evolution and the transformation of gas into stars. A search for systems where the molecular lines can be studied in absorption is required. If the molecular gas plays a dominant role in the interstellar media of galaxies at cosmological redshifts, one should expect more absorption line systems in quasars. Important open questions are: Do the differences in infrared luminosity and CO line emission reflect an evolutionary sequence? Is there a valid unified model that links the different source classes? What are ultraluminous objects? Are they powered by a central star burst or a hidden QSO or both? Detailed multi-frequency high spatial resolution investigations of samples of galaxies are required to tackle these questions.

**Acknowledgements:** I thank L. Tacconi, R.Genzel, A. Quirrenbach, and W. Wild for carefully reading the manuscript, as well as for comments and fruitful discussions.

## References

Adler, D.S., Lo, K.Y., Melvyn, C.H., Wright, C.H., Rydbeck, G., Plante, R.L., Allen, R.J., 1992, Ap. J., 392, 497.

Allen, R.J., Lequeux, J., 1994, p.105, in Proc. of a Conference on "Nuclei of Normal Galaxies: Lessons from the Galactic Center" held at Schloß Ringberg (Tegernsee), July 26-30, 1993, (eds.) R.Genzel & A.I.Harris, Kluwer (Dordrecht).

Alloin, D., Barvainis, R., Gordon, M.A., Antonucci, R.R.J., 1992, Astron. Astrophys., 265, 429.

Barvainis, R., Antonucci, R., 1994, A. J., 107, 1291.
Barvainis, R., Tacconi, L., Antonucci, R., Alloin, D., Coleman, P., 1994, Nature, 371, 586.
Böker, T., et al. 1996 in preparation
Braine, J., Combes, F., 1992, Astron. Astrophys., 264, 433.
Braine, J., Combes, F., 1993, Astron. Astrophys., 269, 7.
Brown, R.L., Vanden Bout, P.A., 1992, Ap. J., 397, L19.
Brouillet, N., Schilke, P., 1993, Astron. Astrophys., 277, 381.
Casoli, F., Dupraz, C., Combes, F., 1992, Astron. Astrophys., 264, 55.
Cohen, R.J., 1989, Rep.Prog.Phys, 25,881
Dickey, J.M., Kazès, I, 1992, Ap. J., 393, 530.
Downes, D., Solomon, P.M., Radford, S.J.E., 1993, Ap. J., 414, L13.
Crawford, M.K., Genzel, R., Townes, C.H., and Watson, D.M., 1985, Ap. J., 291, 755.
Eckart, A., Downes, D., Genzel, R., Harris, A.I., Jaffe, D.T. and Wild, W., 1990a, Ap. J., 348, 434.
Eckart, A., Cameron, M., Wild, W., Zinnecker, H., Rydbeck, G., Olberg, M., Wiklind, T., 1990b, Ap. J., 363, 451.
Eckart, A., Cameron, M., Genzel, R., Jackson, J.M., Rothermel, H., Stutzki, J., Rydbeck, G., Wiklind, T., 1990c, Ap. J., 365, 522.
Eckart, A., van der Werf, P.P., Hofmann, R., Harris, A.I., , 1994, Ap. J., 424, 627.
Garcia-Burillo, S., Combes, F., Gerin, M., 1993, Ap. J., 274, 148.
Garcia-Burillo, S., Guélin, M., Cernicharo, J., 1993, Astron. Astrophys., 274, 123.
Goldsmith, P.F. and Langer, W.D., 1978, Astron. Astrophys., 222, 881.
Güsten, R., Serabyn, E., Kasemann, C., Schinckel, A., Schneider, G., Schulz, A., Young, K., 1993, Ap. J., 402, 537.
Guélin, M., Garcia-Burillo, S., Blundell, R., Cernicharo, J., Despois, D., Steppe, H., 1989, Highlighs Astron. Vol.8, p.575
Harris, A.I., Hills, R.E., Stutzki, J., Graaf, U.U., Russell, A.P.G., Genzel, R., 1991, Ap. J., 382, L75.
Henkel, C., Baan, W.A., Mauersberger, R., 1991, Astron. Astrophys. Review, 3, 47.
Irwin, J.A., Avery, L.W., 1992, Ap. J., 388, 328.
Israel, F.P., van Dishoeck, E.F., Baas, F., Koornneef, J., Black, J.H., de Graauw, Th. 1990, Astron. Astrophys., 227, 342.
de Jong, T., Dalgarno, A. and Boland, W., 1980, Astron. Astrophys., 91, 68.
Knapen, J.H., Beckman, J.E., Cepa, J., van der Hulst, T., Rand, R.J., 1992, Ap. J. (Letters), 385, L37.
Malony, P.R., Begelman, M.C., Rees, M.J., 1994, Ap. J., 432, 606.
Morris, M., Rickard, L.J., 1982, Annu.Rev.Astron.Astrophys., 20, 517.
Phillips, T.G., Ellison, B.N., Keene, J.B., Leigthon, R.B., Howard, R.J., Masson, C.R., Sanders, D.B., Veidt, B. and Young, K., 1987, Ap. J., 322, L73.
Phillips, J.P., Mampaso, A., 1989, Astron. Astrophys., 218, 24.
Quillen, A.C., de Zeeuw, P.T., Phinney, E.S., Phillips, T.G., 1992, Ap. J., 391, 121.
Radford, S.J.E., Solomon, P.M., Downes, D., 1991, Ap. J., 368, L15.
Rand, R.J., 1993, Ap. J., 404, 593.
Rand, R.J., Kulkarni, S.R., 1990, Ap. J., 349, L43.
Rickard, L.J., Palmer, P., Morris, M., Zuckermann, B., Turner, B.E., 1975, Ap. J., 199, L75.

Rydbeck, G., Wiklind, T., Cameron, M., Wild, W., Eckart, A., Genzel, R., Rothermel, H., 1993, Astron. Astrophys., 270, L13.
Sage, L.J., 1993, Astron. Astrophys., 272, 123.
Sage, L.J., Isbell, D.W., 1991, Astron. Astrophys., 247, 320.
Sakamoto K., Ishizuki, S., Kawabe, R., Ishiguro, M., 1992, Ap. J., 397, L27.
Sanders, D.B., Soifer, B.T., Elias, J.H., Madore, B.F., Matthews, K., Neugebauer, G., and Scoville, N., 1988, Ap. J., 325, 74.
Schöniger, F., Sofue, Y., 1994, Astron. Astrophys., 283, 21.
Schilke, P., Carlstrom, J.E., Keene, J., Phillips, T.G., 1993, Ap. J., 417, L70.
Sofue, Y., Reuter, H.-P., Krause, M., Wielebinski, R., Nakai, N., 1992 , Ap. J., 395, 126.
Soifer, B.T., Houk, J.R., Neugebauer, G., 1987, Annu.Rev.Astron.Astrophys., 25, 187.
Solomon, P.M., Downes, D., Radford, S.J.E., 1992, Ap. J., 387, L55.
Solomon, P.M., Downes, D., Radford, S.J.E., 1992, Ap. J., 398, L29.
Solomon, P.M., de Zafra, 1975, Ap. J., 199, L79.
Sternberg, A., Dalgarno, A., 1989, Ap. J., 338, 197.
Tacconi, L.J., Genzel, R., Blietz, M., Cameron, M., Harris, A.I., Madden, S., 1994, Ap. J. (Letters), 426, L77.
Telescon C.M., 1987, Annu.Rev.Astron.Astrophys., 26, 343.
Tielens, A.G.G.M., Hollenbach, D., 1985, Ap. J., 291, 722..
Turner, J.L., Hurt, R.L., 1992, Ap. J., 384, 72.
Verter, F., 1991, Ap. J., 375, 95.
Verter, F., 1992, Ap. J., 386, 398.
Vogel, S., Kulkarni, S., Scoville, N., 1988, Nature 334, 402
Wall, W.F., Jaffe, D.T., Bash, F.N., Israel, F.P., Maloney, P.R., Baas, F., 1993, Ap. J., 414, 98.
Wild, W., Harris, A.I., Eckart, A., Genzel, R., Graf, U.U., Jackson, J.M., Russell, A.P.G., Stutzki, J., 1990, Astron. Astrophys., 265, 447.
Wiklind, T., Combes, F., 1994, Astron. Astrophys., 286, L9.
White, G.J., Ellison, B., Claude, S., Dent, W.R.F., Matheson, D.N., 1994, Astron. Astrophys., 284, L23.
Wright, M.C.H., Ishizuki, S., Turnuer, J.L., Ho, P.T.P., Lo, K.Y., 1993, Ap. J., 406, 470.
Xie, S., Young, J., Schloerb, F.P., 1994, Ap. J., 421, 434.
Yamada, T., Aoki, K., Tomita, A., Takata, T., Ohta, K., Yoshida, M., 1995, Ap. J., 438, L5.
Young, J.S., Scoville, N.Z., 1991, Annu.Rev.Astron.Astrophys., 29, 581.
Zhao, J.-H., Goss, W.M., Ho, P.T.P., 1995, Ap. J., 450, 122.

# THE ESO/SEST KEY PROGRAMME: CO IN THE MAGELLANIC CLOUDS

MONICA RUBIO

*Departamento de Astronomía, Universidad de Chile*
*Casilla 36-D, Santiago, CHILE*

## 1. Introduction

The ESO SEST Key Programme was established to investigate molecular gas in the Magellanic Clouds. These galaxies provide a unique opportunity for studying molecular clouds and star formation in galaxies whose environment is different from that of the Milky Way. In the LMC, the gas-to-dust ratio is 4 times higher than in our Galaxy (Koorneef 1982) and the metal abundance is about 4 times lower (Dufour 1984). In the SMC, these values are even larger, ~17 times larger than our Galaxy for the gas-to-dust ratio and a metal abundance of about 10 times smaller (Dufour 1984, Koorneef 1984). Because of their proximity, 50 kpc for the LMC and 63 kpc for the SMC, the SEST telescope with an angular resolution of 43″ at 2.6 mm provides a linear resolution of 10 pc at the LMC distance, well suited for resolving the largest molecular clouds. The main goals of this programme were to determine the physical properties of the molecular gas in the Magellanic Clouds, derive the characteristics of the CO clouds and compare them to those of the Giant Molecular Clouds (GMC) found in our Galaxy, and to address the question of the conversion of CO luminosity, $L_{\rm CO}$, to molecular hydrogen mass, $M(\rm H_2)$, in an environment with dust content and metal abundance different to that of the Milky Way. In this contribution we summarize the results of this programme which started in May 1988 and ended in January 1995. Detailed results are being presented as a series of publications in *Astronomy and Astrophysics.* Further details and goals of the programme can be found in Israel et al. (1993; Paper I).



*W.B. Latter et al. (eds.), CO: Twenty-Five Years of Millimeter-Wave Spectroscopy*, 265–271.

## 2. Observations

The observations were made during several runs with the Swedish-ESO Submillimeter Telescope (SEST). This radiotelescope is located at La Silla Observatory (Chile) and has a diameter of 15 m; it has beams (FWHM) of 43″ and 22″ at the CO($1 \rightarrow 0$) and CO($2 \rightarrow 1$) frequencies, respectively. The backend is an acousto-optical spectrometer with a total bandwidth of 86 MHz and a channel width of 0.043 MHz, corresponding to a velocity range of 230 km $s^{-1}$ and a velocity resolution of 0.11 km $s^{-1}$ at CO($1 \rightarrow 0$) and 112 and 0.056 km $s^{-1}$ at CO($2 \rightarrow 1$), respectively. Observations were done in frequency-switch mode with a throw of 15 MHz for $^{12}$CO($1 \rightarrow 0$), and position-switch mode, with a reference position far from the known CO emission zones, for the three other lines. The system temperature varied between 400 and 800 K for CO($1 \rightarrow 0$) and 900 to 1500 K for CO($2 \rightarrow 1$). The integration time at each position was automatically varied in order to secure a uniform noise level; in general, 0.1 K per channel at all frequencies. Telescope pointing was checked on the strong SiO maser R Dor, and calibration spectra were done on the strongest lines in the LMC (N159) and SMC (LIRS 49). We estimate the maximum pointing errors as about 5″ . The calibration used the usual absorbing chopping wheel method, and observations of blackbodies at two temperatures. The antenna temperature scale was checked sporadically on the cloud OMC1.

## 3. The Small Magellanic Cloud

In the SMC, observations in the $^{12}$CO($1 \rightarrow 0$)line toward the central region of the southwest part of the Bar and some selected HII regions were performed (see Fig. 1 of Rubio 1992). About a dozen molecular clouds were identified and fully mapped in $^{12}$CO($1 \rightarrow 0$) (Rubio et al. 1993a, Paper II; Rubio et al. 1996, Paper V). The CO clouds have sizes ranging from 10 to 20 pc, line widths between 1.6 and 4.9 km $s^{-1}$ , CO luminosities between $0.5 \times 10^3$ to $6.4 \times 10^3$ $L_\odot$, and virial masses ranging from 2 to $9 \times 10^4$ $M_\odot$. The characteristics of the CO clouds in the SMC are different from those of galactic GMCs. They have smaller sizes and are less luminous in CO than galactic GMCs. From the excitation temperatures derived from the modelling of observed line ratios (Lequeux et al. 1994, Paper IV) and the observed antenna temperatures, it is found that the SMC CO clouds are clumpy with filling factors smaller than that of Galactic CO clouds. The relationships between $L_{\rm CO}$and $\Delta v$ and $L_{\rm CO}$vs. $M_{\rm vir}$ for SMC-CO clouds are noticeably different from those of Galactic molecular clouds. At the scales resolved by SEST (10pc), the CO clouds are less luminous by a factor of 2 compared to outer galaxy clouds and at the largest scales (200pc) this factor increases to 10 (Rubio et al. 1993b, Paper III). These differences can be explained by the combined effects of a higher UV radiation field and of a lower abundance of heavy elements and of dust which produces a higher rate of photodissociation of CO (Maloney & Black 1988, Elmegreen 1989). In the SMC contrary to our Galaxy, CO can only survive in dense clumps while it is completely photodissociated in the diffuse medium. Presumably most of the interstellar medium in the SMC is atomic rather than molecular.

To determine the physical properties of the SMC molecular clouds, observations at the peak position of six molecular clouds in four CO lines [$^{12}CO(1 \rightarrow 0)$, $^{13}CO(1 \rightarrow 0)$, $^{12}CO(2 \rightarrow 1)$, and $^{13}CO(2 \rightarrow 1)$] with the same angular resolution were performed. The derived line ratios range from 7 to 25 for $^{12}CO(1 \rightarrow 0)/^{13}CO(1 \rightarrow 0)$; from 0.8 to 1.3 for $^{12}CO(2 \rightarrow 1)/^{12}CO(1 \rightarrow 0)$, and from 1.0 to 2.7 for $^{13}CO(2 \rightarrow 1)/^{13}CO(1 \rightarrow 0)$. The six clouds under consideration can be classified into 3 categories: i) the "hot" clouds, N 88 and N 66 (Henize 1956), which are close to bright HII regions and exhibit large $^{13}CO(2 \rightarrow 1)/^{13}CO(1 \rightarrow 0)$ integrated line intensity ratios; ii) the "warm" clouds, LIRS 49 and LIRS 36 (Schwering & Israel 1989), which are close to less bright HII regions and exhibit smaller $^{13}CO(2 \rightarrow 1)/^{13}CO(1 \rightarrow 0)$ ratios; and iii) the "cold" clouds, SMC-B1#1 (Paper II) and Hodge 15 (Hodge 1984), which are far from HII regions and have low $^{13}CO(2 \rightarrow 1)/^{13}CO(1 \rightarrow 0)$ ratios. These line ratios have been compared with the predictions of a one-side illuminated plane-parallel model which treats consistently radiative, chemical and excitation processes of CO and its isotopes in SMC conditions (Lequeux et al. 1994, Paper IV). The observed line ratios are consistent with models in which the molecular gas has a moderately high density ($10^4 - 10^5$ cm$^{-3}$). The model shows that gas at low densities, (say $n(H_2) = 100$ cm$^{-3}$) is completely photodissociated in the SMC and contains no CO, nor even $H_2$ in extreme cases. This study confirms the conclusions of Rubio et al. (1993b, Paper III): i) that the high observed $^{13}CO(2 \rightarrow 1)/^{13}CO(1 \rightarrow 0)$ ratios imply high excitation temperatures of the emitting gas; and ii) that the surface filling factors of the emission are low (eg., a value of ~0.08 was derived for the hot clouds).

## 4. The Large Magellanic Cloud

In the LMC, observations in the $^{12}CO(1 \rightarrow 0)$ line towards 92 IRAS sources were first reported by Israel et al (1993). During the following observing periods of the Key programme, several regions were fully mapped in the $^{12}CO(1 \rightarrow 0)$ line. These are: a) 30 Dor and N159 regions (Booth 1991; Johansson 1991; Johansson; Booth & Heikkila 1995); b) N11 region (de Graauw et al. 1996; Kutner & Rubio 1996); c) 30 Doradus Central and Southern Complex (Kutner & Rubio 1996; Kutner et al. 1996); and d)

Cloud 37 (Garay et al. 1996). A total of 86 molecular clouds have been identified. The LMC CO clouds have sizes ranging from 10 to 40 pc, line widths between 2.5 and 10 km $s^{-1}$ , CO luminosities between $8 \times 10^2$ to $1.3 \times 10^4$ K $km\ s^{-1}\ pc^2$, and virial masses ranging from $10^3$ to $7 \times 10^5$ $M_\odot$. These clouds, although larger in size and more massive than those found in the SMC, do not look like typical GMCs in the Galaxy. Their characteristics are closer to molecular clouds in the outer Galaxy than in the inner Galaxy (see Figures 1 and 2). The line width versus size relationship, for the clouds within the N11, Cloud 37, 30 Doradus Central, and 30 Doradus Southern regions, follow the power law relation $\Delta v = (6.5 \pm 1.4)\ R^{0.5 \pm 0.1}$. The slope is similar to that of molecular clouds in our Galaxy and also to the slope of SMC clouds. The $\Delta v$ vs. $L_{\rm CO}$and $L_{\rm CO}$vs. $M_{\rm vir}$ relationships are different from those found for Galactic molecular clouds. A linear fit to the $L_{\rm CO}$versus $\Delta v$ relation for the same ensemble of LMC clouds gives $L_{\rm CO}= (170 \pm 3)\Delta v^{1.8 \pm 0.3}$. A linear fit to the CO luminosity versus virial mass for the LMC clouds gives $M_{\rm vir} = (155 \pm 2)\ L_{\rm CO}^{0.78 \pm 0.07}$.

## 5. Comparison to Galactic Molecular Clouds

In this section, we briefly discuss a comparison of the properties of the molecular clouds in the LMC and SMC with those in our Galaxy. As representatives of the latter, we have used recent data of southern hemisphere clouds obtained with the 1.2 m Columbia telescope at CTIO. May et al. (1996) reported the characteristics of 187 molecular clouds in the III Quadrant of our Galaxy. We also used the data of 75 molecular clouds identified in the inner IV Quadrant of the Galaxy (Bronfman, private communication).

Figure 1 shows the $L_{\rm CO}$versus $\Delta v$ relationship for LMC, SMC, and Galactic clouds. Molecular clouds in the Magellanic Clouds appear to be less luminous in CO than molecular clouds in the outer Galaxy in a given range of line widths. Note that, in that plot, inner Galaxy CO clouds concentrate in the range of large line widths where only a few Magellanic Cloud clouds are found, these being much less luminous in CO than their galactic counterparts.

In Figure 2 we plot the ratio $X/X_{\rm gal}$ as a function of $R$, where $X_{\rm gal} = 2.3 \times 10^{20}$ $cm^{-2}$ (K km $s^{-1})^{-1}$. The factor $X$ is defined such that $X = M(H_2)/L_{\rm CO}$. To compute the $X$ factor for molecular clouds we used as mass their virial mass, $M_{\rm vir} = 190\ R\Delta v^2$. The points in this figure show a large spread but some trends are evident. Inner Galaxy clouds tend to concentrate around the canonical value for the Galaxy, $X_{\rm gal}$, determined from other methods (Strong et al. 1988). That is, for inner galactic clouds $X/X_{\rm gal} \approx 1$. Outer Galaxy clouds exhibit a larger spread, but with an

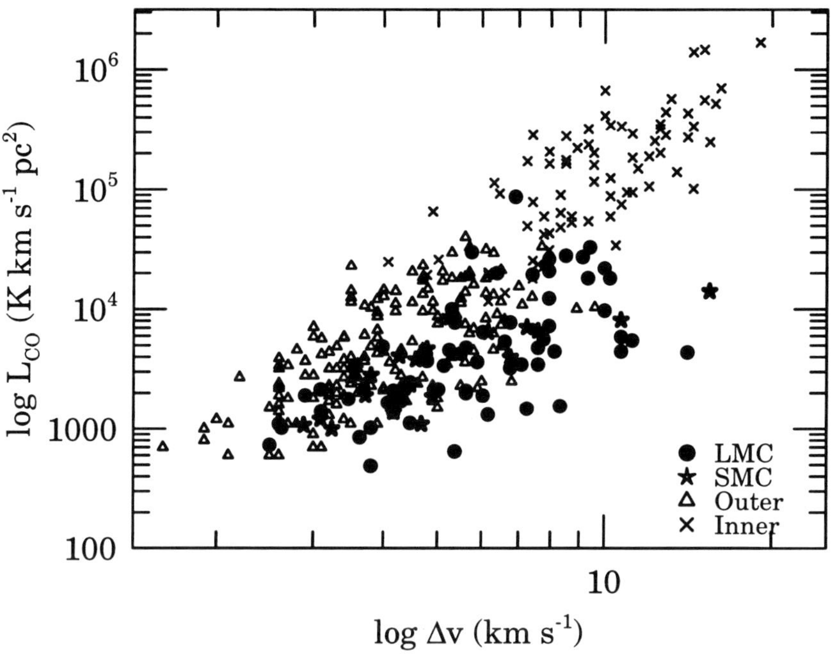


*Figure 1.* CO luminosity versus line width for molecular clouds in the LMC (black dots), SMC (black stars), outer Galaxy (open triangles) and inner Galaxy(crosses).

average value of $\log[X/X_{\rm gal}] = 0.4$, indicating that the conversion factor is about 2.5 times larger than the Galactic value (May et al. 1996). For SMC clouds, Rubio et al. (1993b) found that $X_{\rm SMC}$ is larger than the Galactic value, and that in addition this factor depends on the scale size, $X_{\rm SMC}/X_{\rm gal} = 3.9(R/10{\rm pc})$.

CO clouds in the LMC and SMC cover a similar range of sizes as outer Galaxy clouds, while exhibiting larger values of $X$, with an average value of about 4 times larger than the canonical Galactic value. Johansson (1991) found an $X$ value for the 30 Doradus clouds of about two times larger than the Galactic value, and Israel and de Graauw (1991) found an $X$ value of three times the Galactic one for the N11 clouds. The spread of the $X$ values in the Magellanic Clouds is about one order of magnitude. An explanation of this result would require investigation into the dependence of $X$ on the local ambient conditions of the molecular clouds. LMC and SMC molecular clouds seem to share similar properties. In both Figures 1 and 2, no appreciable difference is seen between LMC and SMC clouds. They follow the same CO luminosity-line width relationship and are indistinguishable in the

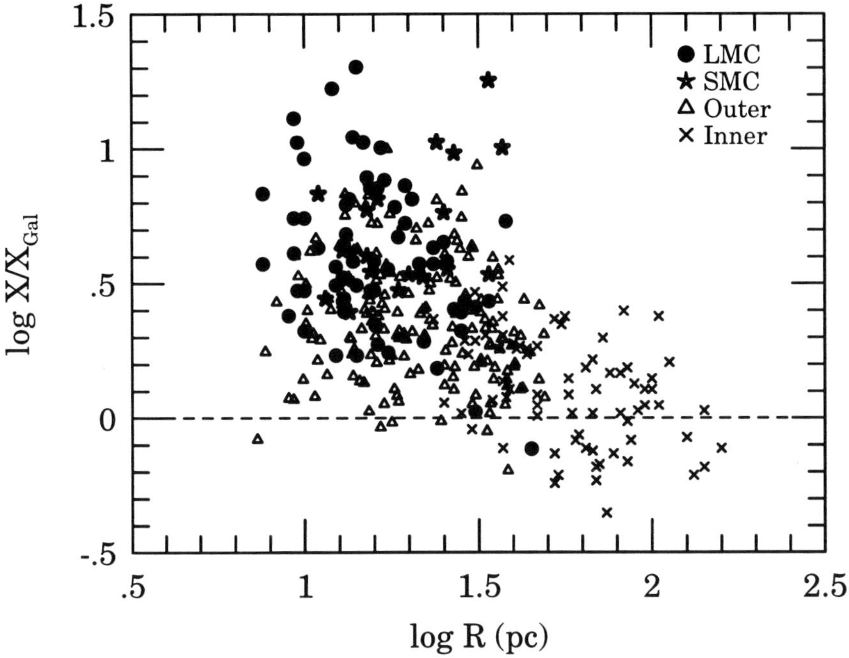


*Figure 2.* Ratio of the conversion factor $X/X_{gal}$ for molecular clouds in the LMC (black dots), SMC (black stars), outer Galaxy (open triangles) and inner Galaxy (crosses) versus size.

conversion factor versus size dependence (see Figure 2). Clearly, more work is needed to determine how local ambient conditions, e.g, photodissociation, metallicity, etc, affect the properties of the CO clouds. The suggestion that the conversion factor depends linearly on metallicity (Cohen at al. 1988) or that the $X$ factor decreases with metallicity (Rubio et al. 1991, Verter & Hodge 1995) is not as straightforward as proposed. This result is currently being investigated and will be one of the main outcomes of this long term project.

We would like to acknowledge support from the Local Organizing Committee of the Symposium. We wish to thank G. Garay for interesting discussions. This research has been supported by FONDECYT (Chile) through Grant#1930928.

## References

Booth, R.S., 1991, New Aspects of Magellanic Cloud Research, in Baschek, B., Klare, G.,& Lequeux, J., (eds.) Springer-Verlag, p.26

de Graauw, Th., et al. 1996, in preparation
Dufour, J., 1984, Structure and evolution of the Magellanic Clouds, in van den Bergh S., de Boer K.S. (eds) proc. IAU Symp. 108. Reidel Dordrecht, p.353
Elmegreen, B.G. 1989,338, 178
Garay, G., et al. 1996, Ap. J., in preparation
Henize, K., 1956, Ap.J. S. 2, 315
Hodge, P. 1974, PASP, 86, 263
Israel, F.P., de Graauw, Th., 1991, in The Magellanic Clouds, R. Haynes & D.E. Milne (eds.) in IAU Symp. 148. Kluwer Dordrecht, p.45
Israel, F.P., Johansson, L.E.B., Lequeux, J., et al., 1993, A&A 276, 25 (Paper I)
Koorneef, J., 1982, A&A 107,247
Koorneef, J., 1984, Structure and evolution of the Magellanic Clouds, in van den Bergh S., de Boer K.S. (eds) proc. IAU Symp. 108. Reidel Dordrecht, p.333
Kutner, M., Rubio M., 1996, CO: Twenty-five Years of Millimeter-wave Spectroscopy, IAU Symp. 170, these proceedings
Kutner, M., Rubio M., et al. 1996, in preparation
Johansson, L.E.B., 1991 in Dynamics of Galaxies and their Molecular Cloud Distributions, IAU Symposium 146, F. Casoli & F. Combes, ed. Kluwer, Dordr p1
Johansson, L.E.B., Booth, R.S., & Heikkila, A., 1995, CO: Twenty-five Years of Millimeter-wave Spectroscopy, IAU Symp. 170, these proceedings
Lequeux J., Le Bourlot J., Pineau des Forts G., et al., 1994, A&A 292, 371 (Paper IV)
Maloney, P.R., Black, J.H., 1988 ApJ 325,389
May, J., Alvarez, H., Bronfman L., 1996 A&A, submitted
Rubio, M., Garay, G., Montani, J., & Thaddeus, P., 1991, ApJ 368, 173
Rubio, M., 1992, New Aspects of Magellanic Cloud Research, in Baschek, B., Klare, G.,& Lequeux, J., (eds.) Springer-Verlag, p.36
Rubio, M., Lequeux, J., Boulanger, F., et al., 1993a, A&A 271, 1 (Paper II)
Rubio, M., Lequeux, J., Boulanger, F., 1993b, A&A 271, 9 (Paper III)
Rubio, M., Lequeux, J., Boulanger, F., 1996, A&AS submitted, (Paper V)
Schwering, P.B.W., Israel, F.P. 1989, A&AS, 79, 79
Strong, A.W., Bloemen, J.B.G.M., Dame, T. M., Grenier, I.A., Hermsen, W., Lebrun, F., Nyman, L., Pollock, A.M.T., Thaddeus, P. 1988, A&A, 207, 1
Verter, F., Hodge, P., 1995, ApJ 446, 616

# CO IN NEARBY NORMAL GALAXIES

YOSHIAKI SOFUE
*Institute of Astronomy, University of Tokyo, Mitaka, Tokyo 181, Japan*

**Abstract.** We review CO line observations of nearby galaxies, particularly those with high-resolution and high-sensitivity using large aperture telescopes such as the Nobeyama 45 m and and IRAM 30 m telescopes. Based on the observations, we discuss kinematics of the molecular gas, including the inner rotation curves of galaxies, and interstellar physics including the galactic scale HI–molecule transition and the CO to $H_2$ conversion factor.

## 1. Introduction

CO observations of galaxies provide information about: (A) Distribution and approximate mass of molecular hydrogen gas. Face-on galaxies show us the radial distribution and spiral arm related density enhancements. High-resolution mapping of edge-on galaxies tells us about the vertical extent of the molecular disk and halo. Accurate information on gas masses cannot be obtained as yet because of the uncertainty in the CO-to-$H_2$ conversion factor. (B) Kinematics of the molecular gas, particularly in the inner region of galaxies, thanks to the high velocity resolutions. This provides detailed rotation curves for the inner 1 kpc, kinematics of non-circular motions related to bar-shocked gas, which is related to triggering enhanced star formation in the nuclei, and density wave kinematics in the disk. And (C) Physical properties of the interstellar molecular gas, such as the temperature, density, or clumpiness from multi-transition observations.

The major constituents of interstellar matter (ISM) in spiral galaxies are HI and $H_2$ gas, which share several to ten percent of the total mass. HI gas distributions have been determined for many galaxies from high-resolution maps. It is, however, only recently that a number of galaxies have been mapped in the CO line with high resolution and sufficient sensitivity by using large filled aperture dishes like the Nobeyama 45 m and



*W.B. Latter et al. (eds.), CO: Twenty-Five Years of Millimeter-Wave Spectroscopy,* 273–281.

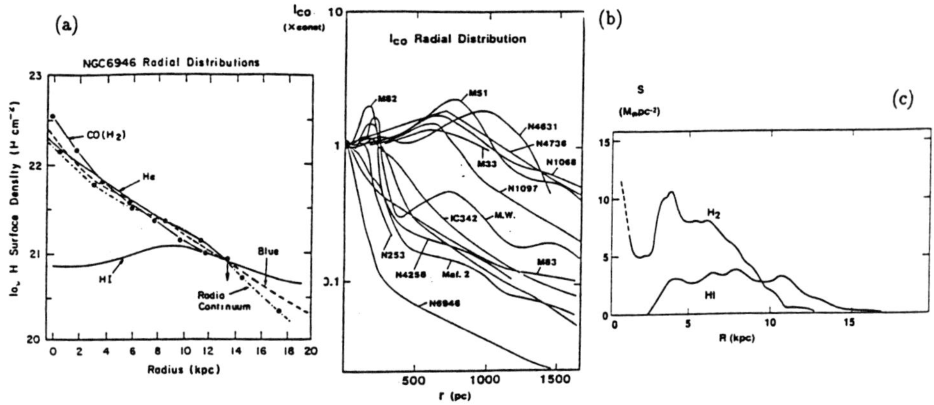


*Figure 1.* (a) $I_{CO}$ distribution in the Sc galaxy NGC 6946. (b) $I_{CO}$ distributions in the central regions of disk galaxies. (c) Radial distribution of $I_{CO}$ in the Sb galaxy NGC 891.

IRAM 30 m telescopes. In this review we concentrate on results from large filled aperture telescopes and on nearby, well studied normal Sb and Sc spirals. Interferometric observations are reviewed by Scoville in this volume. Readers may also refer to reviews on CO in galaxies by Morris & Rickard (1982); Young & Scoville (1991); Combes (1991); and Sargent & Welch (1993).

## 2. Molecular Gas Distribution

### 2.1. RADIAL DISTRIBUTION

The global distribution of the CO line intensity indicates the molecular disks of late type (Sc, Sd) spiral galaxies are characterized as exponential disks with nuclear concentrations and superposed spiral arms. The central region often shows strong non-circular motions caused by a bar induced shock wave and inflow. In addition to these features, Sb and Sa galaxies usually show a large scale ring structure with a radius of a few kpc, similar to the 4 kpc molecular ring in the Milky Way. Fig 1. shows the $I_{CO}$ distribution in the Sc galaxy NGC 6946, indicating an exponential disk (Young & Scoville 1982, 1991). The central region of this galaxy also has a steeper-exponential, high density disk (Fig. 1b) (Sofue 1990). The central regions of other galaxies generally exhibit more ring-like structures with radii of several hundred pc (Fig. 1b). The radial $I_{CO}$ distribution in the Sb galaxy NGC 891 (Fig. 1c) shows a several-kpc molecular ring (Sofue & Nakai 1993).

In higher resolution CO maps, spiral arms show up in positional coincidence with the optical dark lanes inside the spiral arms. Optical arms,

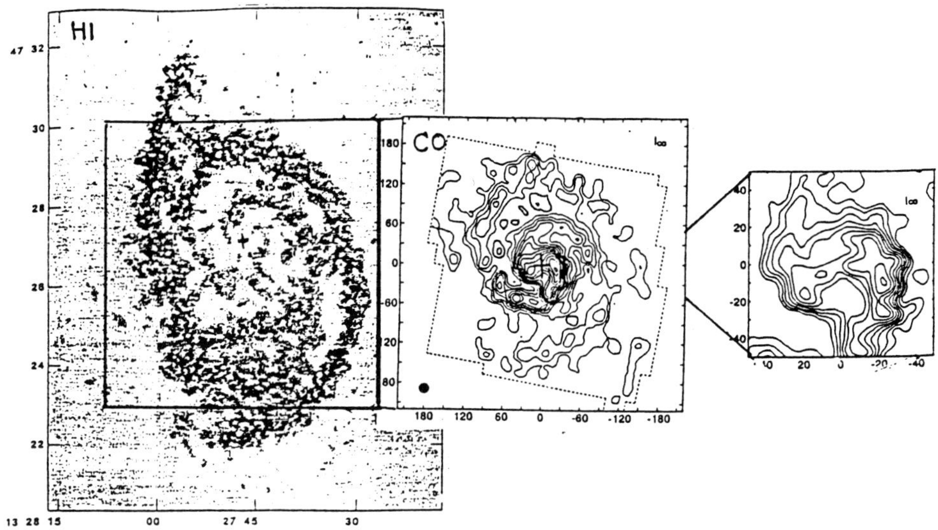


*Figure 2.* CO intensity distribution in the face-on galaxy M51 (Nakai et al. 1994) compared with a high-resolution HI map (Rots 1990).

particularly those seen in H$\alpha$ emission and indicating massive star forming regions, are shifted slightly downstream from the CO spiral arms by 0.5–1 kpc in the azimuthal direction ($\sim$ 100 pc perpendicular to the arms) due to the time lag of $\sim 10^7$ yrs between the galactic shock compression of gas and star formation. Fig. 2 shows the CO intensity distribution in the face-on galaxy M51 observed with the Nobeyama 45 m telescope (Nakai et al. 1994) in comparison with a high-resolution HI VLA map (Rots et al. 1990). It is clear in this figure that the molecular gas dominates in the inner 3 kpc region, while HI dominates in the outer disk.

## 2.2. VERTICAL DISTRIBUTION

High resolution CO images of edge-on galaxies are useful for investigating the vertical extent of the molecular gas disk and halo. Fig. 3 shows the variation of the CO intensity perpendicular to the disk plane in the edge-on galaxy NGC 891 (Garcia-Burillo et al. 1992; Sofue & Nakai 1993). The molecular disk is unresolved, but a deconvolution of the profile using an assumed beam indicates the molecular disk is about 100 pc thick. Further data indicate the thickness increases with the distance from the center. In addition to this thin and dense disk, a thick halo component is detected, extending about $\pm 1$ kpc above the galactic plane. The amount of molecular gas in this extended component amounts to a few tens of percent of the total CO emission, indicating a significant molecular gas in the halo.

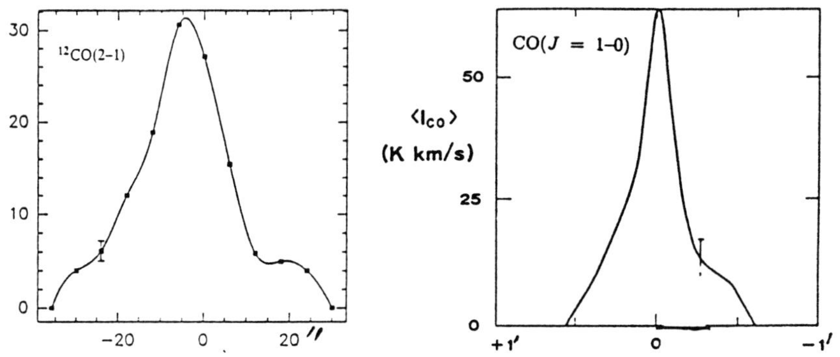


*Figure 3.* Variation of the CO intensity perpendicular to the disk plane in the edge-on galaxy NGC 891 (Garcia-Burillo et al. 1992; Sofue & Nakai 1993)..

## 3. Kinematics and Rotation

### 3.1. INNER CO + OUTER HI ROTATION CURVES

The rotation curves of galaxies have been obtained from optical and HI 21 cm line emission observations along the major axes (e.g., Rubin et al. 1982). Optical measurements are affected by the contamination of the bright bulge light, which increases the uncertainty of the curve near the center. The HI gas distribution has a depression in the central few kpc region, which yields an apparently solid rotation curve for the central few kpc. The CO-line emission on the other hand is concentrated in the central region, so it gives the most accurate rotation curves of the inner few kpc region.

A rotation curve can be derived by using the loci of terminal velocities in a PV diagram. Therefore, the velocity dispersion of the interstellar gas and the velocity resolution of observations must be corrected properly. Fig. 4a shows an example of a composite PV diagram for NGC 891 using CO data from Sofue & Nakai (1993) and HI data from Rupen (1991). The HI gas traces the rotation of the outer disk, whereas the CO emission traces the rotation in the innermost region including the rapidly rotating nuclear disk (see also Garcia-Burillo et al. 1992; Scoville et al. 1993). In Fig. 4b, we present rotation curves obtained this way for several galaxies (Sofue 1996). Generally, the rotation velocity rises steeply within a few hundred pc, indicating the existence of a central compact mass component. Many galaxies (the Milky Way, NGC 891, NGC 3079, NGC 6946) exhibit a sharp maximum at $R \sim$ a few hundred pc, reaching a velocity as high as $\sim 200$ to 300 km s$^{-1}$.

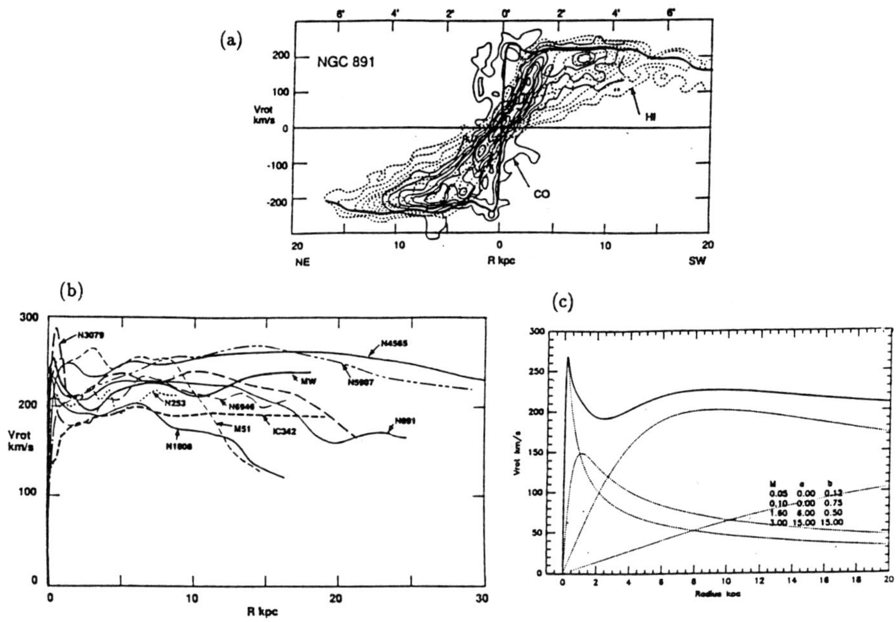


*Figure 4.* (a) CO + HI position-velocity diagram for NGC 891. (b) Most completely sampled rotation curves of galaxies. (c) A calculated rotation curve for a four-component model.

## 3.2. FITTING BY A POTENTIAL WITH FOUR MASS COMPONENTS

The observed rotation curves can be fitted by the Miyamoto-Nagai (MN) (1975) potential with multiple mass components. The rotation velocity is calculated from $V_{\rm rot} = (R\partial\Phi/\partial R)^{1/2}$, where $\Phi$ is the MN potential. The usual three-component model (bulge, disk, and halo) is not sufficient to fit the steep central peak, so we have to introduce a fourth, compact nuclear component. Fig. 4c shows an example of a calculated rotation curve of this four-component model ($n = 4$), where we assumed (1) nuclear compact mass, (2) bulge, (3) disk, and (4) massive halo components. This nuclear mass component has significant implications for the formation and evolution of the galactic bulge and the central mass condensation in galaxies.

## 3.3. VELOCITY FIELDS: NON-CIRCULAR MOTIONS

A velocity field, defined as the distribution of the intensity weighted mean velocity over a galaxy, provides information not only about rotation, but also about non-circular motions. Non-circular motions can arise from density waves, galactic shocks, bar-induced shocks, and inflows. The CO VF is useful to trace these effects in the central region of a galaxy, while the HI

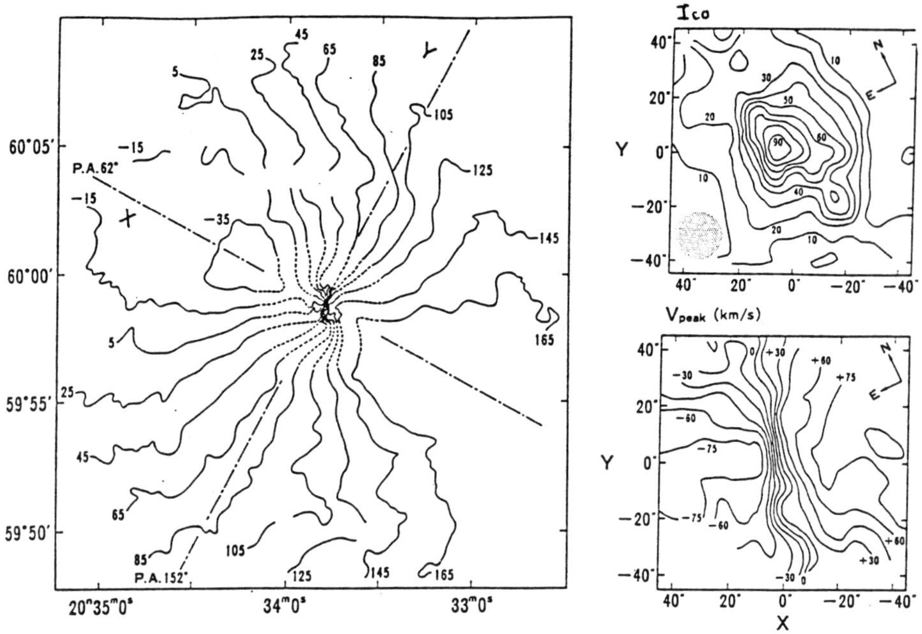


*Figure 5.* HI + CO velocity field for NGC 6946.

VF can be used to study the motion of the outer disk and outskirts (Bosma 1981). Fig. 5 shows an HI + CO velocity field for the face-on galaxy NGC 6946 (Sofue et al. 1988). Spiral arms show up by tracing loci with corrugated isovelocity contours, indicating velocities distorted from the circular motion

by a few to ten percent of the rotation ($\sim 10 - 30$ km s$^{-1}$). In M51, however, a larger non-circular motion has been detected in the spiral arms, exceeding 50 km s$^{-1}$ (Garcia-Burillo et al. 1993). This is too large for a density wave and poses a crucial problem for arm formation theory. In general, the inner regions show larger non-circular motions, amounting to a few tens of percent of the rotation velocity ($\sim 30 - 50$ km s$^{-1}$) and much larger non-circular distortions are observed in the nuclear disks with interferometric measurements (e.g., Ishizuki et al. 1990). Such highly distorted velocity fields have been simulated extensively and suggest existence of a nuclear bar potential (e.g., Wada & Habe 1996 and the literature therein).

## 4. Physics of ISM in Galaxies

### 4.1. GALACTIC SCALE HI–$H_2$ TRANSITION: MOLECULAR FRONT

The HI and the gas density distributions derived from the HI and CO line intensities can be used to obtain distribution of the molecular fraction, de-

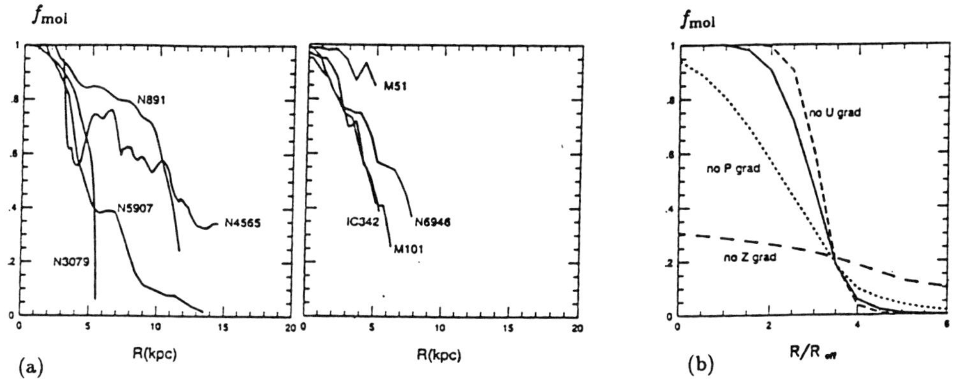


*Figure 6.* (a) Distributions of $f_{mol}$ for several galaxies. (b) Model molecular front.

fined as $f_{mol} = \rho_{H_2}/(\rho_{HI} + \rho_{H_2})$. Fig. 6 shows the derived $f_{mol}$ distributions for several galaxies (Sofue et al. 1994; Honma et al. 1995). The figure shows the ISM in the central few kpc is almost totally molecular, while the ISM is almost totally HI in the outer disk. The HI to $H_2$ transition occurs suddenly in a narrow range of radius, which we call the *molecular front.* The HI and $H_2$ disks are clearly separated by this front. The front will be the place where the phase transition from HI to $H_2$, and vice versa, is taking place on a galactic scale.

The observations can be well explained by a phase-transition theory proposed by Elmegreen (1993). The critical parameters for determining the molecular fraction $f_{mol}$ are the interstellar pressure $P$, the UV radiation field $U$, and the metallicity $Z$. We constructed a model galaxy in which $P$, $U$, and $Z$ are exponentially distributed with the radius $r$, $\propto \exp - r/r_e$, where $r_e$ is the disk scale radius. Fig. 6b plots the calculated results, which indicate the molecular front is a fundamental feature of an exponential disk galaxy. Results for constant $P$ and $Z$ are also shown by dashed and dotted lines, but cannot reproduce the observations. The molecular front will have an important effect on the evolution of the interstellar matter in galaxies, which should be deeply coupled with the chemical evolution of galaxies from their birth.

## 4.2. THE CO-TO-$H_2$ CONVERSION FACTOR AND METAL ABUNDANCE

The mass of molecular gas has been derived from the CO intensity. This is, however, still very crude because of uncertainty in the conversion factor. The column density of $H_2$ molecules is assumed proportional to the integrated intensity of the $^{12}$CO ($J = 1-0$) line, $N(H_2)$ [cm$^{-2}$] $= X \int_{V_{min}}^{V_{max}} T_B dv$ [K km s$^{-1}$]. Here $X = X^* \times 10^{20}$ is the conversion factor, for which we of-

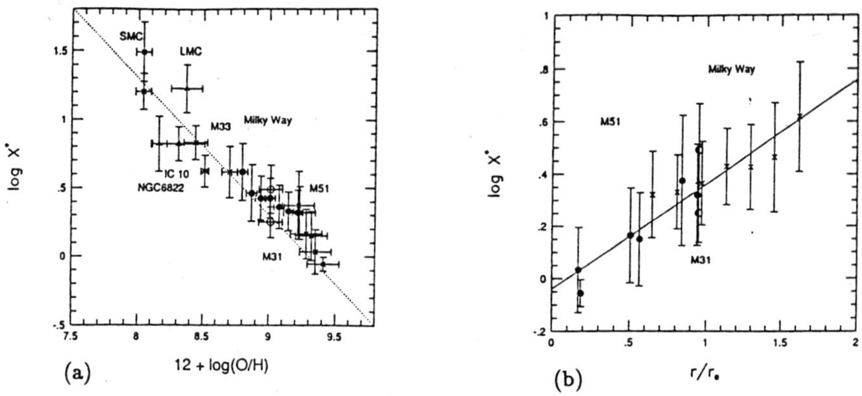


*Figure 7.* (a) Conversion factors for nearby galaxies against metallicity. (b) Conversion factor as a function of the distance from the galactic center.

ten adopt the empirical value obtained for molecular clouds in the Galaxy, $X \sim 2 - 3 \times 10^{20}$ (Bloemen et al. 1986).

The conversion factor varies by more than a factor of several between the Galactic Center and the outer arms and by more than an order of magnitude from normal spirals to metal-poor dwarfs like the Magellanic Clouds to the metal-rich galaxy M51 (Arimoto et al. 1996; Nakai & Kuno 1995). The conversion factor should be a complex function of interstellar conditions (metal abundance, density, temperature, and radiation fields). Arimoto et al. (1996) have examined the dependency of $X$ on metallicity, as it is the best studied quantity among the others in external galaxies. They used $X$ values derived by the virial-mass method. Fig. 7a shows conversion factors obtained in nearby galaxies against the O abundance, $12 + \log \mathrm{O/H}$. The diagram demonstrates that $X$ is proportional to the metallicity and can be fit by $\log X^* = -1.0(12 + \log \mathrm{O/H}) + 9.30$. It is also known that O/H varies with absolute magnitude $M_B$ of galaxies, $12 + \log \mathrm{O/H} = -0.20 M_B + 4.86$ (Roberts & Haynes 1994). We can then estimate a typical $X$ value for a galaxy with known luminosity, $\log X^* = 0.20 M_B + 4.44$. This figure indicates that $X$ is a simple function of metallicity in individual regions (clouds).

### 4.3. IMPLICATION OF THE CONVERSION FACTOR ON THE STUDY OF STAR FORMING EFFICIENCY AND DISK STABILITY

As the metallicity decreases with radius within a disk galaxy, the metallicity dependence of $X$ implies a radial gradient in individual galaxies. In Fig. 7b we plot $X^*$ as a function of radius $r$ normalized by the effective radius $r_e$ of the stellar disk for the galaxies, M51 and M31. The plot can be fit by $\log X^* = 0.37 + 0.40(r/r_e - 1)$. If we apply this radial gradient in deriving molecular hydrogen masses, we may expect a factor of two or three less molecular gas in the central region than previously estimated. For example, Fig. 1 shows the CO intensity is proportional to the blue light intensity. If the conversion factor is smaller at the center, this means the SF efficiency is higher in the central region, which will also affect the current studies of star formation (e.g., Kennicut 1989).

The smaller gas mass in the central region would be crucial for the stability analysis of the nuclear gas disk. If the gaseous mass is by a factor of three or more smaller than that so far estimated, the nuclear gas disk will be no longer self-gravitating as often mentioned in theoretical models. This implies that the gas kinematics of the central regions of galaxies can be a good indicator of the stellar mass potential.

## References

Arimoto, N., Sofue, Y., & Tsujimoto, T. 1996 A&A, submitted.
Bloemen, J. B. G. M. et al. 1986, A&A 154, 25
Bosma, A. 1981, AJ, 86, 1825.
Combes, F. 1991 ARAA, 29, 195.
Elmegreen, B. G. 1993, ApJ 411, 170.
Garcia-Burillo, S., Guélin, M., & Cernicharo, J. 1993, A&A, 274, 123.
Garcia-Burillo, S., Guélin, M., Cernicharo, & J., Dahlem, M. 1992, A&A, 266, 210.
Honma, M., Sofue, Y., & Arimoto, N. 1995 A&A 296, 33.
Ishizuki, S., Kawabe, R., Ishiguro, M., Okumura, S. K., Morita, K. -I., Chikada, Y., Kasuga, T., & Doi, M. 1990 ApJ 355, 436
Kennicutt, Jr . R. C. 1989 ApJ 344, 685.
Miyamoto, M., & Nagai, R. 1975, PASJ **27**, 533.
Morris, M., & Rickard, L.J. 1982, ARAA, 20, 517.
Nakai, N., & Kuno, N., 1995 PASJ, 47, 761.
Nakai, N., Kuno, N., Handa, T., & Sofue, Y. 1994, PASJ 46, 527.
Roberts, M. & Haynes, M. P. 1994, ARAA 32, 115.
Rots, A. H., Bosma, A., van der Hulst, J. M., Athanassoula, E., & Crane, P. C. 1990, AJ 100, 387.
Rubin, V. C., Ford, W. K., & Thonnard, N. 1982, ApJ, 261, 439
Rupen, M. P. 1991, AJ 102, 48.
Sargent, A. I., & Welch, W. J. 1993, ARAA 31, 297.
Scoville, N. Z., Thakker, D., Carlstrom, J. E., & Sargent, A. E. 1993 ApJ 404, L63.
Sofue, Y. 1996 ApJ 458, 120.
Sofue, Y., & Nakai, N. 1993, PASJ 45, 139
Sofue, Y., Honma, M., & Arimoto, N. 1994 A&A 296, 33.
Sofue, Y., Doi, M., Ishizuki, S., Nakai, N., & Handa, T. 1988 PASJ 40, 511.
Sofue, Y., & Nakai, N. 1993, PASJ 45, 139.
Young, J. S., & Scoville, N. 1982, ApJ 258, 467.
Young, J. S., & Scoville, N. Z. 1991, ARAA, 29, 581.
Wada, K., & Habe, A. 1996, MNRAS, in press.

# MOLECULAR GAS AND STAR FORMATION IN GALAXIES

JUDITH S. YOUNG
*Department of Physics and Astronomy and FCRAO*
*University of Massachusetts*
*Amherst, MA 01003*

## 1. Introduction

Using the 14m millimeter telescope of the Five College Radio Astronomy Observatory (HPBW=45″), we have observed the CO J=1–0 emission at 1412 positions in 300 galaxies. These observations, which comprise the *FCRAO Extragalactic CO Survey*, are the most extensive and uniform database on the content and large-scale distribution of molecular gas content in galaxies (Young et al. 1995). Since stars form in molecular clouds, these observations enable us to investigate galaxy evolution through a determination of the star formation efficiency and the molecular-to-atomic-gas ratio within and among galaxies.

The galaxies observed as part of the *Survey* were selected at optical and infrared wavelengths to cover a wide range of morphological type, luminosity, and environment; 260 of the 300 galaxies are classified as spiral (S0 – Sm). Among galaxies that are either brighter than 11th magnitude in the blue or brighter than 15 Jy at 100 $\mu$m, the *Survey* is 80% complete. CO was detected in 236 galaxies (79%) and observations were made along the major axis in 193 galaxies. Since many of the galaxies in the *Survey* are at distances of ~20 Mpc, the regions we are observing are at least a few kiloparsecs across.

The $H_2$ masses used in this work were derived from CO observations assuming a constant CO→$H_2$ proportionality (Bloemen et al. 1986). The global $H_2$ masses derived for luminous spirals have been shown to be statistically accurate to ~ 30% (Devereux and Young 1990) and the absolute value of the CO→$H_2$ proportionality for the MiLky Way has been confirmed in the Local Group spirals M31 (Sb) and M33 (Scd) by observations which resolve individual molecular clouds (cf. Figure 1 in Young and Scoville 1991).



*W.B. Latter et al. (eds.), CO: Twenty-Five Years of Millimeter-Wave Spectroscopy*, 282–285.

## 2. Molecular Gas Distributions

Since both the Milky Way and Andromeda have molecular rings with radii ~5 kpc, it is of interest to determine the frequency of such large molecular rings in other spiral galaxies. In fact, among the galaxies whose CO distributions were determined in the *Survey*, large-scale molecular rings are very unusual. Only 7 out of the 193 galaxies that were mapped exhibit CO rings with radii $\geq$ 5 kpc. The galaxies with molecular rings are primarily early-type unbarred spirals.

For ~150 of the galaxies in the *Survey*, we have characterized the extent of the CO in the disk by determining CO isophotal diameters (for details, see Young et al. 1995). Among these galaxies, we find the mean ratio of CO-to-optical isophotal diameters is 0.52±0.02; that is, the CO typically extends over half of the optical disk. Additionally, we find a trend along the Hubble sequence such that the CO extends over a smaller fraction of the optical disk for the early-type spiral galaxies while it extends over a larger fraction of the optical disk among Sc galaxies.

With regard to the form of the CO radial distributions in galaxies, it has been known for over a decade that the CO, blue light, H$\alpha$, far-infrared, and radio continuum emission all display similar exponentials in Sc galaxies, while the HI distribution is typically flat across the disk (Young and Scoville 1982, 1991). Thus, the ratio of molecular to atomic gas decreases as a function of radius in these galaxies.

## 3. The Star Formation Efficiency (SFE)

One of the most fundamental parameters describing the star formation activity in a galaxy is the ratio of the current star formation rate to the mass of molecular gas available to form stars, which we call the star formation efficiency. We have used both the luminosity in the far-infrared, from IRAS data, and the luminosity in H$\alpha$, from CCD imaging (Young et al. 1996), as measures of the star formation rate and we find similar results using both tracers.

To date, of all the investigations of the star formation efficiency in galaxies, the largest effect that has been found is the factor of ~7 to 10 elevated star formation efficiency in interacting galaxies relative to isolated galaxies (Young et al. 1986; Sanders et al. 1986; Solomon and Sage 1988).

Another global property that appears to correlate with an elevated star formation efficiency is the bar morphology of a galaxy. Among 219 of the *Survey* galaxies for which bar classifications are available in the literature, 141 are barred or mixed, and 78 are unbarred. The mean value of the star formation efficiency for the barred galaxies is a factor of ~2 higher than for the unbarred galaxies, a trend that is present in both the early- and late-type spiral galaxies, independently (Young 1993).

Finally, investigations of the star formation efficiency in galaxies along the Hubble sequence indicate that there is no variation in the mean SFE with Hubble type among spiral galaxies of types Sa-Sc. That is, the mean global star formation efficiency is independent of the spiral morphology of a galaxy (using L(IR) to trace the star formation rate—Rengarajan and Verma 1986; Devereux and Young 1991; using L(H$\alpha$) to trace the star formation rate—Young et al. 1996). Therefore, it appears that spiral disks are similar and that disk morphology has little to do with the global rate at which molecular gas is converted into stars.

## 4. Molecular and Atomic Gas

Since stars form from molecular clouds, and molecular clouds form from atomic gas, both processes are of fundamental importance in the evolution of galaxies. The above results indicate that the star formation rate in spiral disks is proportional to the available mass of molecular gas, independent of Hubble type. The situation is very different for the comparison of the molecular and atomic gas masses in galaxies.

For 178 galaxies of types Sa-Sc, we find that the dominant neutral gas phase in spiral galaxies changes smoothly along the Hubble sequence, with the mean ratio of molecular to atomic gas ranging from 2±1 among Sa galaxies to 0.4±0.2 for the Sc galaxies, as suggested by Young and Knezek (1989). Additionally, the mean neutral gas surface density (molecular+atomic) increases only a factor of 2 from Sa to Sc galaxies. Thus, it appears that the dominant neutral gas phase changes along the Hubble sequence, with more molecular gas in the early type spirals and more atomic gas in the late type spirals.

## 5. Summary and Conclusions

The trends revealed with regard to the star formation efficiency and the molecular-to-atomic-gas ratio in galaxies as a function of morphology and environment provide important constraints for theories of the evolution of galaxies.

First, it is clear that the SFE in spiral galaxies is independent of morphology, so that the global rate of star formation depends primarily on the global quantity of molecular gas. Apparently, once molecular clouds are formed, star formation proceeds in a way that is similar from one galactic disk to another. However, forming the molecular clouds is another matter entirely and there is a significant difference in the molecular-to- atomic-gas ratio as a function of morphology—early-type spirals are predominantly molecular, while late-type spirals are predominantly atomic. In other words, the dominant phase of the neutral gas in galaxies changes with the degree of central mass concentration, or the depth of the central potential.

These results can be simply understood in terms of the effects of gravity in cloud and star formation and the size scale on which gravity must operate. In order for molecular clouds to form from atomic gas, it is not surprising that gravity on the large scale should be important. This is reflected in the $H_2$/HI ratio changing with type or, equivalently, with the mass distribution of the galaxy. In contrast, it appears that star formation in unbarred, isolated, spiral galaxies occurs with the same mean efficiency independent of morphological type. This suggests that once the molecular clouds are formed, star formation within them is a local process, due to gravity on the small scale. These results suggest that spiral disks are similar and that the Hubble sequence is primarily a sequence of changing bulge-to-disk ratio. Only when a large scale perturbation is present, such as a bar or a tidal encounter with another galaxy, does the global star formation rate per unit mass of molecular gas become elevated.

In the galaxies surveyed, the global gas cycling times range between $10^8$ and $10^{10}$ years, although this can vary dramatically with position within individual galaxies as well. In M82, for example, the gas in the center will be cycled into stars in less than $10^9$ years, while the outer parts can sustain star formation for $\sim 10^{11}$ years. Thus, gas depletion may not occur everywhere in a galaxy and may not occur at all if only high mass stars form.

## References


Bloemen, et al. 1986, Astr. Ap, 154, 25.
Devereux, N., & Young, J.S. 1990, Ap.J., 359, 42; 1991, ApJ., 371, 515.
Sanders, D., et al. 1986, Ap.J.(Letters), 305, L45.
Solomon, P., & Sage, L. 1988, ApJ., 334, 613.
Young, J. 1993, in The Environment and Evolution of Galaxies, p. 295.
Young, J., Allen, L., Kenney, J., Kleinmann, S., et al. 1996, A.J., submitted.
Young, J., Kenney, J., Tacconi, L., Claussen, M., et al. 1986, ApJ.Letters., 311, L17.
Young, J., & Knezek, P. 1989, Ap.J.(Letters), 347, L55.
Young, J., & Scoville, N.Z. 1982, ApJ., 258, 467.
— 1991, Ann. Rev. Astr. Ap., 29, 581.
Young, J., Xie, S., Tacconi, L., Knezek, P., et al. 1995, Ap.J.Suppl., 98, 219.

# STAR FORMATION AND GAS CONTENTS IN DISK GALAXIES: COMPLEX RELATIONSHIPS

F. CASOLI
*DEMIRM, Observatoire de Paris*
*61 av. de l'Observatoire*
*75014 Paris (France)*

## 1. Introduction

Star formation activity is highly variable in galaxies. Can we relate it to the gas content of the galaxy? To the gas composition, the relative amounts of available molecular and atomic hydrogen? Or are these variations mainly due to the galaxy environment? To study this problem, we made a statistical study of the links between the star formation indicators and gas tracers in a sample of 469 disk galaxies. Since the molecular phase is the place of star formation, information about the $H_2$ content is crucial. Hence, the sample galaxies have all been observed in the $^{12}CO(1–0)$ line. We use a conversion factor $X = 2.3\ 10^{20}$. We studied the global properties of the sample objects: total gas content and global star formation indicators at various wavelengths from the optical colors to the radio continuum emission.

## 2. The Sample

One third of the C0(1–0) data comes from the FCRAO extragalactic survey and from Sage's survey of nearby galaxies (1993). The other CO data have been obtained by us in several surveys, in particular of the Coma region and of isolated galaxies from the Karachenseva catalog. These surveys have been conducted with the NRAO 12m and the SEST 15m telescopes, with some additional data from IRAM. The beamsizes are then respectively 55″ and 45″. All galaxies with $D_{25} > 4'$ have been mapped in order to get all the emission. 109 objects were not detected down to typical noise levels of 2–3 mK in 10 km $s^{-1}$ channels and we take the upper limits into account using a survival analysis technique.



W.B. Latter et al. (eds.), CO: Twenty-Five Years of Millimeter-Wave Spectroscopy, 286–288.

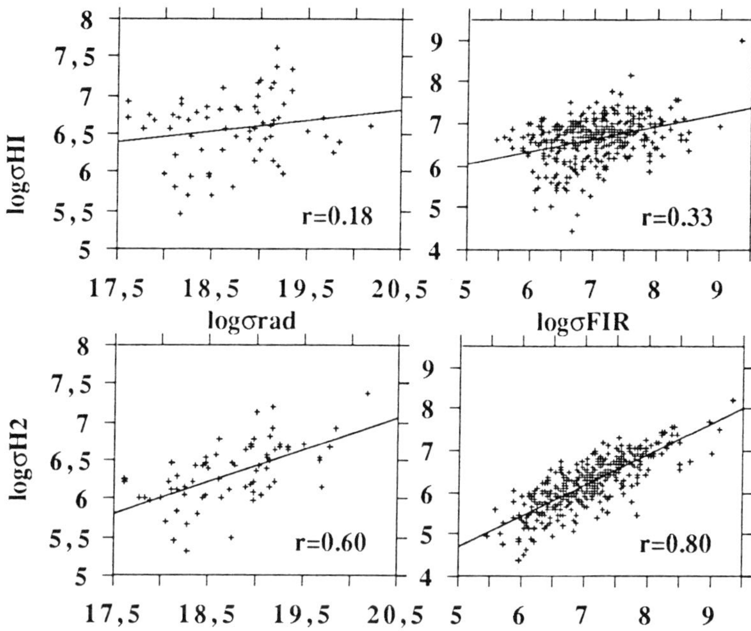


*Figure 1.*

The other available data are the blue magnitudes and diameters (467 galaxies), the 60 and 100 $\mu$m flux densities (425), the HI fluxes (421), the B-V and U-B colors (227), the H-band magnitudes (146), the 21 cm continuum fluxes (69 galaxies in the Coma region), and the H$\alpha$ equivalent widths (68).

These are indicators and not true measurements of the star formation rate. For example, the colors are sensitive to the star formation history integrated over several $10^8$ yr and to a limited range of stellar masses. As for the gas, the $H_2$ masses are derived from the CO line areas with the well-known uncertainty in the conversion factor, $X$. In addition, we have to normalize the observed quantities in order to get rid of the size effect. We have chosen to divide the luminosities and masses by the optical surface of the disk, $D_{25}^2$. We get essentially the same results if we scale by the blue luminosity, but not if we scale by the H-band luminosity.

## 3. Results and Interpretation

The average spiral in the sample has much more HI than $H_2$. For the 322 objects that are not strongly HI-deficient, we find $<\log(M(H_2))/M(HI))> = \log(0.16) \pm 0.04$. We see a slight decrease (a factor 2–3) of $M(H_2)/M(HI)$ for later morphological types.

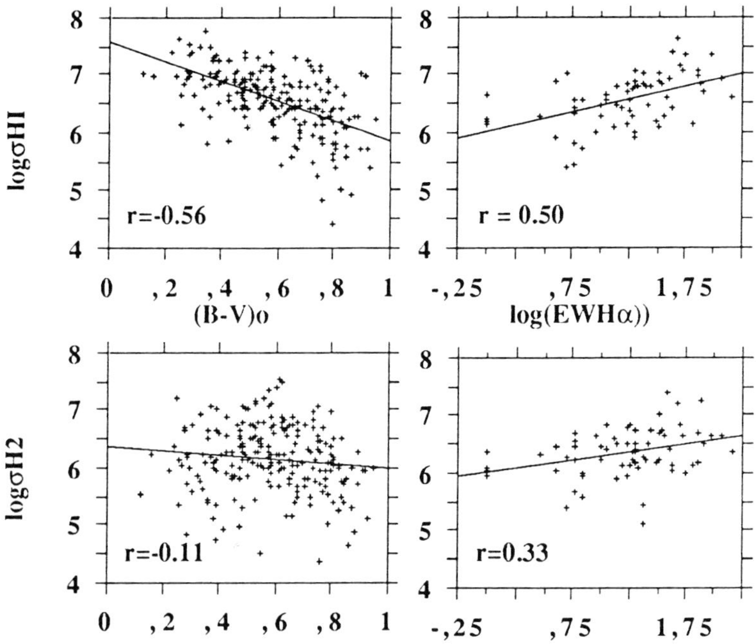


*Figure 2.*

Since stars are born in molecular clouds, we expect a tight relationship between the star formation tracers and the avaiblable $H_2$ link with HI. This is indeed what we see with the FIR and the radio-continuum emission (Figure 1). This also suggests that the CO line traces the available molecular gas well. Things are more complex, however, with the optical star formation tracers. There is little or no relation to the CO emission and only a fair relationship to the HI emission (Figure 2).

Can we explain these unexpected results if the CO emission is not a fair tracer of the $H_2$ mass? But this would explain neither the good CO-FIR and CO-radio relationships, nor the HI-optical relation. The optical data are likely affected by extinction, which could blur the relation with CO. But this cannot explain the correlation with the HI. A third explanation could be that the atomic phase has something to do with star formation: either directly, if for example the transformation of the HI into $H_2$ is a much slower process than the transformation of $H_2$ into stars; or indirectly, if most of the HI present in a galaxy is a by-product of star formation via photodissociation of molecular clouds.

In conclusion, it seems that the relationships between the star formation indicators and the gas tracers in disk galaxies are still poorly understood.

# MOLECULAR GAS IN ULTRALUMINOUS INFRARED GALAXIES

P. M. SOLOMON
*Astronomy Program, SUNY, Stony Brook*
*Stony Brook, N.Y. 11794-2100*

## 1. Introduction

The discovery of CO emission (Wilson, Jefferts & Penzias 1970) 25 years ago opened a major new astronomical window which has led to important advances in many areas of Galactic and extragalactic astronomy. The strength and ubiquity of the CO lines, CO isotopic lines, and other millimeter transitions have provided extraordinary diagnostics for studying molecular interstellar matter in the Milky Way and in external galaxies. All dense interstellar gas is molecular hydrogen, $H_2$ (Solomon & Wickramasinghe 1969), due to the sharp transition from H I to $H_2$ caused by the self-shielding of $H_2$ from photodissociation by absorption in the Lyman bands followed by emission downward into the continuum. The balance between formation on grains and photodissociation yields an optically thick self-shielding layer in all clouds where the total neutral hydrogen density $n(\mathrm{H} + 2\mathrm{H}_2) > 100\ \mathrm{cm}^{-3}$. Since $H_2$ itself has no radio or millimeter transitions and H I is absent, abundant minor constituents like CO become the main tracer of $H_2$. All current star formation takes place in molecular clouds since the first stage of star formation is the formation of high-density interstellar clouds. Molecular clouds and particularly giant molecular clouds (GMC), with masses $> 1 \times 10^5\ \mathrm{M}_\odot$, the most massive objects in the Galaxy are a major component of the interstellar medium.

My assignment from the organizers is to discuss molecular clouds in the context of extragalactic observations. Presumably, this involves an analysis of the differences and similarities between Milky Way GMCs and the sources of millimeter lines from CO and other molecules in galaxies near and far. Within the confines of a 15 minute talk this is a difficult task but the real problem is more fundamental. There are only a few extragalactic



*W.B. Latter et al. (eds.), CO: Twenty-Five Years of Millimeter-Wave Spectroscopy,* 289–298.

molecular clouds observed with sufficient angular resolution to determine their morphology or dynamical mass and these are in a few of the nearest galaxies where the resolution is very marginal compared with the size of individual clouds. Therefore, many important physical properties of molecular clouds, including their mass, size distribution, and average densities cannot be directly determined for extragalactic GMCs. Since most extragalactic clouds cannot be imaged, observations pertain to the ensemble of clouds in a galaxy and indirect measurements including line ratios, use of high-density tracers, and large scale morphology must be used to deduce the physical properties of the molecular ISM. Section 2 discusses the evidence for gravitationally bound molecular clouds in the Milky Way. Milky Way observations are a reasonable foundation for studying clouds in the disks of spiral galaxies. Section 3 presents observations and interpretation of molecular emission from ultraluminous IR galaxies with extremely different physical conditions from clouds in the disk of the Milky Way.

## 2. Gravitationally Bound Giant Molecular Clouds

The physical basis for the use of CO as a quantitative tracer of $H_2$ rests on the origin of the emission from gravitationally bound molecular clouds (Dickman et al. 1986; Solomon et al. 1987). For a single cloud or an ensemble of non-overlapping, virialized gas clouds, the gas mass $M_{gas}$, and CO line luminosity $L_{CO}$, are linearly related by the quantity $\alpha$ given by

$$M_{gas} \,/\, L_{CO} = \alpha = \left(\frac{4\,\mu 1.36}{3\,\pi\,G}\right)^{1/2} \frac{n^{1/2}}{T_b} = 2.6\,\frac{n^{1/2}}{T_b}, \tag{1}$$

where $\mu$ is the mass of an $H_2$ molecule (helium adds 36% by mass), $n[\mathrm{cm}^{-3}]$ is the volume averaged $H_2$ number density in the clouds, and $T_b[\mathrm{K}]$ is the intrinsic (rest frame) brightness temperature of the CO(1–0) line. $M_{gas}$ is given in units of $M_\odot$ and $L_{CO}$ in units of in $\mathrm{K\,km\,s^{-1}\,pc^2}$.

For gravitationally bound clouds the mean density $n$ is determined by the size and velocity linewidth. Maloney (1990) has claimed that the correlation between CO luminosity and virial theorem mass determined directly form size and linewidth measurements is an artifact of the size-linewidth relation and does not prove that molecular clouds are virialized. Maloney repeated the derivation of the observed virial mass from the size-linewidth relation given by Solomon et al. (1987) and turned the argument upside down by suggesting that any size-linewidth relation would produce a virial mass-CO luminosity relation, even if the clouds were not virialized. This argument misses the point and ignores the physics involved in the various methods of mass determination. The strong evidence for gravitationally bound clouds is not the simple correlation between virial mass and CO luminosity; it is (Solomon et al. 1987) the agreement between the absolute value of the virial mass and the mass determined by completely different and totally independent techniques. The dynamical (virial) mass determined for Galactic molecular clouds agrees well with the "luminosity mass" determined from CO line luminosities and integrated intensities calibrated by $\gamma$ ray measurements and also by comparison of $^{13}$CO intensities with extinction in nearby dark clouds. Indeed, for massive molecular clouds the dynamical mass and the luminosity mass agree in general to within a factor of 2. One of these methods involves the use of Newton's gravitational constant G, the other two involve star counting techniques combined with optically thin column densities and the comparison of $\gamma$ rays from molecular and atomic regions. It is not an accident that these 3 techniques basically agree. If molecular clouds were not gravitationally bound the virial mass would turn out wildly different from the CO luminosity mass, as is the case for unbound sources such as outflows from young stars. There is no "dark matter" in molecular clouds. Most of the mass in the molecular ISM is at the high end of the mass spectrum. While there are obviously many small clouds, including "high latitude clouds," and some of them are not gravitationally bound, they do not contribute significantly to the total molecular mass (Heithausen et al. 1993).

## 3. CO Emission from Ultraluminous Infrared Galaxies

We have carried out a systematic CO(1–0) survey (Solomon et al. 1996) with the IRAM 30 m telescope of a large sample of ultraluminous galaxies out to $z = 0.27$. This is part of a multi-line study with a goal of understanding the difference between the star-forming environment in ultraluminous galaxies and spiral galaxies like the Milky Way. A second stage of this investigation involves imaging about 10 galaxies with the IRAM Interferometer (Downes & Solomon 1996). Ultraluminous galaxies, with IR luminosities $> 10^{12}\,L_\odot$ (cf. Sanders et al. 1988), are the most luminous objects in the local universe. The brightest (nearest) ten ultraluminous galaxies (Sanders et al. 1988) are all either merging systems or have tails indicating a recent merger. Besides the very large IR luminosity, the molecular interstellar medium in such galaxies differs from that in normal spiral galaxies in several fundamental respects. First, although these galaxies have high CO luminosities and molecular masses, the ratio of far-infrared to CO luminosity is about an order of magnitude greater than for normal spiral galaxies (e.g., Sanders et al. 1986; Solomon & Sage 1988). In star formation models, this implies a much higher star formation rate per solar mass of molecular gas than in normal galaxies, even gas rich spirals. Second, a large fraction of their molecular mass is at densities much higher than in ordinary giant molecular clouds, as shown by HCN observations which trace gas at densities $> 10^5\,\mathrm{cm}^{-3}$ (Solomon, Downes & Radford 1992). In nearby ultraluminous galaxies, direct interferometric measurements (e.g., Scoville et al. 1991; Radford et al. 1991) show most of the CO, HCN, and molecular mass is concentrated in a small central region (less than 1 kpc). Indeed, estimates of the molecular mass in the central condensation often equal the dynamical mass (e.g., Scoville et al. 1991). This conflict between dynamical mass and $H_2$ mass derived from CO luminosity leads to a new interpretation of the CO luminosity (Downes, Solomon & Radford 1993; hereafter DSR) for a medium which may fill a disk or sphere in the central few hundred parsecs of a galaxy.

## 3.1. CO SURVEY OBSERVATIONS

The sample of 37 sources includes a large fraction of the most luminous IRAS galaxies in the northern sky. Of the ten most luminous galaxies in the optical redshift survey (Strauss et al. 1992) above $\delta > -10°$, our sample includes eight. Twenty galaxies in our sample are more luminous than the prototype IR galaxy Arp 220. The most luminous object in the sample, 14070+0525 with $4 \times 10^{12}\,\mathrm{L}_\odot$ in the far IR, is also the most distant. Figure 1 shows some sample CO spectra and R band images (full data are presented in Solomon et al. 1996).

All of the candidate galaxies, without exception, have high CO luminosities and a high infrared to CO luminosity ratio. Figure 2 is a diagram of FIR luminosity vs. CO luminosity for the galaxies in our sample. For comparison, a well-known galaxy and the trend for normal and weakly interacting spirals (Solomon & Sage 1988) are also shown. For ultraluminous galaxies (but not in general for other or normal galaxies) the CO integrated line intensity, $I_{\mathrm{CO}}$, is tightly correlated with $100\,\mu$m flux density (Fig. 3). In order to explain the high IR to CO luminosity ratio and the tight CO – $100\,\mu$m correlation we (DSR; Solomon et al. 1996) have developed a "black body" model which predicts a high ratio and shows there is a clear upper limit to $L_{\mathrm{FIR}}/L_{\mathrm{CO}}$. The most luminous galaxies are very near this limit. This implies the dust is optically thick even in the far IR (see also Condon et al. 1991), and that the ratio is not a simple indicator of luminosity to molecular mass. The model also leads to a re-interpretation of the molecular mass determination for ultraluminous galaxies.

## 3.2. THE BLACK–BODY MODEL

We assume the CO traces a huge cloud of area filling factor unity and the dust is opaque for $\lambda \leq 100\,\mu$m. The source may be a disk, a torus, a bar, or have an irregular shape. For illustration, we derive the luminosity ratio for

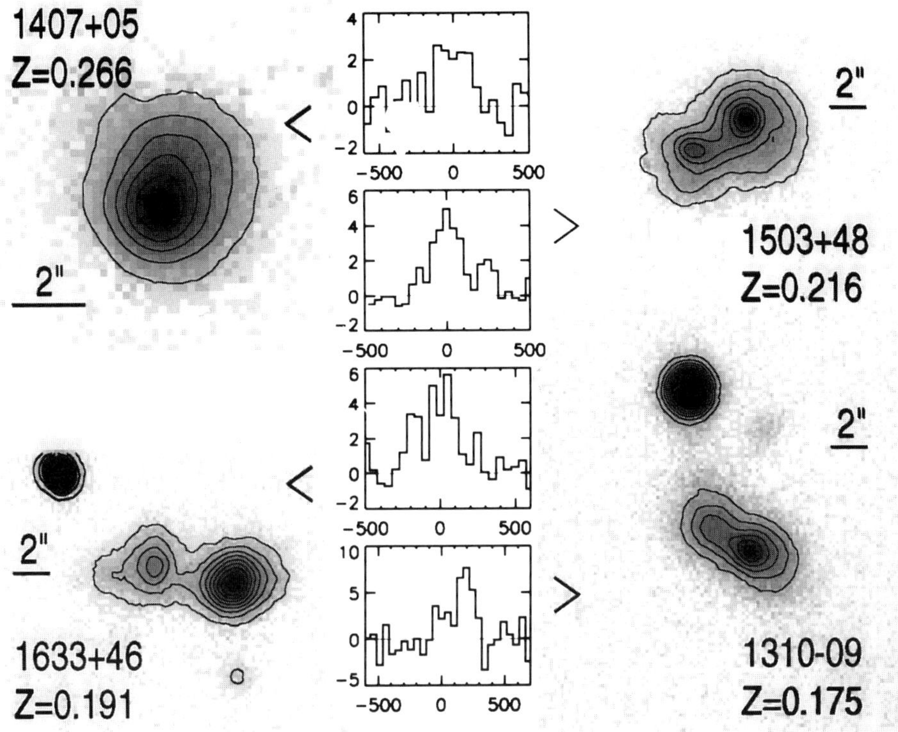


*Figure 1.* Sample IRAM 30 Meter CO Spectra and R band CCD images of ultraluminous galaxies. North is at the top, East is to the left. Next to each image is source name, redshift and CO(1–0) spectrum, with intensity in milliKelvin (from Solomon et al. 1996). The images were taken at the Univ. of Hawaii 3.5 m telescope by Sanders & Kim (1993, priv. communication).

a sphere, but as long as the CO and far IR sources cover the same area, the FIR/CO luminosity ratio is independent of source shape. For an optically thick sphere of temperature $T_d$ at its outer radius $R$,

$$L_{\rm FIR} = 4\pi R^2 \sigma T_d^4 \quad , \tag{1}$$

where $\sigma$ is the Stephan-Boltzmann constant. Similarly, the CO(1–0) luminosity

$$L_{\rm CO} = 4\pi^2 R^2 \frac{2k}{\lambda^2} \int T_b \, d\nu \quad , \tag{2}$$

where $T_b$ is the intrinsic brightness temperature, $\nu$ is frequency, and $L_{\rm CO}$ is in the same units as $L_{\rm FIR}$ (e.g., Watts or $L_\odot$). We assume that regardless of how the gas is clumped in volume, the region has an area filling factor of unity. The temperatures are determined from the far IR colors. If $T_b = T_d$,

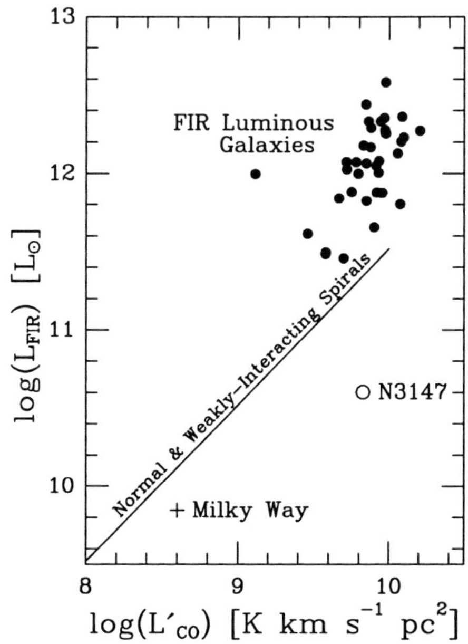


*Figure 2.* FIR luminosity vs. CO(1–0) luminosity (lower scale) and molecular gas mass (top scale). Solid circles indicate galaxies in our sample. Normal and weakly interacting spirals are scattered about the solid line. Open circles show locations in this diagram for one well-known galaxy (from Solomon et al. 1996).

$$\frac{L_{\rm FIR}}{L_{\rm CO}} = \frac{\sigma T_d^3 c^3}{2\pi k \nu_{\rm CO}^3 f_V \Delta V} \quad , \tag{3}$$

where $\nu_{\rm CO}$ is the CO(1–0) rest frequency, $f_V$ the velocity filling factor, and $\Delta V$ the linewidth. For typical values,

$$\frac{L_{\rm FIR}}{L_{\rm CO}} = 4.8\times 10^6 \left(\frac{T_d}{50K}\right)^3 \left(\frac{300}{f_V \Delta V}\right) \quad . \tag{4}$$

$$\frac{L_{\rm FIR}}{L_{\rm CO}} = 4.8\times 10^6 \left(\frac{L_{\rm FIR}}{10^{12}\ L_\odot}\right)^{3/4} \left(\frac{300\ {\rm pc}}{R}\right)^{3/2} \left(\frac{300}{f_V \Delta V}\right) \quad . \tag{5}$$

With the CO luminosity in $\rm K\,km\,s^{-1}\,pc^2$,

$$L'_{\rm CO} \equiv \pi R^2 T_b f_V \Delta V \quad , \tag{6}$$

the ratio becomes

$$\frac{L_{\rm FIR}}{L'_{\rm CO}} = 224 \left(\frac{T_d}{50K}\right)^3 \left(\frac{300}{f_V \Delta V}\right) \quad , \tag{7}$$

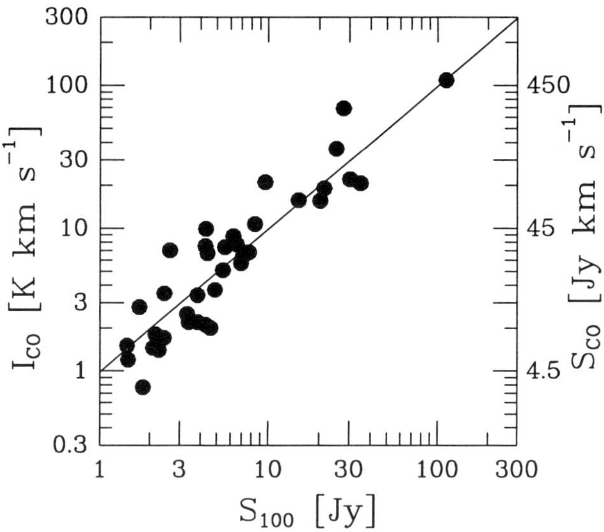


*Figure 3.* Integrated CO(1–0) line intensity vs. IRAS 100 $\mu$m flux density for the ultraluminous galaxies in our sample. The solid line corresponds to the relation $I_{CO} = 1.0 \times S_{100}$ (from Solomon et al. 1996).

where the $L_{FIR}/L'_{CO}$ ratio now has units of $L_\odot$ $(\mathrm{K\,km\,s^{-1}\,pc^2})^{-1}$.

Figure 4 shows that the most luminous galaxies are within a factor of 2 of the black body limit. We can use this model to derive sizes of the FIR and CO emitting regions and the enclosed dynamical masses. Much of the IR and CO emission originates in regions several hundred parsecs in radius. The entire ISM is a scaled up version of a normal galactic disk with the ambient densities a factor of 100 higher, making even the intercloud medium a molecular region. Normal GMCs are not the source of CO emission in this environment.

The dynamical masses are typically a factor of 3 below the $H_2$ mass obtained from the use of a Milky Way conversion factor. Under conditions where the CO emission is from a filled medium in the center of a galaxy and not from an ensemble of individual bound clouds, the CO luminosity measures the geometric mean of the dynamical and $H_2$ (gas) mass

$$M_{dyn} M(H_2) = (\alpha L'_{CO})^2 \quad . \tag{8}$$

$H_2$ masses determined from the assumption of optically thin CO emission yield lower limits about a factor of 3 below the dynamical mass.

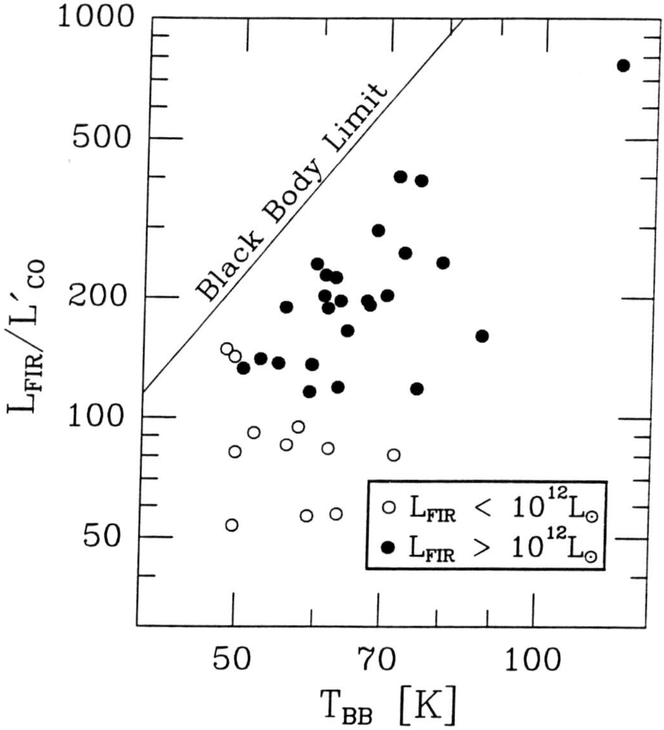


*Figure 4.* The distant-independent quantity $L_{\rm FIR}/L_{\rm CO}$ vs. the far IR black body dust temperature derived from the IRAS flux ratio $S_{60}/S_{100}$. The line indicates the upper limit for the luminosity ratio for the black body model in eq.(7), with $f_V\,\Delta V = 300\,{\rm km\,s^{-1}}$. Solid circles: galaxies in our sample with $L_{\rm FIR} > 10^{12}\,{\rm L_\odot}$. Open circles: galaxies with $L_{\rm FIR} < 10^{12}\,{\rm L_\odot}$ (from Solomon et al. 1996).

## 4. HCN and CS: Dense Gas in Ultraluminous Galaxies

The centers of ultraluminous galaxies exhibit several other important differences from normal spiral galaxies. Most important is the extremely high HCN luminosity. HCN traces $H_2$ at a much higher density ($\sim 10^5$ cm$^{-3}$) than CO ($\sim 500$ cm$^{-3}$). The HCN luminosities of the ultraluminous galaxies Mrk 231, Arp 193, Arp 220, and NGC 6240 are greater than the CO luminosity of the Milky Way (Solomon, Downes & Radford 1992). They have $\sim 10^{10}$ M$_\odot$ of $H_2$ at a density $> 10^4$ cm$^{-3}$, about 300 times the mass of dense $H_2$ in the Milky Way. The ratio of HCN to CO luminosity is 1/7 to 1/4 for ultraluminous galaxies, but only 1/30 or less in the disks of normal spiral galaxies. CS emission is also present but at an intensity about a factor of 3 less than HCN. A large fraction of the molecular gas in ultraluminous galaxies, perhaps 50%, is in very dense regions similar to star forming cloud cores, rather than in the envelopes of giant molecular clouds. Thus, the density of molecular clouds in ultralumious galaxies is much higher than Milky Way GMCs. The ratio of far-infrared to HCN luminosity is similar for ultraluminous galaxies and for normal spirals, including the Milky Way, which suggests that the star formation rate per mass of *dense* gas is independent of the infrared luminosity or the state of interaction. The molecular gas density in the central regions of the ultraluminous galaxies ($\sim$500 $M_\odot$ $pc^{-3}$) is similar to the stellar density in the centers of elliptical galaxies, consistent with the idea that some mergers may eventually become ellipticals.

## 5. $^{13}$CO Observations of Ultraluminous galaxies

The brightness temperature ratio of $^{13}$CO to $^{12}$CO in ultraluminous galaxies is extremely low (Solomon, Downes & Radford 1996). Our observations show that for Arp220, NGC 6240 and Mrk231 this ratio is in the range of 1/50 to 1/30, instead of the typical 1/10 in spiral galaxy disks. This indicates that CO emission has lower optical depth than in galactic disks, or $^{13}$CO is less abundant relative to $^{12}$CO, or both. Again, the physical conditions are very different for Milky way GMCs and a full interpretation requires a broad based multi-line study. The cores of ultraluminous galaxies are the sites of the most extreme molecular environments with very dense clouds, and an intercloud medium emitting much of the CO and $^{13}$CO emission.

## 6. IRAS F10214+4724: A Lensed Ultraluminous Galaxy at z=2.3

In the past few years, through the combination of improved technology and a boost from gravitational lensing, it has become possible to observe CO emission from ultraluminous galaxies at high redshift. An example is F10214+4724 shown in Figure 5 (Downes, Solomon & Radford 1995). By combining information from Keck telescope K band images and IRAM interferometer CO images, it is possible to deduce the magnification factor for the CO emission. The resulting intrinsic CO luminosity, IR luminosity and size are very similar to that of local ultraluminous IR galaxies.

## References

Condon, J.J., Huang, Z.P., Yin, Q.F., & Thuan, T.X. 1991, *ApJ*, **378**, 65.
Dickman R.L., Snell, R.L., & Schloeb, F.P. 1986, *ApJ*, **309**, 326.
Downes, D. & Solomon, P.M. 1996 (to be published).
Downes, D., Solomon, P.M., & Radford, S.J.E. 1995, *ApJ*, **453**, L65.
Downes, D., Solomon, P.M., & Radford, S.J.E. 1993, *ApJ*, **414**, L13 (DSR).

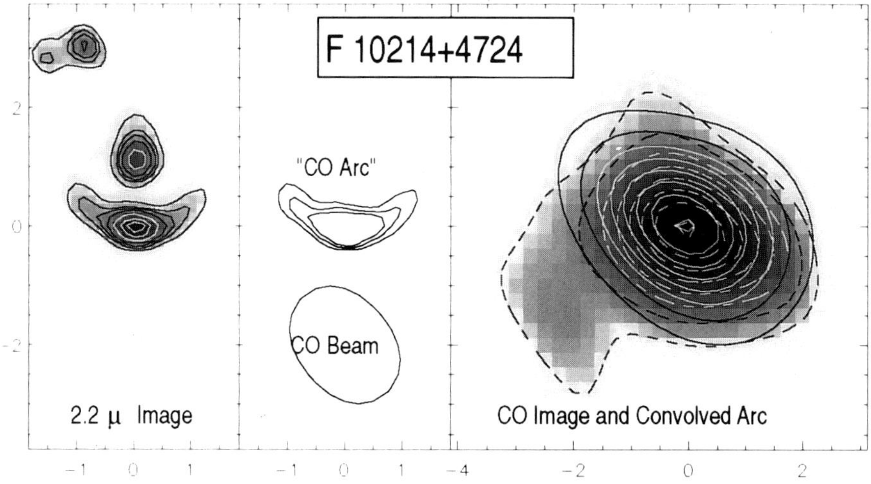


*Figure 5.* CO arc and image from 10214+4724 at z = 2.3. Offsets are in arcseconds. Left: Keck telescope 2.2 $\mu$m image. The feature north of the arc is the lensing galaxy and the other features are unrelated (Matthews et al. 1994). Middle: Truncated 2.2 $\mu$m image used to model the CO arc and the half power shape of the IRAM interferometer beam. Right: IRAM interferometer CO map in gray scale and dashed contours. Solid contours are the model image obtained by convolving the beam with the arc. The fit is very good, and the difference between the observed image and the model has the same rms noise as the original map. The CO magnification is a factor of 5 to 10 (Downes, Solomon & Radford 1995).

Heithausen, A., Stacy, J.G., deVries, H.W., Mebold, U., & Thaddeus, P. 1993, *A&A*, **268**, 265.
Maloney, P. 1990, *ApJ*, **348**, L9.
Matthews, K., et al. 1994, *ApJ*, **420**, L13.
Radford, S.J.E., et al. 1991, in Dynamics of Galaxies and Their Molecular Cloud Distributions, ed. F. Combes & F. Casoli (Dordrecht: Kluwer), 303.
Sanders & Kim 1993 (priv. comm.)
Sanders, D.B., Soifer, B.T., Elias, J.H., Madore, B.F., Matthews, K., Neugebauer, G., & Scoville, N.Z. 1988, *ApJ*, **325**, 74.
Sanders, D.B., et al., 1986, *ApJ*, **305**, L45.
Scoville, N.Z., Sargent, A.I., Sanders, D.B., & Soifer, B.T. 1991, *ApJ*, **366**, L5.
Solomon, P.M., Downes, D., Radford, S.J.E., & Barrett, J.W. 1996 (to be published in ApJ).
Solomon, P.M., Downes, D., & Radford, S.J.E. 1992, *ApJ*, **387**, L55.
Solomon, P.M., & Sage, L.J. 1988, *ApJ*, **334**, 613.
Solomon, P.M., Rivolo, A.R., Barrett, J.W., & Yahil, A. (1987), *ApJ*, **319**, 730.
Solomon, P.M., & Wickramasinghe, N.C. 1969 *ApJ*, **158**, 449.
Wilson, R.W., Jefferts, K.B., & Penzias, A. 1970, *ApJ*, **158**, L43.

# MOLECULAR CLOUDS AND CO EMISSION IN LOW–METALLICITY GALAXIES

PHILIP R. MALONEY
*CASA, University of Colorado*
*Boulder, CO 80309-0389*

AND

MARK G. WOLFIRE
*NASA/Ames Research Center*
*Mail Stop 245-3*
*Moffett Field, CA 94035*

## 1. Introduction

Irregular and dwarf galaxies are probably the most numerous actively star-forming galaxies. They are generally gas-rich (in neutral hydrogen) and metal-poor, with gas-phase metallicity $Z \lesssim 0.25$ solar and similarly low dust-to-gas ratios. The nearest examples are the Large and Small Magellanic Clouds, but there are many others within a few Mpc (*e.g.*, NGC 1569, NGC 4449). Irregular and dwarf galaxies are notoriously poor in CO emission and the resulting high ratios of infrared to CO luminosities have been interpreted as a high star formation efficiency. Here we show that CO emission is not generally a reliable tracer of molecular mass in low-$Z$ objects, and discuss alternative tracers of the molecular gas content.

## 2. Low–Metallicity Cloud Models

The key parameter controlling the physical and chemical state of gas exposed to FUV photons is $G_o/n$, where $G_o$ is the scaling factor for the FUV radiation field (relative to the solar neighborhood value) and $n$ is the total hydrogen density. If $G_o/n$ is high, then attenuation of the FUV photons by dust determines where molecules survive; if $G_o/n$ is low, molecular survival is determined by self-shielding. The transition between these regimes (for $H_2$) occurs at $(G_o/n)_{cr} \sim 10^{-2} Z^{1/2}$ (assuming that the $H_2$ formation



W.B. Latter et al. (eds.), CO: Twenty-Five Years of Millimeter-Wave Spectroscopy, 299–301.

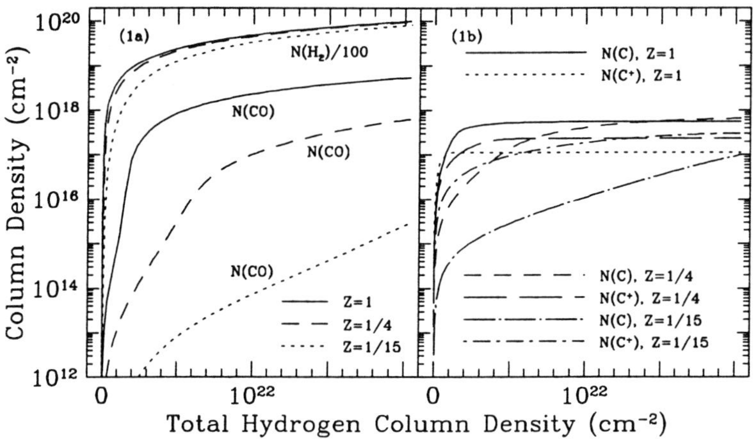


*Figure 1.* $H_2$ and carbon species column densities for $G_o = 1$ models.

rate on grains scales with grain surface area $\propto Z$). We have constructed chemical and physical models (including thermal balance) for a series of cloud models, all with $n = 10^3$ cm$^{-3}$. These include models with parameters appropriate to Galactic molecular clouds and models that approximate the physical conditions in the LMC and SMC, with the gas-phase metallicity and dust-to-gas ratios reduced by factors of 4 and 15, respectively, compared to the Galactic model.

$G_o = 1$ *Models.* These are appropriate to quiescent molecular clouds exposed only to the ambient FUV flux. Figure 1a shows the cumulative molecular hydrogen and CO column densities as functions of total hydrogen column density $N_H$. The differences in $H_2$ between the different models are small. The differences in carbon chemistry, however, are pronounced, as these models are all in the self-shielding regime. Figure 1b shows the C and $C^+$ columns as functions of $N_H$. The column density of CO drops by more than three orders of magnitude as $Z$ declines from 1 to 1/15 (relative to solar), while the $C^+$ column is nearly constant, *increasing* by about a factor of 3. Neutral carbon is also roughly constant, with $N_C \sim 6 \times 10^{17}$ cm$^{-2}$ for $Z = 1$ and $Z = 1/4$, and about a factor of 6 smaller for $Z = 1/15$ (here $C^+$ is the dominant carbon species). The observable consequences for CO are dramatic: for the LMC ($Z = 1/4$) model, the emergent intensity is smaller than in the Galactic model by about 35% (due to the smaller CO column), but the size of the optically thick CO core has decreased by more than a factor of two, leading to a drop in the CO luminosity of more than a factor of six. CO in the SMC model would be undetectable. For realistic (clumpy) cloud models, the effects on the CO luminosity depend on the clump column density spectrum, but are generally severe.

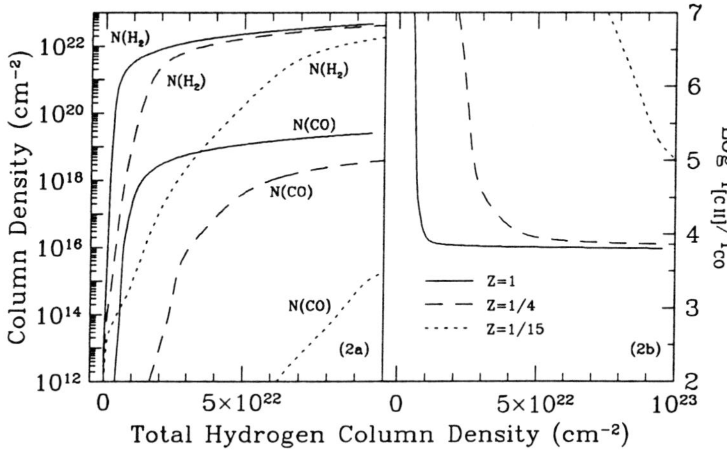


*Figure 2.* (a) $H_2$ and CO column densities for $G_o = 10^3$ models. (b) Ratio of emergent integrated intensities in the $C^+$ 158$\mu$m and CO $J = 1 - -0$ lines for the same models.

$G_o$ = 1000 *Models.* These are appropriate to star-forming molecular clouds exposed to intense FUV fluxes from OB stars, *e.g.*, photodissociation regions or PDRs. Here the survival of molecules is determined by dust; due to the lower dust-to-gas ratios in the low-$Z$ models, formation of $H_2$ and CO requires a larger $N_H$ to reach the same extinction. The dramatic differences between $Z = 1$ and low-$Z$ PDRs are shown in Figure 2a. At the end of the integration (to $N_H = 10^{23}\,\mathrm{cm}^{-2}$, or $A_v \approx 60$ for the Galactic model) the CO column in the SMC model is still only $N_{CO} \approx 2 \times 10^{15}\,\mathrm{cm}^{-2}$, while the $C^+$ column is nearly constant in all three models at $N_{C^+} \approx 2 \times 10^{18}\,\mathrm{cm}^{-2}$. As shown in Figure 2b, this leads to enormous differences in the $C^+$ 158$\mu$m to CO $J = 1 - 0$ line ratio, with the observed ratio for the low-$Z$ models depending strongly on the total column through the PDR. This is in agreement with observations of low-$Z$ galaxies, which show $I_{[CII]}/I_{CO} \sim 5 - 15$ times larger than typical for IR-bright galactic nuclei or Galactic HII regions.

## 3. Conclusions

Clouds in low metallicity galaxies, such as irregular and dwarf galaxies, can have most of their hydrogen in molecular form with little or no CO.

The 158$\mu$m/CO $J = 1 - 0$ ratio in low-$Z$ PDRs depends on the total column through the PDR and the degree of clumping and is in general considerably higher than in $Z \sim 1$ PDRs.

On global (kpc) scales, photodissociation processes dominate the ISM of low metallicity galaxies. Neutral carbon and $C^+$ emission will be more useful as a tracer of molecular gas mass than CO.

# $N(H_2)/I_{CO}$ IN GALACTIC BULGES

R. MAUERSBERGER
*Steward Observatory*
*University of Arizona*
*Tucson, AZ 85721 USA*

**Abstract.** Gas masses derived for the bulge regions of the Milky Way and the nearby starburst galaxy NGC 253 using the $C^{18}O$ rotational lines are 5–10 times lower than those from $^{12}CO$ and a standard $X$. Also, the distributions of both isotopomers differ. It is suggested that due to the stellar potential, $^{12}CO$ arises in part from optically thin intercloud gas.

Rotational lines of CO can be easily mapped over large regions of the Galactic plane (e.g., Dame et al. 1987) and can be detected toward more external galaxies than other molecular tracers (e.g., Young et al. 1995). It is, therefore, wishful to think that these measurements have a simple meaning. A correlation between the velocity integrated intensity of CO and the column density of $H_2$ was first found empirically for a number of clouds in the Galactic disk. Theoretical approaches, which explain and quantify this correlation remarkably well (e.g., Radford et al. 1991; Downes et al. 1993), require that the CO emission is a) optically thick, b) subthermally excited, and c) arises from virialized clouds. In particular, Mauersberger et al. (1996) stress that if a) is not fulfilled, $X = N(H_2)/I_{CO}$ will be 40 times smaller than the commonly assumed value by Strong et al. (1988). In the optically thick case, $X$ is not very sensitive to the assumed CO abundance.

Optically thin tracers, such as the $C^{18}O$ rotational lines or thermal FIR/(sub-)mm continuum, are alternative tracers of $H_2$ (see Mauersberger et al. 1996 for a compilation of useful relations) and can be used to check the value of $X$. A map in the $C^{18}O$ $J = 1$–$0$ line of the inner 500 pc of the Galactic center has recently been obtained with the 1.2 m mini radiotelescope on Cerro Tololo (beam width: 9′) (Dahmen et al. 1996; Fig.1).

In Table 1, total gas masses from CO are compared with those from dust and $C^{18}O$ in the centers of the Milky Way and the nearby starburst galaxy NGC 253 (Mauersberger et al. 1996). The dust emission as well as



*W.B. Latter et al. (eds.), CO: Twenty-Five Years of Millimeter-Wave Spectroscopy*, 302–304.

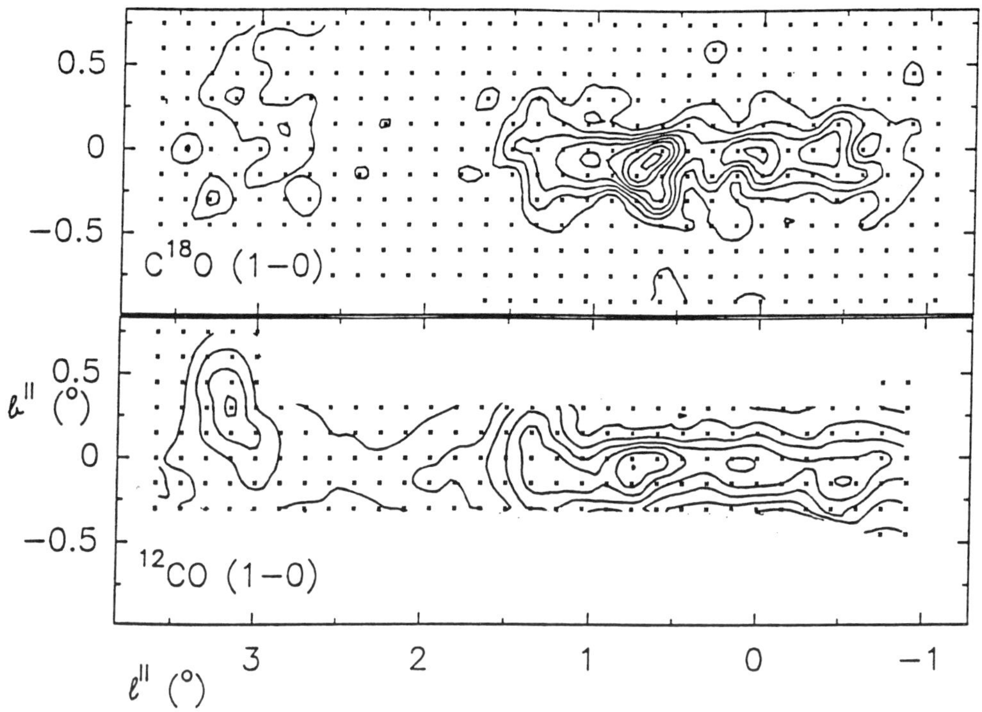


*Figure 1.* Contours of $^{12}$CO (200, 400, 600... K km s$^{-1}$) and C$^{18}$O $J = 1$–0 (2.5, 5, 7.5... K km s$^{-1}$) emission from the Galactic center (Dahmen et al. 1996).

TABLE 1. Gas mass in two galactic bulges

| Source | Method[a] | | | |
|---|---|---|---|---|
| | $^{12}$CO | C$^{18}$O | cold dust | $M_{\rm dyn}$ |
| | | ($10^8$ M$_\odot$) | | |
| Milky Way (inner 500 pc) | 1.7[b] | $\sim$ 0.15 | 0.15–0.5[d] | 60[g] |
| NGC 253 (inner 150 pc) | 3[b,c] | $\sim$ 0.6[e] | 0.4[f] | 10[h] |

a) See Mauersberger et al. (1996) for relations used; b) $X$ from Strong et al. (1988) corrected for calibration error (Bronfman et al. 1988); c) $I_{21}/I_{10} =$ 0.9 assumed; d) Sodroski et al. (1995); e) data: Sage et al. (1991); f) flux density: Krügel et al. (1990) corrected for the $^{12}$CO (2–1) contribution; dust conversion: $b = 1.9$, $Z = 3{\rm Z}_\odot$, $T_{\rm d} = 30$K (see Mezger 1990). g) Güsten et al. (1989) h) from the CO rotation curve.

the $C^{18}O$ emission lead to values for the gas mass that are 5–10 times lower than the masses from $^{12}CO$. The different spatial distributions of $^{12}CO$ and $C^{18}O$ in the Galactic center show that those cannot both be good tracers of $H_2$ in that region. Presumably the line widths are not only determined by the clouds' virial masses but by the stellar potential (Stacy et al. 1989; Solomon 1996). In some cases, line widths become large enough that $^{12}CO$ is no longer optically thick (Stark & Bania 1986). That a "standard" $X$ determined from disk clouds can probably not be applied to clouds in the galactic bulges is also supported by the detection of galaxies where the dynamical mass based on stellar velocities is lower (!) by a factor of 4–10 than the $H_2$ mass obtained from $^{12}CO$ (Shier et al. 1994).

Any model of the bulge molecular emission has to take into account that it arises from a multicomponent medium with a range of kinetic temperatures and densities (Mauersberger et al. 1995; Aalto 1996). The Southern mini and the new 60 cm southern mm-wave radiotelescope (Hasegawa 1996) may yield important clues about the more excited gas component in the Galactic bulge.

**Acknowledgements.** The $C^{18}O$ map of the Galactic center was obtained by a collaboration of the MPIfR (G. Dahmen, T.L. Wilson, et al.), Univ. de Chile (L. Bronfman, J. May, et al.), and CfA (S. Hüttemeister, T. Dame, P. Thaddeus). The project was supported by a Bennigsen-Foerder research award from Northrhine-Westphalia.

## References

Aalto, S., 1996, these proceedings
Bronfman, L., Cohen, R.S., Alvarez, H., May, J., Thaddeus, P., 1988, ApJ 324, 248
Dahmen, G., et al., 1996, in prep.
Dame, T.M., Ungerechts, H., Cohen, R.S. et al., 1987, ApJ 322, 706
Downes, D., Solomon, P.M., Radford, S.J.E., 1993, ApJ 414, L13
Güsten, R., 1989, in: The Center of the Galaxy, ed.: M. Morris, IAU Symp. 136, Kluwer, p.89
Hasegawa, T., 1996, these proceedings
Krügel, E., Chini, R., Klein, U., Lemke, R., Wielebinski, R., Zylka, R., 1990,
Mauersberger, R., Henkel, C., Chin, Y.-N., 1995, A&A 294, 23
Mauersberger, R., et al., 1996, A&A, 305, 421
Mezger, P.G., 1990, in: Physics and Composition of Interstellar Matter, Inst. of Astronomy, Nicolaus Copernicus University, Torun, eds.: J. Krelowski and J. Papaj
Radford, S.J.E., Solomon, P.M., Downes, D., 1991, ApJ 368, L15
Sage, L. J. , Mauersberger, R., Henkel, C., 1991, A&A 249, 31
Shier, L.M., Rieke, M.J., Rieke, G.H., 1994, ApJ 433, L9
Sodroski, T., Odegard, N., Dwek, E., et al., 1995, ApJ, 452, 262
Solomon, P.M., 1996, these proceedings
Stark, A.A., Bania, T.M., 1986, ApJ 306, L17
Stacy, J.G., Bitran, M.E., Dame, T.M., Thaddeus, P., 1989, in: The Center of the Galaxy, ed.: M. Morris, IAU Symp. 136, Kluwer, p.157
Strong, A.W., Bloemen, J.B.G.M., Lebbrun, F., et al., 1988, A&A 207, 1
Young, J.S., Xie, S., Tacconi, L., et al., 1995, ApJS 98, 219

# MOLECULAR GAS IN GALACTIC NUCLEI

N.Z. SCOVILLE, M.S. YUN, AND P.M. BRYANT
*Owens Valley Radio Observatory*
*Caltech 105-24, Pasadena, CA 91125*

**Abstract.** Dense molecular clouds are critical to the activity occurring in galactic nuclei. They are the active component from which starbursts arise and this dense interstellar gas may be the fuel for AGNs. In both the nearby infrared bright galaxies (e.g., M82, NGC 253, and IC342) and the ultraluminous IRAS galaxies, high resolution millimeter line mapping has shown extremely high gas surface densities in the central kpc. Often a significant fraction (> 25%) of the total molecular line emission from the galaxy arises from these central regions. New maps at resolutions down to 0.8″ of the ultraluminous IRAS galaxies reveal kinematic gradients parallel to the major axis of the CO intensity distribution, suggesting that the gas is situated in a central rotating disk. The most extreme central concentrations are seen in Arp 220 and Mrk 231 (Sey 1), which have now been mapped in both the 2.6 and 1.3 mm CO transitions. In both galaxies, the high observed CO brightness temperatures indicate large area filling factors with mean $H_2$ densities exceeding $10^4 cm^{-3}$. To produce the observed luminosities, the star formation rates must be ~100 $M_\odot$ $yr^{-1}$ within the central 500 pc radius. Estimated time scales for both the dynamical evolution and the exhaustion of the observed central ISM are typically $2 \times 10^8$ years.

## 1. INTRODUCTION

Many of the most active starburst galaxies are subject to strong dynamical perturbations; they are either interacting galaxy systems or show evidence of a central bar-like mass distribution. In both instances there will be strong non-axisymmetric gravitational forces that disturb the circular rotation. The interstellar gas is fundamentally different from the stars—the ISM is extremely dissipative (because the gas clouds have much larger cross-sections relative to their mean separation and in cloud collisions the gas radiates strongly in shock fronts). The dissipation of orbital energy (and transfer of angular momentum) leads to transfer of gas towards the galac-



*W.B. Latter et al. (eds.), CO: Twenty-Five Years of Millimeter-Wave Spectroscopy*, 305–313.

tic center, resulting in high central concentrations of active star-forming molecular clouds.

In the last few years there have been major advances in observations of molecular gas in the nuclei of galaxies. There are now a number of galaxies that have been mapped at resolutions of 2–4″ in CO, $^{13}CO$, and high dipole moment molecules (CS and HCN which trace the higher density molecular gas). As a result of continued improvements in receiver sensitivity and expansions of the mm interferometer arrays, capabilities have improved to the point where such observations can now be done out to systemic velocities of 10,000 km $s^{-1}$. The most spectactular observational results have been achieved on the infrared bright galaxies.

The IRAS survey provided a complete sample of luminous starburst galaxies out to moderate cosmological distances ($z < 0.1$); however, high resolution observations of that gas are vital to understand the differences in starburst activity from one galaxy to another. Without such molecular line data, it is not known if the high luminosities are simply due to a large quantity of gas, a more concentrated distribution of molecular gas, or a more efficient trigger for stimulating gas to form stars. It now appears as though all three factors may be at work to varying degrees. Single dish molecular line surveys of samples of IRAS bright galaxies (e.g., Sanders, Scoville, & Soifer 1991; Tinney et al. 1988; Sage & Solomon 1988) clearly indicated that the more luminous infrared bright galaxies have greater CO luminosities and thus greater overall masses of dense molecular gas. On the other hand, at the highest luminosities ($> 10^{11}L_\odot$) the ratio $L_{IR}/M(H_2)$ is enhanced by a factor of 5–50 over that measured in normal spiral galaxies like the Milky Way or M51. At high $L_{IR}$, there is also a higher frequency of interacting and merger systems and an increasing fraction of galaxies with optical emission line ratios indicative of a relatively hard ionizing spectrum, possibly from an active galactic nucleus (see Sanders et al. 1990). The increasing luminosity-to-mass ratio can be interpreted as a greater efficiency for forming stars in those galaxies or an additional non-stellar source of luminosity.

## 2. Nearby Galactic Nuclei

### 2.1. M82

The M81/M82/NGC 3077 group is at a distance of only 3.3 Mpc and is thus the closest interacting/starburst galaxy group. The entire group has been mapped in HI at 60″ resolution by Yun, Ho, & Lo (1994) and extensive tidal tails are seen extending well beyond the optical disks of the galaxies. In M82, the molecular gas is strongly concentrated towards the center with the total molecular gas mass being as much as 25% of the dynamical mass (Young & Scoville 1984). The nucleus of M82 has been mapped at high resolution in

CO, HCN, and $HCO^+$ by Lo et al. (1987); Nakai et al. (1987); Carlstrom (1988); and most recently by Shen & Lo (1995). The earlier studies indicated that a substantial fraction of the molecular gas was contained in a torus at radius $\sim$ 150 pc and mass $\sim 10^8 M_\odot$; the relatively strong emission from high dipole moment molecules indicates that the physical conditions are quite different than those of typical Galactic giant molecular clouds. Derived kinetic temperatures are 50 K and the volume densities within the clouds are approximately $10^5 cm^{-3}$. The most recent observations at 2.5″ (38 pc) resolution by Shen & Lo (1995) resolve the molecular gas into spiral arms at 125 and 390 pc radii. They also indicate outflow motion at 230 km $s^{-1}$ normal to the plane of the galaxy; however, the galaxy is nearly edge on and thus there is a large inclination correction for motions perpendicular to the disk. The earlier investigations had noted that the bulk of the starburst activity was interior to the molecular gas ring/arms and hence concluded that the starburst had propagated outwards in the central disk of the galaxy. However, Shen & Lo find substantial molecular gas within the starburst region and thus propose that the starburst may be propagating inwards. Modelling of the infrared, optical, and radio data from M82 by Rieke et al. (1993) suggests two bursts of star formation occurring $5 \times 10^6$ and $3 \times 10^7$ years ago.

## 2.2. IC 342 AND BARRED SPIRALS

At a distance of 3 Mpc, IC 342 is one of the closest barred spiral galaxies. In fact, the barred nature of this galaxy was first discovered by low resolution aperture synthesis observations (Lo et al. 1984). High resolution CO and $^{13}CO$ 2.6 and 1.3 mm maps have been obtained by Turner, Hurt, & Hudson (1993) and Turner & Hurt (1993). They find the strongest molecular emission arising from two ridges or spiral arms (along the leading edges of the bar), radial streaming towards the center, and a molecular gas mass fraction $\sim$ 50% in the inner 100 pc. The high CO (2–1)/(1–0) ratios at the interfaces of the molecular and ionized gas suggest substantial external UV heating of the molecular clouds.

Kenney et al. (1992) have mapped four galaxies in which the bar potential is probably stronger than that in IC 342. In all cases, the molecular gas is strongly concentrated towards the nucleus of the galaxy. In three of the galaxies (M101, NGC 3351, and NGC 6951), the central gas concentration is resolved into twin peaks near the center but positioned perpendicular to the direction of the large-scale stellar bar. In general, the orbits in a bar potential will be elongated parallel to the bar; however, if there is a Lindblad resonance within the bar, orbits inside this ILR are elongated perpendicular to the bar. Kenney et al. (1992) suggest that the twin concentrations of

CO emission are due to orbit crowding where the inner perpendicular orbits meet the outer parallel orbits. At these locations, cloud-cloud collisions and gravitational instabilities could lead to massive accumulations of gas.

### 2.3. NGC 1068

NGC 1068 is the closest (18 Mpc) galaxy in which both starburst and AGN activity are occurring in the central regions. A total luminosity of $3 \times 10^{11} L_\odot$ is approximately equally divided between an extended starburst and a point-like Seyfert 2 nucleus. In addition, near infrared mapping revealed a strong stellar bar extending out to radius 1 kpc (Scoville et al. 1988). High resolution maps of the CO, $^{13}$CO, and HCN emission have been obtained by Planesas, Scoville, & Myers (1991); Kaneko et al. (1992); Tacconi et al. (1994); Helfer & Blitz (1995); and Papadoupolus, Seaquist & Scoville (1996). The major features seen in the molecular gas are two strong spiral arms at radius $\sim 15''$(1.5 kpc) that originate from the ends of the stellar (2$\mu$m) bar and a compact source ($\sim 3''$) coincident with the Seyfert 2 nucleus. The maps by Heifer & Blitz (1995), which included both interferometric and single dish data, also reveal molecular gas flowing into the nucleus along the leading edges of the stellar bar. The total mass of molecular gas in the inner disk of NGC 1068 is approximately $4 \times 10^9 M_\odot$ and approximately 50% of this gas is in the discrete structures mapped interferometrically. Within the spiral arms the gas structures range in size up to 500 pc and their masses are in the range $2 \times 10^7 - 7 \times 10^8 M_\odot$ (Planesas et al. 1991). At present it is not clear if these are gravitationally bound structures or simply transient associations of gas clouds similar to those seen in M51 (Rand & Kulkarni 1990). The gas mass in the nuclear concentration is estimated to be $8 \times 10^7 M_\odot$ (within 130 pc radius, Planesas et al. 1991); however, relatively strong HCN emission is seen here (e.g., Tacconi et al. 1994) and it is, therefore, doubtful that the standard Galactic conversion ratio should be applied. The mass of gas and dust in this source is clearly sufficient to obscure the broad line Seyfert 1 nucleus as originally suggested by Antonucci & Miller (1985). NGC 1068 also contains a one-sided nuclear radio jet (Wilson & Ulvestad 1987) extending from the nucleus to the northeast at approximately the same position angle as the stellar/molecular bar. This alignment is surprising given the general model that the radio jets are oriented along the rotation axis of the galaxies. Clearly, investigations of the interaction of the radio jet with the molecular gas might be illuminating.

## 3. LUMINOUS IRAS GALAXIES

A major direction of extragalactic molecular line research has been measurement of the molecular gas properties and distributions in luminous

IRAS galaxies. The IRAS bright galaxy survey (Soifer et al. 1987) included 238 objects in the northern hemisphere and approximately 150 of these have had single-dish CO measurements. The resulting $H_2$ masses, assuming the same CO emissivity per unit mass as for the Milky Way GMCs, show a clear trend for increasing masses of molecular gas with increasing far infrared luminosity. For the most luminous objects, the total $H_2$ masses are in the range 2–50 $\times 10^9 M_\odot$), that is 1–20 times the $H_2$ content of our Galaxy. CO aperture synthesis mapping has been done on approximately 30 of these galaxies. Early aperture synthesis maps at 4–6″ resolution showed that this gas is also much more concentrated in the galactic nuclei than in normal galaxies (see Scoville et al. 1991; Okumura et al. 1991). Typically 50% of the molecular gas was found within radii less than or equal to 1 kpc and the gas mass fractions in the central regions ($M(H_2)/M_{dyn}$) often exceed 50% (as compared with typical values of 5% near the centers of normal galaxies). Recently, several of these galaxies have been observed at 2″ resolution in CO (1–0) and at 1″ resolution in the CO (2–1) transition. We describe three of these objects (VV 114, Mrk 231, and Arp 220) in more detail below.

### 3.1. VV 114

VV 114 is an early merger system with $L_{IR} = 6 \times 10^{11} L_\odot$ and a very large gas content, $M(H_2) = 5 \times 10^{10} M_\odot$. The two galaxy nuclei are separated by 6 kpc and CO (1–0) images show the molecular peak lies between the optical and infrared nuclei (Yun, Scoville, & Knop 1994). The CO is shown superposed on an r-band image in Figure 1. The molecular gas is extended over $\sim$ 5 kpc between the two galaxy nuclei. In fact, the CO distribution is quite similar to that of the non-thermal radio continuum. In this system, the gas is clearly more concentrated toward a central peak than are the two stellar systems. This is interesting because more rapid merging is expected for the gas than for the stars in interacting systems due to the greater dissipation in the gas.

### 3.2. MRK 231

Mrk 231 (at 174 Mpc) is one of the few ultraluminous IRAS galaxies in which there is general consensus on the existence of a dust-embedded high luminosity Sey 1 AGN (Boksenberg et al. 1977). New CO $(1-0)$ and $(2-1)$ images at 0.8–2″ resolution resolve the CO source and reveal a velocity gradient parallel to the major axis of the CO distribution (Bryant & Scoville 1996). Virtually all of the single-dish CO emission is contained within a radius of 420 pc and the CO brightness temperatures exceed 30K, indicating nearly complete area filling factors for the molecular gas within this radius.

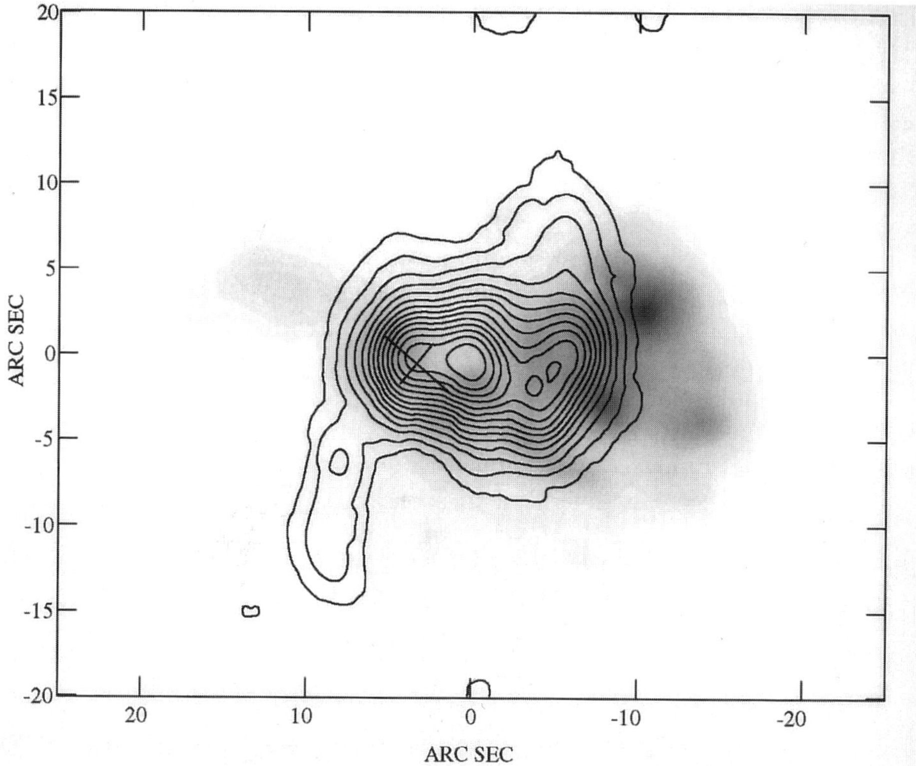


*Figure 1.* The integrated CO (1–0) emission (3″ resolution) is shown superposed on an r-band image of the interacting system VV 114 (Yun, Scoville, & Knop 1994). The position of the eastern nucleus, which is brightest at 2$\mu$m, is indicated by the cross. The molecular gas is distributed in a bar-like configuration between the two galaxy nuclei.

The CO conversion factor in such galactic nuclei was extensively analyzed (Bryant & Scoville 1996) and was found to be within a factor of a few of the Galactic value for Mrk 231. Assuming the Galactic value, the nuclear disk of Mrk 231 must be within 22° of face-on in order to reconcile the dynamical mass with the mass from the CO line emissivity. This is consistent with the modest extinction measured for the Seyfert/quasar nucleus.

### 3.3. ARP 220

Arp 220 at 77 Mpc has been imaged in CO (1–0) (Scoville et al. 1991), in HCN (Radford et al. 1991), and most recently in CO (2–1) (Scoville, Yun, & Bryant 1996). The CO (2–1) maps at 1″ clearly reveal three components in the dense molecular gas: peaks corresponding to each of the double nuclei (separated by 0.9″ at pa = 101°) seen in the near infrared (Graham et al. 1990) and radio continuum and a more extended disk-like structure

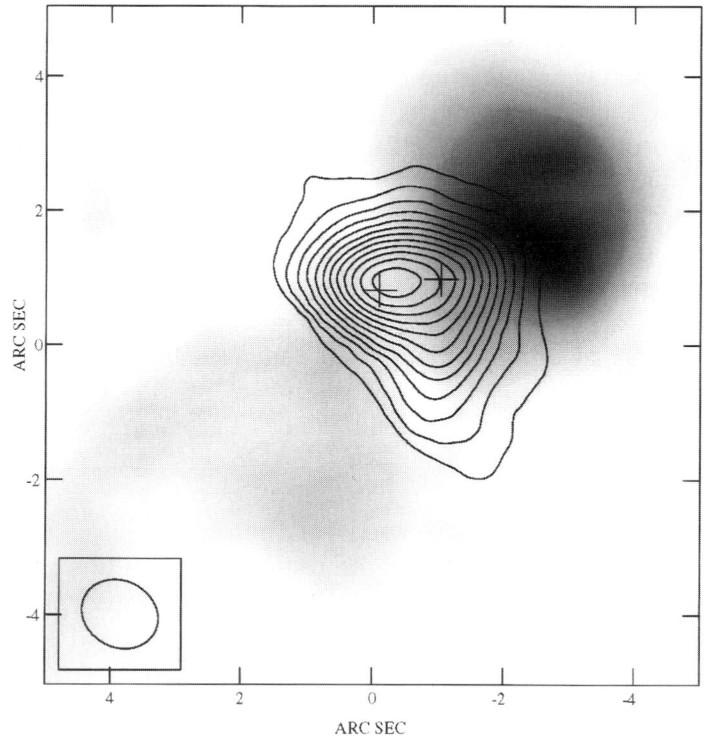


*Figure 2.* CO (2–1) line emission (1″ resolution) integrated over all velocities in Arp 220 is shown superimposed on an H$\alpha$ image (Scoville, Yun, & Bryant 1996). The CO emission in the central 500 pc of the galaxy peaks between the two infrared nuclei (spacing 0.95″) and is elongated in the outer contours parallel to the dust lane crossing the H$\alpha$ emission distribution.

elongated southwest-northeast, similar to the dust lane seen in optical images. In Figure 1, the CO (2–1) line emission integrated over all velocities is shown superimposed on a H$\alpha$ image of Arp 220. The elongated disk feature exhibits a monotonic velocity gradient parallel to the major axis of the CO intensity distribution and the dynamical mass determined from the CO is approximately $3\times10^{10}M_\odot$ within 0.9″ (360 pc) radius. Bright peaks are also seen in the channel maps coincident with the locations of the two infrared nuclei. The observed kinematics of the molecular gas are consistent with the radial distribution of the CO emissivity in the sense that a rotation curve derived, assuming that the total mass is dominated by the molecular gas (with little contribution from the stars), is consistent with the observed CO line profiles (Scoville, Yun, & Bryant 1996).

## 4. SUMMARY

Important general characteristics of the molecular gas in starburst and high infrared luminosity galactic nuclei include: (1) extremely high gas surface densities exceeding $10^3 M_\odot pc^{-2}$; (2) strong non-circular gas motion often accreting radially along a stellar bar potential or two merging galactic nuclei; and (3) high molecular gas mass fractions in the central 500 pc with the gas constituting greater than 50% of the total mass.

The strong non-circular gas motions and high central gas concentrations may be driven by non-axisymmetric gravitational forces from stellar bars or from the merging of comparable mass galactic nuclei. Large nuclear gas concentrations can then arise and set off the starburst activity that becomes non-linear due to cloud-cloud collisions and stimulated star formation effects such as high pressure HII regions and supernova blast waves.

The statistics from the IRAS survey indicate that approximately 0.2% of the present epoch spiral galaxies are undergoing high luminosity global starburst; however, given the short duration of this activity ($2 \times 10^8$ years) and the fact that standard cosmological evolution would indicate that the interaction rate was higher in the past, nearly all galaxies would be undergoing such merger-induced starbursts at $z = 1$–$2$ (see Scoville & Soifer 1990). Although only a small fraction of present-day galaxies are highly luminous in the infrared, the star formation rates in these galaxies are typically enhanced by factors of 10–100 so a significant fraction of the overall star formation may occur during such brief periods of activity.

Our research is supported in part by NSF Grant AST 93-14079.

## References

Antonucci, J., & Miller, J.S. 1985, Ap.J., 297, 621

Boksenberg, A., Carswell, R. F., Allen, D.A. Fosbury, R.A.E., Penston, M.V., & Sargent, W.L.W. 1977, M.N.R.A.S., 178, 451

Bryant, P.M., & Scoville, N.Z. 1996, Ap.J., 457, 678

Carlstrom, J. 1988, PhD Thesis, University of California, Berkeley

Graham, J. R., Carico, D.P., Matthews, K., Neugebauer, G., Soifer, B.T., & Wilson, T. D. 1990, Ap.J. Letters, 354, L5

Helfer, T. & Blitz, L. 1995, Ap.J., 450, 90

Kaneko, N., Morita, K., Fukui, Y., Takahashi, N., Sugitani, K., Nakai, N. & Morita, K. 1992, PASJ, 44, 341

Kenney, J.D.P., Wilson, C.D., ScoviUe, N.Z., Devereux, N.A. & Young, J.S. 1992, Ap. J. Letters, 395, L79

Lo, K.Y., Berge, G., Claussen, M., Heiligman, G., & Leighton, C. 1984. Ap.J. Letters, 282, L59

Lo, K.Y., Cheung, K.W., Masson, C.R., Philips, T.G., Scott, S.L. & Woody, D.P. 1987, Ap.J., 312, 574

Okumura, S.K., Kawabe,R., Ishiguro,M., Kasuga, T., & Morita, K.I. 1991 in IAU Symposium No. 146, 425

Matthews, K., Neugebauer, G., & Scoville, N.Z. 1988, Ap.J., 325, 74
Nakai, N., Hayashi, M., Handa, T., Sofue, Y., Hasegawa, T., & Sasaki, M. 1987. Publ. Astron. Soc. Jpn., 39, 685
Papadopoulos, P., Seaquist, E.R., & Scoville, N.Z. 1996, Ap.J., in press
Planesas, P., Scoville, N.Z. & Myers, S.T. 1991, Ap.J., 369, 364
Radford, S.J.E., Delannoy, J., Downes, D. Guelin. M. & Guilloteau, S. 1991, in IAU Symposium, 146, 303
Rand, R.J., & Kulkarni, S.R. 1990, Ap. J. Letters, 349, L43
Rieke, G.H., Loken, K., Rieke, M.J. & Tamblyn, P. 1993, Ap.J., 412, 99
Sage, L., & Solomon, P.M. 1989. Ap.J., 334, 613
Sanders D.B., Scoville, N.Z., & Soifer, B.T. 1991, Ap.J., 370, 158
Sanders, D.B., Soifer, B.T., Elias, J.H., Madorel, B.F., Matthews, K., Neugebauer, G., & Scoville, N.Z. 1988, Ap.J., 325, 74
Scoville, N.Z., Matthews, K., Carico, D.P. & Sanders, D.B. 1988, Ap.J., 327, L61
Scoville, N.Z., Sargent, A.I., Sanders, D.B., & Soifer, B.T. 1991, Ap.J., 366, L5
Scoville, N. Z., & Soifer, B.T. 1990, in Massive Stars in Starbursts, ed. C. Leitherer (Cambridge: Cambridge Univ. Press), 233
Scoville, N.Z., Yun, M.S., & Bryant, P.M. 1996, Ap.J., (in preparation)
Shen, J., & Lo, K.Y. 1995, Ap. J. Letters, 445, L99
Soifer, B.T., & Wilson, T.D. 1990, Ap. J. Letters, 354, L5
Tacconi, L.J., Genzel, R., Blietz, M., Cameron, M., Harris, A.I., & Madden, S. 1994, Ap.J., 426, L77
Tinney, C.G., Scoville, N.Z., Sanders, D.B., & Soifer, B.T. Ap.J., 362, 473
Turner, J., & Hurt, R. 1993, Ap.J., 384, 72
Turner, J., Hurt, R., & Hudson, D.Y. 1993, Ap.J., 413, L19
Wilson, A.S., & Ulvestad, J.S. 1987, Ap.J., 319, 105
Young, J.S., & Scoville, N.Z. 1984, Ap.J., 287, L65
Yun, M.S., Ho, P.T.P., & Lo, K.Y. 1994, Nature, 372, 530
Yun, M.S., Scoville, N.Z., & Knop, R.A. 1994, Ap. J. Letters, 430, L109

# MODELING OF MOLECULAR EMISSION FROM CENTRES OF GALAXIES

S.AALTO
*Steward Observatory*
*University of Arizona*

## 1. Introduction

Evidence is mounting that the molecular clouds pervading the inner regions of galaxies are rather different from the cold Giant Molecular Clouds (GMCs) dominating the emission in the disk of our Galaxy.

The differences may be especially profound for starburst and interacting galaxies, which have a large fraction of their total $^{12}$CO in the inner kpc. Due to the bright central molecular emission, multi-line mm- and submm observations are possible—shedding light on the structure and physical conditions of the molecular clouds.

## 2. Molecular Line Ratios

The unusual character of molecular clouds in the centres of galaxies is apparent in the measured molecular line ratios for these systems. The $^{12}$CO/$^{13}$CO 1–0 intensity ratio ($=\mathcal{R}_{(1-0)}$) in galaxy centres is usually about a factor of two larger than the mean $\mathcal{R}_{(1-0)} \approx$ 6-7 (Polk et al. 1988) found for the disk of the Milky Way (e.g. Young and Sanders 1986; Aalto et al. 1991, 1995). Aalto et al. (1995) observed $^{12}$CO and $^{13}$CO 1–0 emission in 31 starburst galaxies; the mean $\mathcal{R}_{(1-0)}$ is 13 $\pm$6 for the centres of these galaxies while a few luminous mergers have $\mathcal{R}_{(1-0)} > 20$. The mean $\mathcal{R}_{(2-1)}$ (for 11 galaxies) was also 13. Out of 9 studied galaxies (NGC 3256, NGC 1808, NGC 5055, Arp299, NGC 660 NGC 2146, NGC 4048/39, NGC 6221 and NGC 7552), Aalto et al. found that $\mathcal{R}_{(1-0)}$ and $\mathcal{R}_{(2-1)}$ vary significantly with position in 6 of them. $\mathcal{R}_{(1-0)}$ is usually highest in the centre of the galaxy. The southern merger NGC 3256, for instance, has a large $\mathcal{R}_{(1-0)} =$ 35 in the centre but 5 kpc north-east of the nucleus $\mathcal{R}_{(1-0)}$ is $\approx$ 10.



*W.B. Latter et al. (eds.), CO: Twenty-Five Years of Millimeter-Wave Spectroscopy*, 314–316.

In general, the $^{12}$CO 2–1/1–0 intensity ratio ($=r_{12}$) is close to unity in centres of galaxies. Aalto et al. (1995) find a mean $r_{12}$ of 0.93±0.22 for a set of 19 starburst galaxies, in good agreement with Braine et al. (1993) who find a mean of 0.89 for the centres of nearby spirals. The $r_{12}$ for the disk of the Milky Way is 0.5 in the solar neighbourhood increasing to 0.75 4 kpc from the Galactic centre (Hasegawa, this volume). Aalto et al. (1995) find a mean value of 1.3 for the $^{13}$CO 2–1/1–0 ratio measured in 11 galaxies, but the uncertainties are large.

In galaxy centres, the emission from CS, HCN, and HCO$^+$ is bright relative to $^{12}$CO, indicating the presence of significant amounts of gas with densities $n > 10^4$ cm$^{-3}$. Aalto et al. (1995) find a mean $^{12}$CO/HCN 1–0 intensity ratio of 16 for 11 starburst galaxies. Helfer and Blitz (1993) find a mean ratio of 20 for the bulges of 19 spirals.

## 3. Cloud Properties

A simple model consisting of a mixture of dense "cores" ($n > 10^4$ cm$^{-3}$) surrounded by lower density "envelopes" ($n = 10^2 - 10^3$ cm$^{-3}$) will, in general, reproduce the above line ratios. Although the lower-density component will have the highest volume filling factor, the bulk of the molecular mass may reside in the cores. In this scenario the $^{12}$CO 1–0 emission will often be dominated by emission from the "envelopes" (large filling factor combined with high temperature) and *the optical depth of the $^{12}$CO 1–0 line is moderate*, $\tau \approx 1$ rather than optically thick ($\tau >> 1$), as is often assumed for the $^{12}$CO 1–0 line. The $^{13}$CO 1–0 emission, on the other hand, will be characterised by a mixed contribution from the "envelopes" and cores. The $^{13}$CO 1–0 line is optically thin in the "envelope" component, but may be saturated in the core clouds. Although having a smaller filling factor, the contribution from the saturated emission could be significant due to higher brightness. The dense "core-component" will be bright in high-density tracer emission.

Individually, the line ratios discussed above are not usually good indicators of temperature. The $^{12}$CO 2–1/1–0 line ratio, for instance, is governed by a combination of density, temperature, and optical depth. The ratios combined into one model, however, are consistent with rather high temperatures, $T_k$=50-100 K (e.g. Martin and Ho 1986; Mauersberger et al. 1990), for the molecular gas—at least in the "envelope" component.

The causes of spatial variations remain poorly understood, although some new clues have recently emerged. Aalto et al. (1995) report a tentative correlation between $\mathcal{R}_{(1-0)}$ and gas surface density: $\mathcal{R}_{(1-0)}$=5-10 for disks with $\Sigma = 10 - 10^2$ M$_\odot$pc$^{-2}$, $\mathcal{R}_{(1-0)}$=10–20 for galactic centres with $\Sigma = 10^2 - 10^4$ M$_\odot$pc$^{-2}$, and $\mathcal{R}_{(1-0)} > 20$ for galactic centres with $\Sigma \gtrsim 10^4$

$M_\odot pc^{-2}$.

## 4. The Effect of the Environment

The properties of the gas clouds are likely to respond to the surrounding conditions. Clouds in the inner regions of starburst galaxies experience vigorous star-forming activity and feel squeezed by the deep central potential well. Moreover, as indicated above, the gas surface density is considerably higher than that of the Galactic disk.

Using the estimates of gas surface density, one can make a simple estimate of the hydrostatic pressure associated with that gas column. Not surprisingly, this pressure is several orders of magnitude higher than that of the Milky Way disk. Given that the clouds are in equilibrium through $P_c \propto n(\Delta V)^2$, they must either be very dense or have large individual line-widths. Helfer and Blitz (1993) suggest the mean density of molecular gas is large, possibly as a result of a high ambient pressure. However, a large fraction of the $^{12}$CO 1–0 emission may originate in gas at a density of only $n \approx 10^3$ cm$^{-3}$. These structures should have large individual line-widths in a high-pressure environment and may not be self-gravitating clouds. Even if there is a significant column density in these structures, their large line-widths (and high kinetic temperatures) will prevent the $^{12}$CO 1–0 optical depths from becoming large. Starburst activity and the dissipation of turbulence are possibly important sources for heating the gas in the inner region of galaxies (e.g. Güsten 1989).

Multitransition observations of $^{12}$CO and its isotopomers as well as the high-density tracers will explore the limitations of the above model. It is essential to combine data on both low and high transitions with data from various molecules for a more complete picture of the molecular ISM. Interferometric observations allow further exploration of gas distribution and kinematics and their relation to the physical conditions of the molecular gas.

## References

Aalto, S., Black, J.H., Johansson, L.E.B., and Booth, R.S., 1991, A&A 249, 323
Aalto, S., Booth, R.S., Black, J.H., Johansson, L.E.B., 1995, A&A in press
Braine, J., Combes, F., Casoli, F., Dupraz, C., Gerin, M., Klein, U., Wielebinski, R., Brouillet, N., 1993, A&AS 97, 887
Güsten, R., 1989, In: *The Center of the Galaxy*, Morris M., (ed.) (Dordrecht.Kluwer) p. 89
Helfer, T.T., Blitz, L., 1993, ApJ 419, 86
Martin R.N, Ho, P.T.P., 1986, ApJ 308, L7
Mauersberger, R., Henkel, C., and Sage, L.J., 1990, A&A 236, L63
Polk, K.S., Knapp, G.R., Stark, A.A., and Wilson, R.W.,1988, ApJ 332, 432

# STAR FORMATION AND DYNAMICS OF MOLECULAR GAS IN CIRCUMNUCLEAR REGIONS OF BARRED GALAXIES

SUMIO ISHIZUKI
*Institute of Astronomy, Division of Science, Tohoku University*
*Aoba, Aramaki, Aoba-ku Sendai, 980-77, JAPAN*

## 1. Introduction

To investigate gas dynamics in bar or oval potentials of galaxies and the relation to star formation in the nuclear and circumnuclear regions, the CO ($J = 1 \rightarrow 0$) emission from nine nearby galaxies (NGC 1530, NGC 2782, NGC 2903, NGC 3504, NGC 4369, NGC 5383, NGC 7479, NGC 7541, & NGC 7714) was mapped using the Nobeyama Millimeter Array with angular resolutions of 4–13 arcseconds. Observations of M 83 by Handa et al. (1994) are included in the following discussion.

## 2. Categories of CO Morphology

The observations revealed a variety of CO distributions in barred galaxies: in category [1], the CO distribution is elongated parallel to the large-scale stellar bar (NGC 7479 & NGC 2903); in category [2], there is an inner CO bar inscribed in a pair of dust lanes that lie along the large-scale stellar bar (NGC 5383, M83, NGC 1530, & NGC 2782); and in category [3], the CO is concentrated in the nucleus (NGC 3504).

## 3. Interpretation in Terms of Orbit Resonance Theories

This variation can be interpreted in terms of orbit resonance theories. Category [1] is the case where the $x_2$-to-$x_1$ coverage ratio, the ratio of the coverage by orbits in the $x_2$-family (orientations perpendicular to the large-scale stellar bar) to the coverage by orbits in the $x_1$-family (parallel to the large-scale stellar bar) is very small. In this case, the radius of the inner Lindblad resonance (ILR) is much smaller than that of the corotation resonance or there is no ILR. The narrow CO distribution parallel to the large-scale stel-



*W.B. Latter et al. (eds.), CO: Twenty-Five Years of Millimeter-Wave Spectroscopy*, 317–320.

lar bar in NGC 7479 resembles the results of numerical simulations with small $x_2$-to-$x_1$ coverage ratios (Athanassoula 1993). In NGC 2903, the CO isovelocity contours are parallel to the large-scale stellar bar, also suggesting a small $x_2$-to-$x_1$ coverage ratio.

Category [2] is the case where the $x_2$-to-$x_1$ coverage ratio is larger than in the category [1]. In particular, the deviation angles of the inner bars in NGC 1530, M 83, and NGC 5383 with respect to the main stellar bars are close to 45°. This agrees with the prediction of the damped-orbit-resonance theory (Wada 1994).

## 4. Dynamical Evolution of Inner Bars

There is a variety of CO distributions in the inner bars of galaxies in category [2]. The inner gaseous bar of NGC 5383 (Figure 1) has local peaks of integrated CO intensity at the apocenters and H$\alpha$ emission spread over the inner gaseous bar. On the other hand, the inner bars of M 83 (Handa et al. 1994; Cowan & Branch 1985) and NGC 1530 (Figures 2,3) have leading ridges of the CO integrated intensity and star forming regions inscribed in the leading ridges. In addition, NGC 2782 (Figure 4) has a pair of circumnuclear CO peaks between the ends of the dust lanes that run along the large-scale stellar bar. The southern CO peak is elongated west toward the inner end of the straight dust lane. This variation suggests that the galaxies are in different stages of dynamical evolution as molecular gas is transported toward the nucleus. NGC 5383 is in the earliest stage and NGC 2782 is in the latest stage. Note the near-infrared image of NGC 2782 shows a weak elongation in the east-west direction (Pompea & Rieke 1990) and the southern CO elongation is on the leading edge of the near-infrared elongation. It is plausible there are oval potentials inside the outer ILRs of the inner-barred galaxies. The gas in the leading ridges may be affected by the gravitational torques exerted by the inner bar and driven toward the nucleus. If dynamical evolution of this kind is real, NGC 3504 may be in the stage after the molecular gas has been transported toward the nucleus by the torques exerted by the inner gaseous bar. This means high density gas in the nuclear region can be realized due to the dual-bar system, the large-scale stellar bar, and the inner gaseous bar.

## 5. Star Formation in the Circumnuclear Kiloparsec Regions

CO and 20 cm radio continuum data compiled from the literature (Figure 5) reveal galaxies are divided into two classes in terms of the gas depletion time due to star formation, $\tau_{SF}$, the reciprocal of the radio-to-CO intensity ratio (Wright et al. 1988). Galaxies with higher radio-to-CO ratios have $\tau_{SF} \sim 10^8$ years while galaxies with lower ratios have $\tau_{SF} \sim 10^9$ years. The shorter $\tau_{SF}$ is similar to that for well-known starburst galaxies like M 82 and NGC 253. The longer $\tau_{SF}$ is similar to that for the normal galaxies (Young et al. 1989; Young this Symposium). Note many hot spot galaxies have long $\tau_{SF}$, indicating their apparent star formation enhancement is a simple result of the piling up of a large amount of molecular gas, not of a starburst. Starbursts are divided into two categories: the "ring starbursts" which occur in a ring with a diameter comparable to the width of the large-scale stellar bar (e.g., NGC 1068, NGC 1097, & NGC 3310), and "nuclear starbursts" which occur inside this diameter. The inner bar mentioned above is a plausible mechanism for fueling the nuclear starburst.

## 6. Evolution of Kiloparsec Rings and AGNs

After the nucleus is fueled with matter, the resulting mass concentration may disrupt the inner bar and form a ring (Friedli & Martinet 1993). NGC 1097 (Hummel et al. 1987), NGC 3310 (Duric et al. 1986), and NGC 4736 (Pogge 1989) possess rings resembling such a morphology. NGC 1068 has a similar ring with a starburst, although there is a small bar inside the ring (Thronson et al. 1989; Planesas et al. 1991). Optical spectroscopy of the nuclei and starburst rings of NGC 1097 (Talent 1982; Phillips et al. 1984) and NGC 3310 (Terlevich et al. 1990; Kikumoto et al. 1993) suggests star formation occurred in the nuclei before the rings, supporting the scenario of dynamical evolution. It is also suggested (Friedli & Martinet 1993) the morphology of a galaxy changes to an early-type spiral in the final stage of the dynamical evolution. This is consistent with statistics (Balzano 1983) indicating starburst galaxies are often early-type spirals. Note NGC 1068 and NGC 1097 have active nuclei. This is consistent with other statistics that Seyfert galaxies tend to have outer and/or inner rings (Simkin et al. 1987) and to be early-type spiral galaxies (Terlevich et al. 1987).

## References

Athanassoula, E. 1993, MNRAS, **259**, 345
Balzano, V.A. 1983, ApJ, **268**, 602
Cowan, J.J., & Branch, D. 1985, ApJ, **293**, 400
Duric, N., Seaquist, E.R., Crane, P.C., & Davis, L.E. 1986 ApJ, **304**, 82
Friedli, D., & Martinet, L. 1993, A&A, **277**, 27
Handa et al. 1994, in IAU Colloq. **140**, 341
Hummel, E., van der Hulst, J.M., & Keel, W.C. 1987, A&A, **172**, 32
Kikumoto, T., Taniguchi, Y., Suzuki, M., & Tomisaka, K. 1993, AJ, **106**, 466
Phillips, M.M., Pagel, B.E.J., Edmunds, M.G., & Diaz, A. 1984, MNRAS, **210**, 701
Planesas, P., Scoville, N., & Myers, S.T. 1991, ApJ, **369**, 364
Pogge, R.W. 1989 ApJS, **71**, 433
Pompea, M., & Rieke, G.H. 1990, ApJ, **356**, 416
Puxley, P. J. (private communication)
Simkin, S.M., Su, H.J., & Schwartz, M.P. 1980, ApJ, **237**, 404

Talent, D. L. 1982, PASP, **94**, 36
Terlevich et al. 1987, IAU Symp 121, 499
Terlevich, E., et al. 1990, MNRAS, **242**, 48P
Thronson, H.A., Jr., et al. 1989 ApJ, **343**, 158
Wada, K. 1994, PASJ, **46**, 165
Wilson, A.S., & Willis, A.G. 1980 ApJ, **240**, 429
Wright, G.S., et al. 1988, MNRAS, **233**, 1
Young, J.S., Xie, S., Kenney, J.D.P., & Rice, W.L. 1989, ApJS, 70, 699
Young, J.S. (this Symposium)

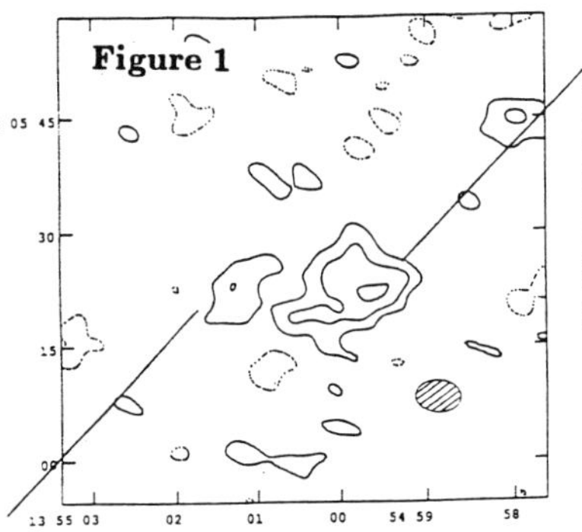


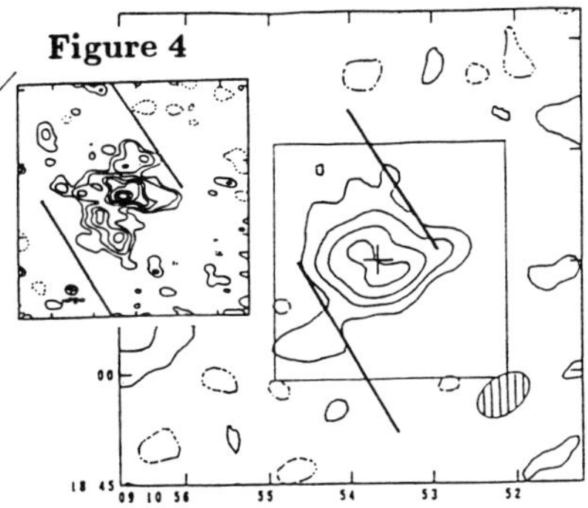


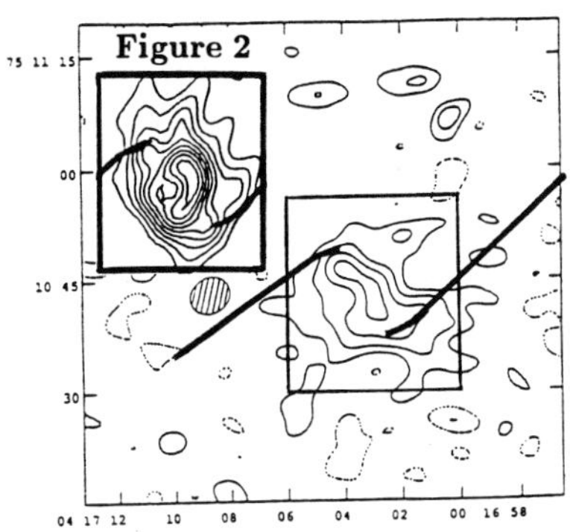


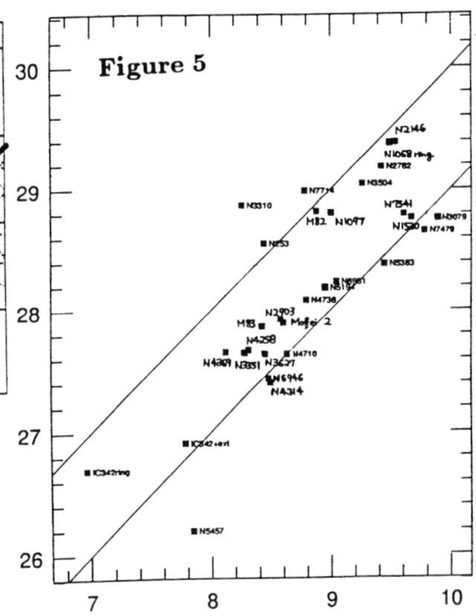


Figure 3

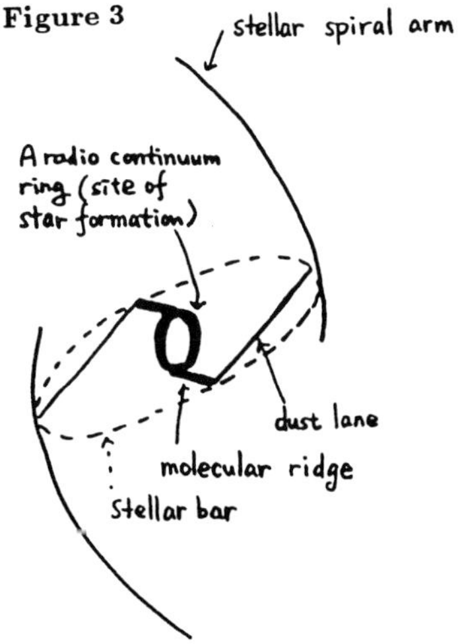


**Figure 1** Map of CO integrated intensity in NGC 5383. The thick lines represent dust lanes. **Figure 2** Same as Fig. 1 for NGC 1530. The inset is a 6 cm map (Puxley). **Figure 3** The schematic for NGC 1530. **Figure 4** Same as Fig. 2 for NGC 2782. The 6 cm map is from Wilson & Willis (1980). **Figure 5** log[20 cm radio spectral power / erg s$^{-1}$ Hz$^{-1}$] (ordinate) vs. log[CO derived $H_2$ mass/$M_\odot$] (abscissa). The values are only for the circumnuclear kiloparsec regions. The two lines are for $\tau_{SF} = 10^8$ yr (upper) and $\tau_{SF} = 10^9$ yr.

# STRUCTURE OF CIRCUMNUCLEAR REGIONS OF AGNS: STAR FORMATION AND DENSE GAS

L. J. TACCONI, A. ECKART, R. GENZEL, A. KRABBE, AND L. E. TACCONI-GARMAN
*MPE, Postfach 1603*
*D-85748 Garching, Germany*

## 1. Introduction

Our group is undertaking a large scale program of studying the dynamics and energetics of active galactic nuclei (AGN) through high spatial and spectral resolution observations in the millimeter and near infrared wavebands. With millimeter interferometric CO and other molecular line observations, we are studying the physical characteristics and kinematics of the dense, possibly obscuring gas in the circumnuclear environments of these objects. With near-infrared imaging spectroscopy, we are determining the frequency of intense starbursts associated with the AGN and also estimating the age and stellar composition of these bursts. Taken together, these data sets allow us to identify the possible fueling mechanisms for the AGN and/or the starbursts. In the conference talk we discussed our results for the 3 Seyfert/starburst galaxies NGC 1068, NGC 7469 and Mkn 231. Many of the results for our work on NGC 1068 have already been published elsewhere (Tacconi *et al.* 1994; Sternberg, Genzel and Tacconi 1994), and we refer the reader to those papers for more details. Here we present short descriptions for NGC 7469 and Mkn 231.

## 2. NGC 7469

NGC 7469 contains a classical Seyfert 1 nucleus with a region of prodigious star formation located in a ring or tightly wound spiral arms at a radius of $\sim$1.5$''$ (500 pc at 40 Mpc). This starburst ring is clearly delineated in a 0.4$''$ resolution NIR K-band image and produces $\geq$0.5 $L_{\rm bol}$ of the entire galaxy (Genzel *et al.* 1995). Various K-band line diagnostics observed with



*W.B. Latter et al. (eds.), CO: Twenty-Five Years of Millimeter-Wave Spectroscopy,* 321–323.

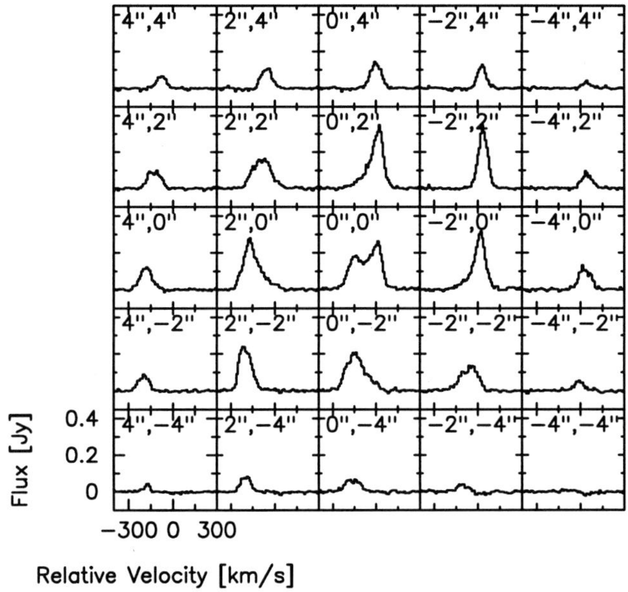


*Figure 1.* Sample spectra from the central 8″ of the CO map. Spectral resolution is 10 km s$^{-1}$. The numbers in each box are offsets from the nucleus in arcsec.

the MPE 3D camera (Genzel *et al.* 1995) show the existence of both an AGN and a starburst component. The AGN is clearly seen in the compact emission of the highly excited coronal line [SiVI], while the 2.3$\mu$m CO bandhead absorption feature, which delineates young stars, peaks in the ring. The Br$\gamma$ emission line has evidence for both a compact component arising from the AGN, and a more extended ring component, which is likely from the hot ionized gas associated with recent star formation. The 2.12$\mu$m S(1) line of $H_2$ also contains emission from a bright central component, showing there is a substantial concentration of hot (T$\geq$10$^3$ K) molecular gas in the nucleus of NGC 7469, as well as a fainter, extended component. Assuming this $H_2$ emission traces the distribution of the bulk molecular gas in the central few hundred pc, as is the case for the nearby Seyfert 2 galaxy NGC 1068 (Tacconi *et al.* 1994), these data suggest there is a dense concentration of molecular material contained within the narrow line region, similar to NGC 1068 (*e.g.* Blietz *et al.* 1994; Tacconi *et al.* 1994; Helfer and Blitz 1995).

Our CO map made with the IRAM interferometer has 2.5″ (800pc) resolution. The total flux in the map of 280 Jy km s$^{-1}$ is consistent with that observed by Meixner *et al.* (1990) and is roughly half of the single dish flux. The very high S/N of the data and the spectral resolution of 10 km s$^{-1}$ (Figure 1) allow us to glean kinematic details on spatial scales smaller than the spatial resolution. We see evidence for a rotating ring or disk of molecular gas. Following the method outlined in Tacconi *et al.* (1994), we have constructed kinematic models to explain features observed in the CO maps. Kinematic major and minor axis position-velocity maps are well fit by a model where the intensity distribution has both a compact central component and a ring or disk at a radius of 1.5″. The velocity distribution is a slowly rising (not solid-body) rotation curve out to a radius of 4″. The gas lies in a plane with a low inclination angle of ~25°. From the above kinematic parameters we have derived a dynamical mass of $\sim 1.4\times 10^{10}$ $M_\odot$ in the central kpc. This is ~3 times greater than the mass in stars over the same region (Genzel *et al.* 1995) and implies that the ISM dominates the mass and kinematics in the circumnuclear region of this Seyfert galaxy.

## 3. Mkn 231

At a distance of 170 Mpc, this ultra-luminous IRAS galaxy (ULIRG) ($L_{IR} = 3.3\times 10^{12}$ $L_\odot$) with a Seyfert 1 spectrum is a factor of 4 farther away than NGC 7469. Our aim in studying this galaxy is to determine the main powering mechanisms for the extreme IR luminosity and to estimate the properties of the dense gas in the extreme environment of a ULIRG. MPE 3D NIR imaging spectroscopy shows evidence for both an AGN component and a circumnuclear region of intense starburst activity. The 2.3$\mu$m CO bandhead absorption feature peaks in the circumnuclear starburst, while the 2.12$\mu$m $H_2$ emission exhibits a bright central peak and a fainter extended component.

We have also recently detected CO J=6→5 emission from Mkn 231 with the MPE submm heterodyne spectrometer FANATIC on the JCMT. The 6→5 line is roughly half as strong as the 1–0 line (Radford, Solomon, & Downes 1991), assuming all CO lines arise from the same size emitting region. We estimate that the average kinetic temperature of the gas is ≥30 K in our 6→5 beam (7″ or 3 kpc).

We thank A. Harris, J. Howe, M. Heyer, and A. Clapp for taking the CO 6→5 data for Mkn 231 and H. Kroker, M. Tecza, and N. Thatte for all the hard work that made the NIR imaging spectroscopy possible.

## References

Blietz, M., et al. 1994, *Ap.J.*, **421**, 92
Genzel, R., et al. 1995, *Ap.J.*, **444**, 129.
Helfer, T.T., & Blitz, L. 1995, *Ap.J.*, **450**, 90.
Meixner, M., Puchalsky, R., Blitz, L., Wright, M. & Heckman, T. 1990, *Ap. J.* **354**, 158.
Radford, S.J.E., Solomon, P.M., & Downes, D. 1991, *Ap. J.* **368**, L15.
Sternberg, A., Genzel, R., & Tacconi, L. 1994, *Ap.J.*, **436**, L131.
Tacconi, L.J., et al. 1994, *Ap.J.*, **426**, L77.

# BIMA OBSERVATIONS OF MOLECULAR GAS IN NGC 1068

TAMARA T. HELFER
*Department of Astronomy*
*University of Maryland*
*College Park, Maryland 20742*

## 1. Introduction

NGC 1068 is the best studied example of a galaxy with prolific nuclear activity as well as a kiloparsec scale starburst; the bolometric luminosity of this Seyfert galaxy is roughly evenly divided between the two components (Telesco et al. 1984). The starburst region and the nucleus appear to be linked by a 2.2 $\mu$m stellar bar, the linear dimension of which matches the scale of the starburst (Scoville et al. 1988; Thronson et al. 1989). Since molecular gas is a plausible fuel for the active galactic nucleus (AGN), and since molecular gas must be intimately involved with the starburst activity, it is natural to look to the molecular emitters for clues of a possible connection between these two energetic processes. In this paper, I discuss recent CO and HCN observations of NGC 1068 made with the BIMA array. A more complete description of previous observations of this galaxy as well as the results of the BIMA study is presented in Helfer and Blitz (1995).

## 2. Distribution of Molecular Gas

The BIMA images of CO and HCN are presented in Figure 1; these maps include zero-spacing data from the NRAO 12m telescope so they are sensitive to structure on large as well as small size scales.

The CO emission is distributed in the form of a molecular bar from which emanate two spiral arms. The linear extent ($r \sim 1$ kpc) and position angle (P.A. $\sim 63°$) of the molecular bar are in good agreement with those of the 2.2 $\mu$m stellar bar. As is common with dust lanes within bars in spiral galaxies, the CO emission within the bar is at the leading edge and is probably associated with shocked gas.



*W.B. Latter et al. (eds.), CO: Twenty-Five Years of Millimeter-Wave Spectroscopy,* 324–326.

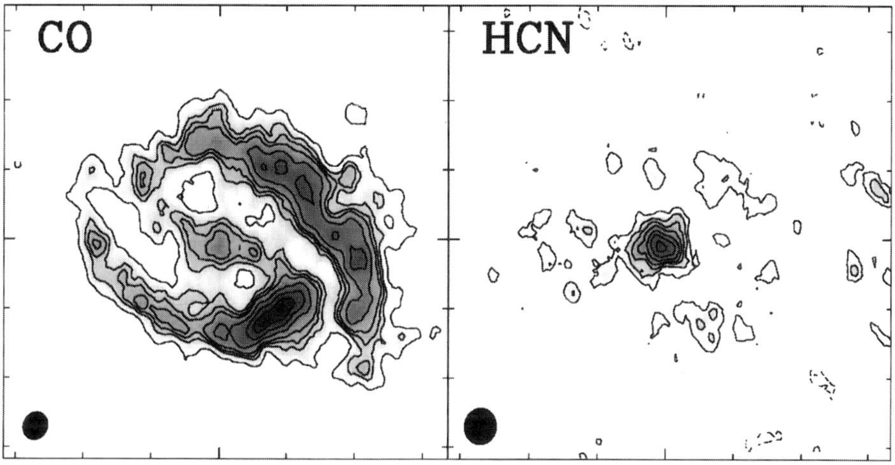


*Figure 1.* BIMA maps of CO emission (left) and HCN emission (right) in NGC 1068. The linear scale is such that 1 kpc $\approx 14''$ at $d = 16$ Mpc; the tick marks are separated by $10''$.

While the bulk of molecular gas traced by CO is associated with the starburst at kiloparsec size scales, there is also a compact distribution of CO emission around the nucleus of NGC 1068. By contrast, *most* of the HCN emission is strongly concentrated towards the nucleus of NGC 1068, though there is low-level emission detected in the region of the starburst. HCN traces gas at densities $> 10^5$ cm$^{-3}$, which is about two orders of magnitude greater than the critical density required to excite CO; the strong HCN emission suggests that there is a substantial quantity of high density gas near the nucleus.

## 3. Kinematics

The excellent signal-to-noise ratios of the individual 8 km s$^{-1}$ CO channel maps allow for a detailed study of the kinematics in the inner kiloparsec of NGC 1068. In the region of the spiral arms, the velocity field is well ordered; the gas is in simple rotation with a kinematic line of nodes of 84°. Interior to the bar, the isovelocity contours twist and run parallel to the axis of the bar in the signature kinematic response of gas to a bar potential.

Simple kinematic modeling shows that the bar's influence on the gas kinematics in the spiral arms is modest, with typical noncircular motions of $< 30$ km s$^{-1}$ in the plane of the galaxy. The noncircular residuals show an ordered spiral structure that lie along the spiral arms; the magnitude and distribution of the noncircular motions are consistent with spiral density wave streaming. Interior to the spiral arms, noncircular motions dominate the gas kinematics.

## 4. Extreme Physical Conditions in the Central ~ 200 pc

The ratio of HCN to CO integrated intensities is about 0.6 over the central $r \approx 175$ pc; this is the highest ratio measured in the center of any galaxy. The physical conditions in the molecular gas in this region are well modeled by gas at a kinetic temperature of about 50 K and a density of about 4 $\times 10^6$ cm$^{-3}$. The pressure implied by this modeling is $P/k \approx 2 \times 10^8$ cm$^{-3}$K. This pressure is about four orders of magnitude greater than that of the local interstellar medium (Bloemen 1987) and it is some three orders of magnitude greater than the pressure of the self-gravity of a typical giant molecular cloud in the solar neighborhood (Blitz 1993).

## 5. Comparison With the Milky Way

NGC 1068 shows all the kinematic and morphological characteristics of a classic barred spiral galaxy. What distinguishes this source from the Milky Way, another barred galaxy with a central concentration of molecular gas? Both galaxies have an appreciable quantity of dense gas in their centers (as do most spiral galaxies, Helfer & Blitz 1993 and references therein). Both galaxies have similar gas surface densities ($\sim 600 M_\odot$ pc$^{-2}$) averaged over their inner $r \sim 200$ parsecs. Both galaxies show evidence for $m =$ 1 modes (Blitz, these proceedings). An important difference between the two galaxies may be that in NGC 1068, there is a massive ($> 10^9$ M$\odot$) reservoir of molecular gas in spiral arms that lies within $r = 0.6$–2 kpc of the nucleus; this reservoir may be the source of fuel for the AGN. While the Milky Way has a similar concentration of molecular gas at its center and the same potential mechanism for channeling gas inward from large radii (i.e., bar streaming), there is a distinct lack of molecular gas at radii between 0.3–3.5 kpc in the Milky Way.

## References

Blitz, L. 1989, in Physical Processes in Interstellar Clouds, ed. G.E. Morfill and M. Scholer (Dordrecht:Reidel), 35
Bloemen, J.B.G.M. 1987, ApJ, 322, 696
Helfer, T.T. and Blitz, L. 1993, ApJ, 419, 86
Helfer, T.T. and Blitz, L. 1995, ApJ, 450, 90
Scoville, N.Z., Matthews, K., Carico, D.P., and Sanders, D.B. 1988, ApJLett, 327, L61
Telesco, C.M., Becklin, E.E., Wynn-Williams, C.G., and Harper, D.A. 1984, ApJ, 282, 427
Thronson, H.A., Hereld, M., Majewski, S., Greenhouse, M., Johnson, P., Spillar, E., Woodward, C.E., Harper, D.A., and Rauscher, B.J. 1989, ApJ, 343, 158

# CO IN HIGH REDSHIFT GALAXIES

SIMON J. E. RADFORD
*National Radio Astronomy Observatory, Tucson, AZ, USA*

## 1. Introduction

Study of molecular gas in distant galaxies during the last twenty years has followed the steady progress in mm wave receiver sensitivity. In 1975, CO was detected in M 82, NGC 253, and several other galaxies with redshifts of a few hundred km $s^{-1}$ (Rickard et al. 1975; Solomon & de Zafra 1975). Over the next fifteen years, the CO detection horizon increased steadily, reaching $z \approx 0.22$ by 1990 (Downes et al. 1991). The discovery that the large population of infrared luminous galaxies detected by IRAS are very gas rich (e. g., Sanders, Scoville, & Soifer 1991) was especially significant. In the last few years, CO has been detected in two high redshift objects, IRAS FSC 10214+4724 at $z = 2.3$ (Brown & Vanden Bout 1992b) and the Cloverleaf quasar (H 1413+117) at $z = 2.6$ (Barvainis et al. 1994; Barvainis 1996). These objects offer glimpses of galaxies' properties when the Universe was only about 15% of its present age. The presence and conditions of molecular gas in galaxies at such an early epoch are clues to understanding galaxy formation and evolution in the early Universe.

Despite much observational effort, however, no other high redshift sources have confirmed detections of CO. Indeed molecular gas in both 10214+4724 and the Cloverleaf is visible only because they are gravitationally lensed. In intrinsic molecular content, gas distribution, and IR luminosity, these galaxies resemble gas rich, ultraluminous IR galaxies in the local universe. They may be powered by ongoing bursts of massive star formation in the gas rich interstellar medium or by active nuclei fueled by gas accretion. Whatever the power source, substantial processing by massive stars must have already taken place to create the observed heavy elements, dust, and molecular gas. These galaxies indicate the existence of a high redshift parent population similar to nearby ultraluminous galaxies.



*W.B. Latter et al. (eds.), CO: Twenty-Five Years of Millimeter-Wave Spectroscopy*, 327–334.

## 2. IRAS FSC 10214+4724

### 2.1. APPARENT PROPERTIES

The apparent luminosity of 10214+4724, about[1] $10^{14}h^{-2}\,L_\odot$ emitted primarily in the far infrared (Rowan-Robinson et al. 1991, 1993), places it among the most luminous objects known. It exceeds the quasar luminosity threshold by two orders of magnitude. Although 10214+4724 is not strictly speaking a primordial galaxy, because there has been sufficient processing of material through stars to produce metals, it obviously has not had much time to convert its gas to stars. It clearly formed at least a few $\times 10^7$ yr, and more probably $> 10^8$ yr, earlier.

To date, CO in 10214+4724 has been detected with six different telescopes. Since the first observations (Brown & Vanden Bout 1991), however, the CO line flux, precise redshift, and source size have all been disputed. Follow up observations with the IRAM and Nobeyama telescopes (Fig. 1) have all indicated a much smaller line flux, $4 \pm 1$ Jy km s$^{-1}$, than first measured with the NRAO 12 m telescope, $21 \pm 5$ Jy km s$^{-1}$ (Brown & Vanden Bout 1991). The lower measurements imply an apparent mass $M(H_2) \approx 10^{11}h^{-2}\,M_\odot$ (Solomon, Downes, & Radford 1992), as much gas as the total mass of stars in a large spiral galaxy or the core of a giant elliptical galaxy. The discrepancy with the original measurement led to suggestions (Brown & Vanden Bout 1992b; Sakamoto et al. 1992; Tsuboi & Nakai 1992) the molecular gas is extended over a diameter as large as $60''$ ($240h^{-1}$ kpc) with an apparent molecular mass of about $10^{12}\,M_\odot$. With smaller beams, the large telescopes and interferometers would see, therefore, only the central peak, while the smaller telescope would detect the more extended component. Although it cannot be ruled out a priori, such a large source seems unlikely since only a few molecular clouds have been observed $> 20$ kpc from the center of the Milky Way and CO has never been detected $> 100$ kpc from the center of any other galaxy.

To resolve the observational quandry, then, we reobserved CO(3→2) from 10214+4724 with the NRAO 12 m telescope (Radford et al. 1996). We found an integrated line flux of $6.7 \pm 1.4$ km s$^{-1}$. This is $3 \pm 1$ times smaller than the first measurement and is consistent, within its uncertainties, with observations at other telescopes. No evidence remains for an extended source larger than a couple of arcseconds. Molecular gas in 10214+4724 is concentrated in a small central region.

In high redshift objects, submm lines usually difficult to study can be observed at mm wavelengths through the standard atmospheric windows. In 10214+4724, the CO(3→2), (4→3) (Brown & Vanden Bout 1992b), (6→5)

[1] $H_0 = 100h$ km s$^{-1}$ Mpc$^{-1}$, $q_0 = 0.5$

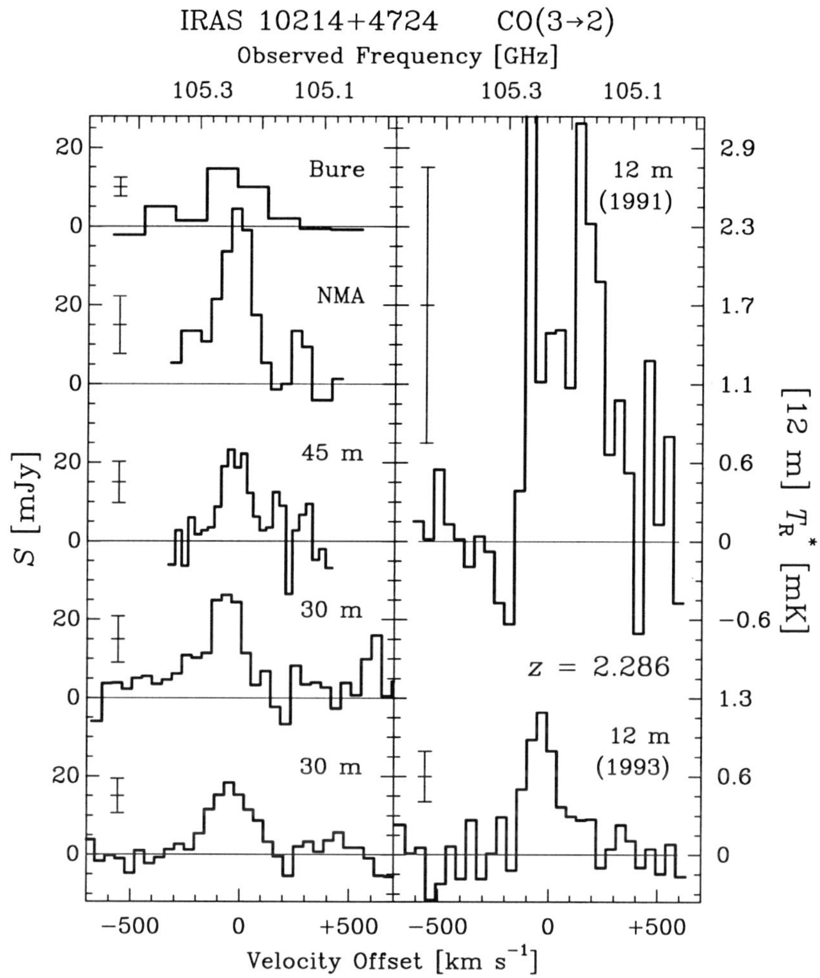


*Figure 1.* Spectra of CO(3→2) emission at $z = 2.286$ from IRAS 10214+4724 observed with the NRAO 12 m telescope at 16 MHz resolution in 1991 (Brown & Vanden Bout 1991) and in 1993 (Radford et al. 1996), the IRAM 30 m telescope at 16 MHz resolution (upper: Brown & Vanden Bout 1992b; lower: Solomon et al. 1992), the Nobeyama 45 m telescope at 10.5 MHz resolution (Tsuboi & Nakai 1992), the Nobeyama Millimeter Array at 32.5 MHz resolution (Kawabe et al. 1992; Sakamoto et al. 1992), and the IRAM interferometer (Bure) at 50 MHz resolution (Radford et al. 1993). The $\pm 1\sigma$ error bars represent the per channel uncertainty.

(Solomon, Downes, & Radford 1992), and, tentatively, (7→6) (Solomon private communication) lines have all been detected. The line ratios are consistent with a single component LVG model that indicates the gas is warmer, $T_{\rm kin} \approx 50\,$K, and denser, $n(H_2) \approx 5000\,{\rm cm}^{-3}$, than the bulk of the gas in the Milky Way (Solomon, Downes, & Radford 1992), where CO(6→5) is only observed in molecular cloud cores near sites of massive star formation (Jaffe et al. 1989). An upper limit to the CO(1→0) line flux (Barvainis 1995) is also consistent with this excitation model.

Searches for C I($^3P_1 \rightarrow ^3P_0$) and ($^3P_2 \rightarrow ^3P_1$) emission from 10214+4724 have been made, but a claimed detection of C I($^3P_2 \rightarrow ^3P_1$) (Brown & Vanden Bout 1992a) remains unconfirmed and controversial. In the Cloverleaf quasar, on the other hand, C I($^3P_1 \rightarrow ^3P_0$) has been detected (Barvainis 1996) with a C I/CO level similar to the nearby galaxy IC 342 (Büttgenbach et al. 1992).

The observed mm and submm spectral energy distribution peaks near 150 $\mu$m, corresponding to a rest frame peak of 46 $\mu$m and a dust temperature of about 80 K. In addition, the 60 $\mu$m flux observed by IRAS corresponds to 18 $\mu$m in the rest frame and indicates the presence of hotter dust at about 200 K. These components each have an apparent luminosity of $4 \times 10^{13} h^{-2}\,L_\odot$. The apparent gas-to-dust ratio, inferred from the CO line flux and the 350 $\mu$m (rest frame) continuum flux, is about 500, similar to the Milky Way and to nearby ultraluminous galaxies. This suggests the metal abundance is already approximately Solar (Downes et al. 1992).

## 2.2. GRAVITATIONAL LENS AND INSTRINSIC PROPERTIES

Optical and infrared images suggest 10214+4724 is magnified 5–50 times by an intervening gravitational lens (Matthews et al. 1994; Elston et al. 1994; Januzzi et al. 1994; Broadhurst & Lehár 1995; Graham & Liu 1995; Eisenhardt et al. 1996). This means, of course, the galaxy's intrinsic luminosity and mass of $H_2$ are proportionally smaller, but it still ranks among the most luminous and gas rich IR galaxies (Downes, Solomon, & Radford 1995). There is still copious molecular gas, comparable to that in an ultraluminous IR galaxy, to fuel vigorous star formation.

The 2.2 $\mu$m (Matthews et al. 1994; Graham & Liu 1995) and red HST (Eisenhardt et al. 1996) images of 10214+4724 show a 1.5″ long arc about 1.2″ south of the intervening galaxy. At the center of the arc is a bright, compact, 0.7″ diameter core that has a faint counterimage just north of the intervening galaxy. This apparent morphology suggests the intrinsic source has at least two components: a compact source almost coincident with a cusp of the lens caustic and an extended envelope or disk that appears as the extended arc (Broadhurst & Lehár 1995). Because the lens magnification depends on the intrinsic source size, the lens may be chromatic since different spectral components are emitted by sources with different intrinsic sizes. Differential magnification may alter, then, the observed spectrum so it no longer accurately represents the intrinsic spectrum.

Current mm wave interferometers do not have the sub arcsecond resolution available in the near IR and optical and necessary to see details of the CO distribution in 10214+4724. Nevertheless, it is possible to gauge its overall extent.

In a CO($3 \rightarrow 2$) image made with the IRAM interferometer, the source is clearly more extended east-west than the beam, but it is no wider north-south. Convolving the $2.3'' \times 1.6''$ synthesized beam with the 15% contour of the $2.2\,\mu$m arc reproduces the observed CO distribution well (Downes, Solomon, & Radford 1995). Convolving the beam with a small compact source, such as the bright, compact $2.2\,\mu$m core, on the other hand, produces a much more condensed distribution than observed in the CO image.

A CO($6 \rightarrow 5$) image made with the OVRO interferometer (Fig. 2) offers better resolution and shows somewhat more directly the east-west extent of the CO distribution. Regardless of the weighting of the visibility data, natural or uniform, this image indicates the apparent CO distribution is $1.5''$ long. We identify the CO distribution with the extended arc in the $2.2\,\mu$m image and, hence, with an extended part of the intrinsic source.

An upper limit to the magnification of the CO image can be determined because the gravitational lens stretches the source image in one dimension, preserving surface brightness, and we can estimate the intrinsic CO brightness temperature from the line ratios (Downes, Solomon, & Radford 1995). The observed luminosity in the CO($3 \rightarrow 2$) line is $L_{\rm CO} = 2.6 \times 10^{10} h^{-2}\,{\rm K\,km\,s^{-1}\,pc^2}$. For the smallest possible intrinsic CO distribution, an optically thick thermal source, $L_{\rm CO} = m_{\rm CO} \pi r^2 T_b \Delta v$, where $r$ is the intrinsic source radius, $T_b$ is the rest frame brightness temperature of the line, $\Delta v = 220 \pm 30\,{\rm km\,s^{-1}}$ is the linewidth, and $m_{\rm CO}$ is the source magnification. Since the magnification is one dimensional, $m_{\rm CO} = a/r$, where $2a$ is the apparent extent of the CO distribution. The observed CO line ratios clearly indicate the molecular gas is warm. If the gas kinetic temperature is 60 K, slightly cooler than the 80 K dust, an LVG excitation model that fits the observed CO line intensities and ratios indicates the CO($3 \rightarrow 2$) and ($6 \rightarrow 5$) brightness temperatures are $43 \pm 7$ and $27 \pm 5$ K, respectively, and the opacities are 6, 37, and 41 for the CO($1 \rightarrow 0$), ($3 \rightarrow 2$), and ($6 \rightarrow 5$) lines. At $z = 2.3$, the apparent extent of the CO distribution $2a = 1.5'' \times (4h^{-1}\,{\rm kpc\,arcsec^{-1}}) = 6h^{-1}\,{\rm kpc}$, so $m_{\rm CO} \approx 10$, independent of $h$.

With this modest lens magnification, 10214+4724 still has a molecular gas content comparable to the most CO rich, IR luminous galaxies (Graham & Liu 1995), but is not extraordinary for that class. The galaxy's intrinsic CO($3 \rightarrow 2$) luminosity is $2.6 \times 10^9 h^{-2}\,{\rm K\,km\,s^{-1}\,pc^2}$ and the true radius of the CO distribution $r \approx 300h^{-1}$ pc. For comparison, Arp 220 has a CO($1 \rightarrow 0$) luminosity of $4 \times 10^9 h^{-2}\,{\rm K\,km\,s^{-1}\,pc^2}$ (Solomon, Radford, & Downes 1990),

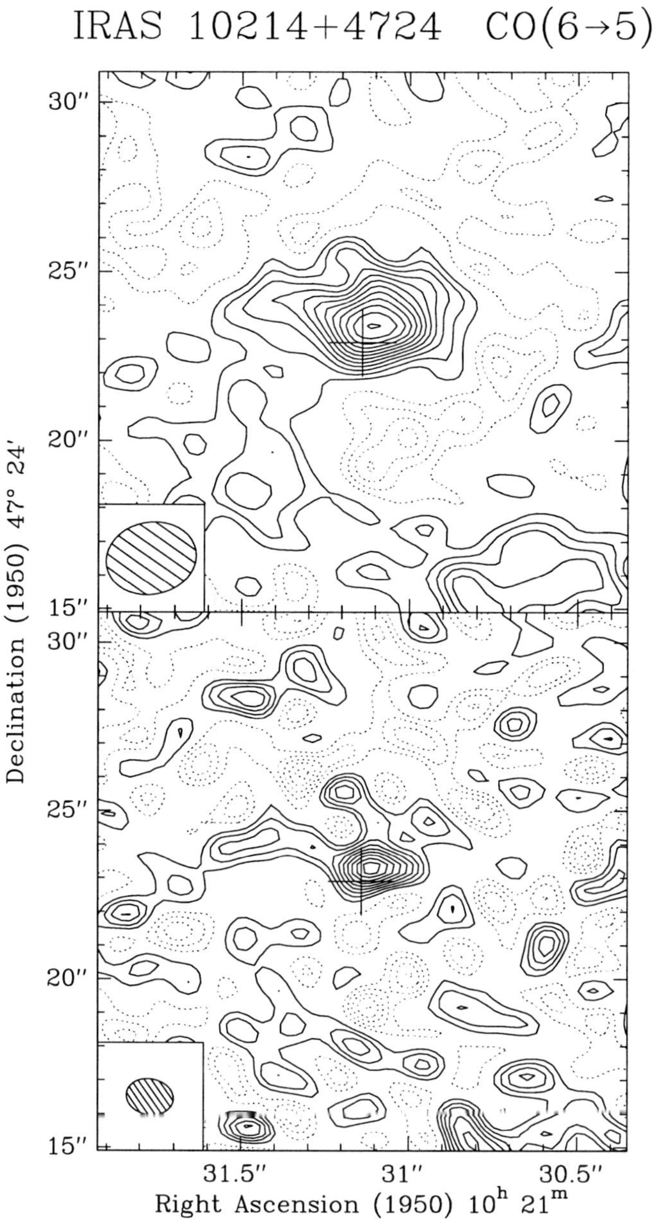


*Figure 2.* Integrated emission maps of CO(6→5) at $z = 2.286$ from IRAS 10214+4724 observed with OVRO interferometer at 210 GHz. The contour interval is 2 mJy beam$^{-1}$ and the synthesized beam (insert) is $2.7'' \times 2.1''$ for natural weighting (top) and $1.4'' \times 1.1''$ for uniform weighting (bottom). The source is more extended than the beam; the deconvolved source size is about $1.5''$ east-west.

with 2/3 of that concentrated within a radius of $240h^{-1}$ pc (Scoville et al. 1991). In a sample of 37 ultraluminous IR galaxies out to $z = 0.27$, the highest CO($1\to0$) luminosity is $9 \times 10^9 h^{-2}\,\mathrm{K\,km\,s^{-1}\,pc^2}$ and the median is $5 \times 10^9 h^{-2}\,\mathrm{K\,km\,s^{-1}\,pc^2}$ (Solomon et al. 1996).

The far IR magnification can be determined by a similar argument, although in this case we have no direct measurement of the source size (Downes, Solomon, & Radford 1995). In 10214+4724, the apparent far IR luminosity of $4 \times 10^{13} h^{-2}\,L_\odot$, observed at 450 to 1300 $\mu$m, is emitted by optically thick dust at $\approx 80$ K (Downes et al. 1992). In nearby, non-lensed ultraluminous galaxies, the far IR source is 0.6–1.0 times the the size of the CO source. This suggests the 80 K dust source in 10214+4724 has a radius of 200–$300h^{-1}$ pc, the far IR magnification is 10–13, and the intrinsic far IR luminosity is 3–4 $\times 10^{12} h^{-2}\,L_\odot$. Again, this is similar to nearby ultraluminous galaxies, albeit at the high end of the distribution. Arp 220, for example, has a far IR luminosity of $6 \times 10^{11} h^{-2}\,L_\odot$. In our sample of 37 galaxies, the most luminous has $2 \times 10^{12} h^{-2}\,L_\odot$. Since the CO and far IR have similar magnifications, their ratio, which indicates the gas-to-dust ratio and the metal abundance, is largely unaffected.

The mid IR radiation, on the other hand, is emitted by a hotter, 200 K source that may be much smaller. If it corresponds to the compact, $0.7''$ diameter core in the 2.2 $\mu$m images, it may be magnified 50 times (Broadhurst & Lehár 1995). Then the intrinsic mid IR luminosity would be $8 \times 10^{11} h^{-2}\,L_\odot$ and the true radius of the core would be $30h^{-1}$ pc. This is a typical radius for an AGN Narrow Line Region (NLR), but is much too small to account for the thermal far IR emission at 80 K. Hence the far IR dominates the intrinsic luminosity of 10214+4724, as it does in nearby ultraluminous galaxies.

Although 10214+4724 has a Seyfert 2 spectrum lacking the usual signatures of star formation (Rowan-Robinson et al. 1991, 1993; Lawrence et al. 1993, Elston et al. 1994; Januzzi et al. 1994; Soifer et al. 1995; Goodrich et al. 1996), this may be an artifact of differential magnification by the intervening lens rather than an intrinsic property. If the galaxy harbors both an extended starburst and an AGN, the small nucleus will be magnified much more than the larger region of star formation where the H II regions are. Since the observed optical spectrum is dominated by the compact $0.7''$ core, which is the magnified image of the nucleus, the AGN characteristics overshadow the starburst signatures.

Even with true bolometric and CO luminosities ten times lower than earlier estimates (Downes et al. 1992; Solomon, Downes, & Radford 1992), the nature of the energy source remains a problem. Is 10214+4724 powered by star formation in the molecular region itself, or are the gas and dust just part of a massive envelope heated by the AGN (Sanders et al. 1989)?

The amount of dense molecular gas in ultraluminous IR galaxies indicates they harbor huge starbursts. To explain even the intrinsic luminosity of 10214+4724 with a starburst, however, time scale constraints (Heckman 1994) imply there must be an IMF of high mass stars only and a high star formation efficiency – 20% of all the gas converted to stars in $10^7$ yr. Nevertheless, in intrinsic molecular content and IR luminosity, 10214+4724 resembles a typical nearby ultraluminous IR galaxy. Whatever its current power source, the heavy elements, dust, and molecular gas must all have been produced by massive stars.

It has been a pleasure to collaborate with Dennis Downes and Phil Solomon. The NRAO is a facility of the National Science Foundation operated under cooperative agreement by Associated Universities, Inc.

## References

Barvainis, R., et al. 1994, Nature 371, 586
Barvainis, R., 1995, AJ 110, 1573
Barvainis, R., 1996, this volume
Broadhurst, T., & Lehár, J. 1995, ApJ 450, L41
Brown, R. L., & Vanden Bout, P. A. 1991, AJ 102, 1956
Brown, R. L., & Vanden Bout, P. A. 1992a, ApJ 397, L11
Brown, R. L., & Vanden Bout, P. A. 1992b, ApJ 397, L19
Büttgenbach, T. H., Keene, J., Phillips, T. G., & Walker, C. K. 1992, ApJ 397, L15
Downes D., et al. 1991, in IAU Symp. 146, ed. Combes & Casoli (Kluwer) p. 295
Downes, D., et al. 1992, ApJ 398, L25
Downes, D., Solomon, P. M., & Radford, S. J. E. 1995, ApJ 453, L65
Eisenhardt, P. R., et al. 1996, ApJ 461, 72
Elston, R., et al. 1994, AJ 107, 910
Goodrich, R. W., et al. 1996, ApJ 456, L9
Graham, J. R., & Liu, M.C. 1995, ApJ 449, L29
Heckman, T. M. 1994, in Mass Transfer ..., ed. Shlossman (Cam. U. P.) p. 234
Jaffe, D. T., et al. 1989, ApJ 344, 265
Jannuzi, B.T., et al. 1994, ApJ 429, L49
Kawabe, R., Sakamoto, K., Ishizuki, S., & Ishiguro, M. 1992, ApJ 397, L23
Lawrence, A., et al. 1993, MNRAS, 260, 28
Matthews, K., et al. 1994, ApJ, 420, L13
Radford, S. J. E., Brown, R. L., & Vanden Bout, P. A. 1993, A&A 271, L71
Radford, S. J. E., et al. 1996, AJ 111, 1021
Rickard, L. J., et al. 1975, ApJ 199, L75
Rowan-Robinson, M., et al. 1991, Nature 351, 719
Rowan-Robinson, M., et al. 1993, MNRAS, 261, 513
Sakamoto, K., Ishizuki, S., Kawabe, R., & Ishiguro, M. 1992, ApJ 397, L27
Sanders, D. B., et al. 1989, ApJ, 347, 29
Sanders, D. B., Scoville, N. Z., & Soifer, B. T., 1991, ApJ 370, 158
Scoville, N. Z., Sargent, A. I., Sanders, D. B., & Soifer, B. T., 1991, ApJ 366, L5
Solomon, P. M., & de Zafra, R., 1975, ApJ 199, L79
Solomon, P. M., Downes, D., & Radford, S. J. E. 1992, ApJ 398, L29
Solomon, P. M., Downes, D., Radford, S. J. E., & Barrett, J. W., 1996, in preparation
Solomon, P. M., Radford, S. J. E., & Downes, D. 1990, ApJ 348, L53
Solomon, P. M., Radford, S. J. E., & Downes, D. 1992, Nature 356, 318
Tsuboi, M., & Nakai, N. 1992, PASJ 44, L241

# CO (AND OTHER) LINES FROM THE CLOVERLEAF QUASAR

RICHARD BARVAINIS
*MIT Haystack Observatory*
*Westford, MA 01886*

## 1. Introduction

The Cloverleaf quasar (H1413+117) at redshift 2.56 derives its name from its optical image, which is gravitationally lensed into a pattern of four points with separations of about 1″. The lensing galaxy is unseen, but models assuming an elliptical potential at intermediate redshift give total magnifications of roughly a factor of 10 (Kayser et al. 1990). The Cloverleaf's optical spectrum has classical broad emission lines, plus broad absorption troughs–it is classified as a BALQ (broad absorption line quasar).

We initially selected the Cloverleaf for molecular line studies because of its strong submillimeter emission, detected during a survey of radio quiet quasars and BALQs at the JCMT (Barvainis, Antonucci, and Coleman 1992). The Cloverleaf was also detected by IRAS and has a far-IR/submm spectrum identical to the luminous infrared galaxy IRAS F10214+4724 at $z = 2.28$ (Barvainis et al. 1995). F10214+4724 was, of course, the first high-$z$ object to be convincingly detected in CO (Brown and Vanden Bout 1992; Solomon, Downes & Radford 1992), with the Cloverleaf being the second (Barvainis et al. 1994). Several new lines of evidence link F10214+4724 closely with the Cloverleaf: spectropolarimetry of F10214+4724 shows broad optical emission lines in reflected light, meaning that it harbors a hidden quasar (Miller 1995); F10214+4724 now appears to be gravitationally lensed (e.g., Broadhurst and Lehár 1995); and similar far-IR and CO emission, but different optical emission, in the two objects suggests that they differ primarily in orientation, with F10214+4724's nucleus obscured by a surrounding dusty torus (Barvainis et al. 1995).

Here I present new millimeter observations of line emission from the Cloverleaf. My collaborators in this work are Ski Antonucci, Danielle Alloin,



*W.B. Latter et al. (eds.), CO: Twenty-Five Years of Millimeter-Wave Spectroscopy,* 335–337.

Linda Tacconi, Paul Coleman, and Phil Maloney.

## 2. Observations and Results

We observed the Cloverleaf using the IRAM 30 m telescope on Pico Veleta in June 1994 and May 1995. Because of the unique ability of the 30 m to collect data in the 3 mm, 2 mm, and 1 mm bands simultaneously, we were able to observe a rather large suite of lines. This has resulted in definite detections in CO $J = 3-2$, $4-3$, $5-4$, and $7-6$ (see Figure 1), and an upper limit on $^{13}CO(3-2)$ emission. Space limitations preclude showing all the spectra here, but a very likely detection of the atomic carbon fine structure line $CI(^3P_1 - ^3P_0)$ was obtained, as was an upper limit on $CI(^3P_2 - ^3P_1)$. We also have a possible detection of HCN$(4-3)$ at the $T_A^* = 1$ mK level, which required some 7 nights of integration. Finally, we have an upper limit on CO$(1-0)$, at $\nu_{obs} = 32.4$ GHz, using the Haystack 37 m telescope in January 1995. This limit is consistent with CO$(1-0)$ being no more than about a factor of two brighter than the thermalized value relative to CO$(3-2)$.

These results will be discussed in detail in a future paper (Barvainis et al. 1996). A few general statements can be made here. First, the CO excitation is interesting, in that CO$(4-3)$ appears to have a higher brightness temperature than either CO$(3-2)$ or CO$(5-4)$. This probably means that conditions are such that the critical density for the CO$(4-3)$ transition obtains, and that the lines are not too optically thick. Second, the (possible) detection of HCN means that there is probably a large quantity of high density gas ($n \sim 10^5$ cm$^{-3}$) present in the Cloverleaf. The HCN/CO brightness temperature ratio is 0.2, very much like the values found in luminous IRAS galaxies. Finally, the CI/CO ratio is also about 0.2, which is similar to the value in IC 342 (the only other extragalactic source where both species have been clearly measured).

## 3. Conclusion

Although there have been many attempts to detect high-$z$ CO emission, only the Cloverleaf and F10214+4724 have been definitely confirmed. It appears that the boosting derived from lensing, uncertain but perhaps an order of magnitude, is essential for studying gas in the early universe, given current instrumentation. Therefore, we need to derive as much information as we can from molecular line studies of the two objects where such studies are now possible. Detailed modeling of the various lines presented here promises to provide interesting constraints on molecular gas in the Cloverleaf quasar at an epoch when the universe was only one-seventh its current age.

## References

Barvainis, R., Antonucci, R., & Coleman, P. 1992, ApJ, 399, L19
Barvainis, R., et al. 1994, Nature, 371, 586
Barvainis, R., et al. 1995, ApJ, 451, L9
Barvainis, R., Maloney, P., Antonucci, R., & Alloin, D. 1996, in preparation
Broadhurst, T., & Lehár, J. 1995, ApJ, 450, L41
Brown, R.L., & Vanden Bout, P.A. 1992, ApJ, 397, L19
Kayser, R., et al. 1990, ApJ, 364, 15
Miller, J. 1995, talk given at the National Academy of Sciences Colloquium "Quasars and AGN, High Resolution Imaging," March 24-25, Irvine, CA
Solomon, P.M., Downes, D., & Radford, S.J.E. 1992, ApJ, 398, L29

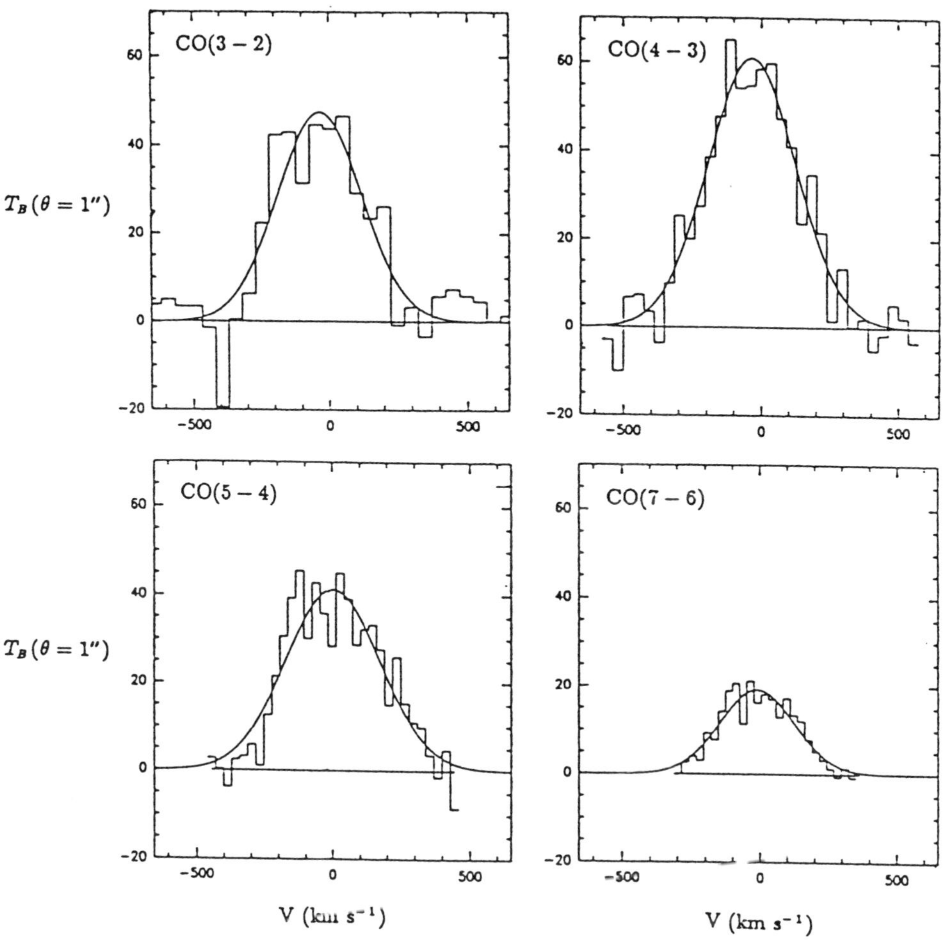


Figure 1. Four transitions of CO in the Cloverleaf quasar. For ease of comparison, the spectra have been put on a brightness temperature scale (in Kelvins), arbitrarily assuming a source size of $0.3''$($\approx$ 1.7 kpc at the source).

# CENTAURUS A:
# THE $^{13}$CO MAP AND MOLECULAR LINE RATIOS

W. WILD
*IRAM, Av. Divina Pastora 7, Granada, Spain*

A. ECKART, M. CAMERON, R. GENZEL AND H. ROTHERMEL
*MPI für Extraterrestrische Physik, Garching, Germany*

AND

G. RYDBECK AND T. WIKLIND
*Onsala Space Observatory, Onsala, Sweden*

## 1. Introduction

Centaurus A (NGC 5128) is the closest radio galaxy (distance about 3 Mpc) and belongs to the class of elliptical galaxies with dust lanes. It is a strong radio source with a milliarsecond nuclear continuum source and two giant radio lobes. Absorption against the nuclear source has been found in HI and several molecular species and transitions. Due to its proximity, it offers the possibility of studying in detail the interstellar medium in an elliptical galaxy.

Here we present the first fully sampled map of $^{13}$CO $J$=1–0 along the dust lane of Centaurus A. The data, obtained with the 15m Swedish-ESO Submillimetre Telescope (SEST) in La Silla, Chile, have a resolution of 44″ (FWHM). In combination with the $^{12}$CO $J$=1–0 and $J$=2–1 maps that we presented earlier (Eckart *et al.* 1990; Rydbeck *et al.* 1993) and newly obtained spectra of $^{13}$CO $J$=2–1 and C$^{18}$O $J$=1–0 at selected positions, study of the excitation conditions of the molecular gas along the dust lane and its variation in the disk is possible. A detailed report of this work will be published elsewhere (Wild *et al.*, in preparation).

## 2. The $^{13}$CO $J$=1–0 map

Fig. 1 (top panel) shows the velocity integrated $^{13}$CO $J$=1–0 emission along the dust lane. The emission is well concentrated along the dust lane and the



*W.B. Latter et al. (eds.), CO: Twenty-Five Years of Millimeter-Wave Spectroscopy, 338–340.*

map shows two peaks of emission symmetrically displaced from the central position. The morphology is quite similar to the $^{12}$CO $J$=1–0 and $J$=2–1 maps.

## 3. Molecular line ratios

Fig. 1 (middle panel) shows the $^{12}$CO/$^{13}$CO $J$=1–0 ratio along the disk of Centaurus A. The $^{12}$CO/$^{13}$CO ratio throughout the disk ranges from 10 to 16, a value found for a large sample of galaxies.

Before calculating the $^{12}$CO $J$=2–1/1–0 ratio, the $J$=2–1 map was smoothed to the resolution of the $J$=1–0 data (44″). The $^{12}$CO $J$=2–1/1–0 ratio (Fig. 1, lower panel) has a value of 0.8–0.9 for the major part of the disk and shows no large variation along the major axis. In the SE the ratio rises to 1.1 .

TABLE 1. CO emission line ratios

| line ratio | center | disk |
|---|---|---|
| $^{12}$CO $J$=2–1/1–0 | 0.8–0.9 | 0.8–0.9 (1.1) |
| $^{12}$CO/$^{13}$CO $J$=1–0 | 14 | 11–14 |
| $^{13}$CO $J$=2–1/1–0 | 1.0±0.4 | 0.85±0.3 |
| $^{12}$CO/C$^{18}$O $J$=1–0 | ~75 | |

Table 1 summarizes the molecular line ratios in the center and disk of Centaurus A. There is no significant variation between the central position and the disk, indicating that the excitation of the molecular gas is not very different between the center and the disk in this galaxy.

Radiative transfer calculations for various CO column densities in the framework of a one component model (taking into account non-LTE effects, clumping, and optical depth dependent filling factors) show that these line ratios are characteristic for gas with a kinetic temperature of 10–15 K and hydrogen densities $n(H_2)$ of $10^4$ cm$^{-3}$ to $3 \times 10^4$ cm$^{-3}$. The inferred CO column density is $\sim 3 \times 10^{17}$ cm$^{-2}$.

## References

Eckart A., *et al.*, 1990, *Ap.J.* **363**, 451.
Rydbeck G., *et al.*, 1993, *Astron. Astrophys. (Letters)* **270**, L13.

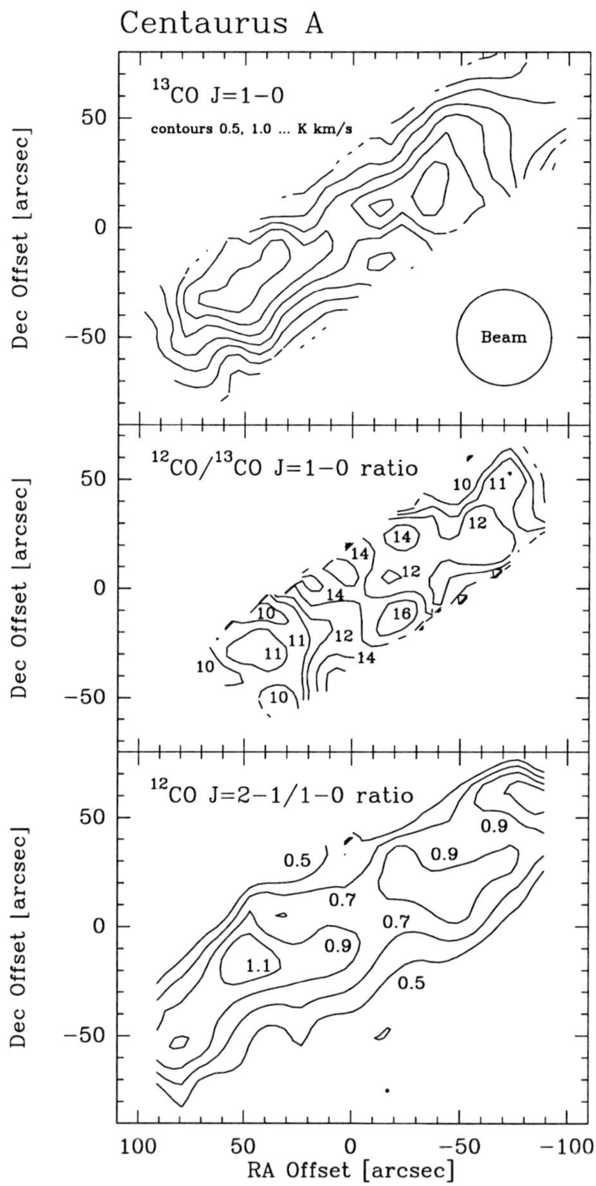


*Figure 1.* Maps of the $^{13}$CO integrated intensity, the $^{12}$CO/$^{13}$CO $J$=1–0 and $^{12}$CO $J$=2–1/1–0 ratios along the dust lane of Centaurus A. The molecular line ratios do not vary significantly across the disk.

# CO OBSERVATIONS OF HIGH REDSHIFT GALAXIES WITH THE OWENS VALLEY MILLIMETER ARRAY

MIN S. YUN AND NICK SCOVILLE
*California Institute of Technology*
*105-24 Robinson, Pasadena, CA 91125*

The peaking of quasar density at $z \sim 2$ as well as the stellar population analysis of galaxies in the local universe suggest that galaxies were formed near or prior to the redshift of 2. The identification of IRAS FSC 10214+4724 with a $z = 2.3$ galaxy (Rowan-Robinson et al. 1991) and the detection of CO emission (Brown & Vanden Bout 1991, Solomon et al. 1992) have substantiated the hypothesized existence of luminous gas-rich galaxies in the early epochs forming stars at prodigious rates. Given that there has been reasonable heavy element production, CO rotational transitions are excellent probes of cold interstellar medium in very distant galaxies. For objects of constant intrinsic CO brightness temperature, the observed line flux is almost independent of redshift at $z = 1$–$6$ due to the factor $(1+z)^2$ in angular size distance. These molecular transitions are also unambiguous tracers of cold, dense molecular gas where stars are forming.

The Owens Valley Millimeter Array is very well-suited for detecting gas-rich high redshift systems since six 10.4 m diameter telescopes add to an effective collecting area of 510 $m^2$ – comparable to a 25.5 m diameter telescope. The use of cross-correlated signals significantly reduces the instrumental effects that often plagues the single dish observations. The OVRO observations also yield spectral map cubes with typical spatial resolution of $2''$ to $5''$, providing both spatial and kinematic information.

The CO (3–2) emission region in $z = 2.3$ IRAS galaxy FSC 10214+4724 is resolved by our $2''$ resolution observations (Figure 1). About 1/3 of the total CO flux originates from an extended structure with a narrow line width while the majority of molecular gas belongs to an unresolved structure (Scoville et al. 1995). The small line width for the extended component may be explained if a kinematically distinct section of the galaxy is strongly amplified to a greater angular extent by gravitational lensing.

The CO (7–6) emission from H1413+117 (the Cloverleaf quasar) is also resolved by our $1''$ beam (Yun et al. 1996). The CO morphology is remark-



*W.B. Latter et al. (eds.), CO: Twenty-Five Years of Millimeter-Wave Spectroscopy,* 341–343.

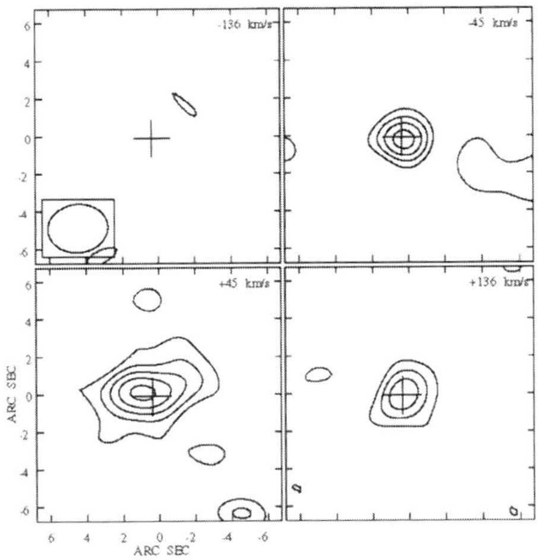


Figure 1

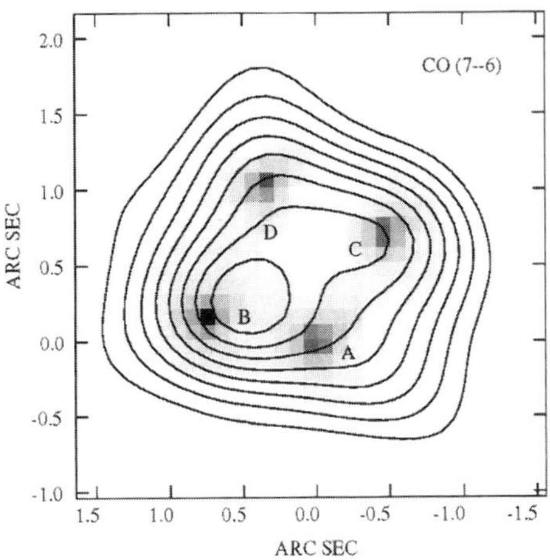


Figure 2

*Figure 1.* New 2″ resolution CO (3–2) maps of FSC 10214+4724 with 90 km $s^{-1}$ velocity channels (Scoville et al. 1995). The cross marks the VLA radio source position. An extended component with a narrow line width is found in V=+45 km/s channel. *Figure 2.* CO (7–6) map of H1413+117 at 1″ resolution in contours overlaid on the grey scale HST F555W image (Falco 1993).

ably similar to that of optical light, suggesting the CO emitting region is very compact and coincides with the quasar (Figure 2). The inferred total $H_2$ mass, $2.3 \times 10^{10} h^{-2}$ $M_\odot$ (assuming magnification factor 10), is nearly identical to the CO (3–2) measurement by Barvainis et al. (1994). The CO emission must arise from dense high excitation gas ($n > 10^5$ $cm^{-3}$, $T > 100$ K). The dust masses estimated from continuum emission are about $10^7$ $M_\odot$ for FSC 10214+4724 and H1413+117, and the inferred gas-to-dust ratios are $\sim 1000$, similar to the IR luminous systems in the local universe (e.g., Sanders, Scoville, & Soifer 1991).

The results of the OVRO CO observations of high redshift systems are summarized in Table 1. In the majority of cases, the $H_2$ mass sensitivity is sufficient to detect the more gas-rich galaxies in the local universe. In all cases, any objects like FSC 10214+4724 should have been detected. Detecting the progenitor of $L_*$ galaxies probably requires an order of magnitude improvement in sensitivity. We note that neither 4C 41.17 (a $z = 3.8$ radio galaxy with dust continuum detection – Chini & Krügel 1994) nor B2 0902+34 (a $z = 3.4$ "protogalaxy" – Eales et al. 1993) were detected in CO. The upper limit on the gas-to-dust ratio is 750 for 4C 41.17. We find the 650$\mu$m (rest) continuum emission from B2 0902+34 dominated by non-thermal emission rather than by dust emission.

TABLE 1. Summary of the OVRO CO Observations

| Name | $z$ | Transition | $S_{CO}$ (mJy) | $\Delta V$ (km s$^{-1}$) | $M(H_2)$ ($h^{-2} 10^{10}$ M$_\odot$) |
|---|---|---|---|---|---|
| H1413+117 | 2.56 | CO (7–6) | $120 \pm 12$ | 340 | $2.3 \pm 0.2$ |
| FSC10214+4724 | 2.29 | CO (3–2) | $14 \pm 2^a$ | 250 | $1.2 \pm 0.2$ |
| 3C 48 | 0.37 | CO (1–0) | $7.7 \pm 1.9^b$ | 250 | $3.9 \pm 0.8$ |
| B2 0902+34 | 3.40 | CO (4–3) | $< 4.1^c$ | $(300)^d$ | $< 3.3^d$ |
| 4C 41.17 | 3.80 | CO (4–3) | $< 4.5$ | (300) | $< 4.2$ |
| 3C 368 | 1.13 | CO (2–1) | $< 8.1$ | (300) | $< 4.5$ |
| TX 1243+036 | 3.58 | CO (4–3) | $< 6.9$ | (300) | $< 5.5$ |
| PC 1033+4750 | 3.68 | CO (4–3) | $< 6.1$ | (300) | $< 5.5$ |
| MG 1019+0535 | 2.77 | CO (3–2) | $< 6.2$ | (300) | $< 6.4$ |
| TX 0211−122 | 2.34 | CO (3–2) | $< 9.1$ | (300) | $< 7.3$ |
| 4C 28.58 | 2.95 | CO (3–2) | $< 7.4$ | (300) | $< 8.5$ |
| H0836+113 | 2.47 | CO (3–2) | $< 13$ | (300) | $< 11$ |
| PC 1643+4631A | 3.14 | CO (3–2) | $< 10$ | (300) | $< 13$ |
| PC 1406+123 | 2.25 | CO (3–2) | $< 12$ | (300) | $< 13$ |

Notes: (a) Scoville et al. (1995); (b) Scoville et al. (1993); (c) all $3\sigma$ upper limits; (d) $\Delta V = 300$ km s$^{-1}$ is assumed.

## References

Barvainis et al. 1994, Nature, 371, 586
Brown, R.L., Vanden Bout, P.A. 1991, AJ, 102, 1956
Chini, R., Krügel, E. 1994, A&A, 288, L33
Eales et al. 1993, Nature, 363, 140
Falco E.E. 1993, in *Gravitational Lenses in the Universe*, eds. Surdej et al. (Universite de Liege, Liege), p.127
Rowan-Robinson et al. 1991, Nature, 351, 719
Sanders, D.B., Scoville, N.Z., Soifer, B.T. 1991, ApJ, 370, 158
Scoville, N.Z., Yun, M.S., Brown, R., Vanden Bout, P. 1995, ApJ, 449, L109
Scoville et al. 1993, ApJ, 415, L75
Solomon, P.M., Downes, D., Radford, S.J.E. 1992, ApJ, 398, L29
Yun et al. 1996, Naure, submitted.

# CO OBSERVATIONS OF ULTRALUMINOUS INFRARED GALAXIES, QSOS, AND POWERFUL RADIO GALAXIES WITH THE 12M TELESCOPE

D. B. SANDERS AND A. S. EVANS
*Institute for Astronomy, University of Hawaii*
*2680 Woodlawn Drive, Honolulu, HI 96822*

## 1. Introduction

During the mid-1980s, the NRAO 12m telescope pioneered the first detections of CO in extragalactic sources at redshifts, $\Delta\lambda/\lambda$, of 0.03–0.2 . These observations strongly suggested that enormous supplies of molecular gas may be directly involved in the genesis of a wide variety of energetic phenomena, including ultraluminous infrared galaxies (ULIGs), quasi-stellar objects (QSOs), and powerful radio galaxies (PRGs). These detections of broad ($\Delta V_{\rm FWZI} \sim 400 - 800$ km s$^{-1}$), weak ($T_R^* \sim 1 - 3$ mK) lines using integration times typically less than 10 hours were made possible by low-noise, dual-channel receivers in the 80–115 GHz band and switching techniques that allowed extremely stable baselines over a relatively wide detector passband. Continued improvements in receivers have recently allowed the possibility of detecting CO luminosities equivalent to the most luminous objects observed locally ($z < 0.2$) at redshifts above 1. However, the required integration times (~60 hrs at the 12m telescope) may be near the limit of current single-dish performance.

## 2. The *IRAS* Galaxy Samples

CO observations at the 12m telescope of luminous infrared galaxies in the *IRAS* Bright Galaxy Survey (BGS) provided the first detections of extragalactic objects at $z > 0.03$. Ground-based optical observations of the most luminous BGS objects showed that the trigger for the enormous infrared luminosities was a strong interaction/merger. The most extreme objects (ULIGs), with infrared luminosities equivalent to the bolometric luminosities of QSOs, appeared to be powered by enormous nuclear starbursts



*W.B. Latter et al. (eds.), CO: Twenty-Five Years of Millimeter-Wave Spectroscopy,* 344–346.

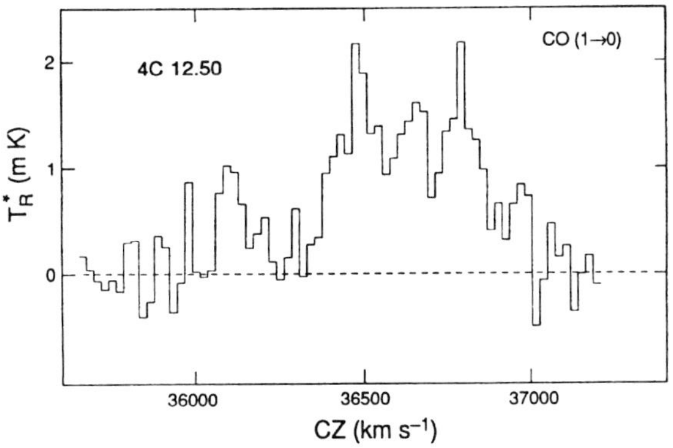


*Figure 1.* CO(1→0) emission from 4C 12.50 (from Mirabel et al. 1989).

and/or AGN (e.g., Sanders et al. 1988). CO observations suggested that the fuel supply was an enormous reservoir of molecular gas, typically 3–10 times the molecular gas content of the Milky Way (Sanders et al. 1991).

CO observations of a complete sample of ULIGs with 'warm' ($S_{60}/S_{100}$) infrared colors provided the first CO detections of objects at $z > 0.1$, including the remarkable detections of abundant molecular gas supplies in the optical QSOs Mrk 1014 and I Zw 1, in the infrared selected QSO IRAS 0759+6508, and in the PRG 4C 12.50 (see Figure 1 and refs. in the review of Sanders 1992). These objects display a clear mixture of powerful nuclear starburst and AGN properties, and they appear to represent a class of transition objects where the circumnuclear dust and gas have been dispersed enough to have a clear line of sight to the central AGN.

Further progress in understanding the role of molecular gas in the origin of QSOs and PRGs and the possible evolutionary connection to ULIGs would seem to depend heavily on our ability to obtain sensitive CO observations of a large sample of optically selected QSOs and radio selected PRGs. The recent detections of CO in 3C 48 (Scoville et al. 1993), in several low-$z$ QSOs (e.g., Alloin et al. 1992), and in a sample of low-$z$ PRGs (Mazzarella et al. 1993) show that molecular gas is indeed important in at least some of these objects. Observations of larger samples at higher redshift are clearly needed. There have been spectacular detections of the lensed object H1413+135 and the 'hyperluminous' infrared object IRAS F10214+4724 (which may also be lensed; see elsewhere this volume) at $z > 2$. Observations of additional high-$z$ objects have been severely hampered, however, by both the lack of accurate systemic redshifts and the lack of sensitivity provided by the current generation of millimeterwave telescopes.

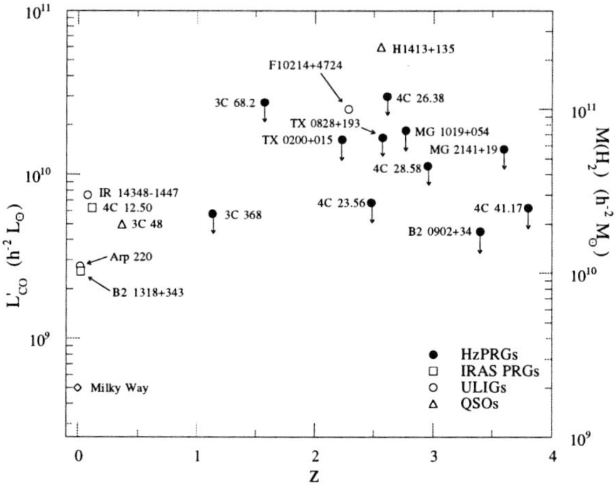


*Figure 2.* $L'_{CO}$ and $M(H_2)$ vs. $z$ (from Evans et al. 1996). (↓) Upper limits.

## 3. High-Redshift Powerful Radio Galaxies

The recent availability of accurate redshifts for a substantial number of high-$z$ powerful radio galaxies (H$z$PRGs) has allowed one of the first sensitive searches for CO at $z > 1$ to be carried out (Figure 2 and Evans et al., these proceedings). Although these observations have not yet yielded any CO detections, they have set meaningful upper limits that show that H$z$PRGs may contain less molecular gas than the most gas-rich objects observed locally. Further comment seems premature without knowing the relative molecular gas content of ULIGs, QSOs, and PRGs at high redshifts. However, CO observations of PRGs at lower redshift (Mazzarella et al. 1993) have already suggested that FR II sources (e.g. Cygnus A) have less CO luminosity than the most gas-rich FR I and core-dominated PRGs (e.g. 3C 120 and 4C 12.50). Our current CO observations of HzPRGs are consistent with the hypothesis that these objects may represent a terminal stage in the evolution of ULIGs and QSOs, one in which the powerful radio jets have helped to dissipate the gas fueling the central AGN.

## References

Alloin, D., Barvainis, R., Gordon, M. and Antonucci, R. (1992) *A&A*, **265**, pp. 429–436.

Evans, A.S., et al. (1996) *Ap.J.*, **457**, 658.

Mazzarella, J.M., Graham, J.R., Sanders, D.B. and Djorgovski, S. (1993) *Ap.J.*, **409**, pp. 170–178.

Mirabel, I.F., Sanders, D.B. and Kazès, I. (1989) *Ap.J.*, **340**, pp. L9–12.

Sanders, D.B. (1992) in *IAU Symp 146*, eds. F. Combes and F. Casoli (Dordrecht: Kluwer), pp. 417–424.

Sanders, D.B. et al. (1988) *Ap.J.*, **325**, pp. 74–91.

Sanders, D.B., Scoville, N.Z. and Soifer, B.T.. (1991) *Ap.J.*, **370**, pp. 158–171.

Scoville, N., Padin, S., Sanders, D., Soifer, B. and Yun, M. (1993) *Ap.J.*, **415**, pp. L75–78

# CO IN PLANETARY SYSTEMS AND IN STELLAR DEATH

# CO IN THE SOLAR SYSTEM

THÉRÈSE ENCRENAZ
*DESPA, Observatoire de Paris*
*F-92195 Meudon, France*

## 1. Introduction

The CO molecule is the most common gaseous species found in the Solar system. In addition to the Earth's atmosphere, CO is found in Mars and Venus; it has been detected in three giant planets, Jupiter, Saturn, and Neptune, and in Titan's atmosphere; CO ices were detected on the surface of Triton and Pluto; CO has been detected, in variable abundances, in most of the comets. The presence of CO in all these objects is remarkable, taking into account the wide diversity of their global atmospheric compositions; in all cases, CO appears to be a minor constituent relative to the dominant atmospheric compound (Table 1). Its abundance, relative to $CH_4$, is an indicator of the conditions and processes which prevailed at the time of the formation of these objects from the surrounding primordial nebula.

## 2. The Formation of Solar-system Objects and the Abundance of CO

### 2.1. THE CONDENSATION SEQUENCE

How do we understand the formation of planets? The currently proposed scenario, based on the condensation sequence of the various species is the following. It is generally accepted that all Solar-system objects were formed after the collapse of the primordial solar nebula into a disk. This nebula, with all its elements in their cosmic abundances, cooled down rapidly, with decreasing temperatures from the inner part to the outer edge of the disk. At the end of this cooling phase, expected temperatures were in the range of a few hundreds K within the first 3 AU (i.e., where the terrestrial planets are found today), but were significantly lower (about 100K or less) beyond 5 AU, in the region where giant planets are now found. Within 3 AU, the species which were able to condense were only the heavy elements,



*W.B. Latter et al. (eds.), CO: Twenty-Five Years of Millimeter-Wave Spectroscopy,* 349–357.

in particular metals and silicates, with, under some conditions, water and carbon dioxide. In contrast, beyond 5 AU, most of the molecules had to be condensed in ices: first $H_2O$ and $CO_2$, then $NH_3$, $CH_4$, $H_2S$, etc.

All planetary bodies are believed to have formed from the accretion of solid particles, the "planetesimals". These planetesimals were made from the species which were present in condensed form. Below 3 AU, the cores which would lead to the formation of the terrestrial planets could only use metals and silicates. In contrast, beyond 5 AU, they could make use of all the ices which, according to the cosmic abundances, constitute most of the mass of the nebula, if we exclude hydrogen and helium. Theoretical models (Mizuno 1980) show that, in this case, the nuclei formed in this process are expected to be about 10 times more massive than the ones formed below 3 AU (which have about 1 terrestrial mass). Once nuclei of 10 terrestrial masses are built, they become able to bind gravitationally the surrounding nebula, mostly composed of hydrogen and helium. This process is likely to have led to the formation of the giant planets. In contrast, the terrestrial planets were not able to retain the surrounding nebula; their mass is basically the mass of their solid body, and their atmospheres were probably later generated, through outgassing and meteoritic impacts. This simple scenario shows that the condensation sequence gives a natural explanation for the two classes of planets which we observe, the terrestrial ones and the giants; we may reasonably expect to find the same basic classification in any other stellar system.

## 2.2. THE $CO/CH_4$ RATIO

In order to interpret the relative abundance of CO in the various Solar-system objects, we need to understand in which form the carbon was present in the primordial solar nebula. This question refers to an earlier step of the Solar system formation, before the cooling of the nebula and the condensation sequence.

Under the conditions of thermochemical equilibrium, the forms in which carbon and nitrogen appear are indicated respectively by the two following reactions:

$$CH_4 + H_2O \quad \leftrightarrow \quad CO + 3H_2 \tag{1}$$

$$2NH_3 \quad \leftrightarrow \quad N_2 + 3H_2 \tag{2}$$

which evolve toward the right-hand side at high temperature, and toward the left-hand side at low temperature. CO, once formed, reacts again with $H_2O$ to form $CO_2$:

$$CO + H_2O \leftrightarrow CO_2 + H_2 \tag{3}$$

In addition, the dynamical factor has to be included in the process (Prinn & Fegley 1988). The solar nebula is not static, but cools rapidly, at a rate which is fast enough for the CO formed at high temperature, at the beginning of the cooling phase, to be quenched. As a result, significant abundances of CO and $N_2$ can be found even at low temperature, at the end of the cooling process. Thermochemical calculations show that an object formed within the solar nebula is expected to have its carbon basically in the form of CO. In contrast, for an object formed within a "Jovian" subnebula (where the temperature is lower), $CH_4$ is dominant. In a similar way, $N_2$ is found to be dominant for objects formed within the solar nebula, while $NH_3$ should be dominant for objects formed in a "Jovian" sub-nebula.

At first order, reactions (1) to (3) seem to account, at least qualitatively, for the observed compositions of Solar-system objects (Table 1). Indeed, at high temperatures CO, $CO_2$, and $N_2$ are formed, as observed in the atmospheres of Venus and Mars; $H_2$ escapes, as it is too light to be gravitationally retained by the terrestrial planets. $CH_4$ and $NH_3$ are formed at low temperatures, as observed in the atmospheres of Jupiter and Saturn; the case of Neptune is more puzzling, as will be discussed below. The CO abundance in comets is also compatible with this scenario, but the apparent depletion of cometary $N_2$ remains to be explained. The case of the outer satellites is also puzzling. In the case of Titan, both $N_2$ and $CH_4$ are present. It has been suggested that both $CH_4$ and $NH_3$ might have been formed first in the vicinity of the Saturnian sub-nebula; in a subsequent evolution, $NH_3$ would have been later transformed into $N_2$ (Atreya 1986). For Triton and Pluto, precise information is still lacking. In conclusion, the relative abundance of CO in the various Solar-system objects is understood at first order, but several questions are still unsolved, especially for the objects of the outer Solar system. In particular, an accurate measurement of the $CO/CH_4$ ratio in Triton, Pluto, and in comets will provide an important constraint upon their formation scenarios.

## 3. Heterodyne Spectroscopy Measurements of CO in the Solar System

In the case of a planetary atmosphere where the molecular lines are pressure-broadened, the typical width of a line (FWHM) is about 3 GHz per atmosphere. As the typical bandwidth of a heterodyne receiver is about 0.5 GHz, such lines are too broad to be observed by this technique in the tropospheres of the planets. They can only be observed at pressure levels in the range 0.01-10 mb.

### 3.1. VENUS AND MARS

The atmospheres of Venus and Mars have similar compositions, but completely different structures. The atmosphere of Venus is extremely dense and hot (surface pressure of 93 bar, surface temperature of 730 K), while the Martian atmosphere is tenuous and cold (surface pressure of 7 mb, average surface temperature of 230 K). As a result, only the Venus mesosphere can be probed ($z = 70 - 110$km), while the Martian atmosphere can be probed down to its surface. As the temperature decreases with increasing altitude, the lines are usually observed in absorption.

In the case of both planets, the CO (1–0) line was monitored in order to retrieve the vertical distribution of CO and the thermal profile in the probed region. In the case of Venus, evidence was found for diurnal variations in the CO mesospheric vertical distribution (Clancy and Muhleman 1985). In the case of Mars, the CO (1–0) and (2–1) transitions were monitored, leading to a temporal monitoring of the Martian thermal profile (Clancy et al. 1990). A remarkable result has been the study of the dynamics in the upper atmospheres of Mars and Venus from the use of CO millimeter lines. Recent observations of the CO (1–0) and (2–1) transitions over the disks of Venus and Mars respectively have led to the determination of wind speeds, through an accurate measurement of the line Doppler shifts (Lellouch et al. 1991; 1994); the precision is about 10–20 m/s. One should mention that there is no other method allowing study of the dynamics of the upper atmospheres of Venus and Mars.

### 3.2. THE GIANT PLANETS

In contrast with the atmospheres of Venus and Mars, the atmospheres of the giant planets are characterized by a temperature inversion, occuring at the tropopause level (pressure level in the range 100–300 mb). As mentioned above, only stratospheric features can be detected by heterodyne spectroscopy; they are expected to appear in emission.

#### 3.2.1. *CO and HCN in Neptune*

A recent exciting result has been the unexpected detection of CO and HCN in the stratosphere of Neptune (Marten et al. 1991; 1993; Rosenqvist et al. 1992). CO was found to be constant with altitude, with a mixing ratio of about $10^{-6}$; for HCN, which has to condense at the tropopause level, a mixing ratio of about $10^{-9}$ was derived in the stratosphere. Another surprise was the lack of detection of CO and HCN in the stratospheres of the three other planets. CO, however, had been previously detected, from infrared observations, in the deep tropospheres of Jupiter and Saturn, with a mixing ratio of about $10^{-9}$. The detection of CO and HCN in Neptune raised two major questions: (1) why are these species so abundant in Neptune? (2) why are Uranus and Neptune different?

A CO mixing ratio close to $10^{-9}$ was found in the case of Jupiter and Saturn, in good agreement with the photochemical models; an external CO origin from meteoritic oxygen was also suggested (Atreya 1986), but more recent data seem to favor the tropospheric origin of CO. In this case, CO, like other disequilibrium species ($PH_3$, $GeH_4$...), is brought up by a strong vertical mixing from the deep levels where it is stable (T = 1100 K) into the upper troposphere where it becomes observable. However, there is a discrepancy by a factor of about 1000, presently unexplained, between the expected and the observed value of the CO mixing ratio in Neptune.

The presence of HCN, probably formed from a reaction between $CH_3$ and nitrogen atoms, requires a source for these atoms. Two origins are possible, either due to an escape of N atoms from Triton, or due to the presence of $N_2$ in the Neptune troposphere. It has been suggested that nitrogen, within the Neptune troposphere, could be in the form of $N_2$ rather than $NH_3$; a $N_2$ mixing ratio of 0.1–1% could also explain the HCN measurement (Gautier et al. 1995).

Regarding the difference between Uranus and Neptune, there is no clear explanation either. However, it has been noticed that this difference could be connected to the absence of internal energy from Uranus, while Neptune, has a strong internal source. In the case of Uranus, convection could be inhibited, leading to the absence of upward transport from the interior (Gautier et al. 1995).

### 3.2.2. *CO in Jupiter after the SL9 Collision*

With a mixing ratio of about $10^{-9}$, CO is not abundant enough to be detectable in the Jupiter stratosphere by millimeter heterodyne spectroscopy. However, this observation became feasible just after the collision of comet Shoemaker-Levy 9 with Jupiter, in July 1994, due to the large amount of CO formed in the Jovian stratosphere after the explosion (Lellouch et al. 1995). The CO millimeter lines, as well as other newly observed transitions (CS, OCS, HCN), became maximum after the brightest impact. These started to decrease a few days later and turned into absorption lines, which remained detectable for more than six months after the impact, yielding information about the evolution of the thermal profile. These observations are the only ones which provided information on the long-term response of the Jovian stratosphere.

## 3.3. THE OUTER SATELLITES AND PLUTO

### 3.3.1. *Titan*

With its nitrogen-dominated composition, its surface pressure of 1.5 bar, and its rich and complex chemistry, Titan's atmosphere might be favorable for the development of prebiotic chemistry. Its strong temperature inversion and the presence of many hydrocarbons make Titan a very good candidate for heterodyne spectroscopy observations. Indeed, several molecules have been detected from this technique: CO, HCN, $HC_3N$ and $CH_3CN$.

CO was first detected in Titan's troposphere, from near-infrared spectroscopy, with a mixing ratio of 6 $10^{-5}$. At the same time, $CO_2$ was detected in Titan's stratosphere with a mixing ratio of about $10^{-8}$, from the Voyager-IRIS infrared spectra. Both detections were unexpected, as $CH_4$ was known to be the most abundant species after nitrogen and, possibly, argon.

The first stratospheric detection of CO was performed in the millimeter range at OVRO by Muhleman et al. (1984), who derived a CO mixing ratio of 6 $10^{-5}$. This implied a constant CO mixing ratio throughout the atmosphere, which was expected in view of the long lifetime of CO. However, subsequent IRAM measurements led to a much smaller mixing ratio (2 $10^{-6}$) for the stratospheric CO. Recent measurements by Gurwell & Muhleman (1995) seem to solve the controversy in favor of Muhleman et al.'s first result. Using the OVRO interferometer and double-side band measurements, the authors were able to obtain an absolute calibration of the two sidebands, something which is not possible with single-dish observations. Gurwell & Muhleman (1995) derived a CO mixing ratio of $5 \pm 10^{-5}$, which implies a uniform vertical distribution of CO in Titan's atmosphere.

What is the origin of CO and $CO_2$ in Titan? The current interpretation is that there is an external source of $H_2O$ coming from meteoritic impacts. $H_2O$ reacts with $CH_4$ to form CO, and CO in turn reacts with $H_2O$ to form $CO_2$. Photochemical models predict that a constant $H_2O$ input corresponding to a mixing ratio of $10^{-9}$ in Titan's stratosphere would be able to account for the observed abundance of CO and $CO_2$. $H_2O$ has not been observed in Titan yet, but this detection should be feasible with the spectrometers of the ISO mission.

### 3.3.2. *Io, Triton and Pluto*

Because of its proximity to Jupiter, Io is subject to strong tidal forces which lead to an intense volcanic activity. $SO_2$, first detected by Voyager-IRIS over a volcanic plume, was later detected in the millimeter range as the component of a stable atmosphere with a surface pressure in the range of 3 to 40 nanobars. Other gases have been searched for without success, including CO. The present upper limit, from the CO(1–0) line, is 5 nbar (Lellouch et al. 1992).

Triton and Pluto both have nitrogen-rich atmospheres, with surface pressures of 14 microbar and 3 microbar respectively. Ices of $N_2$, CO and $CH_4$ have been detected on Triton (Cruikshank et al. 1993) and Pluto (Owen et al. 1983), but the relative gaseous abundances of CO and $CH_4$ are very uncertain (Table 1).

### 3.4. COMETS

As first suggested by Whipple (1950) and confirmed with observations of Comet Halley, the composition of comets is dominated by $H_2O$. Although a minor component, CO plays a key role because, due to its high volatility, it is the first molecule to be outgassed from the nucleus at high heliocentric distance.

CO has been detected on many comets, with variable abundances. Observations were performed in the UV and infrared, and recently in the submillimeter range, with the detection of the (3–2) CO transition in a distant comet, P/Schwassman-Wachmann 1 (Senay & Dewitt 1994). These last measurements allowed to derive the velocity of the CO jet (1 km/s) outgassed from the nucleus.

## 4. Perspectives and Future Plans

CO observations have been of great value for Solar-system studies: they provide clues for constraining formation and evolution models; they are tracers of the dynamics of the terrestrial planets; they are indicators of activity in distant comets. Future progress can be expected in several directions.

First, the extended use of interferometry will provide better spatial resolution for the wind measurements on the disks of Venus and Mars. As discussed above, interferometry is also precious as it provides an absolute calibration of the emission lines of Neptune and Titan. Second, the extension of the spectral range toward higher frequencies will allow the use of stronger CO transitions, leading to improved determination of the CO vertical distributions, and providing, in the case of Io, an improved upper limit. Third, it should be mentioned that the use of millimeter/submillimeter spectrometers with a resolving power of $10^3$–$10^4$, like Fabry-Perot or Fourier-Transform spectrometers, will be very powerful for detecting tropospheric lines, in particular CO in Neptune.

Triton and Pluto appear too weak for CO to be detectable by heterodyne spectroscopy; the detection of gaseous CO, however, might be feasible in the near-infrared, using cryogenically cooled high-resolution spectrometers.

Finally, a millimeter instrument (MIRO) has been submitted to ESA as part of the instrumentation of the ROSETTA mission, devoted to the long-term exploration of a comet. The spacecraft is expected to follow a comet (P/Wirtanen) on its way to perihelion, during the first decade of the next century. The millimeter instrument includes two channels, around 230 GHz and 557 GHz, for the study of several molecules, including two CO transitions. If selected, it will be an ideal instrument for detecting the early activity of the comet, measuring relative abundances of the gaseous species and studying the thermodynamics of the coma.

## References


Atreya, S.K. (1986), "Atmospheres and ionospheres of the outer planets and their satellites", Springer-Verlag

Clancy, R.T. & Muhleman, D.O. (1985) Chemical-dynamical models of Venus mesosphere based upon diurnal microwave CO variations, Icarus 64, 183-204

Clancy, R.T., Muhleman, D.O. & Berge, G.L. (1990) Global changes in the 0-70km thermal structure of the Mars atmosphere derived from 1975-1989 microwave CO spectra, J.Geophys.Res. 95(B9), 14543-14554

Cruikshank, D.P., et al. (1993) Ices on the surface of Triton, Science 261, 742-745

Gautier, D., Conrath, B.J., Owen, T.C., de Pater, I. & Atreya, S.K. (1995) The troposphere of Neptune. In "Neptune and Triton", D.Cruikshank et al (eds), University of Arizona Press, in press

Gurwell, M.A. & Muhleman, D.O. (1995) CO on Titan: evidence for a well-mixed atmosphere, Icarus, in press

Lellouch, E., Goldstein, J.J., Bougher, S.W., Paubert, G., & Rosenqvist, J. (1991) First absolute wind measurements in the middle atmopshere of Mars, Astrophys.J. 383, 401-406

Lellouch, E., Belton, M., de Pater, I., GuLkis, S., & Encrenaz, T. (1992) The structure, stability, and global distribution of Io's atmosphere, Icarus 98, 271-295

Lellouch, E., Goldstein, J., Rosenqvist, J., Bougher, S., & Paubert, G. (1994) Global circulation, thermal structure and carbon monoxide distribution in Venus' mesosphere in 1991, Icarus 110, 315-339

Lellouch, E., et al (1995) Chemical and thermal response of Jupiter's atmosphere to the impact of comet Shoemaker-Levy 9, Nature 373, 592-595

Marten, A., Gautier, D., Owen, T., Sanders, D.B., Tilanus, R.P.J., Deane J.R., & Matthews, H.E. (1991) IAU Circular 5531

Marten, A., et al. (1993) First observations of CO and HCN on Neptune and Uranus at millimeter wavelengths and their implications for atmospheric chemistry, Astrophys.J. 406, 285-297

Mizuno, H. (1980) Formation of the giant planets. Prog. Theor. Phys. 64, 544-557

Muhleman, D.O., Berge, G.L., & Clancy, R.T. (1984) Microwave measurements of carbon monoxide on Titan, Science 223, 393-396

Owen, T.C., et al. (1983) Science 261, 745-748

Prinn, R.G. & Fegley, B. (1988) Solar nebula chemistry: origin of planetary, satellite and cometary volatiles. In "Origin and evolution of planetary and satellite atmospheres", S. Atreya et al. (eds), pp.78-136

Rosenqvist, J., Lellouch, E., Romani, P., Paubert, G. & Encrenaz, T. (1992) Millimeter-wave observations of Saturn, Uranus and Neptune: CO and HCN on Neptune, Astrophys.J. 392, L99-L102

Senay, M.C. & Jewitt, D. (1994) A carbon-monoxide driven coma in periodic comet Schwassmann-Wachmann 1, Nature 371, 229-231

Whipple, F.L. (1950) A comet model. I. The acceleration of comet Encke, Astrophys.J. 111, 375-394. II. Physical relations for comets and meteors, Astrophys.J. 113, 464-474

TABLE 1. Carbon and Nitrogen Atmospheric Composition of Solar-system Objects

| Object | Major Constituent | Minor Constituents (Mixing Ratio) Carbon | Nitrogen | Other |
|---|---|---|---|---|
| Venus | $CO_2$ | CO | $N_2$ | $H_2O,O_2,O_3$ |
| | | $(10^{-3}$–$10^{-4})$ | (0.03) | +... |
| Mars | $CO_2$ | CO | $N_2$ | $H_2O,O_2,O_3$ |
| | | $(8\ 10^{-4})$ | (0.03) | +... |
| Jupiter | $H_2$ | $CH_4$ | $NH_3$ | He |
| | | $(10^{-3})$ | $(2\ 10^{-4})$ | $H_2O,PH_3,$ |
| | | CO | | $GeH_4,AsH_3$.. |
| | | $(1.5\ 10^{-9})$ | | $C_2H_2,C_2H_6$ |
| Saturn | $H_2$ | $CH_4$ | $NH_3$ | He |
| | | $(2\ 10^{-3})$ | $(2\ 10^{-4})$ | $PH_3,GeH_4,$ |
| | | CO | | $AsH_3$ |
| | | $(2\ 10^{-9})$ | | $C_2H_2,C_2H_6$.. |
| Uranus | $H_2$ | $CH_4$ | ? | He |
| | | $(1$–$2\ 10^{-2})$ | | |
| Neptune | $H_2$ | $CH_4$ | $N_2$? | He |
| | | $(1$–$2\ 10^{-2})$ | $(10^{-2}$–$10^{-3})$ | $C_2H_2,C_2H_6$ |
| | | CO | HCN | |
| | | $(10^{-6})$ | $(10^{-9})$ | |
| Titan | $N_2$ | $CH_4$ | HCN | Ar? |
| | | (0.02) | $(2\ 10^{-7})$ | $H_2O$? |
| | | CO | $CH_3CN$ | |
| | | $(5\ 10^{-5})$ | $(10^{-9})$ | |
| | | $CO_2$ | $HC_3N$ | |
| | | $(10^{-8})$ | $C_2N_2$ | |
| Triton | $N_2$ | $CH_4$ | | |
| (* ices) | | $(10^{-4})$ | | |
| | | CO* | | |
| | | $(2\ 10^{-4})$ | | |
| | | $CO_2$* | | |
| Pluto | $N_2$ | $CH_4$* | | |
| (* ices) | | $(10^{-5}$–$10^{-3})$ | | |
| | | CO* | | |
| | | $(2^{-20}\ 10^{-4})$ | | |
| Comets | $H_2O$ | CO | HCN | $H_2S$ |
| | | (0.05) | (0.001) | |
| | | $CO_2$(0.03) | $N_2$? | |
| | | $H_2CO$(0.04) | $(2\ 10^{-4})$ | |
| | | $CH_3OH$ | $NH_3$? | |
| | | (0.01) | $(10^{-3})$ | |
| | | $CH_4$?(0.003) | | |
| | | OCS?(0.002) | | |

# MASS LOSS IN AGB STARS

M. GUÉLIN, R. LUCAS, AND R. NERI
*Institut de Radioastronomie Millimétrique*
*300 rue de la Piscine, F-38406 Saint Martin d'Hères, France*

**Abstract.** The phase of high mass loss, one of the ultimate phases of stellar evolution, is still poorly known. For a long time, large visual extinctions and small angular sizes have impeded the study of the most interesting objects in this stage. The recent development of mm and IR interferometry has allowed a breakthrough in this domain. Expelled molecules and dust are now routinely mapped at arcsecond resolutions, providing a large data base to which model predictions are confronted.

## 1. Mass loss rates

The main component of the expelled gas, molecular hydrogen, is hard to detect. Envelope masses and mass loss rates are therefore derived primarily from the observation of the two most stable trace components, CO and dust: CO, at 4.6 and 2.3 $\mu$m (mostly) in absorption and at mm wavelengths in emission; dust in emission throughout the IR-submm window.

In order to derive the $H_2$ mass from CO, one has to model the CO chemistry. This has been done in particular by Mamon et al. (1988, hereafter MGH), who first introduced accurate CO photodissociation cross sections in their chemical network. In the case of cold red giants ($T_* < 3000$K), the far UV photons come from interstellar space; their energy distribution can be described in terms of an attenuated 'standard interstellar radiation field' (or 'ISRF'). The IS radiation is absorbed by the dust grains as well as by the $H_2$ and CO molecules present in the outer envelope layers. In MGH's model, efficient CO self-shielding and shielding by $H_2$ lines make the CO fractional abundance relative to $H_2$, $x(\mathrm{CO})$, follow almost a step function. It drops from $x(\mathrm{CO}) =$ few $10^{-4}$, to almost zero as soon as the gas overpasses a critical 'photodissociation' radius $R_{\mathrm{ph}}$.

Obviously, $R_{\mathrm{ph}}$ depends on the intensity of the local ISRF and on the mass loss rate, $\dot{M}$. In a more subtle way, it also depends on the dust grain properties. Olofsson et al. (1993), following MGH have derived a simple relation between



*W.B. Latter et al. (eds.), CO: Twenty-Five Years of Millimeter-Wave Spectroscopy,* 359–366.

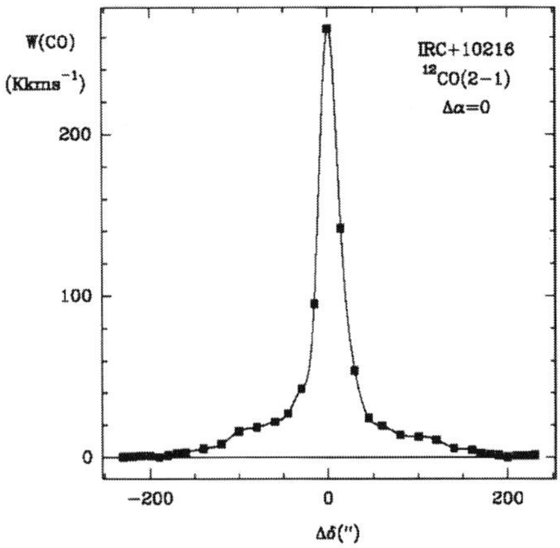


*Figure 1:* $^{12}$CO $J$=2–1 line emission, integrated over the central 13 km s$^{-1}$ velocity interval, along a strip passing through IRC+10216.

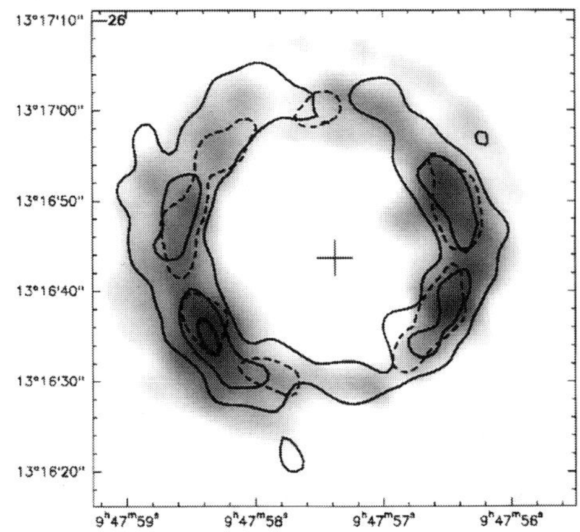


*Figure 2:* The $C_3H$ (dashed) and $C_4H$ (solid) contours of Fig. 4, superimposed on the $C_2H$ brightness distribution (grey scale).

$R_{\rm ph}$ and $\dot{M}$ in the case of a constant isotropic mass loss rate and a 'standard' ISRF:

$$R_{\rm ph} = 4\,10^{16}\left(\frac{\dot{M}}{10^{-6}}\right)^{0.6}\left(\frac{v_{\rm exp}}{15}\right)^{-0.5}\left(\frac{x({\rm CO})}{3\,10^{-4}}\right)^{0.5}\ {\rm cm},$$

where $\dot{M}$ is in $M_\odot$ yr$^{-1}$ and $v_{\rm exp}$, the expansion velocity, in km s$^{-1}$.

Figure 1 shows the $^{12}$CO, $J$= 2–1 line profile along an EW cut passing through CW Leo/IRC+10216. CW Leo, the archetype of carbon stars, located only $D \sim 200$ pc from the Sun, is believed to have reached the very end of the AGB phase and to be just on the verge of evolving toward the PN phase (Guélin et al. 1993 –hereafter GLC– Kastner et al. 1994). It is surrounded by a thick envelope of dust and gas with a very rich chemistry. The CO profile shows a central 'core' and a broad 'plateau' whose intensity drops briskly at $R \simeq 120''$. Setting $R_{ph} = 120'' \simeq 3.5\,10^{17}$ cm, equation (1) yields $4\,10^{-5}$ $M_\odot$ yr$^{-1}$.

The CO profile of Fig. 1 obviously contains more information than the value of $R_{\rm ph}$. Truong-Bach & Rieu (1991) fitted all the CO mm data available for the central $R < 1'$ region of IRC+10216 with a radiative transfer code. They find a mass loss rate of $\dot{M} = 2.5\,10^{-5}$ $M_\odot$ yr$^{-1}$ for the 'core' and $4\,10^{-5}$ $M_\odot$ yr$^{-1}$ for the 'halo': the mass loss rate seems thus to have decreased by about a factor of 2 in the last $10^3$ yr. A modelling of the innermost part of the IRC+10216 envelope has been made by Sahai & Wannier (1985) and Keady et al. (1988) on the basis of the IR data. Keady et al. derive the radial dust and gas temperature profiles as well as a velocity profile (the expansion velocity increases by steps from $V_e = 2$ to 14 km s$^{-1}$). They find for $x$(CO)= 3 $10^{-4}$ a mass loss rate of 4 $10^{-5}$ $M_\odot$ yr$^{-1}$.

The observation of the dust thermal emission offers another way of studying mass loss rates. There, one has to assume a dust absorption coefficient and a dust-to-gas mass ratio (see, e.g., Walmsley et al. 1991). In the case of IRC+10216, Ridgway & Keady (1988) derive from interferometric 1.6–5.0 $\mu$m observations a *dust* mass loss of $2\,10^{-7}$ $M_\odot$ $yr^{-1}$, surprisingly small, in view of the above.

A comprehensive study of IRC+10216's continuum emission was made recently by Winters et al. (1994), who modelled in a self-consistent way: *i)* the formation of the dust grains by nucleation, *ii)* their acceleration by radiation pressure, and *iii)* their thermal equilibrium in the radiation field, taking into account grain-gas collisions. The stellar parameters were adjusted to fit the observed continuum spectrum from the near- to the far-IR, the expansion velocity, and the size (measured by IR interferometry) of the continuum source. Winters et al. arrive at a mass loss rate of $\dot{M} = 8\,10^{-5}$ $M_\odot$ $yr^{-1}$, an envelope mass $>$ 0.5 $M_\odot$ and a carbon-to-oxygen ratio of 1.3–1.5. The central star has presently a mass of 0.7 $M_\odot$, a surface temperature of 2000 K and a peak bolometric luminosity of $2.3\,10^4$ $L_\odot$, which yields a distance $D$ = 170 pc. Of course, these values are only a set of possible solutions. They illustrate, nevertheless, how much can be derived from a well studied source.

There are not many AGB stars where such a wealth of observations is available. In most cases, one has to rely on coarse CO and/or IR continuum observations to estimate the mass loss rates. To date several hundreds of circumstellar envelopes have been detected in CO (see, e.g., the compilation by Loup et al. 1993), but for the vast majority of them only a single spectrum (e.g., $^{12}$CO $J$= 2–1) is available. A crude idea of their mass loss rates can nevertheless be obtained from the relation between the CO integrated line intensity and the mass loss rate derived by Knapp & Morris (1985) and Kastner (1992) for optically thick envelopes. The distances of most objects being unknown, they are crudely evaluated from the IR and FIR fluxes by assuming AGB stars have bolometric luminosities $\simeq 10^4$ $L_\odot$ (see, e.g., Loup et al. 1993 for a more detailed discussion).

With the advent of powerful mm interferometers, it now becomes possible to map the CO emission of many circumstellar envelopes and directly measure their radial profile, size, and degree of asymmetry. This latter parameter is very important (and so far poorly known), since the first step in the evolution from AGB to PPN and PN is a disruption of the spherically symmetrical shell. (Such a disruption has just started in IRC+10216 – see Fig. 2 and GLC.)

The survey of a sample of 50 AGB and post-AGB stars has started at IRAM, and a first atlas, consisting of $^{12}$CO $J$= 2–1 line emission maps, observed with the 30-m telescope, and of $J$=1–0 maps, resulting from a combination of 30-m telescope and the Bure interferometer data, is already available (Neri et al. *submitted*). The basic kinematical parameters, as well as the visibility curves,

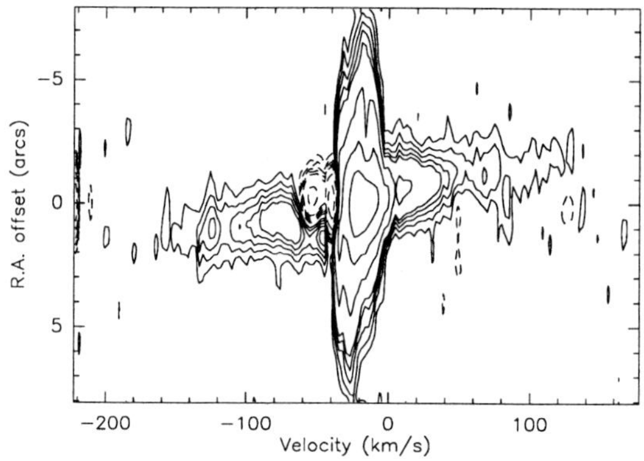


*Figure 3:* Position-velocity diagram along the EW high velocity outflow of the PPN CRL 618, observed in $^{12}$CO $J$=1–0 line with a 2″ angular resolution (IRAM interferometer, Neri et al. *in preparation*).

envelope sizes, and degrees of asymmetry are given for all objects. The CO visibilities are typical of Gaussian CO line brightness radial distributions. They show no evidence of a sharp photodissociation cutoff at the (limited) sensitivity of the present data. Only a few envelopes are found to be strongly asymmetrical. Most stars, however, were only observed with the interferometer in the 'snapshot' mode with poor *uv* coverage, so that weak asymmetries could have been missed.

## 2. Pre-Planetary Nebulae

Among the objects showing strong asymmetries, are the well known post-AGB stars M1–92 and CRL 618, both of which exhibit high velocity (HV) molecular outflows. These outflows and the central cores of the envelopes were mapped in the $^{12}$CO (1–0) line with a good *uv* coverage and high sensitivity using the Bure interferometer. In both objects there is a clear interaction between the outflow and the core material. The results on M1–92 were already reported by Bujarrabal et al. (1994). The core-outflow interaction in CRL 618 is illustrated in Fig. 3, where we show the position-velocity diagram observed along the outflow E–W axis of symmetry. The slowly expanding core appears as a bright, slightly bipolar source, extending between $\pm 7''$ in R.A. and $-40$ and 0 km s$^{-1}$ in velocity. The HV outflow is also bipolar; it is visible from $-180$ to $+140$ km s$^{-1}$. The gas at large positive velocities arises from distinct impact regions, unresolved by the 2″ (FWHP) synthesized beam and located 1–2″ W of the central source. That at negative velocities seems less collimated: it extends from 2″ E until right in front of the continuum source, as attested by the absorption visible between $-40$ and $-80$ km s$^{-1}$. Both the core and the HV CO data can be explained by the schematic model of CRL 618 proposed by Neri et al. (1992) on the basis of HCN (1–0) interferometric observations. Other species mapped in CRL 618 are $^{13}$CO (Yamamura et al. 1994), CS (Hajian et al. 1995b) and $NH_3$ (Martin-Pintado et al. 1995).

## 3. Chemistry

While matter is expelled from the stellar atmosphere and expands into outer space, it assumes a wide range of physical conditions. The density decreases by several orders of magnitude and the temperature drops from thousands of K to a mere 10–20 K. In the process, about every molecule formation/destruction mechanism occurs: 3-body reactions in the hot atmosphere, grain condensation in the upper atmospheric layers, ion–molecule and radical–molecule reactions in the circumstellar envelope, and photodissociation in the envelope outer 'skin'. All of this makes an exceptionally rich chemistry, as attested by the more than 50 molecular species identified in IRC+10216.

The simple geometry and velocity field of AGB star envelopes is another asset. It allows to determine the physical conditions and dynamical timescales and makes it possible to follow in time the chemical processes by measuring the molecular abundances at different radii. Such studies, however, require linear resolutions reachable only in the closest objects. IRC+10216, $\simeq$ 200 pc away, is obviously a prime target.

The chemistry in the envelope is far from thermodynamic equilibrium, as illustrated by the large abundance of radicals (e.g., $x(\mathrm{CP}) > x(\mathrm{HCP})$), and is highly time dependent. The short dynamical time scales (in IRC+10216, it takes $10^4$ yr for the matter ejected by the central star to cross the entire CO envelope) imply that chemical processes must be fast. The formation of dust grains, which removes from the gas most elements heavier than carbon; photodissociation, which breaks out molecules into small radicals and atoms; photoionization and ionization by cosmic rays, which produce $C^+$, $C_2H_2^+$,...; and, finally, fast neutral and ion–molecule reactions shape the molecular composition. It is also likely that grain chemistry plays a role in the final composition, all the more that the dust grains, while drifting outwards, are exposed to an increasingly large UV intensity.

The most recent models of gas phase chemistry in CSEs (with particular emphasis on IRC+10216) are those of Glassgold and co-workers (see Cherncheff et al. 1993 and references therein) and Millar & Herbst (1994). In addition to the photodissociation and ion–molecule reactions already present in previous models, they include neutral reactions, such as radical–radical reactions. Although neutral reactions have smaller rate coefficients than the fast ion–molecule reactions, they turn out to be much faster in the dense envelope cores where the density of ions is particularly low. Gas phase models predict centrally peaked radial distributions for SiS, CO, $C_2H_2$ and HCN, which come from the stellar atmosphere, and hollow shell distributions for $C_2H$, CN, and their products $C_3H$, $C_4H$, $C_3N$, $HC_3N$, etc..., which are formed in the envelope.

The first interferometric maps of molecular line emission in IRC+10216 were made by Bieging & Rieu (1988, 1989) with the Hat Creek mm array

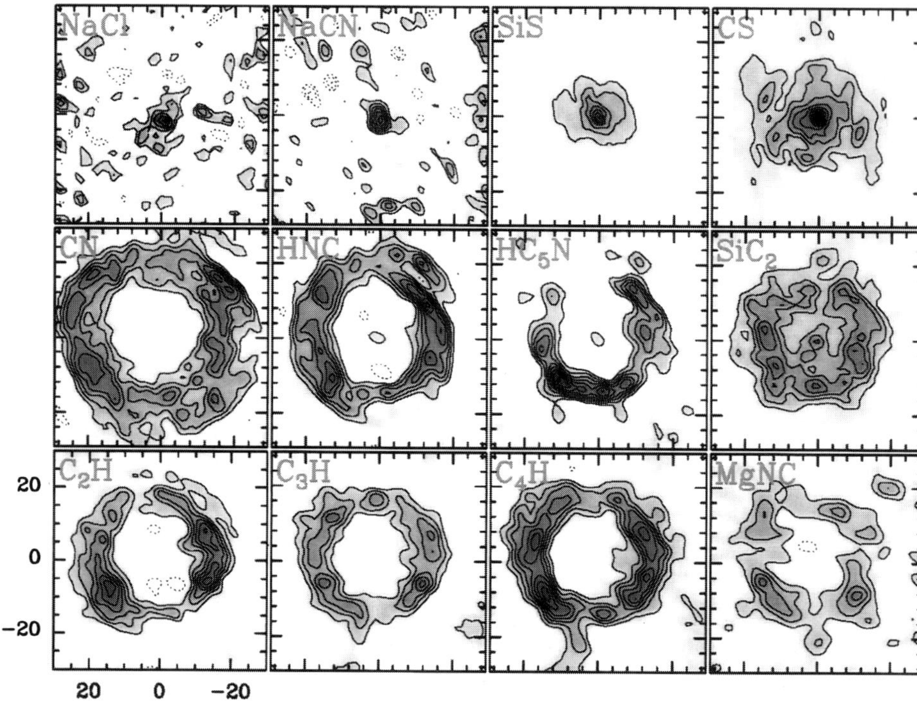


*Figure 4.* The IRC+10216 envelope, observed with an angular resolution of $\simeq 3''$ (FWHP) in the 3 mm lines of different molecules. The line intensities are integrated over the central 5 km $s^{-1}$ interval (IRAM interferometer).

(HPBW 8″). They nicely confirmed that the SiS emission was arising from a small central source and the $HC_3N$, HNC and $C_2H$ emissions from hollow shells. They lacked, however, the resolution and sensitivity needed to resolve either the central SiS source or the structure of the shells. Further observations with the Hat Creek and Nobeyama arrays (Bieging & Tafalla 1993; Takano et al. 1992; Dayal & Bieging 1995) extended these observations to further species ($SiC_2$, CN, and $C_4H$).

The Plateau de Bure interferometer, which is more sensitive than the other mm arrays to small scale structures, is particularly well suited to study IRC+10216's dense core. A programme of mapping the chemically most significant molecules is under way with this instrument (Lucas et al. 1992; GLC; Lucas et al. 1995).

Figure 4 shows some examples of maps observed at Bure with a $\sim 3''$ synthesized beam. The emission is integrated over a 5 km $s^{-1}$–wide interval, centered on the source systemic velocity ($v_{\rm sys} = -26.4$ km $s^{-1}$). This interval was chosen narrow enough that the contours represent roughly the line brightness distribution in the meridian plane perpendicular to the line of sight (more exactly the surface brightness from a thin conical slab, since the envelope expansion velocity is roughly constant). The CN map results from a mosaic of 7 interferometer fields spaced by 20″ steps; the other are derived from single field observations and are not corrected for primary beam attenuation (HPBW$\simeq$ 50″). Some maps (CN, $C_3H$, $C_4H$, CS, $SiC_2$, and MgNC) include short spacing data, obtained by mapping the Bure field of view with the 30-m telescope; the others use only the central 30-m spectrum (HPBW 23″–27″). The continuum emission (essentially, a point-like source of flux 60 mJy) has been removed.

The maps in Fig. 4 confirm that the radicals and the cyanopolyynes have hollow-shell distributions (i.e. ring-like distributions in the velocity interval plotted in the figure) and that the small stable molecules are centrally peaked, as predicted by the gas-phase models. They show much more, however. First, the rings and the central sources show a rich azimuthal structure, in the form of thin arcs and clumps, with a crude axis of symmetry (see, e.g., the $C_2H$ and SiS maps). Second, the rings show a radial structure suggesting the presence of double or even triple shells (e.g., the CN and $C_2H$ maps). Third, some maps show at the same time a central source and a ring (CS and $SiC_2$). Fourth, the rings are off-centered by $\sim$ 2″ to the E of the central star.

Most of the bright clumps are visible on several maps. Since the lines have different excitation conditions and opacities, the clumps cannot just be artifacts of the excitation and must trace an increase in column density. Similarly the decrease in intensity toward the N-NE and the S-SW (see $C_2H$) probably indicate a hole in the envelope.

According to gas phase models, the species in the envelope peak at different radii, since they are formed from different parent molecules through reactions with different time scales. For example, in Cherchneff et al.'s model (1993) $C_4H$ is mainly formed by the neutral reaction $C_2H + C_2H$, whose characteristic time is $t = (k_n x(C_2H) n_{H_2})^{-1} \sim 2\,10^3$ yr, whereas $C_3H$ is formed from $C_2H$, but by radiative association with C, a reaction at least 5 times slower. Neutral carbon, moreover, is abundant only in the outer envelope, further out than $C_2H$, so that the $C_3H$ abundance is predicted to peak at a radius at least two times larger than $C_4H$. $C_4H$, itself, is predicted to peak at a radius $\sim$ 1.5 times larger than $C_2H$ (A. Glassgold, private communication).

This is not observed: the bright blobs in the $C_2H$, $C_3H$ and $C_4H$ maps (as well as in the maps of several species, such as MgNC–see GLC) coincide within 2″ (Fig. 2). This implies a time delay for the formation of $C_3H$ and $C_4H$ from $C_2H \leq 150$ yr, or a few percent of the characteristic reaction times! The peaks do not trace the envelope border, since CN and, mostly, CO extend further out. They do correspond to density clumps, but those are unlikely to be dense enough to reconcile the dynamical and chemical times, since they are not detected in $^{13}CO$. Other examples of discrepancies between observations and predictions are the similarities between the CN and HNC distributions (Fig. 4) and the relatively small radius of the CN shell.

Clearly, the models, which assume spherical symmetry and a smooth density distribution, are much too crude in view of the complexity of the new data and attempts to model line excitation and gas phase chemistry in clumpy, asymmetrical media are required. Nonetheless, the above discrepancies probably ask for a reconsideration of the chemical networks.

It is tempting to see in the remarkable similarities between different species a common formation mechanism, capable of building large molecular abundances in timescales of few $10^2$ yr. The gas density being a few times $10^4$ cm$^{-3}$ at $R = 15''$ and the abundance (relative to $H_2$) of the parent neutral species ($C_2H_2$, $C_2H$, HCN...) $x < 10^{-5}$, the reaction rates should be > a few times $10^{-9}$ cm$^3$ s$^{-1}$. This is faster than the Langevin rate and much faster than expected for neutral gas phase reactions. Photon-related processes, such as photodissociation and photodesorption from dust grains, on the one hand, and shock desorption of molecules weakly attached to grains, on the other hand, can be fast enough. What role they actually play in IRC+10216's complex chemistry, however, remains to be seen.

## References

Bujarrabal, V., Alcolea, J. Neri, R., Grewing, M. 1994, Ap.J. 436, L169
Bieging, J.H., N-Q-Rieu 1988, Ap. J. 329, L107
Bieging, J.H., N-Q-Rieu 1989, Ap. J. 343, L25
Bieging, J.H., Tafalla, M. 1993, A. J. 105, 576
Cherchneff, I., Glassgold, A.E., Mamon, G.A. 1993, Ap. J. 410, 188
Dayal, A., Bieging, J.H. 1995, Ap. J. 439, 996
Guélin, M., Lucas, R., Cernicharo, J. 1993, in *The Cold Universe*, ed. Montmerle et al., Editions Frontières, Gif
Kastner, J.H., Weintraub, D.A. 1994, Ap. J. 434, 719
Kastner, J.H. 1992, Ap. J. 301, 337
Keady, J.J., Hall, D.N.B., Ridgway, S.T. 1988, Ap. J. 326, 832
Knapp, G.R., Morris, M. 1985, Ap. J. 292, 640
Loup, C., Forveille, T., Omont, A. and Paul, J.F. 1993, A&A Supp. Ser. 99, 377
Mamon, G., Glassgold, A.E., Huggins, P.J. 1988, Ap. J. 328, 797 (MGH)
Lucas, R., 1992 I.A.U. Symp. 150, p. 389
Lucas, R. et al. 1995, Ap. and Space Sience, 224, 293
Hajain, A., Phillips, J.A., Terzian, Y. 1995, Ap. J. 446, 244
Millar, T. J., Herbst, E. 1994, A&A 288, 561
Martin-Pintado, J. et al. 1995, Ap. J. 446, 687
Neri, R. et al. 1992, A&A 262, 544
Olofsson , H., Eriksson, K., Gustafsson, B., Carlstrom, U. 1993, Ap. J. *Supl* 87, 267
Ridgway, S.T., Keady, J.J. 1988, Ap.J. 326, 843
Sahai, R., Wannier, P.G. 1985, Ap. J. 299, 424
Shibata, K. et al. 1993, Ap. J. 415, 708
Takano, S., Saito, S., Tsuji, T. 1992, PASJ, 44, 469
Truong-Bach, Morris, D., Nguyen-Q-Rieu 1991, A&A 249, 435
Winters, J.M., Dominik, C., Sedlmayr, E. 1994, A& 288, 255
Walmsley, M. et al. 1991, 248, 555
Yamamura, I., Shibata, K., Kasuga, T., Deguchi, S. 1995, Ap. J. 427, 406.

# UNRAVELING MYSTERIES IN LATE STAGES OF STELLAR EVOLUTION: THE ENIGMA OF THE S-TYPE STARS

WILLIAM B. LATTER
*NASA/Ames Research Center*
*MS 245-3*
*Moffett Field, CA 94035*

AND

JOHN H. BIEGING
*Steward Observatory*
*University of Arizona*
*Tucson, AZ 85721*

**Abstract.** We report the results of an ongoing survey of S stars within 1 kpc of the Sun for emission from the $J = 1 \rightarrow 0$ and $J = 2 \rightarrow 1$ lines of CO. We derive mass loss rates from the CO emission and compare with other estimates. We find that S stars are losing copious amounts of mass at rates comparable to that of the carbon stars. There are indications that the gas-to-dust ratio is *higher* than for carbon stars. To aid in determining the evolutionary status of the S-type stars, we are also searching the brightest CO stars for $J = 1 \rightarrow 0$ HCN and $J = 2 \rightarrow 1$, $v = 0$ SiO emission. We use our results to estimate the chemical composition of the circumstellar envelopes. Comparison is made with thermodynamic equilibrium models for the formation of these and other species under photospheric conditions.

## 1. Introduction

Nearly seventy-three years ago a new spectral type called S was added to the Harvard system of spectral classification (Merrill 1922). At the time little was known about these unusual stars. Recently, the situation has been improving. Today we have the tools required to address some of the fundamental properties of the S-type stars.

S stars are a relatively rare class of luminous red giants undergoing evolution along the asymptotic giant branch (AGB). (We consider here



*W.B. Latter et al. (eds.), CO: Twenty-Five Years of Millimeter-Wave Spectroscopy*, 367–369.

only the "intrinsic" S stars, not the "extrinsic" S stars whose abundances are believed to have been altered by mass transfer effects – cf. Jorissen and Mayor 1992.) Their distinguishing features are a C/O abundance ratio which is unity (to within 5%); and evidence of strong enrichment in s-process elements. Together these properties suggest that the S stars have undergone mixing of the products of nucleosynthesis from the interior to the photosphere. They might be transition objects between the O-rich M giants, with $C/O < 1$, and the carbon stars, with $C/O > 1$. If so, their rarity could indicate a short timescale during which the surface C/O ratio switches from O-rich to C-rich. The details of how this transition occurs, or even if it occurs has been the subject of much debate.

We have undertaken a program to survey as completely as possible the the nearest S-type stars for their mass-loss and chemical properties using single-dish and interferometric techniques. The first results of this work have been reported (Bieging & Latter 1994). A study of CO emission and mass-loss from S stars has been presented by Sahai & Liechti (1995). Dust properties have been studied by Chen & Kwok (1993) and Skinner et al. (1990).

## 2. The study

The goals of our initial survey (Bieging & Latter 1994), and our continued study is 1) to determine the time-averaged mass-loss rates for a complete sample of S stars within 1 kpc of the Sun, by observing emission from the CO molecule as a mass-loss tracer; and 2) to use observations of key molecular species, namely HCN and SiO, as indicators of the C/O ratio in the present envelope gas in order to constrain the possible evolutionary sequence of the surface C/O abundance for the detected stars. To achieve these goals, we are combining data from the NRAO 12-m, SEST, and BIMA.

Chemical models of AGB envelopes with $C/O \approx 1$ are also being developed. Currently, our models only assume photospheric-type conditions (local thermodynamic equilibrium), which is where formation of HCN and SiO is believed to occur. These calculations indicate that HCN and SiO abundances are extremely sensitive to the C/O ratio (see, for example, Fig. 9 of Bieging & Latter 1994).

Our observational results so far suggest that S star circumstellar envelopes have abundances consistent with a chemistry at photospheric densities, $C/O \approx 1$, and gas temperatures $T_{\rm gas} < 1500$ K. Also, the S stars detected appear to have been losing mass for timescales $\tau > 10^4$ years. There is evidence that S stars have a "dust deficit" in that the typical gas-to-dust ratio is $\sim 2$ times larger than the typical gas-to-dust ratio found for carbon stars (see also Sahai & Liechti 1995).

## 3. Implications

The results obtained have several important implications to understanding the evolution of AGB stars in general and suggest further study –

- S stars have been identified as such by observations of photospheric abundances. This means that no fully enshrouded S stars are known. Yet, measured mass loss rates are comparable to the carbon stars, implying heavily obscured objects could exist. If fully enshrouded S stars are present, then they can be identified by observations of both HCN and SiO.
- Observations and modeling show that the chemistry within S star envelopes is extremely sensitive to the C/O ratio. This can give insight into dust formation and other fundamental chemical problems.
- Of key importance is the evolutionary status of the S stars. Are they in transition to C-rich, or are they a stable phase in the evolution of some AGB stars? Molecular abundances found in the envelopes of S stars are consistent with those observed at the photosphere, implying that the S star phase is long-lived and might be the terminal phase of their AGB evolution.

Further studies are required to fully understand the status of S stars in AGB evolution, and are being conducted or planned. These include interferometric mapping of key molecular species in the nearest S star envelopes. Improving the size of the single-dish sample, especially the HCN and SiO detections, is necessary to complete the statistical analysis. Modeling of the molecular and dust contents in circumstellar envelopes with C/O $\approx$ 1 is required to fully understand the observations.

## Acknowledgements

We thank the NRAO 12-m staff for their assistance. WBL thanks the NRAO and the NRC for support. We are grateful to Hans Olofsson and Lewis Knee for their contributions.

## References

Bieging, J. H., & Latter, W. B. 1994, ApJ, 422, 765
Chen, P.S., & Kwok, S. 1993, ApJ, 416, 769
Jorissen, A., & Mayor, M. 1992, A&A, 260, 115
Merrill, P. 1922, ApJ, 56, 457
Sahai, R., & Liechti, S. 1995, A&A, 293, 198
Skinner, C. J., Griffin, I., & Whitmore, B. 1990, MNRAS, 243, 78

# NEW DETECTIONS OF METAL-BEARING MOLECULES IN IRC+10216: FROM CHEMISTRY TO NUCLEOSYNTHESIS

L. M. ZIURYS

*Chemistry Dept., Arizona State Univ., Tempe, AZ 85287-1604*

## 1. Introduction

The carriers of the metallic elements in interstellar gas are at present uncertain. If these elements exist as atoms, they are likely to be a major contributor to the electron density and the main source of neutralization for complex molecular ions. On the other hand, metals might be condensed onto grains. Certainly there are large depletions of these elements in diffuse gas, but do they apply to dense material as well? Alternatively, the metallic elements could be contained in simple molecules. It has been postulated that these elements should be inert in cold gas, but in actuality, few gas-phase reactions involving metals have been studied experimentally.

Detection of metal-containing molecules in interstellar gas was first accomplished in 1987, when the halide species AlCl, KCl, NaCl, and AlF were observed towards the circumstellar shell of the late-type carbon star IRC+10216 by Cernicharo and Guélin (1987). Very recently, MgNC was also discovered in this source (Guélin et al. 1986; Kawaguchi et al. 1993), as well as NaCN by Turner et al. (1994). These limited detections have already raised questions concerning metal chemistry. For example, all six species have been found only in IRC+10216, where they have a curious distribution. The halide species are found in the inner envelope, where the temperatures and densities are quite high and equilibrium chemistry occurs. In contrast, MgNC, a free radical, is found only in the *outer* envelope (Guélin et al. 1993). In this region, temperatures and densities are much lower and are similar to those found in molecular clouds. Under these conditions, the metallic elements are predicted to be inert.



*W.B. Latter et al. (eds.), CO: Twenty-Five Years of Millimeter-Wave Spectroscopy*, 370–372.

## 2. New Studies of Magnesium-Bearing Molecules

The curious results for MgNC have prompted new laboratory and observational studies of MgCN, MgCCH, and $MgCH_3$. MgCN is the metastable isomer of MgNC and lies $\sim$ 700 K higher in energy. The MgNC/MgCN system is analogous to the HCN/HNC isomer pair. MgCCH and $MgCH_3$ are relevant because of the large abundances of HCCH, CCH, and other carbon chains, as well as methane, found in IRC+10216. Measurement of the rotational spectrum of these radicals has been recently carried out by Anderson, Steimle, and Ziurys (1994) for MgCN and Anderson and Ziurys for MgCCH (1994) and $MgCH_3$ (1995).

Ziurys, Guélin, and collaborators subsequently have been conducting astronomical searches for these radicals towards IRC+10216. Initially, for MgCN, the $N = 11 \rightarrow 10$ and $N = 10 \rightarrow 9$ transitions were detected towards this object using the NRAO 12m telescope, although the $N = 10 \rightarrow 9$ lines were partially contaminated by $C_6H$. These transitions were subsequently confirmed using the IRAM 30m antenna (Ziurys et al. 1995). The $N = 9 \rightarrow 8$ lines were observed as well, although they are blended with cyclic $C_3H$ (see Fig. 1).

The IRAM spectra have double-horned shaped line profiles, which indicates that MgCN, like MgNC, must arise from the outer envelope of IRC+10216. In this region, the column density of MgCN is $N_{tot} \sim 10^{12}$ cm$^{-2}$, which implies a MgNC/MgCN ratio of about 22/1. This ratio is very low, considering the metastable nature of MgCN, and indicates that non-equilibrium, low temperature chemistry is producing these species.

Searches for the MgCCH and $MgCH_3$ radicals, using the 12m and 30m telescopes, have thus far proved unsuccessful. Upper limits to the column densities of these two species are $N_{tot} < 10^{12}$ cm$^{-2}$.

Magnesium has three stable isotopes with a relative abundance ratio of $^{24}$Mg: $^{25}$Mg: $^{26}$Mg = 7.9:1:1.1. Through a joint laboratory/observational effort, several transitions of both $^{25}$MgNC and $^{26}$MgNC have been identified in the outer envelope of IRC+10216 (Guélin et al. 1995). The astronomical measurements have enabled the first accurate determination of the Mg isotope ratio in an AGB star, which was found to be very close to solar. The three isotopes are produced in the Mg/Al cycle, which also produces $^{26}$Al. The near solar ratio puts constraints on the nucleosynthesis occurring in IRC+10216, as well as suggests that AGB stars may not be a good source of $^{26}$Al.

## 3. Conclusions

The recent detections of metal-bearing molecules certainly have altered the past view of their chemistry. Magnesium-containing molecules appear to be

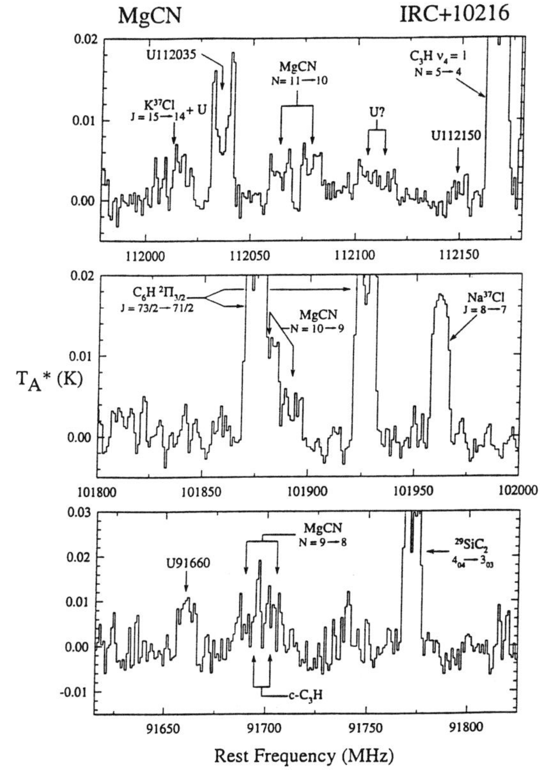


*Figure 1.* Spectra of MgCN observed towards IRC+10216 using the IRAM 30m telescope.

forming via cold, non-equilibrium chemistry. Why they are not observed in molecular clouds is thus surprising. Also, the pattern of detected species is not easily explained. Finally, observation of $^{26}$MgNC and $^{25}$MgNC in IRC+10216 suggests that determination of other metal isotope ratios in AGB stars may be possible, leading to a better understanding of nucleosynthesis.

## References

Anderson, M.A., Steimle, T.C., and Ziurys, L.M. 1994, ApJ, **429**, L41
Anderson, M.A., and Ziurys, L.M. 1994, ApJ., **439**, L25
Anderson, M.A., and Ziurys, L.M. 1995, ApJ, **452**, L157
Cernicharo, J. and Guélin, M. 1987, A & A, **183**, L10
Guélin, M., Cernicharo, J., Kahane, C., and Gomez-Gonzales, J. 1986, A & A, **157**, L17
Guélin, M., Lucas, R., and Cernicharo, J. 1993, A & A, **280**, L19
Guélin, M., et al. 1995, A & A, **297**, 183
Kawaguchi, K., Kagi, E., Hirano, T., Takano, S., and Saito, S. 1993, ApJ, **406**, L39
Turner, B.E., Steimle, T.C., and Meerts, L. 1994, ApJ, **426**, L97
Ziurys, L.M., Apponi, A.J., Guélin, M., and Cernicharo, J. 1995, ApJ, **445**, L47

# CO IN PLANETARY NEBULAE AND PROTO-PLANETARY NEBULAE

SUN KWOK
*The University of Calgary*
*Dept. of Physics & Astronomy*
*Calgary, Alberta, Canada T2N 1N4*

The evolution between the asymptotic giant branch (AGB) and the planetary nebula (PN) phases has long been a "missing link" in our understanding of stellar evolution. Through the study of cool IRAS sources, an increasing number of transition objects, or "proto-planetary nebulae", has now been identified. The CO molecule has been found to be a useful tracer of the evolution from the AGB to PN. The detection of CO emission in PNe represents a key piece of evidence that PNe are created from the circumstellar envelopes of AGB stars through the interacting winds process.

## 1. Introduction

The field of stellar evolution between the asymptotic giant branch (AGB) and planetary nebulae has been very active in the last decade. Since 1986, there have been at least ten conferences devoted to this subject (Kwok & Pottasch 1987; Preite-Martinez 1987; Torres-Peimbert 1989; Mennessier & Omont 1990; Schwarz 1993; Sasselov 1993; Acker & Weinberger 1993; Clegg et al. 1994; Tylenda 1994; Harpaz & Soker 1995). In this talk, I can only give a very brief summary of the recent developments, and the readers are encouraged to get a more comprehensive picture by consulting the above conference proceedings. In line with the theme of this conference, I will emphasize the role that the CO molecule has played in our understanding of the late stages of stellar evolution, in particular in the search for proto-planetary nebulae (PPN).

## 2. Effects of Mass Loss on the Late Stages of Stellar Evolution

One of the major breakthroughs in the field of stellar evolution in the 1970s was the discovery of large-scale mass loss on the asymptotic giant



*W.B. Latter et al. (eds.), CO: Twenty-Five Years of Millimeter-Wave Spectroscopy*, 373–381.

branch. The evidence for such mass loss first came in infrared photometry, where large infrared excesses were found to be common in late-type stars (Gehrz & Woolf 1971). The detection of thermal emission from CO confirms that gas, in addition to dust, are being lost (Solomon et al. 1971), and CO observations have allowed accurate estimates of the mass loss rates in hundreds of stars (Knapp et al. 1982; Loup et al. 1993).

The rate of mass removal by stellar winds from the surface was found to be much larger than the rate of nuclear burning in the core. The AGB is therefore terminated by the complete depletion of the hydrogen envelope by stellar winds, and carbon may never be detonated under degenerate conditions. The observed mass loss rates of $10^{-6} - 10^{-4} M_\odot$ yr$^{-1}$ will allow stars with initial masses as high as 8 $M_\odot$ to end their lives as white dwarfs, and several $M_\odot$ of neutral material can be ejected over the short ($\sim 10^6$ yr) AGB lifetime of a star.

The detection of CO in the planetary nebula NGC 7027 came as a surprise not only because the molecule has survived in the ionized environment of the PN, but also that the observed mass of the molecular gas ($> 1 M_\odot$) is much higher than the ionized masses ($\sim 0.1 M_\odot$) usually associated with PNe (Mufson et al. 1975). Far infrared observations have also revealed that a large fraction of the total PN energy output is from the neutral component (Zhang & Kwok 1991). If the molecular envelopes observed in PN represent the remnants of the circumstellar envelopes of AGB stars, then stellar winds from AGB stars, not a sudden ejection at the end of the AGB, must be responsible for the formation of PN. This has led to the proposal of the interacting winds model, where a fast wind from the central star compresses and accelerates the remnant AGB wind material, to be the mechanism for the formation of PN (Kwok et al. 1978; Kwok 1982). The interacting winds model predicts the following components in a PN system in addition to the high-density shell: (i) a fast wind, (ii) a halo, and (iii) a hot bubble, all of which have been detected since their prediction (Heap et al. 1978; Chu et al. 1987; Chu & Ho 1995).

## 3. CO in Planetary Nebulae

The first extensive survey of CO in PN was done by Huggins & Healy (1989) using the NRAO 12 m. Approximately 20% of the 100 PN observed were detected. The amount of molecular mass derived ranges from $10^{-3}$ to 1 $M_\odot$, and in many cases larger than the amount of ionized mass derived from radio continuum observations. The detection of this previously-unseen mass component conveniently solves the "missing mass" problem in PN (cf. Kwok 1994), where the observed "nebular+stellar" masses of PN are found to be much smaller than the main-sequence masses of their progenitors

(1–8 $M_\odot$). Huggins & Healy (1989) also found that the molecular mass to ionized mass ratio decreases with increasing nebular radius, and suggest that the molecular envelope is gradually depleted by photo dissociation as the nebula evolves.

### 3.1. MOLECULAR GAS DISTRIBUTION IN PN

With the advent of mm interferometry, a number of PN have been mapped in CO (Bachiller et al. 1989a,b; Bieging et al. 1991). The CO emission regions are generally found to lie outside of the ionized nebula, confirming that the molecular gas represents the remnant of the AGB envelope. The ionized nebula of NGC 7027 ($\sim 10''$) fits neatly into the central minimum of the CO distribution (Bieging et al. 1991). Similar molecular cavities corresponding to the size of the ionized region have also been seen in IRAS 21282+5050 (Likkel et al. 1994; Deguchi 1995). IRAS 21282+5050 is a particularly interesting PN for most of its energy is emitted from the neutral component (Kwok et al. 1993). The CO map of M1-7 (Shibata et al. 1994) shows a double-peaked structure, very similar to the radio continuum structure of most PN. An inner bright ring is seen in the CO map of NGC 7027, suggesting that the molecular gas is compressed by the shock front created by the interacting winds process (Bieging et al. 1991). Comparison of the CO and $H_2$ maps of NGC 6720 also suggests that a shock is propagating into the molecular envelope (Greenhouse et al. 1988).

### 3.2. MOLECULAR HYDROGEN EMISSION

The 2.121 $\mu$m v=1–0 S(1) electric quadruple vibration-rotation line of $H_2$ is widely observed in PNe (Latter et al. 1995). It is excited either by shocked gas or fluorescence (Dinerstein et al. 1988). The $H_2$ line is therefore useful as a probe of the dynamics or the radiation distribution in a PN. It has been suggested that PNe of "butterfly" or "bipolar" morphology have a higher probability of having $H_2$ emission (Kastner et al. 1994). This correlation could be explained by PNe of these morphological types having higher progenitor masses, which have higher mass loss rates and shorter transition times between the AGB and PN.

## 4. CO in Proto-Planetary Nebulae

Proto-planetary nebulae, defined as objects in transition between AGB and PN phases, are expected to contain a blueward-evolving central star surrounded by a dust envelope ejected during the AGB phase. Due to the low temperatures of the dust envelopes, PPN should appear as far-infrared sources and can be found among IRAS sources of low color temperatures

(100-200 K). As the central star emerges from the dispersing dust envelope, PPNe will take on a "double-peaked" energy distribution showing both the photospheric and dust components. The central stars of PPNe are found to be mainly of F or G spectral types, and have luminosity classes of I, corresponding to the Class I low surface gravity expected of such objects. Over 30 PPNe have been identified to date (Kwok 1993).

The detection of circumstellar CO emission quickly followed the identification of PPNe (Woodsworth et al. 1990; Likkel et al. 1991; Bujarrabal et al. 1992). The CO observations confirm that these "yellow supergiants" are indeed evolved objects, and their CO envelopes represent the remnant of the AGB mass loss. A summary of the CO observations of PPN is given in Table 1.

Most of the PPN are not resolved in CO, except in the two well-known nearby candidates AFGL 618 and AFGL 2688. In spite of the extremely bipolar morphology of the optical reflection nebulae in these two objects, the CO envelope outside is remarkably symmetric (Deguchi 1995), suggesting that a morphological change occurred during the transition period.

### 4.1. HIGH VELOCITY GAS IN PPN

One of the major uncertainties in post-AGB evolution is when does the fast wind begin? A significant rate of mass loss during the post-AGB (and pre-PN) evolution will greatly shorten the transition time from AGB to PN. While the PN central-star winds can be detected by P Cygni profiles of the UV-resonance lines, they are difficult to detect before ionization. The only evidence for the existence of a central-star wind during the PPN phase is through the detection of extended wings of the CO line.

The development of SIS receivers has allowed the detection of weak, extended line wings of CO in the PPN AFGL 618 (Gammie et al. 1989; Cernicharo et al. 1989) and AFGL 2688 (Young et al. 1992; Jaminet et al. 1992). Molecular winds of velocities of approximately ten times higher than the AGB wind velocities have been found. Such velocities are in the range of expected escape velocities for the spectral types (B and F) of AFGL 618 and AFGL 2688.

### 4.2. VIBRATIONAL-ROTATIONAL TRANSITIONS OF CO IN PPN

The vibrational band of CO is one of the strongest absorption features observed in the near infrared spectra of late-type stars. Recent high-resolution infrared spectroscopic observations have detected first overtone ($\Delta v = 2$) vibrational bands of CO in absorption up to $v = 8$ in many PPNe (Hrivnak et al. 1994). Most interestingly, the CO spectrum of IRAS 22272+5435 was found to change from emission to absorption over a 3-month period, sug-

TABLE 1. PPNe detected in CO

| IRAS name | Line | $T_{\mathrm{mb}}$ [K] | $\int T\,dV$ [K km s$^{-1}$] | $V_e$ [km s$^{-1}$] | $V_{\mathrm{lsr}}$ [km s$^{-1}$] | Telescope | Ref |
|---|---|---|---|---|---|---|---|
| 04296+3429 | 2–1 | 0.4 | ... | 15.6 | −62 | JCMT | 1 |
| | 2–1 | 0.45 | 8.3 | 12.0 | −66 | IRAM | 2 |
| 07134+1005 | 1–0 | 1.3 | 21 | ... | 72 | IRAM | 3 |
| | 1–0 | 0.4 | 6 | 11.0 | 75 | SEST | 4 |
| | 2–1 | 2.2 | 28 | ... | 73 | IRAM | 3 |
| | 2–1 | 3.2 | 44 | 10.2 | 73 | IRAM | 2 |
| 17436+5003 | 1–0 | 0.8 | 16 | 14.9 | −35 | IRAM | 7 |
| | 1–0 | 0.9 | 19 | ... | −35 | IRAM | 3 |
| | 2–1 | 1.8 | 33 | 13.2 | −35 | IRAM | 7 |
| | 2–1 | 2.0 | 36 | ... | −35 | IRAM | 3 |
| 19114+0002 | 1–0 | 1.7 | 76 | ... | 99 | IRAM | 3 |
| | 1–0 | 0.3 | 16 | 30.4 | 99 | SEST | 4 |
| | 2–1 | 3.4 | 171 | ... | 99 | IRAM | 3 |
| | 2–1 | 0.5 | 23 | 28.1 | 99 | SEST | 4 |
| | 3–2 | 2.1 | 110 | 37.1 | 98 | JCMT | 6 |
| 19500−1709 | 1–0 | 1.0 | 21 | ... | 24 | IRAM | 3 |
| | 2–1 | 0.2 | 4 | 13.0 | 25 | SEST | 4 |
| | 2–1 | 1.9 | 48 | ... | 23 | IRAM | 3 |
| | 3–2 | 0.4 | 7 | 13.2 | 26 | SEST | 4 |
| 22223+4327 | 1–0 | 0.25 | 2 | 15 | −25 | IRAM | 7 |
| | 2–1 | 0.64 | 12.3 | 14.0 | −30 | IRAM | 2 |
| 22272+5435 | 1–0 | 1.3? | 18 | 10.5 | −28 | OSO | 5 |
| | 2–1 | 3.1 | ... | 9.6 | −28 | JCMT | 1 |
| | 2–1 | 6.27 | 80 | 11.8 | −28 | IRAM | 2 |
| | 3–2 | 3.2 | 38 | 9.8 | −28 | JCMT | 6 |
| 22574+6609 | 1–0 | 0.3 | 10 | 24 | −62 | IRAM | 7 |
| | 2–1 | 0.5 | 11 | 15 | −64 | IRAM | 7 |
| | 3–2 | 0.7 | 8.3 | 20 | −62 | JCMT | 6 |
| 23304+6147 | 1–0 | 0.6 | 10 | 8.0 | −17 | IRAM | 7 |
| | 2–1 | 0.7 | ... | 15.5 | −16 | JCMT | 1 |
| | 2–1 | 1.6 | 25 | 11.5 | −16 | IRAM | 7 |
| | 3–2 | 0.8 | 14 | 17.4 | −13 | JCMT | 6 |
| 23321+6545 | 1–0 | 0.6 | 14 | 15.2 | −56 | IRAM | 7 |
| | 2–1 | 1.2 | 28 | 17.0 | −55 | IRAM | 7 |

1. Woodsworth et al. (1990); 2. Omont et al. (1993); 3. Bujarrabal et al. (1992); 4. van der Veen et al. (1993); 5. Lindqvist et al. (1988); 6. Kwok et al. (1995); 7. Likkel et al. (1991)

gesting recent episodes of mass loss. High-resolution infrared spectroscopic observations have resolved the rotational lines in the first overtone band of CO, and comparison between the observed and the predicted line velocities can be used to study the kinematics of the mass loss process.

### 4.3. DUST IN PPNE

While most of the dust in AGB stars can be identified as silicates (for oxygen-rich stars) or SiC or graphite for carbon-rich stars, PNe are known to show a family of infrared features attributed to Polycyclic Aromatic Hydrocarbons (PAH). The detection of PAH features in PPNe suggests that these molecules are either synthesized during the evolution to PNe, or produced in the AGB stellar envelopes but excited when the star becomes hotter. Most interestingly, nine carbon-rich PPNe are found to show a strong unidentified emission feature at 21 $\mu$m (Kwok et al. 1989; Kwok et al. 1995). This 21 $\mu$m feature is often accompanied by unusually strong 3.4–3.5 $\mu$m features (Geballe et al. 1994), as well as by an unidentified emission feature at 30 $\mu$m (Omont et al. 1995). The strength of the 21 $\mu$m feature suggests that it probably originates from a large carbon-based molecule.

## 5. The Interacting Winds Process as a Cause of "Bipolar" Morphology Observed in PN

It has been long known that most PNe are not spherically symmetric but have a "double-peaked" (sometimes referred to equivocally as "bipolar") morphology. The more extreme cases take on the form of a "butterfly" shape. This is true for the most evolved PNe (Balick 1987) to the youngest PN (Aaquist & Kwok 1990). Balick (1987) suggests that a diversity of PNe morphologies can be explained by the interacting winds process, where small degrees of asymmetries in the AGB envelope are amplified by the fast wind from the central star. Recent 2-D hydrodynamical models based on this idea have been very successful in simulating the observed morphologies of PNe (Mellema et al. 1992, Frank et al. 1993).

## 6. Planetary Nebula as an Evolving System

Recent multi-wavelength observations have found PNe to be active in almost in all wavebands between the X-ray and radio While the mass of the nebula is provided by the AGB progenitor star, it is constantly being shaped and pushed by a fast wind from the central star. As the star becomes hotter, both the mechanical output (mass loss rate and the wind velocity) and the radiative output (Lyman continuum photons) change, which results in the gradual acceleration and ionization of the nebula. The fact that PN is a

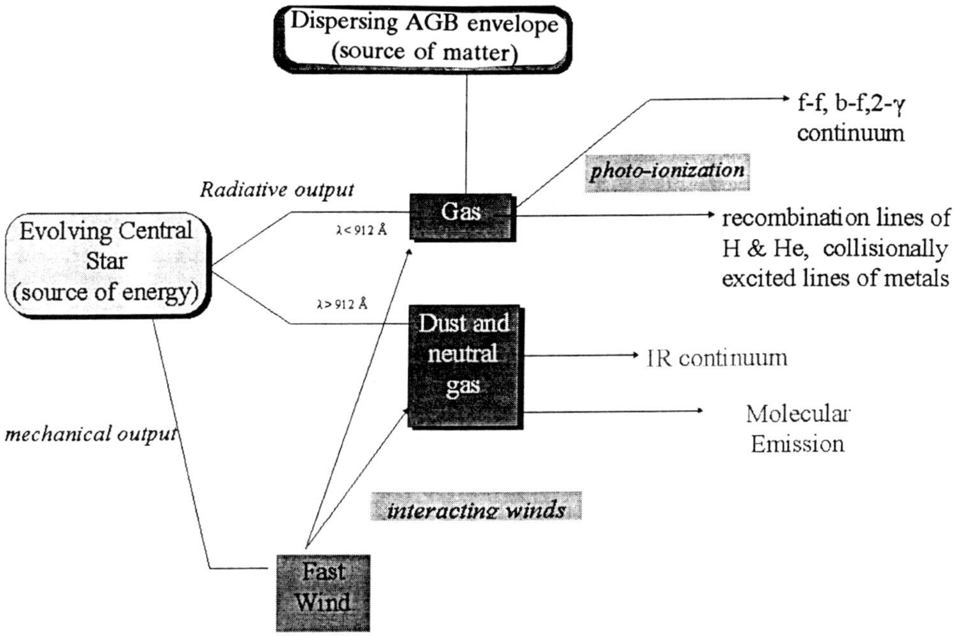


*Figure 1.* A schematic diagram of planetary nebula evolution.

time-dependent system, consisting of two tightly coupled components (star and nebula) evolving at different time scales, is a new concept with implications we are just beginning to appreciate (cf Kwok 1994). A schematic diagram showing the time-dependent radiative and dynamical interactions between the central star and the AGB wind is shown in Fig. 1.

## 7. Summary

The CO molecule has been shown to be a very effective tool in tracing stellar evolution from the AGB through the PPN phase to the PN stage. Millimeter-wave spectroscopy of the rotational lines of CO have revealed the presence of massive neutral halos in PNe. The discovery of this previously-undetected component has led to a revolution in our understanding of the origin and evolution of PNe. By interpolating the infrared and CO properties of AGB stars and PNe, we are able to search for and identify objects in transition between these two phases. The discovery of PPNe not only fills a "missing link" in the late stages of stellar evolution, but has also led to new challenges to our understanding of interstellar chemistry.

## References

Aaquist, O.B., & Kwok, S. 1990, A&AS, 84, 229
Acker, A., & Weinberger, R. 1993, IAU Symp. 155: Planetary Nebulae, Kluwer: Dordrecht
Bachiller, R., Planesas, P., Martin-Pintado, J., Bujarrabal, V., & Tafalla, M. 1989a, A&A, 210, 366
Bachiller, R., Bujarrabal, V., Martin-Pintado, J., & Gómez-González, J. 1989b, A&A, 118, 252
Balick, B. 1987, AS, 94, 671
Bieging, J.H., Wilner, D., & Thronson, H.A. 1991, ApJ, 379, 271
Bujarrabal, V., Alcolea, J., & Planesas 1992, A&A, 257, 701
Bujarrabal, V., Alcolea, J., Neri, R., & Grewing, M. 1994, ApJ, 436, L169
Cernicharo, J., Guélin, M., Martin-Pintado, J., Peñalver, J., & Mauersberger, R. 1989, A&A, 222, Ll
Chu, Y., & Ho, C.-H. 1995, ApJ, 448, 127
Chu, Y., Jacoby, G.H., & Arendt, R. 1987, ApJS, 64, 520
Clegg, R.E.S., Stevens, I.R., & Meikle, W.P.S. 1994, Circumstellar Media in the Late Stages of Stellar Evolution, Cambridge University Press
Deguchi, S. 1995, in Asymmetric Planetary Nebulae, eds. Harpaz, A., & Soker, N., Ann. Israel Phys. Soc., Vol. 11, 132
Dinerstein, H.L., Lester, D.F., Carr, J.S., & Harvey, P.M. 1988, ApJ, 327, L27
Frank, A., Balick, B., Icke, V., & Mellema, G. 1993, ApJ, 404, L25
Gammie, C.F., Knapp, G.R., Young, K., Phillips, T.G., & Falgarone, E. 1989, ApJ, 345, L87
Geballe, T.R., Tielens, A.G.G.M., Kwok, S., & Hrivnak, B.J. 1994, ApJ, 387, L89
Gehrz, R.D., & Woolf, N.J. 1971, ApJ, 165, 285
Greenhouse, M.A., Hayward, T.L., & Thronson, H.A. 1988, ApJ, 325, 604
Harpaz, A., & Soker, N. 1995, Asymmetrical Planetary Nebulae, Ann. Israel Phys. Soc., Vol. 11.
Heap, S.R. et al. 1978, Nature, 275, 385
Hrivnak, B.J., Kwok, S., & Geballe, T.R. 1994, ApJ, 420, 783
Huggins, P.J., & Healy, A.P. 1989, ApJ, 346, 201
Jaminet, P.A., Danchi, W.G., Scndell, G., & Sutton, E.C. 1992, ApJ, 400, 535
Kastner, J.H., Gatley, I., Merrill, K.M., Probst, R., & Weintraub, D.A. 1994, ApJ, 421, 600
Knapp, G.R., Phillips, T.G., Leighton, R.B., Lo, K.Y., Walmier, P.G., & Wootten, H.A. 1982, ApJ, 252, 616 Knapp, G.R., et al. 1982, ApJ, 252, 616
Kwok, S. 1982, ApJ, 258, 280
— 1993, ARAA, 31, 63
— 1994, PASP, 106, 344
Kwok, S., & Pottasch, S.R. 1987, Late Stages of Stellar Evolution, Reidel: Dordrecht
Kwok, S., Purton, C.R., & FitzGerald, P.M. 1978, ApJ, 219 L125
Kwok, S., Volk, K., & Hrivnak, B.J. 1989, ApJ, 34S, L51
Kwok, S., Hrivnak, B.J., & Langill, P.P. 1993, ApJ, 408, 586
Kwok, S., Hrivnak, B.J., & Geballe, T.R. 1995, ApJ, 454, 394
Kwok, S., Woodsworth, A.W., & Purton, C.R. 1995, in preparation
Latter, W.B., Kelly, D.M., Hora, J.L., & Deutsch, L.K. 1995, ApJS, 100, 159
Likkel, L., Forveille, T., Omont, A., & Morris, M. 1991, 246, 153
Likkel, L., Morris, M., Kastner, J.H., & Forveille, T. 1994, A&A, 282, 190
Lindqvist, M., Nyman, L.A, Olofsson, H., & Winnberg, A. 1988, A&A, 205, L15
Loup, C., Forveille, T., Omont, A., & Paul, J.F. 1993, A&AS, 99, 291
Mellema, G., Eulerink, F., & Icke, I. 1992, A&A, 252, 718
Mennessier, M.O., & Omont, A. 1990, From Miras to Planetary Nebulae: Which Path for Stellar Evolution?, Editions Frontieres, Gif-sur-Yvette
Mufson, S.L., Lyon, J., & Marionni, P.A. 1975, ApJ, 201, L85

Omont, A., Loup, C, Forveille, T., te Lintel Hekkert, P., Habing, H., & Sivagnanam, P. 1993, A&A, 267, 515
Omont, A. et al. 1995, ApJ, 454, 819
Preite-Martinez, A. 1987, Planetary Nebulae and Proto-Planetary Nebulae: From IRAS to ISO, Reidel: Dordrecht
Sasselov, D. 1993, Luminous High-Latitude Stars, ASP Conf. Ser. 45
Schwarz, H.E. 1993, Mass Loss on the AGB and Beyond, ESO Conf. Proc. No. 46
Shibata, K.M., Deguchi, S., Kasuga, T., Tamura, S., Hirano, N., & Karneya, O. 1994, in Astronomy with Millimeter and Submillimeter Wave Interferometry, eds. M. Ishiguro and W.J. Welch, ASP Conf. Ser. 59, 143
Solomon, P., Jefferts, K.B., Penzias, A.A., & Wilson, R.W. 1971, ApJ, 163, L53
Torres-Peimbert, S. 1989, IAU Symp. 131: Planetary Nebulae, Kluwer: Dordrecht
Tylenda, R. 1993, Acta Astron., 43, 4
van der Veen, W.E.C.J., Trams, N.R., Waters, L.B.F.M. 1993, A&A, 269, 231
Volk, K., Kwok, S., & Woodsworth, A.W. 1993, ApJ, 402, 292
Woodsworth, A.W., Kwok, S., & Chan, S.J. 1990, A&A, 228, 503
Young, K., Serabyn, G., Phillips, T.G., Knapp, G.R., Güsten, R., & Schulz, A. 1992, ApJ, 385, 265
Zhang, C.Y., & Kwok, S. 1991, A&AS, 250, 179

# PRE–SHOCK AND POST–SHOCK ABUNDANCE RATIOS OF ATOMIC CARBON TO CO IN IC 443 G


JOCELYN KEENE AND T. G. PHILLIPS
*Caltech 320-47*
*Pasadena, CA 91125, USA*

AND

EWINE F. VAN DISHOECK
*Sterrewacht Leiden, P.O. Box 9513*
*2300 RA Leiden, The Netherlands*


## 1. Introduction

IC 443 is a prominent, well-studied supernova remnant. Optical pictures show multiple filaments through the midst of which runs a lane of extinction. Bright $H_2$ emission marks the region of interaction between the SNR and the foreground cloud. Clump G is at the western side of the SNR and is heavily obscured. At its position the velocity of supernova shock is nearly in the plane of the sky and the radial velocity of the shocked material is displaced very little from that of the ambient foreground material.

We have mapped clump G of IC 443 in the $^3P_1 - {}^3P_0$ line of atomic carbon (C I) with the CSO, as well as in $^{12}$CO lines from $J = 2 - 1$ up to $6 - 5$, $^{13}$CO $2 - 1$ and $3 - 2$, and have obtained a few C$^{18}$O $2 - 1$ spectra (Figs. 1 & 2). The $^{12}$CO lines all exhibit strong wide lines which emanate from the shocked region and, up to $^{12}$CO $4 - 3$, strong absorption by very cold foreground gas. The C I lines are much stronger than the $^{13}$CO $2 - 1$ lines, which is unusual. The C I and CO isotopomeric lines also show the wide lines from the shocked region, but, rather than exhibiting absorption by cold gas, they have narrow strong emission lines at the same velocity as the $^{12}$CO absorption, evidently emanating from warm pre–shocked gas. Because of the low temperature of the foreground gas, it does not strongly absorb the $^{12}$CO $6 - 5$ line, so at most map positions (but not at the CO peak position with spectra shown in Fig. 2) the CO $6 - 5$ line also shows the narrow emission line in addition to the wide emission from the shock.



*W.B. Latter et al. (eds.), CO: Twenty-Five Years of Millimeter-Wave Spectroscopy*, 382–384.

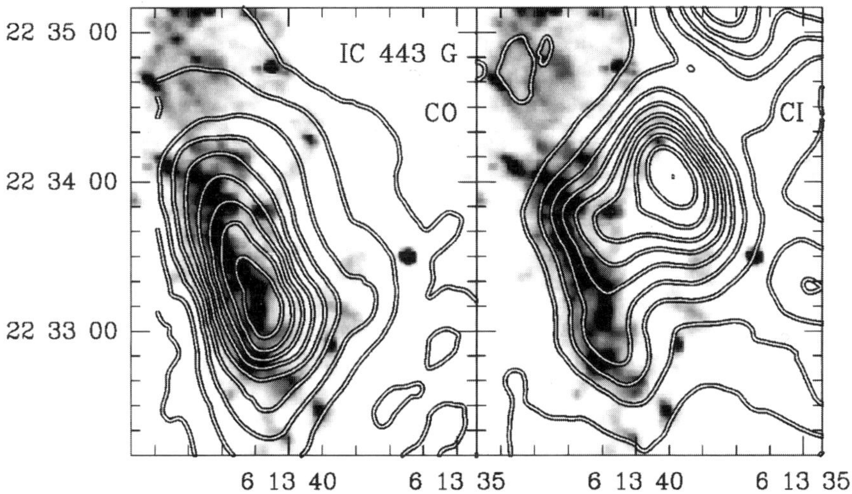


*Figure 1.* *left*) Contours of integrated $^{12}CO$ 2−1 emission superposed on an image of the 2.12 $\mu$m $H_2$ emission (Merrill, Gatley, & Probst 1995, in preparation). *right*) Contours of peak C I temperature superposed on $H_2$ image. The direction of shock propagation is from east to west.

The $^{12}CO$ lines up to $J = 4$ peak strongly near the shock front as delineated by the $H_2$ emission (Fig. 1, left). However, the C I emission shows a more complex behavior – while the broad-line emission has a similar distribution to that of $^{12}CO$, the narrow-line emission peaks well to the west of the front, *upstream* from the shock (Fig. 1, right). The same distribution is seen in the $^{12}CO$ 6 − 5 line and the CO isotopomeric lines.

## 2. Interpretation

We have employed a Monte–Carlo radiative–transfer code (Bernes, 1979, A&A, 73, 67) to model the data. We have found that the emission regions can be satisfactorily represented as a superposition of three layers: 1) an extremely cold ($T_K < 10$ K) and/or low density foreground cloud of normal abundances ($N_1$[C I]/$N_1$[CO] $\approx 0.1$), 2) an intermediate narrow–line layer of extremely carbon–rich material ($N_2$[C I]/$N_2$[CO] $\approx 10$) with $25 < T_K < 60$ K and moderate density ($n \approx 10^4$ cm$^{-3}$), the presence of which indicates that most of the CO molecules have been dissociated *before* the arrival of the shock front, and 3) a background carbon–rich ($N_3$[C I]/$N_3$[CO] $\approx 1$) shocked region with $T_K \approx 100$ K and high density ($n \approx 10^5$ cm$^{-3}$) in which molecule re–formation must be taking place. Table 1 gives the calculated column densities and abundance ratios for the position of the $^{12}CO$ emission peak with the observed and model spectra shown in Fig. 2.

The Caltech Submillimeter Observatory is supported by the National

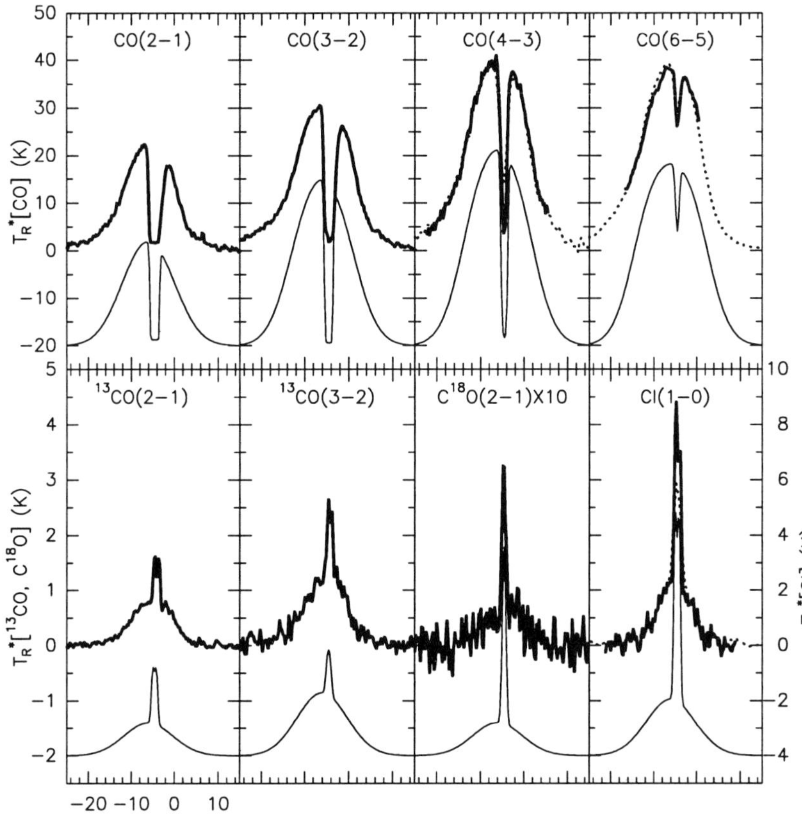


*Figure 2.* C I and CO spectra at the $^{12}$CO peak [6:13:40.9 +22:33:10 (1950)]. Below the observations (heavy lines) are spectra calculated from the three–layer Monte Carlo model (light lines). Low spectral–resolution observations are shown with dotted lines.

TABLE 1. Pre– and Post–shock C I and CO Column Densities at the CO Peak

| | 1) Cold Pre–shock | 2) Warm Pre–shock | 3) Post–shock |
|---|---|---|---|
| *N*[C I] | 0.3 (17) cm$^{-2}$ | 6.0 (17) cm$^{-2}$ | 3.0 (17) cm$^{-2}$ |
| *N*[CO] | 2.0 (17) cm$^{-2}$ | 0.5 (17) cm$^{-2}$ | 3.3 (17) cm$^{-2}$ |
| *N*[C I]/*N*[CO] | 0.15 | 13 | 0.9 |

Science Foundation under grant AST 93–13929. We thank K. M. Merrill for permission to use the $H_2$ image before publication.

# REFLECTIONS

# TWENTY-FIVE YEARS OF CO ASTRONOMY: REVEALING THE COLD UNIVERSE

CHARLES J. LADA
*Smithsonian Astrophysical Observatory*
*60 Garden Street*
*Cambridge, MA 02138*

## 1. Introduction

I suspect that future historians of science will regard the discovery of CO in space as one of the more significant milestones of astronomical research in the 20th century — a period which is likely to be considered the Golden Age of astronomical discovery. It was during this century that such fundamental phenomena as galaxies, the big bang, the expanding universe, the interstellar medium, quasars, pulsars, black holes, extra-solar planets and protostars were first revealed by observational exploration. Indeed, it was also in this century that observation and theory combined to decipher the very nature of the stars themselves as thermonuclear reactors which convert hydrogen, the primary product of the big bang, into the zoo of heavier elements which make up the periodic table. These discoveries were the direct result of the dramatic evolution of observing technology which among other things, opened the entire electromagnetic spectrum to astronomical investigation and enabled observations at optical, radio, infrared, UV, X and $\gamma$ ray wavelengths with large telescopes on the ground, in the air and in space.

The discovery of interstellar CO emission in 1970 was a consequence of the quest to open the millimeter-wave portion of the electromagnetic spectrum to astronomical research and it ushered in a whole new era in the study of the interstellar medium and star formation. Its significance to astronomy lies in the fact that observations of CO (more than any other tracer) revealed to astronomers a new component of the cosmos: *the cold universe.* This is the universe of birth and creation. It is the universe where planets, stars and perhaps even galaxies come into being. It is the universe where the interesting physical mechanisms are not those which produce



*W.B. Latter et al. (eds.), CO: Twenty-Five Years of Millimeter-Wave Spectroscopy,* 387–396.

ever increasing amounts of energy, but those which produce ever decreasing amounts of entropy. It is the astrochemical universe of molecules and dust. Before the discovery of CO the cold universe was an unknown and uncharted frontier of astrophysics. With the discovery of CO, the cold universe became the subject of detailed astronomical scrutiny and CO observations have proved to be the primary probe for its exploration.

## 2. State of Affairs Prior to 1970

By the middle of the twentieth century, such observations as: 1) the discovery of stationary atomic absorption lines in the spectra of spectroscopic binary stars (Hartmann 1907; Plaskett and Pearce 1933), 2) deep photographs of dark nebulae within the Milky Way (e.g., Barnard 1919) and 3) the general extinction of starlight toward open clusters (Trumpler 1930), had convinced most astronomers of the existence of an interstellar medium (ISM) of gas and dust. The discovery of HI emission at $\lambda$21 cm in 1951 revealed in dramatic fashion just how pervasive the interstellar gas was in the disk of the Milky Way. By the 1960s radio and optical observations of atomic clouds and HII regions, as well as of dust polarization and extinction contributed to an extensive body of literature (both theoretical and observational) dealing with the subject of interstellar gas and dust in the galaxy (e.g., Dufay 1957). Although the basic physical processes (e.g., cooling, heating, ionization, interaction with stellar radiation fields, etc.) seemed well understood (e.g., Spitzer 1968), the basic structure of the interstellar medium was not at all clear. In particular, it was not certain whether or not the interstellar medium was in fact a medium. That is, whether it was made up of smoothly distributed gas and dust, or of a collection of discrete clouds. By 1970, however, an elegant theoretical picture of the ISM was developed. Simple considerations of the equilibrium heating and cooling balance in atomic clouds suggested that the interstellar gas likely existed in only one of two stable phases: a high density cool phase which could be identified with discrete clouds and a low density warm phase which could be identified with an "intercloud" medium (Field, Goldsmith and Habing 1969). On another front the development of the theory of stellar structure and evolution by the middle of the century led to an understanding of the basic nature of stars and their evolution on and off the main sequence. Although this theory is one of the greatest achievements of 20th century astronomy, by 1970 (and even at present) it was incomplete in one very fundamental aspect: it did not predict or explain the origin of stars. As a result the problem of star formation became a prime focus of ISM research in the latter part of the century. In the two-phase ISM picture, star formation presumably occurred in the cool atomic clouds through some process of global collapse and hierachical fragmentation (e.g., Hoyle 1953; Spitzer 1968).

Unfortunately, this tidy picture of the ISM and star formation quickly came into conflict with numerous observations. For example, HI clouds were observed over a wider range of temperatures than appeared permitted by the theory. In addition, optical and X-ray observations uncovered a major and very hot ($\approx 10^6$ K) component of insterstellar gas not predicted by the two phase theory. Moreover, the youngest known low mass stars were found intimately associated with dark nebulae (e.g., Herbig 1962) which were not, as it turns out, the HI clouds envisioned by the theory of the two phase ISM. Although HI column densities were generally proportional to dust column densities (e.g. Lilley 1955), HI was either weak or entirely absent in dark nebulae (e.g., Bok, Lawrence & Menton 1955; Heiles 1969), the regions of highest visual extinction, suggesting that the gas that made up these nebulae was molecular not atomic in nature. With the discoveries of radio transitions of OH, $H_2O$ and $NH_3$ in space, the significance of a molecular phase of the ISM became increasingly apparent. By 1970 the stage was set for a major shift in paradigm.

## 3. CO and the Cold Interstellar Medium

When it was discovered in 1970, both the strength and the ubiquity of CO emission in space were a complete surprise (Wilson et al. 1970; Penzias et al. 1971). Indeed, the fundamental rotational transition of CO was detected in emssion toward HII regions, dark nebulae, Bok Globules, reflection nebulae, supernova remnants, planetary nebulae and even in the extended atmospheres of certain luminous infrared giant stars. Moreover, in regions where it was found, emission from CO was more widespread than any other molecular tracer known at that time or subsequently discovered. It was found essentially everywhere there was a significant amount of extinction (i.e., $A_v > 1$ magnitude). The excitation of this transition by collisions required sufficiently high densities (100-1000 $cm^{-3}$) that it was clear that CO clouds had to be composed primarily of hydrogen. Furthermore, the lack of observable HI emission in such regions indicated that the hydrogen had to be in molecular form. It became quickly apparent that emission from the $J = 1 \rightarrow 0$ transition of CO was very effectively tracing the (cold) molecular component of the interstellar medium. And this was the component of the ISM where star formation occurred. It was the cold, dense gas that proved to be so impenetrable to the optical observations of the first half of the century.

Surveys of CO emission in the disk of the Milky Way soon revealed the unexpected result that cold molecular gas constituted a significant fraction of the ISM: in the vicinity of the Sun, as much as half the mass of the interstellar medium was in the form of molecular clouds while a considerably higher fraction of the interstellar gas was molecular in form in the inner and central regions of the Galaxy (e.g., Combes 1991 and references therein). Besides this significant finding, CO mapping of molecular clouds led to two other discoveries of fundamental importance for astronomy: Giant Molecular Clouds (GMCs) and bipolar molecular outflows. These latter two discoveries helped to revolutionize the study of star formation and thus merit further brief discussion.

### 3.1. GIANT MOLECULAR CLOUDS (GMCS)

With sizes between 50–100 parsecs in extent, GMCs are the largest objects in the Galaxy and with masses between $10^5$–$10^6$ $M_\odot$, they rival globular clusters as the most massive objects in the Milky Way. With temperatures seldom exceeding 10 K, GMCs are the coldest objects in the universe, yet their existence was unknown prior to 1970. Over the last twenty years, observations have established that GMCs are the sites of all star formation currently taking place in the Milky Way. In the 1970s CO mapping of GMCs revolutionized the way astronomers viewed the problem of star formation (e.g., Tucker, Kutner and Thaddeus 1973; Lada 1975; Loren 1975; Elmegreen and Lada 1977; Blitz 1980). During the preceding twenty years observations of giant HII regions, compact HII regions, $H_2O$ and OH maser sources, infrared stars and Herbig-Haro objects provided a menagerie of astronomical phenomena that were thought to be somehow intimately related to the processes of star formation and early stellar evolution. Yet, the true natures of most of these objects as well as their precise relation to each other and to the actual process of stellar birth were obscure. GMCs provided a unifying context within which these phenomena could be linked together. The underlying physical connection was that all these interesting objects were inhabitants of GMCs. Moreover, within a given GMC the various objects were often simultaneously present. This gave rise to the notion that these seemingly disparate phenomena could be understood as manifestations of differing stages in the development of young stars within a star forming molecular cloud complex. And, this notion gained considerable support from subsequent observations and theoretical modelling of the infrared energy distributions of embedded young stellar objects (e.g., Adams, Lada and Shu 1987).

Yet after twenty-five years of research, many basic questions concerning the nature of GMCs remain unanswered. These include the questions of their origin, evolution, death, and physical structure. For example, we do not know if GMCs are formed from HI clouds, the intercloud medium, smaller molecular clouds, or all of the above. It is likely that cloud formation involves spiral density waves, supernova explosions, the galactic dynamo, photoionization by galactic starlight and maybe other phenomena (e.g., Elmegreen 1991), but, unfortunately, observations and theoretical understanding of many of these processes are still lacking and the origin of the largest and most massive objects in the Galaxy remains a complete mystery. Likewise, we are not sure how GMCs die. They are probably destroyed by OB associations. But are they converted into ionized and/or atomic gas or are they shredded into smaller molecular cloud chaff? What are the timescales for their formation, evolution and destruction? The critical observations needed to answer these questions either don't yet exist or are not fully recognized, a circumstance which indicates how deep our lack of understanding of these problems actually is. Much of this uncertainty surely results from the fact that there is not as yet a clear picture of molecular cloud structure. Molecular-line observations certainly indicate that GMCs are highly structured, but finding a meaningful way to characterize this structure has proved to be surprisingly elusive. Are clouds composed of (round) clumps, (elongated) filaments, or is their structure best described by chaotic complexity? Do we even know what we really mean by any of these terms? It is difficult to see how progress will be made in this area since existing molecular-line data are about as detailed as one could hope to obtain and furthermore are unlikely to be surpassed by any future observations. A fundamental revision of the manner in which we think about cloud structure may be required to make appreciable progress in understanding these issues.

### 3.2. BIPOLAR MOLECULAR OUTFLOWS

About a decade after the discovery of CO, CO observations of the molecular gas surrounding certain young stellar objects led to the discovery an unanticipated phenomenon of great significance for understanding the process of star formation. In mapping the velocity field around the infrared source known as L1551 IRS 5, Snell Loren and Plambeck (1980) discovered the first example of a bipolar outflow of cold molecular gas emanating from a buried protostellar object. This discovery came as a complete surprise because up until that time it was generally believed that if there were any systematic bulk motion of material in a molecular cloud, especially around a protostellar object, that motion should be collapse, not outflow (e.g., Goldreich and Kwan 1974). Subsequent CO observational surveys revealed many more bipolar outflow sources (e.g., Bally and Lada 1983) so that at the present time well over 100 molecular outflows are known, most within a kiloparsec of the sun (e.g, Fukui et al. 1993). Their properties have been extensively and thoroughly reviewed in the literature (e.g., Lada 1985, Bachiller & Gomez-Gonzalez 1992). Briefly, the masses of such outflows are substantial, containing anywhere between 0.1 and 100 solar masses. Consequently, the outflowing molecular gas consists of swept-up ambient cloud material rather than original ejecta from the driving source. More significantly, the corresponding kinetic energies of the flows are enormous, ranging between $10^{43}$ and $10^{47}$ ergs! The dynamical timescales of the flows are estimated to be between $10^3$ and $10^5$ years and their local formation rate is estimated to be roughly comparable to the formation rate for stars of a solar mass or greater. Taken together, these facts suggest that molecular outflows play a critical role in the star formation process. In particular, bipolar molecular outflows are individually energetic enough to disrupt cloud cores and it is suspected that they are the likely agent that drives the evolution of a young stellar object from the deeply embedded protostellar stage to the optically revealed pre-main sequence stage of early stellar evolution.

At the same time the coincidence of bipolar outflows with protostellar objects presented an interesting paradox for star formation researchers: How can a protostar—an embryonic stellar core in the process of accreting matter and building up a stellar mass—be simultaneously a source of outflow and inflow? How can a star form by losing mass? The answer to these questions lie at the heart of the star formation problem. Because of its conservation, angular momentum presents a barrier which prevents the contraction of interstellar gas to stellar dimensions. Although magnetic breaking during the early stages of cloud formation and the process of viscous accretion through a circumstellar disk can remove a significant amount of the initial angular momentum of an accreting element of interstellar gas, the gas still retains too much specific angular momentum to reach a configuration of stellar dimensions. If generated close enough to an embryonic star, bipolar outflows can, in principle, carry away significant amounts of angular momentum and thus enable the accretion process to build up a protostellar embryo to stellar mass. Evidently, bipolar outflows may be the physical process employed by nature to solve the angular momentum problem of star formation and in doing so thus resolve the paradox of the simultaneity of outflow and infall in protostellar evolution (e.g., Shu et al. 1988).

In many respects more may be understood about the fundamental aspects of bipolar outflows than is understood about the fundamental aspects of GMCs. For example, the origin of the bipolar outflow phase is thought to be a natural consequence of the formation of a star while the cessation of the outflow phase a result of the termination of the accretion process which occurs at that point when the build up of a young star comes to an end. However, there are still many basic aspects of the bipolar outflow phenomenon that are mysterious. For example, the physical mechanism that accelerates the primary wind driving the molecular outflow is not understood, neither is the mechanism that produces the bipolar, often jet-like morphology observed in CO maps. In addition, it is not clear how the primary wind actually sweeps up the large masses of cold ambient molecular gas that make up the bipolar lobes of an outflow. Nevertheless, the discovery of bipolar molecular outflows by CO observations has provided a most critical piece to the ultimate solution of the star formation puzzle.

## 4. CO and the Cold Component of Galaxies

In the limited space I have remaining I will briefly outline the tremendous impact of CO observations on the studies of the Milky Way and extragalactic astronomy. Not long after its discovery, CO observations were employed to investigate the structure of the Milky Way, and as receivers, spectrometers and telescopes improved, the structure and properties of the cold component of external galaxies. Early studies showed that the distribution of the molecular gas in the Milky Way differed significantly from that of atomic gas. Cold molecular gas was found to be primarily located in a ring between 4 and 8 kiloparsecs from the Galactic center with an additional strong concentration in the inner region of the Galaxy (e.g., Scoville and Solomon 1975, Burton et al. 1975). In these two regions of the Galaxy the ISM is primarily molecular, while outside the solar circle it is primarily atomic. Although not perfect, CO also turned out to be a much better tracer of spiral structure than HI, revealing strong spiral features in the outer as well as inner Galaxy (e.g., Cohen et al. 1980; Dame et al. 1986). CO observations also enabled the rotation curve of the Galaxy to be determined well beyond the solar circle and thus provided a more accurate weighing of the Milky Way than previously possible (e.g., Blitz 1979). The resulting increase of about a factor of 2 in its mass led to the recognition of "... A bigger and better Milky Way..." as Bart Bok once put it. More recently, examination of the kinematics of molecular gas in the inner Galaxy has helped lead to the recognition that our Galaxy is likely a barred spiral (Liszt and Burton 1980, Binney et al. 1991), and this is something that can perhaps account for the absence of molecular gas between 1–4 kpc, a feature which is not present in the molecular component of most normal spiral galaxies. CO observations also revealed the very unusual nature of the ISM in the inner regions of the Galaxy, where GMCs reach masses in excess of $10^6$ $M_\odot$ and average densities two orders of magnitude higher than those characteristic of GMCs in molecular ring (e.g., Bally et al. 1988)!

The significance of CO studies for external galaxies lies in the fact that star formation drives galactic evolution. Star formation takes place in GMCs and CO is by far the best tracer of cold molecular gas that we have at our disposal. Over the last twenty years or so, CO observations have been successfully used to probe the cold component of hundreds of external galaxies (e.g., Young and Scoville 1991 and references therein). In nearby galaxies such as M31 and M33, interferometric and single dish observations have been able to resolve individual GMCs enabling estimates of their sizes, masses and densities. The molecular component of spiral arms has been very clearly traced out in at least one galaxy, M51. Such observations enable the investigation of such issues as GMC formation and destruction in a manner not possible for Milky Way clouds and may be the best way to investigate these problems. In more than 100 spiral galaxies the radial distribution molecular gas has been determined from CO mapping. Among the interesting results that have been found is the fact that the radial distributions of CO in most spiral galaxies have central peaks and then fall off monotonically with increasing radial distance from the center. This differs from the Milky Way which, as mentioned earlier, has a depression between 1-4 kpc and a ring-like appearance. On the other hand, these galaxies are similar to the Milky Way in that the distributions of HI and CO in them are distinctly different. The ratio of atomic-to-molecular hydrogen has also been measured in many galaxies and found to be dependent on galactic type. Although such determinations as well as the overall $H_2$ mass determinations for galaxies appear to still be plagued by the uncertainty in the conversion of CO luminosities to hydrogen masses.

Some of the most interesting results have come from studies of the nuclear regions of galaxies. CO observations have enabled measurements and in some cases mapping of the molecular contents of active nuclear regions of nearby galaxies such as M82 and NGC 1068 where similar to the nuclear region of our galaxy, gas densities are significantly higher than found in the GMCs of the disk. Particularly intriguing are the observations of ultraluminous infrared galaxies. Many of these systems appear to be interacting or merging galaxies and all appear to be rich in molecular gas which is highly concentrated in their nuclear regions. Moreover, evidence suggests that the mass of molecular material is comparable to the dynamical mass (Scoville et al. 1991) suggesting that whatever is being formed in these regions (e.g., the nuclear bulge of a merged galaxy) is in a very early stage of development. The detection of the lensed, ultraluminous galaxies F10214+4724 and the Cloverleaf seem to indicate that ultraluminous starbursts are important even at relatively large look back times (i.e., $z \approx 2$). Clearly, observations of the cold molecular component of external galaxies are telling us something about the evolution and perhaps even the formation of galaxies. What exactly that is remains to be determined by future research.

## 5. Concluding Remarks

The preceeding brief discussion of the impact on astronomy of 25 years of CO observations would take hundreds of additional references and many figures to properly document. Even so, it would be still far from comprehensive and would require many more pages of text to do justice to the wide range of exciting research which has resulted from the opening of the cold molecular universe to astronomical investigation a quarter century ago. The reader has only to skim the table of contents of this volume to appreciate both how much was left out of this short review and the true scope of present day research on this topic. However, I hope I have conveyed here some sense of the remarkable progress achieved as a result of the early exploration of the cold universe by CO observations. This conference marks not only the end of the first 25 years of CO astronomy but also the beginning of a new quarter century of vigorous exploration of a universe which has not yet revealed all its secrets. Young astronomers take heart, there is still much left to decipher.

## References


Adams. F.C., Lada, C.J. and Shu 1987, ApJ, 312, 788.

Bachiller, R. and Gomez-Gonzeles J. 1992, Astr.Ap.Rev., 3, 257.

Bally J. and Lada, C.J. 1983, ApJ, 265, 824.

Bally J., Stark, A. A., Wilson, R. W., and Henkel C. 1988, ApJ, 324, 223.

Barnard, E. E. 1919, ApJ, 49, 1.

Binney, J. J., Gerhard, O. E., Stark, A. A., Bally, J. and Uchida, K. I. 1991, MNRAS 252, 270.

Blitz, L. 1979, ApJL, 231, 115.

Blitz, L. 1980, in "Giant Molecular Clouds in the Galaxy", eds P.M. Solomon and M.G. Edmunds, (Pergamon: Oxford), p. 1.

Bok, B. J., Lawrence, R. S. and Menon T. K. 1955, PASP, 67, 108.

Burton, W.B., Gordon, M.A., Bania, T. and Lockman F.J. 1975, ApJ, 202, 30.

Cohen, R. S., Cong, H., Dame, T. M., and Thaddeus, P. 1980, ApJL, 239, 53.

Combes, F. 1991, ARAA, 29, 195.

Dame, T. M., Elmegreen, B. G., Cohen, R. S. and Thaddeus, P. 1986, ApJ, 305, 892.

Dufay, J. 1957 "Galactic Nebulae and Interstellar Matter," (Philosophical Library: New York).

Elmegreen, B. G. 1991 in "The Physics of Star Formation and Early Stellar Evolution", eds. C. J. Lada and N. Kylafis, (Kluwer: Dodrecht), p. 35.

Elmegreen, B. G. and Lada C. J. 1977, ApJ, 214, 725.

Field, G. H., Goldsmith, D.W. and Habing, H. 1969, ApJL, 155, 49.

Fukui, Y. et al. 1993, in "Protostars and Planets III", eds., E.H. Levy and J.I. Lunine, (Arizona: Tucson), p. 603.

Goldreich, P. and Kwan J. 1974, ApJ, 189, 441

Hartmann, J. F. 1904, ApJ, 19, 268.

Herbig, G. H. 1962, Advan. Astron. Ap., 1, 47.

Heiles, C. 1969, ApJ, 156, 493.

Hoyle, F. 1953, ApJ, 118, 53.

Lada, C.J. 1975, PhD Dissertation, Harvard University.

Lada, C.J. 1985, ARAA, 23, 267.

Loren, R.B. 1975, PhD Dissertation, University of Texas.
Lilley, A. E. 1955, ApJ, 121, 559.
Liszt, H. S. and Burton B. 1980, ApJ, 236, 779.
Penzias, A.A., Solomon, P.M., Wilson, R.W. and Jefferts, K.B. 1971, ApJL, 168, 53.
Plaskett, J. S. and Pearce, J. A. 1933, PDAO, 5, 169.
Scoville, N. Z., Sargent, A. I., Sanders, D. and Soifer, B. T. 1991, ApJL, 366, 5.
Scoville N. Z. and Solomon P. M. 1975, ApJL, 199, 105.
Shu, F. H., Lizano, S., Ruden, S. and Najita, J. 1988, ApJL, 328, 19.
Snell, R. L., Loren, R.B. and Plambeck R. 1980, ApJL, 239, 17.
Spitzer L. 1968, "Diffuse Matter in Space", (Interscience Publishers: New York).
Trumpler R. 1930, Lick Obs. Bull., 14, 154.
Tucker, K.D., Kutner, M.L. and Thaddeus 1973, ApJL, 186, 13.
Wilson, R.W., Jefferts, K.B. and Penzias A.A., 1970, ApJL, 161, 43.
Young, J. and Scoville, N. J. 1991, ARAA, 29, 581.

# POSTER ABSTRACTS

# POSTER ABSTRACTS

Only abstracts of the posters presented at the Symposium are included in these Proceedings because of space limitations. A companion volume, *IAU Symposium 170: Poster Presentations,* edited by Jeffery G. Mangum. Simon J. E. Radford, and William B. Latter (1996 NRAO), contains expanded versions of many of these presentations.

## Using Herbig-Haro Objects to Probe the Interclump Medium of Molecular Clouds

V. P. Alten (University of Colorado)

A recent observational survey of the Perseus molecular cloud complex has resulted in the discovery of a large number of new Herbig-Haro (HH) objects. This implies that the density of shock waves in star forming molecular clouds is much higher than previously thought. Outflows associated with some HH objects are many parsecs in length. It is possible that they pass through the molecular cloud interclump medium so regularly that the gas therein is reprocessed. The short time interval between the passage of subsequent shocks has implications for cloud support and the chemical ages of molecular clouds.

## The Chemical Evolution of an Interstellar Cloud Model

M. Y. Amin (Cairo University, Egypt) and M. S. El-Nawawy (UAE University, United Arab Emirates

The aim of the present work is to integrate the chemical rate equations and the hydrodynamic equations simultaneously, where we have used a one-dimensional program to simplify the treatment of contraction as that of a spherically symmetric cloud. The atomic, molecular, and ionic species included in our chemical scheme are altogether 104 species from the families



*W.B. Latter et al. (eds.), CO: Twenty-Five Years of Millimeter-Wave Spectroscopy,* 399–483.

of H, C, O, S, Si, Cl, metals and grains. The involved network contains 622 reactions including 111 reactions for negative ions. The rate coefficients of our system are updated with its most recent values. The central density increases from $n = 10$ cm$^{-3}$ in the diffuse initial cloud model to a dense cloud with central density number of $n \geq 10^5$ cm$^{-3}$ after a time of $1.2 \times 10^7$ years.

The results predict that the different molecular species increase in abundance as the contraction proceeds. The species which enhance significantly are CO, HCO, CS and NO. The fractional abundances of many of the other molecular species increase distinctly with contraction, e.g., CH, $C_2$, CN, $SO_2$, $CO_2$, $H_2O$, $C_2$, $NH_3$, HCN, SO, OCS, and SN. The transformation of the initial diffuse cloud model with small abundances of molecular species to a dense molecular cloud with enhancement of the different molecular species is confirmed. Our results are compared with both observations and some recent theoretical works. The comparison shows good agreement of our results with both the observations and the other theoretical studies. Molecules such as the hydrocarbons and related species increase in abundances distinctly. This is because there is a significant abundance of carbon atoms which drive the hydrocarbon chemistry.

## The Nitrogen/Oxygen Chemical Network: Observations of NO, HNO, $N_2O$, $NO_2$, and $NO^+$ Towards Sgr B2

A. J. Apponi (Arizona State University)

Unlike that of carbon or oxygen, the chemistry of interstellar nitrogen has yet to be understood even on a qualitative level. One problem is that the observed abundances of several simple nitrogen-bearing molecules such as $NH_3$ and $N_2H^+$ cannot be reproduced by the usual "early time" calculations. Another puzzle has been the supposed lack of compounds containing an N-O bond. Here we present recent observations towards the galactic center of N-bearing molecules in the nitrogen/oxygen chemical network that shed light on these problems. The species studied include NO, HNO, $N_2O$, $NO_2$, and $NO^+$. Recently, we have confirmed the presence of HNO in several molecular clouds, including the Sgr B2 complex, by observing transitions of this species at 2 and 1.2 mm. We also have detected $N_2O$, a new nitrogen-bearing species, in the Galactic center. In conjunction with past work on NO, these observations show that N-O bonds cannot be so rare as previously thought. We additionally have mapped NO and $N_2O$ emission across the Sgr B2 complex. Interestingly, NO, HNO, and $N_2O$ all appear to have peak abundances at the same position, Sgr B2N. Moreover, their abundances are all well-predicted by ion-molecule models, at early times, in contrast to other N-containing species. Such data suggest that nitrogen chemistry involving oxygen differs from that with hydrogen. Other chemical compounds participating in the N/O network are $NO^+$ and $NO_2$. We are conducting searches for these species in molecular clouds. The results of this work were reported.

## Balanced Finline Mixer in the Millimeter Wave Band

D. S. Armağan (Marmara Research Centre, Turkey)

Rapidly expanding activities in millimeter wave hardware developments have created an urgent need for broad-band mixers for the receivers used in radioastronomy, remote sensing, electronic warfare, surveillance, meteorology, radiometer, and communication systems. Significant improvements have been made in millimeter wave mixers recently. Integrated circuit technologies provide the advantages of low cost, light weight, and small size. These also have the potential of direct translation into monolithic circuits and even large-scale integration. The use of beamlead diodes in integrated circuit mixers avoids a mechanical diode contact in whisker-contacted mixers and, therefore, has better reliability and needs less assembly. This paper describes a balanced finline mixer at 34 GHz frequency. The mixer is the first stage of a radiometer which is aimed to investigate the vegetation canopy. Using the same design technique, it is possible to implement the mixer at the other millimeter wave frequencies. The mixer circuit is etched from a $4 \times 6$ cm Duroid-5880 substrate of 0.38 mm thickness and it is placed in a waveguide mount. The signal is fed to the diodes via a symmetrical unilateral finline. The local oscillator power reaches the diodes via symmetrical unilateral finline and transition from finline to coplanar line. This transition works in a similar way to that of a probe type transition from waveguide to coaxial line. For proper operation, a short is placed a quarter wavelength away from the transition, allowing some tuning of the LO input impedance. In a mixer, RF and LO signals produce $nf_{\mathrm{RF}} \pm mf_{\mathrm{LO}}$ products. The desired IF output is either $f_{\mathrm{RF}} + f_{\mathrm{LO}}$ or $f_{\mathrm{RF}} - f_{\mathrm{LO}}$ (upper and lower sidebands) and the IF filter is usually selected to pass only one of these outputs. If it is the desired $\mathrm{f_{RF}} - \mathrm{f_{LO}}$ output, usually the low-pass filter is designed at the IF port. For this finline mixer, the microstrip low-pass filter at 2.8 GHz is connected to the coplanar line. The combination of finline and coplanar line forms a frequency-independent 18° hybrid, giving the basis for this balanced mixer. For the mixer, commercially available GaAs Schottky-barrier beam-lead diodes were used. A first set-up was tested with

AEI diodes DC1306. These diodes have a high noise figure at 34 GHz frequency, so the DSB system noise figure was not good enough. The next step in this study will be to use the HSCH-9101 GaAs Schottky beam-lead mixer diodes.

## The Embedded Pre-main Sequence Population in NGC1333-S: $C^{18}O$ Observation

C. Aspin (Joint Astronomy Centre)

We present new JCMT $C^{18}O$ line data towards selected regions in the NGC1333-S molecular cloud/star formation complex. Combined with existing near-IR photometry, we study the relationship between molecular column density (and implied visual extinction) and similar values derived from de-reddening studies. We use the data to provide additional diagnostics for the extractions of the true embedded pre-main sequence stellar population from the distribution seen in the near-IR.

## Physical and Chemical Evolution of Dense Cores from Isotopic CO, $HCO^+$, $NH_3$, and $HC_3N$ Observations

Peter Barnes (Haystack Observatory), Phil Myers (Center for Astrophysics)

The complex comprising IC 5146 and associated clouds was studied with a variety of molecular line tracers to investigate the properties of the dense cores. Cores associated with the optical cluster (cluster cores) are significantly hotter and somewhat denser than cores far from the cluster (dark cores). The dark cores also have a fairly uniform excitation temperature. Where excitation temperatures are high, optical depths in these low-lying transitions are low, but total CO column densities are comparable in the different cores. Thus the main physical difference between the cluster and dark cores is that molecular populations have been excited to higher $J$ levels. $C^{18}O$ emission correlates well with IR extinction, $HCO^+$ emission correlates well with star-forming activity, HCO emission occurs only near the HII region, and $NH_3$ and $HC_3N$ emissions are brightest in the dark cores but are anticorrelated with each other. This anticorrelation is consistent with early-time chemical evolution. Strong fractionation of deuterium has occurred in the dark cores, and fractionation of $^{13}C$ in the less extincted parts of cores. The intrinsic $^{12}C/^{13}C$ ratio is somewhat less than the terrestrial value (~60 vs. 89), the double ratio $(^{16}O/^{18}O)/(^{12}C/^{13}C)$ is more than twice terrestrial (~12 vs. 5.5), and so the $^{16}O/^{18}O$ ratio is also somewhat larger than terrestrial (~700 vs. 500) in this complex. The warm cluster cores exhibit distinctly different chemical and physical properties from the cool dark cores, but there are also differences among the dark cores.

### CO in the Atmospheres of Pluto and Triton

Peter Barnes (Haystack Observatory)

Despite the fact that Pluto has not been visited by a spacecraft, an impressive amount of information has been deduced about its surface, atmosphere, and moon Charon, due in large measure to the mutual occultation events between the two bodies and Pluto's occultation of a star. The latter revealed that Pluto has a thin atmosphere, and subsequent work has suggested that it is composed mainly of nitrogen. In this and other respects, then, Pluto resembles Neptune's moon Triton, as revealed by the Voyager 2 encounter. In addition, both atmospheres contain CO as a minor constituent so, especially in the case of Pluto, CO could be used as a unique probe of atmospheric conditions. Here I give a set of predictions for CO emission from these atmospheres; these can be used, once CO has been detected using ground-based mm-wave radiotelescopes, to derive atmospheric abundances and other properties of these bodies.

### Potential Protostars in Cloud Cores: $H_2CO$ Observations of Serpens

Mary Barsony, Robert L. Hurt (UC Riverside), Alwyn Wootten (NRAO)

The Serpens cloud is an active star-forming site. The recent discovery in Serpens of submillimeter continuum sources lacking near-infrared counterparts (a defining characteristic of the youngest observable protostellar phase), has motivated our study of the gas properties of these unique sources. We have chosen $H_2CO$ (formaldehyde) as our molecular probe, because it can trace exceptionally dense gas more sensitively than most other commonly-used molecular species. In the four submm continuum sources which were observed in at least three $H_2CO$ transitions, we present evidence for warm, kinematically active regions surrounded by cooler, more quiescent gas. The temperatures of the embedded cores are warm, ranging from 40–190 K. The gas densities are also high, in the range of $n(H_2) \sim 10^6$–

$10^{6.5}$ cm$^{-3}$, in general agreement with densities derived from dust emission. Some of the objects exhibit strong self-reversal features which may indicate infall of the cooler outer layers of the core, and two of them possess exceptionally broad linewings which may be associated with molecular flows. Several of these sources may represent extremely young protostars that are in the process of accreting the bulk of their mass from their surrounding envelopes.

## Multitransition Submillimeter CO Spectra of Ultraluminous IRAS Galaxies

Dominic J. Benford, Min S. Yun, Todd R. Hunter, Peter M. Bryant, Thomas G. Phillips (California Institute of Technology)

We present submillimeter CO observations of 8 ultraluminous IRAS galaxies as preliminary results from a multitransition survey at the Caltech Submillimeter Observatory. To date, high signal-to-noise spectra have been obtained for all 8 galaxies in the $^{12}$CO $J = 2 \rightarrow 1$ and $J = 3 \rightarrow 2$ transitions. Also, we have recently detected the CO $4 \rightarrow 3$ and $6 \rightarrow 5$ transitions in some of these objects. Observations of the higher frequency lines have been limited by the AOS backend bandwidth (500 MHz = 220 km s$^{-1}$ at CO $6 \rightarrow 5$). As of July 1995, we have begun to extend the high frequency survey using a new AOS that covers the entire 1 GHz receiver bandwidth. We are in the process of preparing an LVG analysis of the multitransition data to better constrain the temperature and density of the molecular gas in these luminous galaxies.

## A Comparative Chemical Study of Three Giant Molecular Cloud Cores

E. Bergin (FCRAO, University of Massachusetts)

We discuss the results of a comparative chemical survey of three cloud cores in Orion A, M17, and Cepheus A. The observations were obtained using the FCRAO 15-element array to map the emission of 21 molecules and isotopic variants in $>$ 100 positions in each source. To ascertain if differences in emission are due to abundance variations, we have developed a model of the physical conditions (density and temperature) that determine the molecular excitation. The densities and temperatures are used to derive accurate column densities for each source. The relationship between the physical and chemical properties for each source will be discussed as well as a discussion of the differences and similarities in the chemical abundances between the three sources. This research is a systematic effort to examine whether current gas phase chemical models can account for observed abundances in GMC cores and, in cases where it fails, to examine other possibilities.

## Molecular Oxygen: A New Astrophysical Maser?

P. Bergman (Steward Observatory, University of Arizona)

New excitation calculations for molecular oxygen, which predict maser emission for plausible interstellar conditions, are reported. Collisional rate coefficients for the $O_2$–He collision system were calculated for several kinetic energies in the range 20 – 500 K. Excitation calculations have been performed for a typical dark cloud and the core and envelope parts of a giant molecular cloud. It is found that for hot ($T_{\rm kin} \geq 200$ K), dense ($n(H_2) \approx 10^5$ cm$^{-3}$) interstellar clouds the $N_J = N_{N-1}$ transitions ($1 < N \leq 23$) of $O_2$ may become significantly inverted, provided that the $O_2$ abundance (per velocity gradient) is high, $X[O_2]/(dv/dr) > 10^{-4}$ (km s$^{-1}$ pc$^{-1}$)$^{-1}$. All these transitions occur at frequencies between 54 and 63 GHz. The physical conditions needed for the masers to exist are similar to, but less severe than, what is required for the 22 GHz water maser line.

## CO in the 30 Dor Region of the Large Magellanic Cloud

R. S. Booth (Onsala Space Observatory)

Within the SEST Key Programme an area of approximately $20' \times 23'$, centered on the 30 Dor nebula, has been fully sampled in the $^{12}$CO (1–0) line. More than 20 clouds are discernible. A sample of seven clouds have been further investigated in the (2–1) transitions of $^{12}$CO and $^{13}$CO. Their physical properties are discussed and compared with those observed in other areas of the LMC.

### A Highly Collimated Outflow in a Southern Bok Globule

Tyler Bourke (Center for Astrophysics & Dept. of Physics, ADFA, Univ. of NSW), Kimmo Lehtinen (Helsinki University Observatory)

We report the discovery of a highly collimated molecular outflow from the nearby (200pc) low mass young stellar object IRAS 11590–6452 ($L_\star \sim 10\ L_\odot$), located in the isolated southern Bok globule BHR 71 (Bourke et al. 1993, Proc. Astron. Soc. Aust., 10, 236; Bourke et al. 1996, MNRAS, 276, 1052. CO (1–0) observations imply a linear size of $\sim$ 0.5 pc and a mass of 0.5 $M_\odot$ for the outflow, with a degree of collimation of 5 : 1. In CO the location of peak emission is not located at the IRAS position, but in the outflow lobes, indicating probable temperature enhancements due to the interaction of shocks and the ambient gas. At these locations, strong emission from CS, $HCO^+$, and SiO is present, indicating the probable destruction of grains by the passage of the shocks. The widespread presence of vibrational $H_2$ emission supports this view of shock activity as the cause of these features. This outflow thus belongs to the select group of young outflows from low mass 'Class 0' objects showing a high degree of collimation and enhanced SiO emission.

### CO Observations of HII Galaxies

Elias Brinks (NRAO, P.O. Box O, Socorro, NM 87801), Christopher L. Taylor (NRAO/Socorro and Univ. of Minnesota, Minneapoils, MN 55455)

HII galaxies are gas rich, actively star–forming dwarf galaxies. We know very little about their molecular gas content. Still, they must harbour substantial amounts of $H_2$ as witnessed by their star formation activity. Several authors have attempted to detect CO in dwarfs (see e.g., Sage et al. 1992, A&A, 265, 19) mostly reporting upper limits or in some cases contradictory results. Several explanations have been put forward why HII galaxies have been so elusive. Because of their low heavy element abundance, the abundances of C and O are down. Moreover, as the dust–to–gas ratio is low, shielding of the UV radiation field will be less effective, suggesting that molecular clouds in irregular galaxies will be more compact than in the Galaxy and that the CO luminosity as a whole can be down by factors of at least 20.

In order to shed more light on this situation we have started a survey in the line of $^{12}CO$ with the NRAO 12 m telescope of a statistically well defined sample of HII galaxies (Taylor et al. 1995, ApJS, 99, 427). None of our targets were detected during our run in January 1995 at sensitivity levels substantially better than previous studies. One object, Haro 2, which was detected previously with the 30 m IRAM telescope was not seen by us. This could indicate that beam dilution plays a role and would suggest that the CO is indeed highly concentrated in small, dense clumps.

## A CS(2–1) Survey of IRAS Point-Sources with Color Characteristics of Ultra-Compact H II Regions

L. Bronfman (Universidad de Chile), L. Nyman (SEST-Onsala Space Observatory, Sweden), J. May (Universidad de Chile)

We have made a complete survey of the CS (2–1) emission toward IRAS point-like sources in the Galactic plane. The observed sources were selected according to their far infrared (FIR) colors, which are characteristic of UC H II regions. They have 25 $\mu$m/12 $\mu$m flux ratios larger than 3.7 and 60 $\mu$m/25 $\mu$m flux ratios larger than 19.3. The survey covered from $b = -2°$ to $b = 2°$ in the range $l = 300°$ through $l = 60°$, and from $b = -4°$ to $b = 4°$ elsewhere. The observations were made between 1989 and 1992 with the 15m SEST Telescope at La Silla, Chile, and with the OSO (Onsala Space Observatory) 20m Telescope, Sweden, in May 1993. The angular resolution at 98 GHz is of $50''$ for SEST and of $39''$ for OSO (main beam FWHM), similar to the IRAS survey angular resolution. Out of 1427 observed sources, 843 were detected. In most cases the observed CS(2–1) profile consisted of a single component, with the VLSR of a known molecular cloud in the line of sight. The IRAS point sources detected in the CS(2–1) line concentrate in the inner Galaxy, forming a "massive star" annulus similar to the molecular annulus that has been previously derived from CO large scale surveys of the Milky Way. The decrease that was then found in the peak molecular gas surface density from the first to the fourth quadrant is compensated by an increase in the FIR surface luminosity due to embedded massive stars.

## High Resolution CO Observations of Luminous IR Galaxies

Peter M. Bryant, N. Z. Scoville (Caltech)

We report $2'' - 3''$ resolution CO ($1 \rightarrow 0$) observations of six galaxies with infrared luminosities ranging from $3.2 \times 10^{11}$ to $3.5 \times 10^{12}$ $L_\odot$; we also report $0.8''$ CO ($2 \rightarrow 1$) observations of the most luminous of these objects, Mrk 231. In the five objects previously classified as mergers by their optical morphology, $10^{10}$ $M_\odot$ or more of molecular gas is contained within a radius of 1.3 kpc or less; in three of these, this gas is within a radius of 0.44 kpc. The only non-merger in our sample — NGC 7674, the larger member of a pair of interacting spirals — shows a massive CO bar of radius 5.6 kpc, roughly coincident with its optical bar in both length and position angle.

In all six objects, the derived nuclear molecular gas masses are comparable to or somewhat greater than the derived dynamical masses. The most serious discrepancy occurs for Mrk 231, where the derived gas mass within a 420 pc radius (0.5″) of the nucleus is over seven times larger than the dynamical mass. We have performed a detailed analysis of the behavior of the CO luminosity to gas mass conversion factor in a variety of geometries and physical conditions, and we conclude the molecular gas mass is unlikely to be overestimated by over a factor of seven in Mrk 231; thus, inclination effects are the probable cause of the discrepancy. This implies the mass of the core of Mrk 231 is dominated by molecular gas and the average gas surface density in the core is roughly $7 \times 10^4$ $M_\odot$ $pc^{-2}$ — over one hundred times greater than in the center of the Milky Way.

**Far-Infrared – CO Study of Star Formation in the Large Magellanic Cloud**

Douglas A. Caldwell, Marc L. Kutner (NRAO, Rensselaer Polytechnic Institute)

We present a study of star forming giant molecular clouds (GMC) in the LMC. We have mapped 21 GMCs in both far–infrared (FIR) and CO ($J=1 \rightarrow 0$). FIR observations are from the IRAS high resolution data set. CO observations were made on the SEST (15 m). The resolutions of both data sets correspond to a linear resolution of $\sim$ 10 pc at the LMC. FIR luminosities and virial masses of these clouds are comparable to Milky Way GMCs. The star formation activity, defined here as $L_{\rm FIR}/M_{\rm VIR}$ is comparable to that found for Milky Way clouds, and shows a similar large range for any given $M_{\rm VIR}$. GMCs in the LMC exhibit a range from little to no star formation to robust, massive star star formation, virtually independent of $M_{\rm VIR}$. Despite different physical conditions, some GMCs in the LMC have star formation activities equivalent to all but the most active Milky Way GMCs.

**Cosmic Evolution of CI, CII, and CO Luminosity of Galaxies**

C. Campbell (Department of Physics, Univ. of Colorado) and John Bally (Center for Astrophysics and Space Astronomy, Univ. of Colorado)

We calculate the gas-phase abundances of neutral carbon (CI), ionized carbon (CII), and carbon monoxide (CO) as functions of cosmic evolutionary time and redshift. In a galactic evolution model using low, intermediate, and high mass stars which includes both Type I and Type II supernovae, infall, and various star formation rates, the interstellar gas phase abundances of carbon and oxygen are computed as functions of the gas mass fraction. For models without infall, the gas phase abundance of carbon bearing species peaks when about 50% of the baryonic mass of a galaxy has been consumed by star formation. The fraction of gas-phase carbon in CI, CII, and CO phases is then determined from models of photon dominated regions. We include the attenuation of radiation by dust grains and self-shielding of CO to find the average column density of CII, CI, and CO. The luminosity of the CO lines is then estimated from an excitation model. Galactic evolution models with varying star formation rates, merging, and the effect of Hubble type are discussed. The analysis of the millimeter and sub-millimeter wavelength lines of all the carbon species may prove to be a powerful probe for studying galactic evolution and cosmology.

**CO at $z = 0.724$**

F. Casoli, P. Encrenaz, P. Boissé, B. Fort (Observatoire de Paris), Y. Mellier (Observatoire de Toulouse)

We present the probable detection of the CO(2–1) line in the giant arc in the cluster A370 with the IRAM 30m telescope. This very blue arc, 26″ in length, is the image magnified about 15 times of a $z = 0.724$ spiral galaxy located behind the cluster. Without magnification, the galaxy image would have a size of about 0.5″. The line is redshifted to 133.7 GHz and the total integration time was 8 hours. The estimated $H_2$ mass in this galaxy is a few times $10^9$ $M_\odot$. We have searched for CO in 3 other arcs (CL2236–04, CL2244–02, and A2218) but detected none. These objects are more distant so we had to observe high-$J$ lines, which are likely to be weak in normal spirals.

The two other objects detected at cosmological distances, IRAS 10214+4724 and the Cloverleaf, are also amplified by gravitational lensing. Thus, the "gravitational telescope" is likely the best way to study objects in the early Universe, either extreme or normal galaxies.

## The Structure of NGC 1333 from $^{13}$CO and C$^{18}$O Observations

A. Castets, S. Warin (Laboratoire d'Astrophysique, Observatoire de Grenoble), and W.D. Langer (Jet Propulsion Laboratory)

This paper presents a study of a dense molecular region, NGC 1333, and the relationship of the embedded infrared sources and young stellar objects to the structure of its molecular core. We use $^{13}$CO and C$^{18}$O $J = 1$–0 and $J = 2$–1 observations, along with those of CS $J = 2$–1 to characterize the structure of the core and to calculate densities and masses. We find from our C$^{18}$O maps that the core consists of a cavity surrounded by a compressed shell of gas. Many of the infrared and outflow sources in NGC 1333 are located in the cavity. The outflow winds from IRAS2, SSV 13, and several other sources appear to be responsible for creating the cavity and forming the shell. We suggest that sequential star formation is taking place in NGC 1333 with successive generations of young stellar objects providing the energy to expand the cavity, sweep up the material, and compress the shell.

## Interferometric Observations of Protostellar Outflows

L. Chernin (University of California, Berkeley)

We present recent high resolution ($3'' - 5''$) $^{12}$CO 1–0 interferometer images of the protostellar outflows in the Orion IRc 2, Lynds 1448, NGC 2071, and NGC 2024 regions. We discuss the structures in the individual outflow maps and make comparisions with recent outflow models. The Orion outflow, which is driven by a massive protostar(s), is best described as a wind driven shell, while the other outflows, which are from lower mass stars, appear to be driven by collimated jets. In the low mass outflows, the outflow lobes are conical near the driving star and in some cases there is no known infrared object at the apex of the cone. Thus, we find new evidence of a fundamental difference between outflows from low and high mass stars.

## Theoretical Maps of Stellar-wind-powered HII Regions

F. Comeron (Universidad de Barcelona)

HII regions often lie in zones with strong density gradients, such as the interfaces between the cloud and the intercloud medium in a molecular cloud. Numerical simulations in 2-D carried out in the past by other authors have outlined the process of expansion of the ionized gas when such density gradients are encountered, establishing the framework of the so-called "champagne model."

In this poster, we present the results obtained in a similar scenario when one takes into account the pressure of the shocked stellar wind from the ionizing star in addition to photoionization as a driving agent for the expansion of the HII region. It is shown how the dynamical evolution is modified by the action of a high-pressure, low-density gas inside the photoionized region. In addition, theoretical maps in optically thin wavelengths are presented to allow comparison with radio and infrared observations.

## CO Observations of Planetary Nebulae

Aditya Dayal, John H. Bieging (Steward Obs.)

We present NRAO 12-meter observations of CO in 3 planetary nebulae, including the first confirmed detection of a molecular outflow in K 3–35 and the first detection of $J$=1–0 emission from NGC 2440. We also report a ($\geq 5\sigma$) detection of $J$=1–0 emission towards IC 418. The $J$=2–1 spectrum towards K 3–35 shows a strong peak at $\sim$ 10 km s$^{-1}$ ($V_{\rm lsr}$) and $\sim$ 50 km s$^{-1}$ wide high velocity wings. The strong peak is narrow ($<$ 5 km s$^{-1}$) and is probably associated with a Galactic molecular cloud; the high velocity emission feature indicates an outflow velocity of about 25 km s$^{-1}$. The CO ($J$=1–0 and $J$=2–1) lines towards NGC 2440 are broad and double-peaked, and indicate an expansion velocity of $\sim$ 25 km s$^{-1}$. The $J$=1–0 emission towards IC 418 shows a parabolic profile which is indicative of optically thick emission; the expansion velocity is $\sim$ 30 km s$^{-1}$. The $J$=2–1 emission is undetected. This suggests that the gas has a fairly low excitation temperature and is consistent with an extended halo of emission, as is also seen in 21 cm observations of neutral H.

## Luminosity-to-Mass [$L_{\rm ir}/M({\rm H_2})$] Ratios within GMCs: NGC 7538 and W3

J. R. Deane, D. B. Sanders (Institute for Astronomy, Univ. of Hawaii), E. F. Ladd (Five College Radio Astronomy Observatory, Univ. of Massachusetts)

The star-forming properties of the NGC 7538 and W3 GMCs have been investigated using infrared and millimeter-wave molecular line data. Total molecular gas masses, $M(H_2)$, for W3 ($3 \times 10^5$ $M_\odot$) and NGC 7538 ($4 \times 10^5$ $M_\odot$) were determined from $^{12}CO(1 \rightarrow 0)$ and $^{13}CO(1 \rightarrow 0)$ images of each GMC. After careful subtraction of the Galactic plane far-infrared background emission from the IRAS Infrared Sky Survey Atlas (ISSA), total infrared luminosities, $L_{ir}(8$ - $1000$ $\mu m)$ for W3 ($2.3 \times 10^6$ $L_\odot$) and NGC 7538 ($1.1 \times 10^6$ $L_\odot$) were also determined.

The total cloud-integrated $L_{ir}/M(H_2)$ ratios (W3: 8 $L_\odot/M_\odot$; NGC 7538: 3 $L_\odot/M_\odot$) are similar to the mean ratios found for HII-region GMCs from previous surveys of the Milky Way. But when each GMC ($1.5° \times 1.5°$) is mapped at $6'$ resolution, the $L_{ir}/M(H_2)$ ratios vary by two orders of magnitude. In W3, the three most massive and active cores in the cloud contain 37% of the cloud mass but contribute 85% of the total luminosity ($L_{ir}/M(H_2) = 19$ $L_\odot/M_\odot$). However, most of the cloud, by mass (62%) and by area (75%), is comparatively quiescent ($L_{ir}/M(H_2) = 2.0$ $L_\odot/M_\odot$). In NGC 7538, the most massive cores, which comprise only 30% of the cloud area, contain 60% of the mass and contribute 73% of the total GMC luminosity ($L_{ir}/M(H_2) = 3$ $L_\odot/M_\odot$).

## $^{12}$CO (3–2) and (1–0) Emission Line Observations Of Nearby Starburst Galaxy Nuclei

Nick Devereux (New Mexico State University), Yoshi Taniguchi (Tohoku University), Dave Sanders (University of Hawaii), N. Nakai (Nobeyama Radio Observatory), J.S. Young (FCRAO/UMASS)

New measurements of the $^{12}$CO (1–0) and (3–2) line emission are presented for the nuclei of seven nearby starburst galaxies. The new observations capitalize on the coincidence between the beam size of the 45m Nobeyama telescope at 115 GHz and that of the 15m James Clerk Maxwell Telescope at 345 GHz to measure the value of the $^{12}$CO (3–2)/(1–0) emission line ratio in a $15''$ ($\leq 2.5$ kpc) diameter region centered on the nuclear starburst. In principle, the $^{12}$CO (3–2)/(1–0) emission line ratio provides a measure of temperature and optical depth for the $^{12}$CO gas. The weighted mean value of the $^{12}$CO (3–2)/(1–0) emission line ratio measured for the seven starburst galaxy nuclei is $0.64 \pm 0.06$. The $^{12}$CO (3–2)/(1–0) emission line ratio measured for the starburst galaxy nuclei is not as high as would be expected if the molecular gas were hot, $\geq 20$ K, and optically thin, $\tau \ll 1$. The total mass of molecular gas contained within the central region of the starburst galaxy nuclei ranges between $10^8$ and $10^9$ $M_\odot$. While substantial, the molecular gas mass represents only a small percentage, $\sim 9$ – 16%, of the dynamical mass in the same region. Details in AJ, 107, 2006.

## DR21: Highlights over 25 Years

Hélène R. Dickel, Douglas A. Roberts (University of Illinois), W. Miller Goss (National Radio Astronomy Observatory - Very Large Array)

CO observations of the DR 21 region in 1973–4 with the NRAO 36 foot radio telescope at a resolution of 1′ revealed a giant molecular cloud associated with this continuum source and large CO line wings toward the continuum peak (Dickel *et al.* 1978). CO observations at 15″ resolution with the Nobeyama 45 m telescope (Garden *et al.* 1991) clearly show this to be the most massive and energetic young stellar bipolar outflow in our Galaxy. Recently we observed the HI absorption with the VLA at 5″ resolution. DR21 exhibits one of the widest HI absorption lines in our Galaxy. The continuum emission of DR21 is nearly totally absorbed with $\tau > 6$ over about 20 km s$^{-1}$; the line FWHM $\sim$ 35 km s$^{-1}$. At least three components are present: two narrow ones at $v$(lsr) = 5 km s$^{-1}$ and 10 km s$^{-1}$ and a negative-velocity wing whose opacity can be measured out to $v$(lsr) = -30 km s$^{-1}$ in some places. No positive velocity wing is detected. The negative-velocity wing is detected over most of the HII region core except in the very southwest corner, near the position of the DR 21 $H_2O$ maser. The HI gas appears to be accelerating from this SW position which is therefore taken as the **origin** of the bipolar outflow. Measurements of Zeeman splitting of the HI line reveal a magnetic field of several hundred microgauss in this outflowing gas close to the molecular core; no magnetic field ($B < 50\mu$G) is detected in the narrow line component at $v$(lsr) = 10 km s$^{-1}$.

## Multitransition Observations of Molecular Clouds in the Far Outer Galaxy

Seth W. Digel (NASA/Goddard Space Flight Center, Hughes STX), Eugéne J. de Geus (California Institute of Technology), Christian Henkel (Max-Planck-Institute für Radioastronomie), Susanne Hüttemeister, Patrick Thaddeus (Harvard-Smithsonian Center for Astrophysics)

Millimeter-wave observations of three molecular clouds in the far outer Galaxy, all with apparent Galactocentric distances $R > 20$ kpc, are presented, along with a preliminary analysis of their properties. These clouds, discovered in a survey of H I concentrations in the far outer Galaxy with the CfA 1.2 m telescope, are among the farthest from the Galactic center of any known. Although they are near or beyond the edge of the star-forming disk, one has been found to contain an H II region. The extensive CO maps and observations of lines of $^{13}$CO and CS with the NRAO 12 m and IRAM 30 m telescopes presented here indicate that all of the clouds are clumpy at 1–2 pc scales and the dense clumps in the star-forming cloud are distinctly different from those in the other two. The cores of this cloud only are detected in CS. These cores are also apparently warmer and more massive than the others, although the latter distinction may not be significant if the cloud is appreciably closer than its kinematic distance, as a recent spectroscopic study of the star exciting the H II region indicates.

## Dust Driven Winds of Pulsating Red Giants

Ernst A. Dorfi, Susanne Höfner (Institut für Astronomie, A-1180 Wien, Austria)

From numerical calculations of dust driven winds from carbon-rich long period variables (LPVs), it has become clear that both a full treatment of the coupled system of gas and radiation and also time-dependent dust formation and growth is essential to describe the properties of this accelerated and expanding stellar outflow. The stellar pulsation causes shock waves to propagate outwards through the stellar envelope and to lift the stellar atmosphere. The density enhancements in the shock waves followed by a radiative cooling zone provide the physical conditions for an effective nucleation and growth of dust particles. The CO molecule plays a crucial role for these winds since the less abundant element of carbon or oxygen is bound in the CO molecule. This process determines which chemical species are available for the dust condensation. In order to determine the temperature and density structure controlling the dust formation process, the coupled system of radiation, dust, and gas dynamics is solved and provides, e.g., also the modulation of the radiation field due to the compressive motions near the dust-forming region. The mass loss rates and outflow velocities obtained are in agreement with the observed values. The zones of dust formation and growth can be traced in various frequencies leading to a rather complicated geometry within the radiation field because regions of large optical depths alternate with almost transparent zones depending on the grain size and the optical properties of the dust particles.

**Compact Objects and Dense Clouds Association**

P. Durouchoux (CE-Saclay-DAPNIA-SAP)

Some compact objects, emitting in the gamma-ray range, seem not to be associated with a binary system. The alternative for the accretion is a dust cloud which feeds the disk. We review recent observations conducted at the mm wavelengths that support this idea.

**The Evolution of Charged Particles in a Contracting Interstellar Cloud**

M. S. El-Nawawy (UAE University, United Arab Emirates), M. Y. Amin (Cairo University, Egypt)

The objective of the present work is to study the evolution of charged particles with time of contraction. This work is part of the calculations carried out on a model of a contracting interstellar cloud. In the first paper by Amin and El-Nawawy (appearing in this volume), we have discussed the results on the neutral species. In this paper, we follow the evolution of the charged particles during contraction. Special care is given to the details of the negative ions. In addition, we have tested the ambipolar diffusion according to the results of the ion density. We have given attention to the chemical reactions of negative ions in three models in steady state before including them in the network of the contracting model.

The results predict the importance of atomic ions in the diffuse regions. $H^+$ and $C^+$ are distinctly enhanced in the beginning of contraction but decrease as contraction proceeds. Molecular ions enhance as contraction proceeds and becomes important in dense regions. The most enhanced molecular ions are $HCO^+$, $O_2$, $C_2H_3$, and $SO^+$. $H_3^+$ is less abundant. The atomic ions (except metallic ions) decrease noticeably as density increases. In general, the negative ions are of negligible fractional abundances. It has also been found that the time of ambipolar diffusion is shorter than the dynamical time, hence, the magnetic field should be weakened in the central core as the central density increases to $n > 10^4$ cm$^{-3}$.

**Search for CO Emission in High Redshift Powerful Radio Galaxies**
Aaron S. Evans (Institute for Astronomy, University of Hawaii)

We present upper limits to the molecular gas masses of 11 high redshift ($1 < z < 4$) powerful radio galaxies (HzPRGs). These observations include some of the most sensitive extragalactic CO observations to date. Our four best studied radio galaxies (3C 368, 4C 23.56, B2 0902+34, and 4C 41.17) have less than one-third the CO luminosity of the hyperluminous infrared galaxy *IRAS* F10214+4724 and, assuming a galactic CO-to-$H_2$ conversion factor, they have molecular gas masses comparable to or less than the most gas-rich, ultraluminous infrared galaxies (ULIGs) in the local ($z \leq 0.3$) Universe (i.e., $M(H_2) < 3 \times 10^{10} h^{-2} M_\odot$; $H_0 = 100h$ km s$^{-1}$ Mpc$^{-1}$ and $q_0 = 0.5$). Spectral energy distributions of 8 HzPRGs show they have UV-to-optical luminosities ranging from $\sim 0.1$ to 1.0 times that of F10214+4724. Recent submillimeter observations of 3 HzPRGs suggest that one source, 4C 41.17, may have an intense submillimeter-far infrared excess. However, coadded *IRAS* observations suggest that HzPRGs as a class are not hyperluminous at far-infrared wavelengths. We also consider how the possibility that F10214+4724 is a gravitationally lensed galaxy affects our interpretation of the data, as well as what it may mean for gas-rich galaxies in general.

**The Bipolar Outflow NGC2264G**
Michel Fich (University of Waterloo), Charles J. Lada (Smithsonian Astrophysical Observatory)

Detailed observations in CO(2–1) have been obtained of the highly collimated bipolar outflow NGC2264G. A complete map with 20 arcsec resolution shows rich structural detail in both the spatial morphology and the velocity field of the outflow. The blue outflow lobe extends considerably further than realized by previous authors. The blue lobe appears to interact with a molecular cloud in the foreground that is kinematically distinct from the cloud where the outflow originates. The low velocity gas in the red lobe is contained in a large structure that appears similar to a bow shock. Despite striking differences between the two lobes, essentially every point in either lobe is described by the same kinematic time scale.

Both lobes show an elongated structure at the highest velocities, extending beyond the lower velocity gas and unresolved in the direction perpendicular to the flow direction. These structures may be the molecular component of the jet driving this outflow. The low and intermediate outflowing material exhibits a power-law variation of mass with velocity with index of −1.7, similar to what is seen in several other similar outflows. However, the highest velocity gas that looks like a jet in the position-position plots has a steeper power law variation with the identical index in both lobes.

New estimates of the mass, momentum, and energetics of the outflow indicate that a very high efficiency is required for the source driving the outflow.

## Dynamical Models of Circumstellar Dust Shells around Long–period Variables: The Effect of Multiperiodicity

Axel J. Fleischer, Jan Martin Winters, Erwin Sedlmayr (Institut für Astron. & Astrophys., TU Berlin), Andreas H. Gauger (Los Alamos National Laboratory))

We present dynamical models of circumstellar dust shells around carbon–rich long-period variables which include time-dependent hydrodynamics and a detailed treatment of dust formation, growth and evaporation as well as a treatment of chemistry and radiative transfer.

Important effects due to the complex, nonlinear interaction between the dynamics of the pulsating atmosphere and the dust complex are demonstrated. It is shown that in a certain range of the parameter space the dust shell develops its own time scale which is different from the time scale of the interior pulsation. The key parameter which tunes this eigenfrequency of the dust shell is the overabundance of carbon to oxygen. For specific values of the carbon overabundance, the eigenfrequency of the dust shell is an integer multiple of the pulsational period; for other values, irregular or chaotic behavior results from the interaction of the pulsation and the circumstellar shell.

This work has been supported by the DFG (Grant Se 420/8–1), the BMBF (Grant 2BT12A), and the DOE.

## CO(1-0) Emission at $z = 0.437$ toward 3C 196

D. T. Frayer, R. L. Brown (NRAO, Charlottesville)

We report on CO(1–0) observations at a redshift of $z = 0.437$ corresponding to the 21 cm absorption line system toward the background QSO 3C 196. We have systematically searched several positions at and around the QSO line of sight and have found evidence for CO emission at two positions. At 1 arcmin south of 3C 196, we detect CO(1–0) emission representing approximately $10^{11}$ solar masses of molecular gas. We also observe weaker CO emission at the QSO position, but we are currently unable to distinguish between the possibilities that this emission arises from the source which is approximately 1 arcmin south of the 3C 196 or that the CO emission arises from two or more emission regions spread over 200 $h^{-1}$ kpc.

**CO Images of IR-luminous Ring Galaxies: Arp 118 and Arp 119**

Yu Gao, P.M. Solomon (Astronomy Program, SUNY Stony Brook), D. Downes (IRAM, Grenoble), Min Su Yun (OVRO, Caltech), S.J.E. Radford (NRAO, Tucson)

We present high resolution CO images of Arp 118 and Arp 119 obtained with the IRAM and OVRO mm interferometers, respectively. Both systems are ring-like interacting/merging galaxies with extremely broad CO/HI linewidths of FWZI $\sim$ 1,000 km s$^{-1}$and high molecular masses of 2–5$\times 10^{10}$ M$_{\odot}$. In both systems, most of the molecular gas is associated with the "rings/arcs" or "spiral arms" extending up to $\sim$ 15 kpc around the Seyfert nuclei. Arp 118 shows no nuclear molecular gas and even its strongest extranuclear association, a Giant HII Region that is the dominant 10 $\mu$m source, contributes only $\sim$ 15% of the total CO emission. Arp 119 has only $\sim$ 40% of the CO emission from the nuclear region. This is in sharp contrast with most IR-luminous galaxies, which have high nuclear molecular gas concentrations.

Both ring galaxies have high-mass-star formation efficiencies, $L_{\rm IR}/M(\rm H_2) \sim$ 5 L$_{\odot}$/M$_{\odot}$, comparable to the Milky Way's and much lower than those for most IR-luminous interacting galaxies. Arp 119 may be explained as due to a pole-on collision between a gas-rich disk spiral and an elliptical. Arp 118 either involves a collision of an elliptical with an extremely massive spiral of $> 10^{12}$ M$_{\odot}$ or a merger of two disk spirals followed by a collision with an elliptical. Its molecular morphology demonstrates that not all mergers lead to gas infall into the center. There is no connection between the molecular gas–starburst region and the AGN in this galaxy system.

**Dense Molecular Gas and High Mass Star Formation in Galaxies**

Yu Gao & Phil M. Solomon (Astronomy Program, SUNY at Stony Brook)

We have carried out systematic HCN (1–0) observations (complementary with CO observations) of a sample of $\sim$ 40 IR/CO bright/luminous galaxies. HCN luminosity, $L_{\rm HCN}$, is a measure of the mass of dense molecular gas and the ratio $L_{\rm HCN}/L_{\rm CO}$ is an indicator of the fraction of dense molecular gas. The principal goal is to compare the dense molecular gas content with the OB star formation rate traced by far infrared emission. We show that the galaxies observed here follow the correlations between $L_{\rm IR}$ and $L_{\rm HCN}$ and between $L_{\rm IR}/L_{\rm CO}$ and $L_{\rm HCN}/L_{\rm CO}$ proposed by Solomon et al. (1992). In particular, the $L_{\rm IR}/L_{\rm HCN}$ ratio is independent of the infrared luminosity, whereas $L_{\rm IR}/L_{\rm CO}$ increases with IR luminosity. This suggests the star formation rate depends upon the amount of dense molecular gas, not the total molecular content measured by CO emission. For a few nearby galaxies, we present HCN maps along the major axes so we can determine the total HCN luminosity, map the distribution with radius of HCN emission, and obtain the surface brightness ratio $I_{\rm HCN}/I_{\rm CO}$. Dense molecular gas is not confined to $\sim$ 1 kpc nuclear regions. A significant fraction of dense molecular gas is distributed in the inner disks outside the nuclear or inner ring starburst regions, and can be detected to radii as large as a few kpc, perhaps to diameters of $\sim D_{25}/4$. The starburst cores have the highest fraction of dense molecular gas with $I_{\rm HCN}/I_{\rm CO}$ ratios comparable to those of ultraluminous IR galaxies, whereas this ratio falls off in the galaxy inner disks at larger radii to be comparable to those of normal galaxies.

**Observed and Synthetic CO Line Profiles and the Structure of the IRC+10216 Circumstellar Dust Shell**

A. Gauger, J.J. Keady (Los Alamos National Laboratory, Theoretical Division), J.M. Winters, A.J. Fleischer, E. Sedlmayr (Institut für Astronomie und Astrophysik, Technische Universität Berlin)

High resolution infrared spectroscopy of molecular absorption lines, especially of CO, in circumstellar dust shells around AGB stars, such as the prominent carbon-rich Mira IRC+10216, provides a powerful diagnostic probe able to reveal details of the outflow from the photosphere out to more than a hundred stellar radii.

We have performed non-LTE radiative transfer calculations for CO fundamental and $\nu = 2 - 0$ overtone lines using the structure of a dynamical model for the dust shell around IRC+10216, which results from a consistent treatment of hydro- and thermodynamics, radiative transfer, chemistry and carbon grain formation. Compared to high resolution CO line spectra obtained at the KPNO Mayall 4m telescope, the P–Cygni emission components and the position of the absorption cores of the fundamental lines agree well, implying that the model reproduces the outflow velocity and the dust optical depth of IRC+10216. However, the calculated fundamental line absorption cores are broader and the first overtone line strengths are too high, indicating that the density and thus the mass loss rate of the model is too high and should be closer to the value of $\approx 10^{-5}$ $M_\odot$/yr given for IRC+10216 in the literature.

Research at LANL has been performed under the auspices of the U.S. Department of Energy.

## CO Observations of Molecular Gas around a Giant HII Region in M101

J. Giannakopoulou and M. Fich (University of Waterloo, Canada)

Using the James Clerk Maxwell Telescope, we observed two rotational transitions of CO (J=2–1 and J=3–2) towards NGC 5461, the brightest giant HII region in the M101 galaxy. The peak temperature of the CO J=2–1 emission line at the centre of the HII region is a factor of two stronger than the corresponding peak seen in the previously observed J=1–0 line. Furthermore, it is intriguing that no CO J=3–2 signals are observed from two positions directly southeast and northwest from the centre of the HII region, even though CO J=2–1 emission has been detected there. The observed large change in the ratio of intensities of the two lines could indicate significant temperature variations. Another interesting feature in the lines is their double-peaked structure. The velocity peaks appear to be spatially separated. Finally, we obtained a marginal detection of $^{13}$CO J=2–1.

## Cold Dust Emission from the Spiral Arms of M 51: A Reappraisal of the X Conversion Factor

M. Guélin (IRAM), R. Zylka, P.G. Mezger, C.G.T. Haslam, and E. Kreysa (MPIfR)

Using the IRAM 30 m telescope equipped with the MPIfR 7 channel bolometer array, we have mapped the $\lambda$ 1.2 mm continuum emission of the inner part of M 51. This emission traces the central molecular disk and the inner spiral arms and correlates remarkably well with the $^{12}$CO (2–1) line emission. The $\lambda$ 1.2 mm continuum to CO line intensity ratio, averaged over the 60 GHz wide bolometer band, is 6. The $\lambda$ 1.2 mm emission arises mostly from cold dust ($T_d \leq 20$K) associated with molecular clouds. Using the local Galactic value of the dust absorption cross section, corrected for the larger metallicity observed near M 51's center, we derive from this emission the gas distribution. The derived molecular masses are 3–5 times smaller than those obtained from the CO luminosity using Strong et al.'s Galactic CO to $H_2$ conversion factor, but are close to the masses estimated from $^{13}$CO and from visual extinction. Considering the approximations inherent to each method, the molecular masses derived from dust, CO and $A_V$ can still be considered as consistent. We wish to stress, however, that the actual gas mass of M 51 may very well be a factor of 4 lower than usually assumed. A similar conclusion was reached for NGC 891, another nearby spiral whose cold dust emission was mapped with the IRAM 30 m telescope (Guélin et al. A&A, 279, L37). It probably also applies to many other galaxies.

**The Twin 60 cm CO (J=2–1) Survey Telescopes**

Toshihiro Handa, Tetsuo Hasegawa, Seiichi Sakamoto, Tomoharu Oka, Masumichi Seta, Kumiko Sato, Kazuo Sorai, Jun Ichi Morino (Univ. of Tokyo), Hiroko Shinnaga (Tokyo Gakugei Univ.), Masahiko Hayashi (National Astron. Obs., Japan), L. -ÅNyman (Swedish ESO Submillimetre Telescope), R. Booth (Onsala Space Obs.), P. Shaver (European Southern Obs.), L. Bronfman (Univ. of Chile)

We have constructed twin 60 cm radio telescopes in order to make a survey of the Galactic plane in the CO (J=2–1) line. The scientific targets of the project are to figure out the large-scale density distribution of molecular gas in the Galaxy and the large-scale density structure of giant molecular clouds. These basic observational studies will allow us to establish a link between Galactic and extra-galactic multi-line studies. The telescopes are located at Nobeyama, Japan and at La Silla, Chile. They have the same basic structure and are originaly designed for our survey project. The aperture size is 60 cm and beamsize is 9 arcmin, which is well designed to make direct comparisons with the Columbia Survey (Dame et al. 1987). The optics are offset Cassegrain coudé with 6 mirrors, which gives us a high beam efficiency of 0.93. The receivers are SIS with $T_{\rm sys}$ = 1100 K(SSB) and cooled Schottky type with $T_{\rm sys}$ = 700 K(DSB). The spectrometers are AOS with 0.25 and 1 GHz bandwidth. The telescopes are controlled by the original system named ASTROS - Automated Small Telescope Remote Operation System, which is designed based upon the distributed processing concept.

### A Protoplanetary Gas Disk Search around T Tau Stars – Detection of the CO Emission around DM Tauri

Toshihiro Handa (Univ. of Tokyo), Shoken M. Miyama, Takuya Yamashita (National Astron. Obs., Japan), Toshihiro Omodaka, Yoshimi Kitamura (Kagashima Univ.), Masahiko Hayashi (U. Tokyo), Toshikazu Onishi (Nagoya Univ.), Ronald. L. Snell (Univ. of Mass., FCRAO), Stephen E. Strom, Karen M. Strom, Micheal F. Skrutskie (U. Mass.), Suzan Edwards (Smith College), Nagayoshi Ohashi, Kazuyoshi Sunada, Masao Saito (Nobeyama Radio Obs.), Yasuo Fukui, Akira Mizuno (Nagoya U.), Jun-ichi Watanabe (NAO, Japan), Hirokazu Kataza (U. Tokyo)

Sensitive molecular line observations with the Nobeyama 45 m telescope have resulted in the detection of the CO and $^{13}$CO emission centered on the young, classical T Tauri star, DM Tau. No C$^{18}$O emission was detected. The systemic velocity of the emission coincides well with the reported radial velocity of the star. The $^{13}$CO profile shows a symmetric double peak, suggesting a Keplerian rotating disk. The derived radius of the disk is about 1000 AU. The mass is between $7 \times 10^{-4}\ M_\odot$ and $1 \times 10^{-3}\ M_\odot$ from the $^{13}$CO intensity and the upper limit of C$^{18}$O intensity. Our results suggest that molecular gas is depleted from the standard gas-to-dust mass ratio or that mass distribution in the disk is steeper than the standard law, $r^{-1.5}$, although our observations cannot detect emission from an optically thick gaseous component inward of $r < 100$ AU, if the emission comes from a Keplerian rotating disk.

### Evidence for Selective Photodissociation of CO and an Enhanced C/O Elemental Ratio in the Gas Phase in NGC 253

Andrew Harrison, Peter Brand (University of Edinburgh), Adrian Russell, Phil Puxley (Royal Observatory of Edinburgh)

In the central ~300 pc of the circum-nuclear region of the starburst galaxy NGC 253, the bulk of the molecular gas is hot ($T$ ~100 K) and dense ($n > 10^4$ cm$^{-3}$) and the chemistry of this gas is dominated by the large UV field from newly born massive stars. Furthermore, in NGC 253:

- standard models of UV photon dominated chemistry can only account for <5% of the observed [CI] emission.
- there is a high C/O elemental ratio in the gas phase.
- C$^{18}$O is a poor tracer of the bulk of the molecular gas.
- selective photodissociation of CO and its isotopomers is efficient.

## A Search for High-Latitude CO

Dap Hartmann (Harvard-Smithsonian Center for Astrophysics), Loris Magnani (University of Georgia), Patrick Thaddeus (Harvard-Smithsonian CfA)

To determine the quantity of molecular gas at high Galactic latitudes, we are surveying CO (1–0) emission at $|b| \geq 30^\circ$. Using the CfA 1.2-meter telescope (beam = $8.7'$ we sampled the sky (at $\delta \geq -17^\circ\!.5$) at these high latitudes, on a $1^\circ \times 1^\circ$ (locally Cartesian) grid, to a sensitivity ($3\sigma$) of 0.3 K. The spectra are observed in position-switching mode, where the *off* position is a grid point 1° closer to the nearest Galactic pole. The planned grid has 15,650 points, 44% of which have currently been observed. We have made 53 detections of which some appear to be towards directions never before observed. The grid in the region bounded by $30^\circ \leq l \leq 210^\circ$ and $b \geq +30^\circ$ was fully completed; it comprises 37% of the planned grid. We detected CO emission in 15 of the 5742 directions in this part of the sky, amounting to a detection rate of 0.26%. No detections were found for latitudes higher than 43°. Monte Carlo simulations of a random distribution of clouds, ranging in size from 0.02 to 2.0 square deg, indicate that our search would typically detect 73.6±4.5% of all clouds. The undetected clouds fall in the gaps of our sampling pattern. We have only just started sampling the grid at $b \leq -30^\circ$, but the results so far are encouraging. Already we have newly detected CO emission in about a half dozen directions, although it remains to be seen whether this emission originates from large cloud complexes similar to the Polaris Flare in the northern Galactic hemisphere. The region around MBM 54–56 seems to be very rich in CO emission.

## Constraints on the Physical and Chemical Conditions of the Molecular Gas in the Magellanic Clouds

Arto Heikkilä, Lars E.B. Johansson (Onsala Space Observatory), Hans Olofsson (Stockholm Observatory)

Physical and chemical conditions of the molecular gas associated with the HII region complex N159 ($\alpha(1950) = 05^h40^m03.0^s$, $\delta(1950) = -69^\circ47'03''$) in the Large Magellanic Cloud and the IRAS source LIRS 49 ($\alpha(1950) = 00^h46^m32.9^s$, $\delta(1950) = -73^\circ21'50''$) in the Small Magellanic Cloud were derived from results of mm-wave observations of isotopomers of CO, CS and $HCO^+$ with the SEST by performing statistical equilibrium excitation and radiative transfer calculations. The radiative transfer was treated in the mean escape probability approximation.

Two sets of solutions were found: $T_k = 10 - 25$ K, $n(H_2) = 10^4 - 10^6$ cm$^{-3}$ and $T_k \geq 30$ K, $n(H_2) \approx 3000$ cm$^{-3}$. Both sets are appropriate for each of the two studied regions.

The derived column densities of all studied molecules are about a factor of 10 lower in LIRS 49 compared to N159, where they are another factor of 10 lower than in Galactic clouds (see Johansson et al. 1994, A&A 291, 89).

**The IRAM Key Project: Small-scale Structure of Pre-star Forming Regions Preliminary Analysis of the Structure**

A. Heithausen, J. Stutzki (Universität zu Köln), E. Falgarone, J.-F. Panis (Ecole Normale Supérieure, Paris), J.-L. Puget (Université de Paris), and F. Bensch (Universität zu Köln)

Clumpiness is a general property of molecular clouds and is expected to play a key role in the triggering of gravitational instability. The structure found in the three fields selected for the IRAM key project (see their description in the companion poster by Panis et al.) has been decomposed into individual clumps using the Gaussian clump fitting procedure by Stutzki & Güsten (1990, ApJ 356, 513). We analyse the mass spectra and size-line width relation of these clumps and compare the results obtained in each of the three CO isotopes and in each of the fields, which, at large scale, do not present the same level of gravitational binding.

**The Dynamical State of High-latitude Molecular Clouds**

A. Heithausen (Physikalisches Institut der Universität zu Köln)

The size-line width relation and the size-density relation are analyzed for the molecular clouds from the CfA 1.2m telescope CO survey at high latitudes in the second quadrant. It is found that these clouds follow similar relations as giant molecular clouds. It is also shown that the virial mass is correlated to the CO luminosity, although the high-latitude clouds are not gravitationally bound. The influence of the atomic gas surrounding the high-latitude clouds is analyzed, to study if the clouds are pressure bound.

**CO in AGB Stars with Interrupted Mass-Loss**

Kenneth Hinkle (NOAO), H. Alwyn Wootten (NRAO), Raghvendra Sahai (JPL)

We have surveyed the CO $J$=1–0 emission (using the NRAO 12-m) in an IRAS-selected sample of "60 micron excess" AGB stars, in order to look for evidence of interrupted mass-loss. In prototype sources described in the literature (e.g. Olofsson et al. 1988, AA, 196, L1), millimeter wave CO line shapes (doubly-peaked) and maps show the presence of a cold extended hollow shell with radius of a few times $10^{17}$ cm, implying that mass-loss ceased in these stars a few times 1000 years ago. It has been suggested that mass-loss has been interrupted in these stars by a He-shell flash. For a few stars in our CO $J$=1–0 survey, we have obtained high resolution spectra of the 4.6 micron CO vibration-rotation fundamental (using the KPNO 4 m FTS). We have used the circumstellar components of the 1–0 lines as a complementary approach to the microwave spectra. By combining both techniques on spectra of the prototype source S Sct, we are able to derive excitation temperatures, velocities, and mass loss rates. These show that the hollow shell model is correct. The excitation temperature of the hollow, outer shell is 10±5 K. The inner circumstellar shell consists of two regions: one at 75 K expanding by 3 km $s^{-1}$ (relative to the stellar center-of-mass velocity) and a second at 200-1000 K contracting by 5 km $s^{-1}$.

**The Distribution of CO in Sa Galaxies**

D. E. Hogg, M. S. Roberts (NRAO), J. N. Bregman (U. Michigan)

A recent study of the Sa galaxy NGC 1291 has shown that there is a striking spatial anticorrelation between hot (X-ray) and cold (HI,CO) gas, supporting the suggestion that the hot gas is a bulge phenomenon and the cold gas is concentrated to the disk. Additional observations of the cool gas in the X-ray galaxies NGC 2775 and NGC 3623 have now been completed. The CO in NGC 2775 is concentrated in a ring of approximate radius 5 kpc, a region of the disk which lies outside the prominent central bulge. Both CO (1–0) and CO (2–1) are detected over a large radial range in NGC 3623. There is CO emission from the direction of the nucleus, but a significant part of the emission arises at distances of 6 kpc beyond the bulge.

**Atomic and Molecular Gas in Ring Galaxies**

Cathy Horellou (Onsala Space Observatory), Fabienne Casoli, Françoise Combes (Observatoire de Paris)

We present a survey of the $^{12}CO(1\text{–}0)$ and HI emission of a sample of 16 collisional ring galaxies (see Horellou et al. 1995, A&A, 298, 767 for more details). Ring galaxies are believed to form when a compact galaxy plunges through the center of a rotating disk system, triggering a set of radially expanding waves in the target disk. We have found that ring galaxies are bright in CO compared to normal galaxies. Their mean HI content, however, is not significantly different. If the standard conversion factor from CO emissivities into $H_2$ column densities holds for ring galaxies, this indicates that a large amount of molecular gas is available for star formation and that ring galaxies are currently actively forming stars. This result is in agreement with their high far-infrared luminosities. The Cartwheel, often considered as the prototype of ring galaxies, has not been detected. We suggest that the weak CO emission of this galaxy is due to its low metallicity.

**Molecular Gas in the Irregular Galaxy NGC 4449**

D. A. Hunter (Lowell Observatory), H. A. Thronson (Univ. Wyoming)

We present observations of CO ($J$=1–0) emission from the irregular galaxy NGC 4449 obtained with the NRAO 12 m. All but 2 of 8 observations coincided with HI complexes with differing levels of star formation. We found the following: (1) The beam (=1.7 kpc at the galaxy) included molecular cloud complexes with $1\text{–}5\times10^7$ $M_\odot$ of $M(H_2)$. (2) The molecular mass, molecular and atomic column densities, and $M(H_2)/M(HI)$ ratios are similar in spite of large variations in the amount of star formation activity. (3) The molecular complexes are associated with considerable atomic gas, and observations are consistent with the picture that the molecular material is protected from UV radiation by atomic material.

## The Molecular Disk of IC342: On-The-Fly Mapping of CO with the NRAO 12m Telescope

Robert L. Hurt (UC Riverside), Deborah A. Levine (UCLA/IPAC), Robert N. Martin (Steward Obs/SMTO), Jean L. Turner (UCLA), Paul P.T. Ho (CfA)

We present a map of the $^{12}$CO (1–0) emission in the inner disk of the spiral galaxy IC342. This galaxy, at a distance of 1.8 Mpc, is the site of a moderate nuclear starburst and the molecular emission in the central arcminute has been extensively studied interferometrically. Using the newly-implemented spectral line On-The-Fly (OTF) mapping mode of the NRAO 12m telescope, we have observed the extended molecular emission in a $15' \times 15'$($8 \times 8$ kpc) region at a resolution of $1'$ (560 pc). In our sensitive maps (RMS $\sim$ 10 mK) we find the CO to be ubiquitous across the entire mapped region, extending to the edges of the optical disk. The emission is strongly peaked at the nuclear starburst while the disk CO tapers off gradually towards the map edges. There is close correlation between the molecular gas and 20 cm continuum emission tracing the nucleus and spiral arms. However, we find a region of reduced star formation efficiency just outside of the inner starburst where the large concentrations of molecular gas show little associated radio continuum emission. This region corresponds to the 'boxy bulge' seen at near infrared wavelengths in IC342.

## CO Twin Peaks and Centrally Concentrated $H_2$ Emission at the Center of the cD Seyfert Galaxy NGC 1275

Motoko Y. Inoue, Seiji Kameno, Ryohei Kawabe (Nobeyama Radio Observatory, National Astronomical Observatory, Minamimaki, Minamisaku, Nagano, 384-13, Japan), Masuo Tanaka, Tetsuo Hasegawa (Institute of Astronomy, University of Tokyo, 2-21-1, Osawa, Mitaka, Tokyo, 181, Japan)

We report interferometric observations of 2.6 mm CO and 2 $\mu$m spectroscopy of $H_2$ emission in NGC 1275. The CO emission shows a non-axisymmetric distribution in $r \leq 10$ kpc and a ring-like structure with *twin peaks* at $r \approx 1.4$ kpc. On the other hand, the $H_2$ emission is concentrated within $r \leq 340$ pc and its velocity-resolved line profile is fitted by a Gaussian with FWHM $\approx 250$ km s$^{-1}$. All these molecular structures are well explained by a non-axisymmetric potential whose existence is implied by our data. Similarities with NGC 1068 propose the importance of non-axisymmetric potential in a process of AGN formation.

## CO in the Starburst Nucleus of NGC 3628 – NMA Results

Judith A. Irwin (Queen's University, Canada), Yoshiaki Sofue (University of Tokyo)

The edge-on starburst galaxy, NGC 3628, displays spectacular X-ray plumes along its minor axis, indicative of bipolar outflow from the nucleus. We have used the NMA to obtain 3.8″ (123 pc) resolution $^{12}$CO (J=1–0) observations of the nuclear molecular disk in this galaxy. The CO distribution can be characterized by three main components, each of which displays a very clumpy distribution. a) A dense, rapidly rotating nuclear disk several hundred pc in radius with a significant velocity dispersion. The kinematic center of the nuclear disk coincides with the radio continuum peak, rather than the offset optical center, confirming that the radio continuum peak is at the galaxy nucleus. b) A low velocity gradient "ridge" that is outside of the rotating disk but within about 1–2 kpc of the nucleus and that may represent part of an inner ring or spiral arm. c) Various arcs/bubbles, most of which appear to originate in the ridge and which collectively contribute to wide line profiles. Expansion energies of order $10^{52}$–$10^{54}$ erg are suggested. Extraplanar CO arcs and extensions out to 430 pc from the midplane are also observed.

## Mass Loss from OH26.5+0.6

K. Justtanont (NASA/Ames Research Center), C.J. Skinner (Space Telescope Science Institute), A.G.G.M. Tielens (NASA/Ames Research Center), M. Meixner (University of Illinois), F. Baas (Leiden Observatory and Joint Astronomy Center, Hawaii)

We present a self-consistent model for the dust and gas outflows from the extreme OH/IR star, OH26.5+0.6. The source has a deep 10 $\mu$m absorption, inferring a very high mass loss rate ($\dot{M} = 5.5 \times 10^{-4}$ $M_{\odot}$/yr) while the CO J=1–0 and 2–1 lines suggest a more tenous mass loss rate in the past ($\dot{M} = 10^{-6}$ $M_{\odot}$/yr). From these observations plus the recent observations of CO J=3–2 and 4–3 transitions, we have concluded that the star has entered the superwind phase when the mass loss rate increases dramatically during the past 200 yrs.

## The Molecular Outflows in NGC 1333

Lewis B.G. Knee (Swedish-ESO Submillimetre Telescope, Chile, and Onsala Space Observatory, Sweden), Göran Sandell (Joint Astronomy Centre, Hawaii)

We present the results of CO($J$=3→2) mapping, using the James Clerk Maxwell Telescope, of the active star formation region immediately south of NGC 1333, supplemented by CO($J$=1→0) mapping using the Onsala Space Observatory 20 metre telescope.

These spectral line maps provide the first clear overview of the complex cluster of overlapping molecular outflows associated with the Herbig-Haro objects and the far-infrared and (sub)millimetre sources in the region. At least eight molecular outflows are identified, for seven of which we can confidently identify the driving source. Among these are probable "Class 0" protostellar objects driving highly collimated CO jets. Two objects appear to drive outflows which are precessing: IRAS 4A and SSV 13b.

The nature and exciting source of the HH 12 complex is identified for the first time. HH 12 is the working surface of a jet driven by the (sub)millimetre source SSV 13b. This outflow is almost in the plane of the sky, hence the counter-jet also appears blue-shifted.

All of the youngest known members of the current generation of star formation in NGC 1333 are proto-binary or multiple systems. It appears that the hierarchical organisation of multiple systems is present at an extremely early stage.

A cloud-cloud collision model for NGC 1333 is consistent with the structure of the cloud, the arrangement of young objects and their outflows, and the apparently high rate of contemporary star formation activity.

## Molecular Absorption Toward the Low-Latitude Radio Source 2023+336

Henry A. Kobulnicky and John M. Dickey (U. Minnesota), Rachel Akeson (Caltech)

We have conducted millimeter-wave and centimeter-wave absorption observations toward the low-latitude radio source 2023+336 ($l = 73°$, $b = -2°$) in order to search for cold molecular gas, to compare directly the optical depths of the cool atomic and molecular phases of the ISM, and to characterize the chemical composition of molecular clouds in the Galaxy. We find: 1) Multiple-component, translucent CO, HCN, $HCO^+$ absorption in the inner Galaxy near 6 km $s^{-1}$; 2) Only marginal ($3\sigma$) evidence for cold gas (CO absorption without emission) in the Galaxy near −55 km $s^{-1}$;

3) CO absorption line widths as broad as the corresponding emission lines, even though the emission beam diameter subtends 0.2 pc and the absorption column subtends $\sim$ 1 AU! We conclude that the emission preferentially arises in denser clumps with small filling factor while the absorption comes from more diffuse regions of subthermal excitation; 4) $^{12}$CO absorption and emission spectra indicate low CO excitation temperatures of 4 to 6.0 K; 5) The CO absorption implies an optically thin line of sight, whereas $^{12}$CO/$^{13}$CO emission ratios of 13:1 imply a higher optical depth. One resolution to this apparent discrepancy would be that the background source happens to be seen through a "hole" in the molecular clouds so that our measured absorption optical depth is lower than the emission beam average. This solution demands even lower CO excitation temperatures. The alternative solution is a non-standard $^{12}$CO/$^{13}$CO ratio.

**NMA Observations of Dense Molecular Gas in Messier 51**

Kono Kotaro (University of Tokyo), Kawabe Ryohei, Tosaki Tomoka, Okumura K. Sachiko (Nobeyama Radio Observatory)

We have made aperture synthesis HCN (J=1$\rightarrow$0) observations of M51 using the Nobeyama Millimeter Array with spatial resolutions of 7$''$ (340 pc at D=9.6 Mpc). A 4$''$ resolution image was also made to resolve the structure of the circumnucleus.

Intense HCN emission was observed in the central $r < 6''$ region, where CO emission is depressed. There is distortion of the HCN velocity field within $r < 4''$ in the nucleus, where a velocity gradient runs from east to west. If this velocity gradient represents rotation, the axis (P.A.$\sim$20°) is just aligned with the direction of the radio jet and bubble. The $H_2$ column density toward the nucleus is estimated to be $1\times10^{23}$ cm$^{-2}$ from our HCN data, which plausibly accounts for the large absorption reported by X-ray satellites. Considering the large column density toward the nucleus and the alignment of the rotation axis, our observational results imply the presence of a rotating dense molecular disk that obscures the weak Seyfert activity of M51.

In the spiral arm regions, the HCN/CO integrated intensity ratio is in the range of 0.02 – 0.03, whereas the ratio in the central $r < 4''$ area exceeds 0.3, *i.e.*, a factor of 10 larger than those of spiral arm regions. The high HCN/CO ratio means that all the molecular gas in the nuclear region is dense enough to excite HCN (J=1$\rightarrow$0) emission ($n_{H_2} > 10^5$ cm$^{-3}$).

## Structure and Excitation Conditions of the Southern Part of the Orion B Molecular Cloud: A CO Multiline Study

Carsten Kramer (IRAM, Granada, Spain), Jürgen Stutzki, Gisbert Winnewisser (Universität zu Köln, Germany)

We present extended maps of $^{12}$CO and $^{13}$CO ($J$=2→1 and $J$=3→2) of the southern part of the Orion B region covering a ~20′ × 60′ area in $^{13}$CO(2→1). They encompass the NGC 2024 H II region, the interface region between NGC 2024 and the optical nebula IC434, the reflection nebula NGC 2023, and the Horsehead Nebula B33. The physical conditions in these different regions vary significantly due to the influence of the OB association lying to the west and due to the embedded stars. The CO emission originates from many clumps, most apparent in individual velocity channel maps. The observed low-$J$ CO and isotopic CO line ratios are inconsistent with the standard interpretation in terms of a single temperature gas component along the line of sight in LTE. They require an outward temperature gradient on the cloud surface, as is naturally expected to arise from external UV heating. A consistent interpretation for both absolute line brightnesses and the intensity ratios can be reached in terms of PDR models. These indicate densities within the bulk of Orion B of at least close to $10^5$ cm$^{-3}$. In addition, we analyse the structure by subdividing the $^{13}$CO(2→1) emission of Orion B into clumps using an automated procedure. The shape, orientation, LTE mass, virial mass, and stability of these clumps is examined. The mass spectrum of the clumps follows a power law, $dN/dM \propto M^{-\alpha}$, with $\alpha = 1.65 \pm 0.24$.

## Molecular Clouds in the Large Magellanic Cloud

M. L. Kutner (NRAO) and M. Rubio (Universidad de Chile)

We present results of observations of CO(J=1–0) emission form the Large Magellanic Cloud (LMC). These observations were carried out as part of a Key Programme on the Swedish-ESO Submillimeter Telescope (SEST) on La Silla, Chile. The beamwidth at 2.6 mm is 40, or 10 pc at the distance of the LMC, and the maps are fully sampled. The molecular emission in the LMC is dominated by clouds with giant molecular cloud (GMC) dimensions and masses. Cloud properties in two regions of the LMC will be discussed. One is a dark complex extending 2 kpc south of 30 Dor, and the other is associated with the bright HII region N11. We will discuss the cloud size distributions, size-linewidth relations, and the conversion from CO luminosity to virial mass in both regions.

**CO Observations of the Carbon-rich Proto-Planetary Nebula IRAS 22272+5435**

Sun Kwok (University of Calgary), Typhoon Lee, Shudong Zhou (Academia Sinica, Taiwan), A.W. Woodsworth (Dominion Astrophysical Observatory)

We report recent single-dish and interferometric observations of the PPN IRAS22272+5435, which is one of the few stars that are known to have the unidentified emission feature at 21 $\mu m$ (Kwok, Volk, & Hrivnak 1989, ApJ, 345, L51). The CO 2–1 and 3–2 lines were detected at JCMT with a beam sizes of 21 and 14 arcsec respectively. The observed profiles are fitted with a LVG model with an inverse-square density distribution, an inner radius of 0.0077 pc, a kinetic temperature law of the form $T(r) = 50K(r/r_in)^{-0.7}$, a mass loss rate of $10^{-5} M_\odot \ yr^{-1}$, and a distance of 1050 pc. IRAS 22272+5435 has one of the strongest CO emissions among post-AGB stars, and offers an excellent opportunity to study its morphology with millimeter interferometry. A CO (1–0) image has been obtained with BIMA. The object is found to be marginally resolved by the 3 arcsec beam, with the blue-shifted component to the NE, and the red-shifted component to the SW of the infrared position.

**Vibration-Rotation CO in Intermediate Population Miras**

Thomas Lebzelter, Josef Hron (Institute for Astronomy, University of Vienna, Austria), Kenneth Hinkle (Kitt Peak National Observatory)

We have observed time-series IR spectra (2.0 – 2.5 $\mu$m) of four intermediate population Miras (high velocity, early spectral type, short periods, low metallicity) at high resolution. The covered spectral region includes the CO vibration-rotation first overtone region. The high excitation CO and atomic lines follow the radial velocity curve expected during a pulsation cycle. Amplitude and shape of the rv curves are similar to the results found for disk Miras with longer periods. The low excitation CO lines show only small velocity variations in all objects investigated except in R Vir. We have no definite explanation for the strange behavior of this object at the moment.

For three of our program stars, the carbon isotopic ratio is comparable to the values found for thin disk Miras.

Our results allow two conclusions: (1) The amplitude of the radial velocity variations is almost independent from the period for periods between 150 and 350 days. (2) The amplitude does not depend on metallicity.

## PDFs of Centroid Velocities and Centroid Velocity Increments in Supersonic Compressible Turbulence

D.C. Lis, J. Pety, T.G. Phillips (California Institute of Technology), and E. Falgarone (Radioastronomie Millimétrique, École Normale Supérieure)

We have calculated probability density functions (PDFs) of centroid velocities and centroid velocity increments of line profiles computed from the output of a $512^3$ simulation of compressible turbulence (Porter, Pouquet and Woodward 1994, *Phys. Fluids*, 6, 2133). The PDFs of centroid velocities calculated over the whole data cube are roughly Gaussian. On smaller scale, non-Gaussian PDFs are observed in some cases. However, this is far from being the rule, and most of the distributions show only minor deviations from a Gaussian. By contrast, PDFs of centroid velocity increments clearly show non-Gaussian wings for separations comparable to the dissipation scale in the simulation, probably assignable to intermittency. In a recent observational study (Miesch and Scalo 1995, *ApJ Letters*, in press), PDFs of centroid velocities of $^{13}$CO line profiles in star-forming regions were shown often to have near exponential wings. We conclude that these wings are unlikely to be a manifestation of intermittency of turbulence. Investigations of PDFs of centroid velocity increments in non-starforming regions, however, would seem a promising avenue for studying the intermittency in the interstellar medium, as an alternative to the line-shape discussion of Falgarone and Phillips (1990, *ApJ*, 359, 344).

## A Large-scale High-resolution Survey of CO in the Andromeda Galaxy

Laurent Loinard, Enrico Koper, Patrick Thaddeus (Harvard-Smithsonian Center for Astrophysics), James Lequeux (DEMIRM, Observatoire de Paris)

Although surveys of most of the components of the ISM in M31 (H I, IR, radio-continuum emission, etc.) have already achieved an angular resolution of $\sim 1'$, the only complete CO survey of this galaxy (the CfA survey) has a resolution of 8.′7. Several small selected areas have been observed at higher resolution, but the overall small-scale structure of the molecular gas in M31 is largely unknown.

We present here a high-resolution CO(1–0) survey of the southern part of the major axis of M31 done with the QUARRY array of receivers on the 14-m FCRAO telescope (angular resolution $\sim 45''$). This survey consists of a 5 arcminute-wide strip along the major axis from the center to about 16 kiloparsecs, covering an area of 360 square arcminutes. Roughly 10 molecular clouds or cloud complexes are fairly well resolved in our data and nearly all are appropriate for higher resolution CO studies. The distribution of the CO emission along the major axis agrees well with previous determinations and shows the same "ring-like" structure as the H I and other tracers of population I, although the H I shows a peak at 13 kpc from the center that only has a very weak conterpart in CO. The velocity distribution of the CO emission is in good agreement with the velocity distribution of the H I.

## The Variation of the X-factor in Two Translucent Clouds

Loris Magnani (University of Georgia), Dap Hartmann, Patrick Thaddeus (Harvard-Smithsonian CfA)

We determine the "X-factor" [ratio of $N(H_2)$ to $W(CO)$] for two translucent clouds at high Galactic latitude. The method employed to determine the X-factor involves observations of the hyperfine, ground state, main line transition of CH at 3335 MHz as a tracer for $N(H_2)$. Magnani & Onello (1995) discuss the variation of $X_{CO}$ in a sample of 28 lines of sight in 18 different translucent clouds. Values of $X_{CO}$ vary from $0.3 - 6.8 \times 10^{20}$

We determined $X_{CO}$ for 8 lines of sight in MBM40 and found values ranging from $0.8 - 4.8 \times 10^{20}$, with an average of $1.8 \times 10^{20}$. However, the $4.8 \times 10^{20}$ value is uncertain because the CH spectrum from which $N(H_2)$ is derived is noisy. Discarding this measurement, the range of $X_{CO}$ in MBM40 is rather constrained: $0.8 - 2.0 \times 10^{20}$. The case of MBM16 is quite different. $X_{CO}$ was determined for 20 lines of sight and varies more markedly, from 1.2 to $14.9 \times 10^{20}$, with an average value of $5.0 \times 10^{20}$. For MBM16, determining $X_{CO}$ on the basis of an observation along a single line of sight could lead to a highly misleading result.

Future work will determine whether translucent clouds taken as an ensemble display a variation of $X_{CO}$ similar to MBM40 or MBM16.

**High Altitude Molecular Clouds: Properties.**

S. Malhotra, G. R. Knapp (Princeton University Observatory), G. Wright (AT&T Bell Laboratories), R. W. Wilson (CfA)

We describe $^{12}CO$ observations and morphology of a new population of diffuse molecular clouds found far away from the Galactic plane. These high altitude clouds (HACs) are drawn from a sample of 23 clouds identified by Malhotra(1994) as being more than 2.6 scale heights away from the plane. Since these clouds are at tangent point velocities, we know their distances and height above the plane. New observations were carried out in the winter of 1994–95 with the 7 meter telecsope at Bell Laboratories to map these clouds to a noise level of 0.15 K. These observations show that HACs are large—comparable in size to GMC complexes; and arranged in complexes with complicated velocity structure. The HACs have much weaker CO emission than do GMCs, and are not gravitationally bound.

**The Dense Gas Structure and Kinetic Temperature in the Serpens Molecular Cloud**

Jeffrey G. Mangum (Submillimeter Telescope Observatory), William B. Latter (NASA/Ames), Joseph P. McMullin (Submillimeter Telescope Observatory)

Recent high spatial resolution ($\theta_B \simeq 20''$) measurements of the 800 and 1100 $\mu$m continuum emission toward the Serpens molecular cloud (Casali, Eiroa, & Duncan 1993) have uncovered six dust continuum cores, four of which have no near-infrared counterparts. With $L_{bol} \lesssim 30\ L_\odot$, Casali *et al.* suggest that these four newly-discovered submillimeter continuum sources are part of a second, younger phase of star formation in Serpens. To investigate the dense gas properties within the Serpens molecular cloud, we have used the newly-developed on-the-fly (OTF) observing technique at the NRAO 12m telescope to map the $H_2CO$ $3_{03} \rightarrow 2_{02}$ and CS $3 \rightarrow 2$ emission toward this region. Our results are the following: (1) With the exception of SMM2, the 800 and 1100 $\mu$m continuum sources mapped by Casali *et al.* (1993) have CS and $H_2CO$ emission counterparts. (2) Using the analysis described by Mangum & Wootten (1993) for the determination of kinetic temperatures using the $3_{03} \rightarrow 2_{02}$ and $3_{22} \rightarrow 2_{21}$ transitions, we find kinetic temperatures $< 55$ K for the SMM1 through SMM6 sources. (3) An analysis of the velocity structure in our CS map indicates that the S68N and SMM4 sources possess dense-gas kinematic motion. Given the added properties of high gas density, ($n(H_2) > 1.0 \times 10^5$ cm$^{-3}$), moderate gas kinetic temperature, and lack of near– or far–infrared emission, we suggest that S68N and SMM4 are Class 0 objects.

**Molecular Clouds in the Southern Outer Galaxy**

Jorge May, Hector Alvarez, Leonardo Bronfman (Universidad de Chile)

From a deep CO survey of the third galactic quadrant (May et al. 1993, A&A, 99, 105), we derive the physical properties of 177 molecular clouds in the outer Galaxy, beyond 2 kpc from the Sun. Although power-law relations between line widths and sizes of the clouds, and between their densities and sizes have been found, they do not fulfill the requirements to be in strict virial equilibrium. The derived size and mass distributions of the clouds show that they are smaller and less massive than those in the inner Galaxy. A statistical analysis of their CO luminosities and line widths indicates that they appear to be less luminous than those in the inner Galaxy by a factor of 2.4, on the average. Therefore, for the outer Galaxy clouds a $CO/H_2$ conversion factor 2.4 times larger has been adopted. The warping and flaring of the outer molecular disk is clearly delineated, although no grand design spiral pattern can be visualized from the projection of the clouds on the galactic disk.

**Molecular Depletions in NGC 1333 IRAS 4**

Joseph P. McMullin (SMTO), Lee G. Mundy (University of Maryland), Geoff A. Blake (Caltech)

We present interferometric observations (BIMA, OVRO) of the young stellar objects NGC 1333 IRAS 4A and 4B, focussing on a comparison between the emission distributions of gas and dust. The strong continuum signature of these sources contrasts with generally weak molecular emission on small scales. We explored the possibility of optical depth effects hiding a larger column density. However, the Blake *et al.* (1994) line survey derives low values for the optical depth in our observed lines while the possibility of temperature increases masking column density increases requires unrealistically large changes in $T_K$ to substantially alter the derived column densities. We conclude that the observed emission may be explained in terms of gas phase depletions by factors of greater than 20, even for CO (based on isotopic information).

Based on the magnitude and scales of the observed depletions, we suggest an extremely youthful stage of development for these sources and attempt to place them within a crude chemical evolutionary sequence based on other sources within the literature and basic model predictions.

## Stars and Protostars in NGC 2024

David M. Mehringer, Hélène R. Dickel (Department of Astronomy, University of Illinois), T. L. Wilson (Max-Planck-Institut für Radioastronomie)

Mezger et al. (1988, A&A, 191, 44) reported the detection of six "Protostars" in the high density molecular gas *behind* the HII region NGC 2024. We have imaged this region in $J = 1 \rightarrow 0$ $C^{18}O$ and 2.7 mm continuum emission with a resolution of $\sim 10''$ using the Berkeley-Illinois-Maryland Association Array. Continuum emission was detected from dust associated with the sources FIR 3, 5 and 6 (Mezger et al. 1988). The 1.3 mm/2.7 mm spectral indices of these dust cores are all $\geq 4$. The upper limits for the 2.7 mm continuum flux densities for FIR 1, 2, 4 and 7 indicate similar spectral indices for these cores.

$C^{18}O$ $J = 1 \rightarrow 0$ emission is observed in 3 slightly extended regions. The two southern peaks are offset by $> 15''$ (6000 AU) from the positions of FIR 5 and FIR 6. The northern peak is located $\leq 4''$ (1700 AU) from the position of FIR 2. For the $C^{18}O$ clouds, $H_2$ column densities based on the virial theorem imply ($CO/H_2$) ratios of $\sim 10^{-5}$, about 10% of the normally used value. The low value of the CO fractional abundance can be explained if molecules are depleted onto grains; it is surprising that $X(\mathrm{CO})$ is this *high*, because current models suggest greater depletion in such high density regions. Toward most of the dust cores, no $C^{18}O$ emission is detected and so depletion may be much more severe. On the other hand, if the dust cores contain stars, then molecules may have been destroyed by energetic stellar winds.

## BIMA Array CS J=2→1 Observations of Sagittarius B2

David M. Mehringer (Department of Astronomy, University of Illinois)

Observations of the $J = 2 \rightarrow 1$ transition of CS toward the Sgr B2 star-forming region have been carried out using the BIMA (Berkeley-Illinois-Maryland Association) array. The angular and spectral resolutions are $9'' \times 14''$ ($\alpha \times \delta$) and 1.2 km s$^{-1}$, respectively. Results indicate that a high velocity

(FWZI $\sim$ 60 km $s^{-1}$) molecular outflow is present in Sgr B2(M), a core which contains more than 20 compact and ultracompact HII regions. The mass of the outflow, derived assuming a normal CS fractional abundance of $2\times10^{-9}$, is 8000 $M_\odot$. Because this derived mass is much higher than that derived from previous observations, it is likely that the fractional abundance of CS in the Sgr B2(M) outflow is significantly higher than the usual value. No CS emission is observed toward Sgr B2(N), a core/outflow source $\sim$50″ north of Sgr B2(M). The absence of CS emission toward this core is due at least in part to self-absorption from the extensive molecular envelope. It is also possible that CS is underabundant toward the Sgr B2(N) core. CS absorption is observed toward both the Sgr B2(M) and (N) continuum sources. Much of the absorbing material is probably located in the extensive envelope of the Sgr B2 molecular cloud and other clouds not associated with Sgr B2 that lie along the same line-of-sight.

## Beyond CO – Complex Molecules In Sgr B2(N): The Importance of Grain Chemistry

Y. Miao, D. M. Mehringer, L. E. Snyder, and Y.- J. Kuan (Department of Astronomy, University of Illinois)

The complex molecules methyl formate ($HCOOCH_3$), vinyl cyanide ($CH_2CHCN$), and ethyl cyanide ($CH_3CH_2CN$) were observed in the Sgr B2 star-forming region with the BIMA millimeter wavelength array. A region with diameter $< 0.1$ pc toward the Sgr B2(N) molecular core is found to be the major source of these molecules. Also, this source is coincident with continuum emission from dust and a center of $H_2O$ maser activity. Ultracompact (UC) HII regions are located within 0.1 pc. Strikingly, none of these molecules is detected toward Sgr B2(M), a core located 1′ south of Sgr B2(N). The existence of complex molecules, a large mass of dust, high velocity $H_2O$ masers, and UC HII regions strongly suggest that the Sgr B2(N) region has just begun to form stars, while the absence of strong dust emission and large molecules suggest Sgr B2(M) is more evolved. The detection of large molecules coincident with continuum emission from dust supports the idea found in current chemical models that grain chemistry is of crucial importance for the formation of these molecules.

This paper has been accepted for publication. Interested parties may contact the authors for reprints.

**S140 - Unravelling the Mystery?**

Nigel R. Minchin, Glenn J. White (Physics Department, Queen Mary & Westfield College, University of London, London, E1 4NS, UK)

The complex consisting of the Lynds 1204 molecular cloud and the adjacent S140 HII region, produced by the nearby B0 star HD211880, is somewhat unique. The molecular cloud not only contains one of the best examples of an edge-illuminated PDR, but only 70 – 80 arcsec to the north-east is an embedded cluster of three infrared sources, which lie at the centre of a massive and extensive molecular outflow. Such close association of an archetypical PDR and molecular outflow make this a fascinating region for observational study. Four years ago we began a major observational program to study this region through submillimetre isotopic CO and neutral carbon (CI) emission lines and also broad-band continuum mapping. Selected highlights from our four published/accepted papers are outlined.

Comparison of the continuum emission with high resolution CS, $NH_3$, isotopic CO and CI observations provides evidence that, for the first time, demonstrates the PDR and outflow are intimately linked. We postulate that the outflow has expanded towards the edge of the molecular cloud such that the edge of the blueshifted outflow lobe is bounded by the expanding HII region. The CI emission originates from the outer edge of the cloud (the PDR), with the CS emission tracing compressed high density gas, sandwiched between the expanding outflow and PDR regions. The $NH_3$ and continuum emission emanates from the inner edge of the outflow lobe, shielded from the external UV field.

**The Star-forming Core of the Monoceros R2 Cloud**

George F. Mitchell (Saint Mary's University, Nova Scotia), Jean Giannakopoulou (University of Waterloo, Ontario)

The physical nature of the core of the Mon R2 molecular cloud is investigated using new, high resolution observations. The data include spectral line maps (in CO $J$ = 3–2, HCN $J$ = 4–3, and $H_2CO$ 51,5–41,4), mm/submm continuum maps (at 450, 800, 1100, and 1300 $\mu$m), and an M-band absorption spectrum of IRS 3. In addition, spectra of $^{13}CO$ J = 3–2, $H_2CO$ 50,5–40,4, and $H_2CO$ 52,3–42,2 were obtained at a number of positions. The CO map shows that the cloud core is broken up into numerous clumps (13 in the $3' \times 3'$ field) with masses ranging from 0.01 $M_\odot$ to 3 $M_\odot$. An unusual feature of the clumps is their high velocity dispersion. The formaldehyde line ratio, 50,5–40,4/52,3–42,2, yields a gas kinetic temperature in the range 50 to 70 K. An excitation diagram of 14 rotational lines of the fundamental band of $^{13}CO$ (from the IR absorption spectrum) is used to find the gas temperature and column density toward IRS 3. Two gas temperatures are found, $T$ = 45 K, which we presume to represent the core gas temperature in this line of sight, and $T$ = 310 K, which probably is due to gas very close to IRS 3. The HCN and $H_2CO$ emission line maps indicate that the densest core gas is organized into two structures with different radial velocities. We find that IRS 3 is the most active IR source in the core, in the sense that it is producing high velocity gas in its surroundings. There is, however, no evidence for organized, bipolar flow from IRS 3, which might connect it to the well known and very extended bipolar flow in this region. The compactness of the high velocity region around IRS 3 suggests that it may be a young object which has only recently become active.

**Molecuar Line Analyses of W49N**

Ryosuke Miyawaki (Fukuoka University of Education), Masahiko Hayashi (SUBARU Project Office, National Astronomical Observatory of Japan), Testuo Hasegawa (Institute of Astronomy, The University of Tokyo)

Studies of the W49A molecular cloud complex are difficult not only because it is compact as a result of its large distance but because different molecular lines exhibit different shapes of line profiles. We presented classification of various molecular lines toward W49N using the Nobeyama 45 m telescope. There are three types; deep dip lines, shallow dip lines, and single peak lines. These are charcterized by their optical depths and excitation conditions. From evaluation of physical properties of molecular lines, star formation is taking place between two velocity components. These conclusions are supported by the results of CS (J=1–0) and SiO (J=2–1) emission using Nobeyama Millimeter Array. The 4 kms$^{-1}$ component of CS is surrounded with the 12 kms$^{-1}$ component of CS. SiO has a velocity between two velocities, and its distribution is compact.

**3mm SIS Receiver for the Australia Telescope 22 metre Mopra Antenna**

G.G. Moorey, M.W. Sinclair (Australia Telescope National Facility, CSIRO), J.M. Payne (National Radio Astronomy Observatory)

As the first step towards equipping the Australia Telescope 3-kilometre array at Narrabri with mm wave receivers covering the 70 to 115 GHz band, a prototype receiver for the 86 to 115 GHz band was built and installed on the 22 metre antenna at Mopra Rock. The antenna is located near the Siding Springs Anglo-Australian Optical Observatory at Coonabarabran in the picturesque Warrumbungle National Park about 450 kilometres north-east of Sydney. To obtain the best possible results from the antenna and site, a dual-frequency , dual-polarisation receiver employing SIS mixers was designed and installed. The receiver is capable of operation in the SSB or DSB mode and has an instantaneous IF bandwidth of 600 MHz centred on 1500 MHz. The initial commissioning tests utilising the inner 15.4 metres of the antenna across the 86-to-98 GHz band gave an average DSB system noise temperature on the sky of 125K with a zenith optical depth $T =$ 0.1. The full width to half power beamwidth is $\sim$45 arcsec and the gain is 40 Jy/K which corresponds to an aperture efficiency of $\sim$35%. The initial pointing model gives errors of $< 10$ arcsec rss of both axis. The flexibility of the ATNF Mopra correlator is demonstrated by the diversity of its configuration modes. These range from 1-bit sampled 256 MHz bandwidth, dual-polarisation, 512 channels down to 2-bit sampled 4 MHz bandwidth, dual-polarisation with 16,384 channels. Initial on/off source observations of SIO, HCN, $HCO^+$,and CS molecular lines with 20 second integration times, produced good base line spectra with a signal-to-noise ratio of $> 20$. The initial engineering and observational tests as mentioned above were carried out in the winter months of the last quarter of 1994. For the same period in 1995, the Mopra site will be operated on a shared-risk basis while commissioning tests continue. Commencing in the June quarter of 1996, it is proposed to operate the site as a National Facility. In future development plans we hope to extend the solid surface out to 22 metres, install a nutating subreflector to enable beam switching, improve the pointing and extend the receiver capabilities possibly to cover the 140-to-170 GHz band.

## Observation of the Rosette Molecular Cloud in CO($J$=2–1) line

Jun Ichi Morino, Shigehisa Takakuwa, Tsuyoshi Yoshida, Tetsuo Hasegawa, Toshihiko Handa, Seiichi Sakamoto, Tomoharu Oka, Masumichi Seta, Kumiko Sato, Kazuo Sorai (University of Tokyo), Hiroko Shinnaga (Tokyo Gakugei University), Masahiko Hayashi (National Astronomical Observatory Japan)

The molecular cloud accompanied by the Rosette Nebula have been observed in the $J$=2–1 transition of carbon monoxide with 9′ resolution (4pc at 1600 pc) using the Tokyo-NRO 60 cm telescope. Comparing CO($J$=2–1) line intensity with that of CO($J$=1–0) line (Blitz and Thaddeus 1980; May and Bronfman 1993), we find that CO($J$=2–1) line emission is relatively stronger near the HII region than in other regions. A close comparision with the CO($J$=1–0) data of May et al. (1993) in a position-velocity map confirms the tendency that the line intensity ratio, $R_{2-1/1-0}$, is high near the HII region and declines as we move away from it. The gas component with a steep velocity gradient, which we think evidence of interaction with the HII region, exhibits ratios ranging from 1.0 up to 1.2. An LVG line excitation model shows that $R_{2-1/1-0}$ is a good indicator of gas density although it is also slightly dependent on gas kinetic temperature. A possible interpretation for the observed ratio is that the density is enhanced by interaction of molecular gas with the expanding HII region. An alternative interpretation is that the molecular gas near the HII region is made of small clumps surrounded by photodissociation regions as was modeled by Gierens et al. (1992).

## The CO-to-$H_2$ Conversion Factor in Galaxies

N. Nakai (Nobeyama Radio Observatory)

The CO-to-$H_2$ conversion factor has been evaluated in M51 using visual extinctions and CO data toward 30 HII regions. The derived value is $N(H_2)/I(CO) = (1.1 \pm 0.1) \times 10^{20}$ cm$^{-2}$ [K km/s]$^{-1}$ where $N(H_2)$ is the $H_2$ column density and $I$(CO) the CO integrated intensity. The derived value is comparable to the smallest value in our Galaxy or a factor of three smaller than the mean value in our Galaxy. The $N(H_2)/I(CO)$ ratio tends to be larger in the outer region. At very low values of $I$(CO) of $< 20-40$ K km/s, the ratio shows large values ($> 3$) in M51 and other galaxies, probably due to the small optical depth of CO emission and low metallicity.

**Evidence for Multiple Outbursts from the Cepheus A Molecular Outflow**

G. Narayanan and C. K. Walker (Steward Observatory, University of Arizona

We report evidence for multiple episodes of outflow activity in the Cepheus A star-forming region. We present new, high signal-to-noise CSO observations of $^{12}$CO $J$=3–2, $^{13}$CO $J$=3–2, and CS $J$=7–6 emission. We also present new, high-angular resolution observations of $^{12}$CO $J$=1–0 emission towards the Cepheus A molecular outflow. Using line core velocity centroid maps, we argue that the multiple self-absorption features in the CO $J$=3–2 line profiles are tracing cool shells of material swept up by an episodic outflow. We present the results of a flexible 3-dimensional LTE outflow model and radiative transfer code that best explains the observations as tracing multiple generations of outflow. The difference in the dynamical time-scales between the "old" and "new" swept-up shells gives an estimate of $\sim 1.6 \times 10^4$ years between the two generations of outbursts. The high resolution $^{12}$CO $J$=1–0 maps obtained by combining single-dish observations with interferometric data clearly show a shell-like morphology at low velocities. This cool shell appears to confine the hot, extremely high-velocity (EHV) winds seen in the $J$=3–2 transition. The interferometric observations show that the current generation of outflow is being powered by the object Cepheus A-HW2. There is also evidence for redirection of the blueshifted lobe of the current generation of outflow, possibly to the extended $NH_3$ structure Cep A-3. We present a picture of outflow geometry that can explain most of the observed structures in Cepheus A as the result of the interaction of an intense stellar wind with dense condensations in the parent cloud, and possibly, swept-up shells of previous generation(s) of outflow. The rotating, dense core traced by the CS observations is $\sim$ 0.32 pc in diameter and has an estimated dynamical mass of 330 $M_\odot$. The velocity structure of the core suggests that it is being disrupted by the high-velocity winds driving the molecular outflow. This new technique of extracting information from self-absorbed line profiles could be used to study other deeply embedded protostellar systems. Since outflows are believed to be intimately tied in with accretion, such studies could lead to constraints on mass accretion models for young stellar objects.

**The Dust Content and the Kinematics of NGC 4565**

Nikolaus Neininger, Michel Guélin (IRAM Grenoble, France), Santiago Garcia-Burillo (CAY Yebes, Spain), Richard Wielebinski (MPIfR Bonn, Germany)

Using the IRAM 30-m telescope, we have mapped the $\lambda\,1.2$ mm continuum emission and the $^{12}$CO J= 1–0 and 2–1 lines in NGC 4565. We selected this edge-on Sb galaxy because, contrary to the nearby galaxies mapped so far at 1.2 mm, it is relatively weak in the CO lines, but strong in the HI $\lambda$ 21 cm line.

The CO emission shows a ring-like distribution plus a central nuclear source. It peaks between $1'$ and $1.5'$ from the center and tapers off for $R > 2'$; it is much less extended than the HI emission, which remains flat until $R = 5 \ldots 6'$. The $\lambda\,1.2$ mm continuum, which is mostly thermal emission from dust, has an intermediate distribution. It exhibits the central peak and compact ring observed in CO *plus* a weaker component extending well beyond the limits of the CO map. This is the first time dust emission from extragalactic HI clouds is detected at mm or submm wavelengths.

The velocity field derived from CO can be described in terms of a circular rotation (solid body rotation between 0 and 4 kpc, constant velocity further out) plus non-circular 'streaming' motions near the center and in the molecular ring. The non-circular motions are explained by the presence of a central bar and a spiral density wave. The presence of streaming motions associated with spiral arms was already proposed by Rupen (1991) to explain the HI velocity field. That of the bar could explain the boxy shape of the central bulge.

**A 3-position Spectral Line Survey of the Sgr B2 Molecular Cloud**

Albert Nummelin, Åke Hjalmarson (Onsala Space Obs.), Per Bergman (Steward Obs.), Per Friberg (JAC, Hilo), Tom J. Millar (UMIST), William M. Irvine (FCRAO), Masatoshi Ohishi (NRO), Shuji Saito (IMS, Okazaki)

We are currently, at SEST, performing a spectral line survey towards 3 positions in the Sgr B2 molecular cloud: The hot cores Sgr B2(M) and Sgr B2(N), and the quiescent Sgr B2(NW). By June 1995, the covered frequency band is 218.3 to 263.5 GHz. The spectrum towards the N core, in which we have 1600 identified and unidentified lines, is dominated by emission from complex organic molecules, e.g., $C_2H_5CN$, $CH_3OCHO$, whereas M, with 600 spectral lines, is more prominent in sulphur species like $SO_2$. At present, the number of unidentified lines is 10% in M, but much larger in

N—approximately 25%. Here several U-lines are 1K or stronger. Analysis work is now in progress.

**Balloon Observations of the 557 GHz Line of Water Vapour**

G. Olofsson, L. Nordh (Stockholm Observatory), J. Tauber, G. Pilbratt (ESA Astrophysics Division, Space Science Department, ESTEC), U. Frisk (Swedish Space Corporation)

We report the first results of PIROG 7, a balloon-borne platform that is equipped with a 60 cm telescope and a heterodyne receiver designed for observations of the ground transition at 557 GHz of ortho-water. Gas-phase water is expected to be one of the primary coolants of dense molecular regions, but its presence in the atmosphere has prevented observations of non-masering low-excitation lines up to now. Observations of this species would provide significant constraints on our knowledge of the chemistry of oxygen-bearing molecules in the ISM, and on the dynamics and energetics of molecular clouds.

**The IRAM Key Project: Small-Scale Structure of Pre-Star Forming Regions**

J.-F. Panis, E. Falgarone (ENS, Paris, France), A. Heithausen, J. Stutzki (Universität zu Köln, Germany), J.-L. Puget (IAS, UMR 120 CNRS, Université de Paris XI, Orsay, France), M. Pérault, M. Gerin (ENS, Paris, France)

This project consisted of mapping large regions of nearby clouds around dense cores which have not yet formed stars. It has been conceived to fill a gap between two complementary observational approaches of star formation processes: the study of immediate environments of newly born stars and the mapping of molecular clouds over large dynamical ranges to characterize their self-invariant dynamical properties. The objective of our project is, therefore, to extend our knowledge of the molecular gas down to scales as close as possible to that of the protostellar environment, before the apparition of any star.

We present preliminary maps performed with the IRAM 30 m antenna in five transitions ($^{12}CO(1\text{–}0)$, $^{12}CO(2\text{–}1)$, $^{13}CO(1\text{–}0)$, $^{13}CO(2\text{–}1)$, and $C^{18}O(1\text{–}0)$) of three fields (L1512, L134A, and a field in Polaris flare). Each contains more than 2500 positions and the resolution of the (2–1) maps is about

2000 AU. The fields extend from the core to the nearest edge of CO cloud emission. The striking results are the large amount of small-scale structure in the $^{12}$CO maps, the remarkable differences between the structure seen in $^{12}$CO and in the rare isotopes, and the uniformity of the (2–1)/(1–0) ratio (0.65 in $T_A^*$).

**High Resolution $^{13}$CO, C$^{18}$O Observations of NGC 1068, Molecular Gas Properties and an Unusually High I(C$^{18}$O)/I($^{13}$CO) Intensity Ratio**

Padeli P. Papadopoulos, Ernest R. Seaquist (Department of Astronomy, University of Toronto, Canada), Nick Z. Scoville (Owens Valley Radio Observatory, California Institute of Technology)

We have conducted high ($\approx 4'' \times 4''$) resolution observations of $^{13}$CO, C$^{18}$O $J$=1–0 observations of NGC 1068. We compare the $^{12}$CO emission (published maps) with our new $^{13}$CO maps and we deduce an average optical depth of $\langle \tau$ ($^{13}$CO) $\rangle = 0.07$, which is similar to the ones obtained for disks of quiescent spirals. However, we do find variations of the optical depth for the individual Giant Molecular Associations (GMAs) as a result of varying gas column density, also revealed in recently published UV color-color maps. Finally, we have detected the rare isotope C$^{18}$O in many positions in NGC 1068 and we have found a ratio of $\langle R\rangle = I(^{18}\mathrm{O})/I(^{13}\mathrm{CO})$ $= 0.4 \pm 0.1$. This is the highest value reported up to date for an extragalactic source. We interpret $R$ as an abundance ratio since both $^{13}$CO, C$^{18}$O are likely to be optically thin and with the same $T_{\mathrm{ex}}$. We argue that such a high abundance is due to the high rate of formation of massive stars, which enhance [$^{18}$O/$^{16}$O] without altering [$^{13}$C/$^{12}$C].

**Star Forming Globules in IC1396**

N. Patel (Smithsonian Astrophysical Observatory)

We have mapped the diffuse HII region IC1396 in the J=1–0 transition of CO and $^{13}$CO, using the FCRAO 14-meter telescope with the QUARRY focal plane array receiver. The prominent bright rimmed globules are found to lie on a ring of radius 12 pc, expanding with a velocity of 5 km s$^{-1}$. The dynamical age of the system of globules, as suggested by its expansion, is 2–3 M yrs, consistent with the age estimates of the exciting O6 star, HD206267. Twenty-four IRAS point sources in the mapped field satisfy criteria of young stellar objects; ten of these sources are directly associated with IC1396 globules, suggesting that star formation is occurring within them. In a mass-UV flux parameter space, globules with the IRAS sources are distinct from the globules without IRAS sources, being relatively more massive. The ring of globules itself appears to be embedded in a larger elliptical structure $\sim 9^\circ$ (120 pc) in size, which appears in the IRAS maps of 60 and 100 micron emission. Preliminary results from a CO study of this large-scale region will be presented. We suggest that star formation in this region has proceeded by sequential triggering, and the stars forming in the globules of IC1396 represent the youngest generation of stars in this region.

**A Chemical Survey of the Dark Clouds TMC-1 and L134N**

Preethi Pratap, Edwin A. Bergin, James Dickens, William M. Irvine, Mari Paz Miralles, F. Peter Schloerb, Ronald Snell (FCRAO, Univ. of Massachusetts)

A large scale chemical study of the dark clouds TMC-1 and L134N is being conducted with the FCRAO 14m telescope. Extended regions of both clouds have been observed in 28 different molecular transitions. The species have been chosen to give us physical conditions such as density ($HC_3N$) and temperature ($CH_3CCH$). The conditions derived from these species are used to obtain abundances. Preliminary estimates of the abundances using LTE and assuming optically thin lines show some abundance variations across the sources. However, we find that the LTE approximation may not be valid in these sources. It is clear that we need more accurate estimates of the densities and temperatures in order to determine whether the observed abundance variations are real or due to variations in the physical conditions. In addition, we need to calculate abundances using a full non-LTE analysis.

**BIMA Observations of Molecular Spiral Structure in M100**

Richard J. Rand (Univ. of Maryland)

We present BIMA observations of CO $J$=1–0 emission in the grand-design spiral M100 (NGC 4321). Two 100″ (FWHM) fields have been mapped at resolutions of 7.7″ × 6.2″ and 4.6″ × 3.6″ and mosaicked together. These fields cover the mild nuclear starburst, the large-scale bar seen in the near-IR, and the main arm which emerges from the north-west end of the bar and swings around to the south. A map of the nuclear region at $3.6'' \times 3.1''$ has also been made. Molecular emission is found along the main arm, showing several GMAs with properties similar to those in M51. The central regions show bright, open arms, in response to the large-scale bar, and a smaller-scale CO bar aligned with a similar structure seen in $K$-band.

Streaming motions are observed across the arm, consistent with the flow of gas through a density wave compression, but they are of much lower amplitude than in M51. An offset between centroids of the CO and H$\alpha$ arms is found in the north and west parts of the arm, indicating triggering of star formation, but the offset disappears further along the arm to the south. This location could be near corotation.

An arm-interarm contrast of molecular surface density can be roughly derived from the amplitude of the radial streaming motions, and is found to be 2–3, somewhat lower than in M51. Gravitational instability or collisional agglomeration are both viable ways of building GMAs in the arms. Both theories run into problems, however, if the $X$-factor in M100 is much different from the standard value. The value of $X$ probably does not vary significantly between arm and interarm regions.

**Hydrodynamics and Mass Calibration in the Cepheus Flare**

G. P. Remy, I. A. Grenier (EUROPA/Université Paris 7 and DAPNIA/ SAp/Centre d'Etudes de Saclay)

The eastern molecular cloud in Cepheus ($111° < l < 118°$, $14° < b < 19°$) coincides with a non-thermal radio spur and displays unusually broad CO lines. Using the 'Mini' telescope of Harvard and POM-2 in France, we have observed this nearby cloud in $^{12}CO$ and $^{13}CO$, in 1–0 and 2–1 transitions, at a resolution of $8'$. A uniform (2–1)/(1–0) line ratio is found over the whole cloud for both isotopomers, implying low excitation temperatures of 10 K and small opacities ($0.1<\tau<3$). The $^{12}CO/^{13}CO$ line ratios fluctuate by a factor of 5 from point to point, maybe due to *in situ* fractionation. $H_2$ column densities have been derived using X $\sim 10^{20}$ $H_2$ cm$^{-2}$ K$^{-1}$ km$^{-1}$ s, as recently obtained from EGRET data on the Cepheus Flare. They agree within a factor of 2 with the column-densities derived from an LVG-model applied to $^{13}CO$ data, while LTE assumptions yield values too low by a factor of 20.

The velocity field in the cloud is quite singular: it is highly organized on a 10-pc scale, with gradients of 0.4 km s$^{-1}$ pc$^{-1}$ along a plane and rings in the (l,b,v) space. Simple cloud rotation, expansion, or Galactic shearing do not reproduce the data. The observed field is quite consistent with the 'slow flow' of a perfect fluid around a spherical obstacle: 5500 solar masses are carried at a velocity of 3 km $s^{-1}$, implying a kinetic energy of $5 \times 10^{47}$ erg. Hydrodynamical simulations of the distortion of a large, dense cloud overtaken by a slow shock wave are underway.

**The Bowshocks of the L1582B Molecular Outflow**

J. Richer, S. Bence, and R. Padman (Mullard Radio Astronomy Observatory, Cambridge University)

High resolution CO 3–2 and 2–1 images of the RNO43/L1582B molecular outflow are presented: these data provide some of the best evidence to date that at least one class of flows are driven by high Mach number optical jets. The CO emission appears to arise in a fragmentary parabolic shell around the optical jet/bow shocks seen in H-alpha and [NII] emission; the CO emission is strongest approximately 30 arcsec upwind of the optical emission, with little emission from the head of the jet. We interpret this as direct evidence that the CO is accelerated in the low velocity wings of the optical jet's working surface; there is little CO at the head of the jet due to molecular dissociation at the high normal shock velocities expected there. The global properties of this outflow are also discussed: a consistent source model is presented in which the observed CO structure is generated by a single intermittent jet whose axis is time dependent; precession of the jet in a cone is one possible explanation for this behaviour.

**A Direct Comparison of CO and $H_2$ from the Orion Molecular Shock**

Matthew J. Richter, James R. Graham (UC Berkeley), Gillian S. Wright (Joint Astronomy Centre)

High resolution long slit spectra of the Orion Peak 2 molecular shock simultaneously detect CO $v = 1-0$ P(3) through P(6) and $H_2$ $v = 0-0$ S(9) emission near 4.7 $\mu$m. Since the $H_2$ line has a significantly higher energy upper state, yet is more wide spread than the CO emission, temperature alone is probably not the dominant constraint. More likely is that the CO traces higher densities. The CO lines are displaced further from the cloud rest velocity than the $H_2$ emission, suggesting the CO may arise in cooled post-shock gas. CO line ratios fail to specify a single temperature and density, but further study of this wavelength region should prove critical for understanding molecular shocks.

## Multiple Winds from the Carbon Star V Hydra: Studying Time-variations Using High-resolution Spectroscopy of the 4.6 $\mu$m CO Lines

R. Sahai (JPL/Caltech), B. Sugerman (Occidental College), and K. Hinkle (KPNO/NOAO)

Observations of millimeter-wave CO lines have shown the carbon star V Hydra (V Hya) possesses not only a slowly-expanding ($V_{\rm exp} \approx 15$ km/s), largely spherically-symmetrical "normal" circumstellar envelope (CSE), but also a fast ($V_{\rm exp} \approx 120$ km/s) secondary outflow, which is bipolar. Observations of individual P & R-brand lines in the fundamental vibration-rotation band of CO at 4.6 $\mu$m over two epochs by Sahai and Wannier (1988, *Astr. Ap.*, 201, L9) showed the presence of several absorption components due to very fast-moving (up to $V_{\rm exp} \approx 120$ km/s) gas. The absorption lines showed significant differences between the two epochs, implying changes in the physical properties of the multiple high-velocity outflow over year-long time-scales. This study provided the first direct evidence of the simultaneous presence of a roughly isotropic, slowly expanding envelope and a very high-velocity stellar wind in a mass-losing AGB carbon star. We are currently performing a detailed study of the high velocity outflow using high-resolution spectroscopic data (obtained with the KPNO 4-m/Fourier Transform Spectrometer) covering the 4.6 $\mu$m band of CO (and its isotopomers) taken over 5 epochs. Our analysis, thus far, has yielded several important results. First, the clean separation of the absorption features in our spectra due to the slow and high-velocity outflows, allows us to determine a precise expansion velocity of the slow component ($V_{\rm exp} = 10 \pm 2$ km/s), which is smaller than published results based on mm-wave lines. The high-velocity outflow consists of 4–5 individual kinematic components, covering a velocity range of about 40 km/s. We find structure within the absorption feature due to the slow outflow, indicative of the presence of kinematic perturbations super-imposed on the predominant expansion at 10 km/s. A detailed analysis of the structure of, and variations in, the various absorption components is in progress, aimed at constraining the nature of the high-velocity outflow, and its hydrodynamic interaction with the slow outflow. This study will thus make an important contribution to our understanding of the physical processes mediating the transformation of mass-losing stars like V Hya into bipolar planetary nebulae.

## Bar-driven Gas Structure and Star Formation in the Center of M100

K. Sakamoto, S. Okumura (Nobeyama Radio Observatory), T. Minezaki, Y. Kobayashi (National Astronomical Observatory), K. Wada (Hokkaido University)

High resolution (2.5″) CO(1–0) interferometry and near-IR imaging have been made toward the spiral galaxy M100 (NGC 4321). In the central 1′ (5 kpc), there are two-armed molecular spirals which start at each end of the nuclear stellar bar seen in the NIR images, and a central peak of CO emission with a diameter about 3″ (240 pc). The isovelocity map clearly shows distortions of the velocity field at the arms. We attribute the gas arms to the orbit-crowding at the outer inner Lindblad resonance (OILR), and the central peak to the collapsed self-gravitating gas that once piled up at the inner inner Lindblad resonance (IILR). Analytical and numerical models of gas dynamics with two ILRs well reproduced the observed gas structure and kinematics. The surface densities of gas at the OILR arms ($\Sigma > 500$ $M_\odot$ $pc^{-2}$) are above the threshold for gravitational instability (i.e., Toomre's $Q > 1$), thus the instability may have triggered the active star formation (H$\alpha$ hotspots) at the arms. In addition, frequent cloud-collisions as a result of orbit-crowding may also be a trigger mechanism. In spite of the similarity to NGC 1068 in gas dynamics, M100 does not show conspicuous AGN activity, indicating that the bar-driven gas fueling to the central ~100 pc is not sufficient for the onset of AGN. The results and the discussion are fully described in Sakamoto *et al.* (1995, *AJ* submitted).

## Physical Conditions of Molecular Gas in the Galaxy

Seiichi Sakamoto, Tetsuo Hasegawa, Toshihiro Handa, Tomoharu Oka, Kazuo Sorai (University of Tokyo), Masahiko Hayashi (National Astronomical Observatory)

We carried out an extensive out-of-plane survey of the inner Galaxy ($l$ = 20°–60°, $|b| \leq 1°$, $\Delta l = \Delta b = 0.25°$) in CO ($J$=2–1) using the Tokyo-NRO 60 cm survey telescope with the same angular resolution to the Columbia CO ($J$=1–0) survey.

Distribution of the CO ($J$=2–1) emission mimics that of the CO ($J$=1–0) emission both spatially and kinematically, with an average $J$=2–1/$J$=1–0 ratio ($\equiv R_{2-1/1-0}$) of 0.66. Through closer comparison of the data, however, we find a radial variation in $R_{2-1/1-0}$. The ratio decreases from $\simeq 0.75$

at 4 kpc to $\simeq 0.6$ at 8 kpc, with a local peak near 6.25 kpc where the Sagittarius arm lies. Gas densities derived through a one-zone model analysis are $\simeq 6 \times 10^2$ at 4 kpc and $\simeq 3 \times 10^2$ cm$^{-3}$ at 8 kpc, respectively. The large-scale variation of $R_{2-1/1-0}$ seems to be ascribed to the variation of the mixing ratio of dense cloud cores and low-density molecular gas. Dense molecular gas marked with large $R_{2-1/1-0}$ was found predominantly along the inner Sagittarius and Scutum arms. This result and the dominance of molecular gas over atomic gas in the inner Galaxy imply that dense molecular gas was formed through compression such as galactic shock compression followed by gravitational collapse, and not through dissociative stripping of low-density molecular gas.

## An Isotope Study of Carbon Monoxide in the Edge-on Galaxy NGC 891

Seiichi Sakamoto, Toshihiro Handa, Yoshiaki Sofue, Mareki Honma, Kazuo Sorai (Institute of Astronomy, University of Tokyo)

We carried out high-resolution strip-scan observations along the disk of the edge-on galaxy NGC 891 in $^{12}$CO ($J$=1–0) and $^{13}$CO ($J$=1–0) emissions with very accurate relative calibrations of pointing and intensity scale. Physical properties of molecular gas were examined through an excitation analysis especially with respect to their relation with spiral arms and the galactocentric distance.

There is a radial variation in the $^{12}$CO/$^{13}$CO ratio. The ratio is >10 inside the molecular ring except for the galactic center. It drops to the lowest value of $\simeq$2.5 near the molecular ring, and rises up outward to >5 at 10 kpc from the center. The arm-to-interarm contrast in the $^{13}$CO emission is higher than that in the $^{12}$CO emission. Nuclear disk emission is almost missing in the $^{13}$CO emission. Since the $^{12}$CO/$^{13}$CO ratio in a large scale reflects a fraction of dense gas to total amount of molecular gas rather than opacity of the $^{12}$CO emission, observed large-scale variation in the $^{12}$CO/$^{13}$CO ratio is understood in terms of concentration of dense molecular gas to inner spiral arms. This result and the dominance of molecular gas over atomic gas in the inner part of the galaxy imply that dense molecular gas in the inner spiral arms was formed through grand-design compression processes such as shock compression accompanied with density waves followed by gravitational collapse, and not through dissociative stripping of low-density molecular gas.

## Structure and Physical Conditions of Molecular Gas in the Orion GMCs

Seiichi Sakamoto, Tetsuo Hasegawa, Junichi Morino, Kumiko Sato, Masumichi Seta, Toshihiro Handa, Tomoharu Oka (Univ. of Tokyo), Masahiko Hayashi (National Astronomical Observatory)

We carried out $^{12}$CO ($J$=2–1) mapping observations of the entire extent of the Orion A and B GMCs using the Tokyo-NRO 60 cm survey telescope with the same beamsize and observation grid as the Columbia survey. Systematic variation in the $J$=2–1/$J$=1–0 ratio was found across the cloud. Gas densities extracted by an LVG analysis are $\geq 3 \times 10^3$ cm$^{-3}$ in central ridges and $\sim 2 \times 10^2$ cm$^{-3}$ in peripheries (Sakamoto et al. 1994).

To figure out origins of the systematic variation in physical conditions of molecular gas across the cloud, we conducted an extensive high-resolution imaging of the Orion A cloud in $^{12}$CO ($J$=1–0) emission with the Nobeyama 45 m telescope. We found two distinct populations of molecular gas: "clumps" with larger $T_b$, smaller linewidth and sharp boundary, and "interclump gas" with smaller $T_b$ and larger linewidth. The surface filling factor of molecular gas is approximately unity throughout the cloud, and the fraction of clump emission to the interclump emission decreases outward. Peripheries of molecular clouds have low CO brightness not because the surface filling factor of molecular gas is small, but because low-density interclump gas dominates the emission.

## Effects of Molecular Cloud Properties on the CO-to-H$_2$ Conversion Factor

Seiichi Sakamoto (Institute of Astronomy, University of Tokyo)

Effects of molecular cloud properties on the CO-to-H$_2$ conversion factor are quantitatively examined on the basis of a simple model of radiative transfer and excitation of the CO ($J$=1–0) emission for clumpy molecular clouds. The model deals with the effects of cloud structure such as elongation, viewing angle, mass distribution, and surface and volume filling factors, in addition to the effects of physical and chemical properties of clouds such as gas density, kinetic temperature, and the CO abundance.

The CO-to-H$_2$ conversion factor is strongly dependent on the CO abundance even when the CO emission is optically thick. It is because most of the molecular gas mass in molecular clouds in galaxies is in the form of low density gas in which even the $^{12}$CO ($J$=1–0) emission is subthermally excited and the degree of photon trapping is reflected in the excitation temperature of the CO emission. Properties of molecular clouds other than the CO abundance have minor effect to the conversion factor.

The numerical results reproduce the empirical relation of the conversion factor to the metallicity found in Galactic and extragalactic molecular clouds (Sofue, Arimoto, & Tsujimoto 1995, this conference). Molecular gas mass in metal-poor galaxies will be severely underestimated if the Galactic value of the conversion factor is assumed.

**Observations of the Perseus Arm in CO($J$=2–1) line**

Kumiko Sato, Tetsuo Hasegawa, Toshihiro Handa, Seiichi Sakamoto, Tomoharu Oka, Masumichi Seta, Kazuo Sorai, Jun Ichi Morino (IoA, University of Tokyo), Osamu Kimura, Hiroko Shinnaga (Tokyo Gakugei University), Masahiko Hayashi (National Astronomical Observatory)

We have made extensive observations of the second Galactic quadrant in CO($J$=2–1) line with the Tokyo–NRO (Nobeyama Radio Observatory) 60cm survey telescope. The observations were done over the area $l = 130.0°$ to $151.875°$, $b = -3.0°$ to $+3.0°$, which covers 840pc × 200pc at the distance of the Perseus arm. The previous Galactic plane surveys in the first quadrant show a trend that the intensity ratio between CO($J$=2–1) and CO($J$=1–0), ($R_{2-1/1-0}$), is about 0.6 at the solar vicinity and increases toward the inner region of the Galaxy. In the Perseus arm at a galactocentric distance of 11.5 kpc, however, the ratio $R_{2-1/1-0}$ is $0.88 \pm 0.03$ which is higher than that we extrapolate from the previous surveys. Most of the CO luminosity in this region comes from molecular gas associated with HII regions. These results suggest that most of molecular gas we can detect in the CO emission in the Perseus arm is strongly biased the high ratio gas ($R_{2-1/1-0} \sim 0.7$–$0.8$) which is adjacent to HII regions, and low ratio gas ($R_{2-1/1-0} \sim 0.5$) such as the Taurus dark cloud is hardly detected. A significant amount of cold or low density molecular gas still can exist in the Perseus arm without being detected in CO lines.

**CSO 607-725 GHz Line Survey of the Orion Hot Core**

P. Schilke, D.J. Benford, T.R. Hunter, D.C. Lis, T.G. Phillips and E. Serabyn (California Institute of Technology)

Using the Caltech Submillimeter Observatory (CSO), we performed a 607 – 725 GHz line survey of the Orion hot core between November 1994 and January 1995, using the 650 GHz facility SIS receiver. Since the weather conditions were excellent, eight nights were sufficient to cover the whole spectrum twice. The single sideband system temperatures were in the range 1500 – 3000 K. The double sideband data were subsequently deconvolved using a MEM method to produce a single sideband spectrum. The achieved sensitivity is sufficiently high to securely identify 0.5 K lines over most of the band. A preliminary analysis revealed around 1500 spectral features. Due to the preliminary stage of the line identification, the fraction of unidentified lines cannot be determined yet, but it appears that it is low. The spectrum is dominated by CO, CS, SO, SiO, HCN, $HCO^+$, $H_2CO$, $SO_2$ and $CH_3OH$, with many lines from very highly excited states. The sensitivity to these highly excited lines is partly due to the small beam size (about 12″), well matched to the hot core. Evidence for two new molecules was found (SiH and DF), but the identifications are only tentative.

**Multi-level CO Observations of Sgr B2**

P. Schilke, D.C. Lis and J. Keene (California Institute of Technology)

With the Caltech Submillimeter Observatory (CSO) we mapped a 2′ by 5′ area around SgrB2 in $^{12}CO(2–1)$, $^{13}CO(2–1)$, $C^{18}O(2–1)$, $C^{17}O(2–1)$, $^{12}CO(3–2)$, $^{13}CO(3–2)$, $C^{18}O(3–2)$, $^{12}CO(6–5)$, $HCO^+(3–2)$, DCN(3–2), SiO(5–4), $H_2S(2,2,0–2,1,1)$, SO(5,6–4,5), $SO_2(11,1,11–10,0,10)$, $H_2CO(3,2,1–2,2,0)$, OCS(18–17) and several transitions of HNCO(10–9) and $HCOOCH_3$, and a 8′ by 12′ area in $^{13}CO(2–1)$. The isotopomeric CO maps allow us to model the excitation conditions in the cores and the extended envelope of SgrB2, while the extended $^{13}CO(2–1)$ map provides information on the large scale kinematic structure of the region. The existence of very widespread, broad $^{12}CO(6–5)$ line emission indicates elevated temperatures and densities in a large fraction of the cloud envelope. The data permit us to determine the large scale cooling rate more precisely and thus allow a better understanding of heating processes. The maps of the other molecules give important insights into the chemistry of the region: some are peaking on SgrB2(M) ($H_2S$, SiO, SO, $SO_2$), some on SgrB2(N) (DCN, $HCOOCH_3$, OCS) and some seem to ignore the cores altogether ($H_2CO$ and HNCO).

The transitions of HNCO allow the determination of the far-IR field in the cores and the answers (150–200 K) agree with earlier estimates.

**CO and C$^+$ Observations in the Rosette Molecular Complex**

N. Schneider, J. Stutzki, G. Winnewisser (I. Phys. Institut, Universität zu Köln)

Large scale mapping of the Rosette Molecular Complex (RMC) in the 158 $\mu$m [CII] fine structure line and lower-$J$ isotopomeric CO transitions reveal the close morphological correspondence between the two tracers. However, there is no simple linear correlation between the CO and C$^+$ intensities as it was found for Photon Dominated Regions (PDR's) with high volume density and strong UV-flux, $F$. We found that the C$^+$ intensity scales with $\log F$. Although the [CII] line was detected on a rather low intensity level of about (1–5)$\times 10^{-4}$ erg s$^{-1}$ sr$^{-1}$ cm$^{-2}$, there is still significant emission deep in the molecular cloud. Necessarily, the cloud structure is highly inhomogeneous to account for the deep UV penetration creating the observed C$^+$ intensities.

The large scale molecular cloud structure is traced by 2$'$ angular resolution $^{12}$CO and $^{13}$CO $J$=2$\rightarrow$1 and 1$'$ angular resolution $^{12}$CO and $^{13}$CO $J$=3$\rightarrow$2 maps obtained with the KOSMA 3-m telescope. Selected regions of the RMC were observed in isotopomeric $J$=2$\rightarrow$1 and $J$=1$\rightarrow$0 CO transitions with the IRAM 30 m telescope. The complex velocity structure of the RMC is revealed in CO channel maps and position-velocity maps. The dynamic of the gas results from the influence of the associated HII region and embedded IR sources. The CO line intensities and ratios are compared with the results of a PDR model, considering a molecular cloud with finite size. The observed line ratios can only be explained by thermalized gas with a local density above $10^4$ cm$^{-3}$ and a beam filling factor of about 0.5.

In addition to the CO maps, we took high signal-to-noise (S/N) $^{12}$CO and $^{13}$CO $J$=2$\rightarrow$1 spectra with KOSMA. Previous observations reported the detection of 'molecular interclump gas' by identifying CO line wing emission in high S/N spectra and a positionally averaged spectrum in the RMC. We confirm the detection of line wings in the composite spectrum but not reliably in the individual spectra. In contrast to the former interpretation, we show that the wing emission most likely results from embedded outflows and from the superposition of individual high velocity clumps. This finding is based on a study of the excitation conditions of the wing gas derived from a multiline analysis of one high S/N-spectra.

**Molecular Clouds Interacting with the Supernova Remnant W44**

Masumichi Seta, Tetsuo Hasegawa, Toshihiro Handa, Seiichi Sakamoto, Tomoharu Oka, Kazuo Sorai, Kumiko Sato, Junichi Morino (University of Tokyo), Masahiko Hayashi (National Astronomical Observatory), Hiroko Shinnaga (Tokyo Gakugei Universtiy)

The supernova remnant W44 was observed in $^{12}$CO($J$=1–0),$^{13}$CO($J$=1–0) and HCO$^+$ ($J$=1–0) lines with 45-m radio telescope at Nobeyama Radio Observatory. Nine giant molecular clouds with masses from $3\times10^4$ to $4\times10^5$ solar masses and diameters from 5 to 30 pc are found near the SNR. Each molecular cloud has substructures identified as regions with brighter CO emission. Some of these substructures show morphological anti-correlation with the radio continuum brightness and this suggests that the expansion of supernova remnant is stopped by the molecular clouds. Wings of $^{12}$CO emission were discovered at eight positions and they are tightly correlated with bright filaments of radio continuum emission. Broad emission of HCO$^+$ extending to 55 km/s was detected at the positions with strong $^{12}$CO wings. These CO and HCO$^+$ wings are considered to be a result of an acceleration of molecular gas in shocks induced by the supernova. Enhanced abundance of HCO$^+$ may have resulted from chemical reactions in the shock.

Observations of the $^{12}$CO($J$=2–1) emission with the Tokyo-NRO 60 cm radio telescope show that the CO $J$=2–1/1–0 intensity ratio increases markedly at the GMC/SNR interface. The ratio will offer a new index in identifying regions of GMC/SNR interaction.

**Discovery of $H_2O$ Maser Emission Associated with the Supernova Remnant W44**

Masumichi Seta, Tetsuo Hasegawa (University of Tokyo), Nobuharu Ukita (Nobeyama Radio Observatory)

We have detected 22 GHz $H_2O$ maser emission in the supernova remnant W44. The 45-m radio telescope at Nobeyama Radio Observatory was used for the observations. Three clear components with intensitics from 1.8 to 7.7K were observed in the velocity range from 45km/s to 65 km/s at a position of $l = 34.787°, b = -0.387°$. Line of sight velocities of the maser features fall within the velocity range of the shocked molecular gas observed as broad CO and HCO$^+$ wings (Seta et al. 1996,these proceedings) and the maser spots are adjacent to a radio continuum peak. Because no IRAS point source is catalogued around this position, compact HII regions and late type stars are unlikely to be the source. The possiblity of a young star is also excluded because no point like radio continuum emission nor prominent molecular core is observed. This is possibly the first detection of $H_2O$ maser emission associated with a supernova remnant.

## Molecular Gas in M82: High Resolution CO, HCN, and $HCO^+$ Observations

Jianjun Shen, K.Y. Lo (Department of Astronomy, University of Illinois)

We present high resolution BIMA observations of CO(1→0), HCN(1→0), and $HCO^+$(1→0) emission from the starburst nucleus of M82.

We have identified two molecular rings in the nucleus of M82 and have found outflow motion of part of the molecular gas normal to the plane of the galaxy. The observations suggest that the starbursts took place within the molecular rings (or spiral arms) at some distance from the nucleus; the inner molecular ring is correlated with the ionized gas ring and the starbursts may, in fact, have propagated inwards. The dual molecular ring structure supports the suggestion that M82 may contain a nuclear bar.

We have identified ~60 molecular cloud clumps from the CO emission, with half-power diameters ranging from 20 pc to 100 pc and velocity widths (full-width half power) ranging from 7 km s$^{-1}$ to 65 km s$^{-1}$. The typical $T_b$ ranges from 13 K to 34 K. We found a power law distribution in the mass spectrum, with an index –1.3 ($dN/dM$ = –160 M$^{-1.3\pm0.2}$) for cloud masses greater than $4\times10^5$ M$_\odot$. The large index indicates that most of the molecular mass is contained in GMCs with large masses.

## CO ($J$=2–1) Mapping of Cygnus X Region with the 60-cm Telescope

Hiroko Shinnaga (Tokyo Gakugei University), Tetsuo Hasegawa, Tomoharu Oka, Toshihiro Handa, Seiichi Sakamoto, Masumichi Seta, Kumiko Sato, Kazuo Sorai, Jun Ichi Morino (IoA, University of Tokyo), Masahiko Hayashi (National Astronomical Observatory, Japan), Fumio Sato (Tokyo Gakugei University)

We have observed the Cygnus X region in the CO ($J$=2–1) line using the Tokyo-NRO 60-cm survey telescope. The observed area covers 6 × 4 deg$^2$, $77° < l < 83°$, $-2° < b < 2°$. The beam size of the telescope is about $9'$ and the sampling interval is $7\mskip-3mu'\!.5$. We compared our data with the CO(J = 1–0) data taken with 1.2m telescope of CfA (Leung and Thaddeus 1992).

The main conclusions are as follows:
(1) The average CO($J$=2–1)/CO($J$=1–0) intensity ratio, $R_{2-1/1-0}$, is 0.69± 0.06. This is similar to the values we got for GMCs with massive star forming regions (e.g., $R_{2-1/1-0} = 0.75 \pm 0.01$ in the Orion A cloud).
(2) We found an agglomeration of molecular clouds at $l \sim 78°$ which coincides with the cluster of HII regions. This could be interpreted as a result of the coherent trigger of star formation in a spiral arm.
(3) There are some high ratio clouds not accompained by known HII regions or SNRs. Some other mechanism that compresses or heats the CO-emitting gas is needed there.

## The Discovery of Methanol Maser Lines at 1.9 mm

V.I. Slysh, S.V. Kalenskii, and I.E. Val'tts (Astro Space Center of Lebedev Physical Institute, Russia)

The series of methanol $J_0{-}J_{-1}E$ lines at 157 GHz has been discovered in galactic sources together with the $2_1 - 3_0A^+$ line of $A$-type methanol. The observations were made in 1994 March, using the NRAO 12-m radio telescope at Kitt Peak. A 138-173 GHz SIS-mixer, dual-linear polarization receiver was used with system noise temperatures between 250 and 500 K. A hybrid spectrometer with 150 MHz total bandwidth and 768 channels was used providing frequency resolution of 195 kHz (or 0.37 km s$^{-1}$). Ten methanol transitions were observed by tuning the receiver to appropriate frequencies. In addition to eight lines of the $J_0{-}J_{-1}E$ series, the $2_1 - 3_0A^+$ line was observed, and the $6_2 - 7_1A^+$ was observed in one source W3(OH). The $E$-methanol lines are intense and narrow and are believed to be maser lines. The radial velocities and profiles are similar to those of other class II methanol lines in W3(OH), G345.01+1.79, W48, and Cep A. The intensity of the lines increases with rotational number $J$, in contrast to thermal methanol features which were detected in the same sources at slightly different radial velocities. In the $2_1 - 3_0A^+$ line, the masers were found only in two sources: G345.01+1.79 and W48. It is suggested that the new maser line sources are the Class II methanol masers, but are excited with higher radiation temperature than the lower frequency Class II masers at 6.7 and 12.2 GHz. The presence of a series of several maser lines confirms an assumption that the whole $K$=0 ladder of $E$-methanol is overpopulated in Class II masers, rather than some selected levels. The masers may be related to molecular outflows or cometary bow shocks of ultracompact HII regions as is evidenced by their velocity shift relative to the parent molecular clouds.

## Interstellar Acetone

L.E. Snyder, Y. Miao, D.M. Mehringer, Y.-J.Kuan (Univ. of Illinois), J.M. Hollis (NASA/GSFC), P.R. Jewell (NRAO), F.J. Lovas (Nat. Inst. of Standards & Tech.)

Acetone is one of the most important molecules in organic chemistry and it was the first 10-atom molecule reported in the ISM (Combes et al. 1987, A&A, 180, L13). The emission lines associated with the original detection were very weak. Thus, this identification was not well-established. Recent BIMA Array observations have shown that the complex species $HCOOCH_3$, $CH_2CHCN$, and $CH_3CH_2CN$ peak toward the Sgr B2(N) molecular core (Miao et al. 1995, ApJ, 445, L59; Miao et al. these proceedings). Combes et al. (1987) would not have detected acetone emission from Sgr B2(N) because their single-dish pointing position was 1.3′ away. Because Sgr B2(N) contains other complex organic species, we used the NRAO 12-m to search for new acetone transitions and the BIMA Array to image acetone emission. The blended $J$=8→7 quadruplets were observed with the BIMA B configuration (10″× 5″ resolution) and the images indicate that, like other complex organic molecules, acetone emission peaks toward Sgr B2(N). We observed these transitions in both Sgr B2(N) and Sgr B2(OH) with the 12-m and found that the acetone emission from Sgr B2(N) is about twice as strong as from Sgr B2(OH). Six new transitions are dectected toward Sgr B2(N). Thus, our results confirm the identification of interstellar acetone and support the conclusion of Miao et al. (1995) that grain-surface chemistry produces complex organic species in Sgr B2(N).

## CO-to-$H_2$ Conversion Factor: Metal Dependency and Radial Variation

Y. Sofue, N. Arimoto (Inst. Astro., Univ. Tokyo), and T. Tsujimoto (NAO, Tokyo)

Compiling values of the CO-to-$H_2$ conversion factor from the literature, we have examined the metallicity dependence and its radial variation in galaxies. The $X$ value (in $H_2$ cm$^{-2}$ K km s$^{-1}$) is found to be proportional to the metallicity of galaxies,

$$\log X^* = -1.0(12 + \log \mathrm{O/H}) + 9.30,$$

where $X^* = X \times 10^{-20}$. Since the metallicity is a function of absolute magnitude $M_B$ of galaxies, we obtain a formula to estimate the $X$ value for a galaxy with known luminosity:

$$\log X^* = 0.20 M_B + 4.44.$$

Since the metallicity is an exponentially decreasing function of the radius within a disk galaxy, the metal dependency would represent a radial gradient, and the data can be fitted by

$$\log X^* = 0.37 + 0.40(r/r_e - 1).$$

This formula can be used for any galaxy if the scale radius of its exponential disk $r_e$ is known. (For details: Arimoto, N., Sofue, Y., Tsujimoto, T. 1995 A&A, submitted).

**Inner-CO vs Outer-HI Rotation Curves of Galaxies**

Yoshiaki Sofue (Inst. Astro., Univ. Tokyo)

The optical and HI rotation curves concern the kinematics of the outer galaxy. The inner rotation curves can be most accurately obtained by using CO-line position-velocity diagrams. We have compiled CO-line PV diagrams for nearby Sb and Sc galaxies observed with large-aperture mm-wave telescopes and interferometers. The rotation velocity is defined by $V_{\rm rot} = V_{\rm t} - (\sigma_{\rm obs}^2 + \sigma_{\rm ISM}^2)^{1/2}$. Here, $V_{\rm t}$ is the terminal velocity at 20% of max-intensity, $\sigma_{\rm obs} \sim 10$ km s$^{-1}$ is the resolution, and $\sigma_{\rm ISM} \sim 7$ km s$^{-1}$ is the velocity dispersion of gas.

All the RC, without exception, are found to have a steep rising up near the nucleus, and some are very similar to that of the Milky Way. The RC can be fitted by the Miyamoto-Nagai potential with four mass components: (a) the central mass (scale radius $\sim 100$ pc, mass $3 - 5 \times 10^9$ $M_\odot$), (b) the bulge ($0.5 - 1$ kpc, $\sim 10^{10}$ $M_\odot$), (c) the disk ($5 - 7$ kpc, $\sim 1 - 2 \times 10^{11}$ $M_\odot$), and (d) the massive halo ($15 - 20$ kpc, $\sim 2 - 3 \times 10^{11}$ $M_\odot$). The most striking feature is the steeply rising nuclear peak of RC at $R \sim 100$ to 200 pc observed for all the galaxies, indicating a compact central mass concentration. This nuclear mass, in addition to the bulge, will have an important implication for the formation of the galaxy center and disk. (See for detail, Sofue, Y. 1995, ApJ, in press.)

## Molecular Front in Galaxies: Galactic-Scale HI–$H_2$ Transition

Yoshiaki Sofue, Mareki Honma, Nobuo Arimoto (Inst. Astro., University of Tokyo)

The major ISM in galaxies comprises the HI and $H_2$ gases. The HI gas is distributed in the outer galaxy, while the molecular gas is more concentrated toward the center. We examined the HI vs $H_2$ distributions in galaxies and discuss the galactic scale phase transition of the ISM. We have obtained the radial variation of the molecular fraction, $f_{\rm mol}$, as defined by $f_{\rm mol} = \rho_{\rm H_2}/(\rho_{\rm HI}+\rho_{\rm H_2})$ for nearby Sb and Sc galaxies. (Four face-on galaxies M51, M101, NGC 6946, IC 342; and five edge-on galaxies , NGC 891, NGC 3079, NGC 4565, NGC 5907, MW). The $f_{\rm mol}$-vs-radius diagrams show that the HI to $H_2$ transition occurs suddenly in a narrow range of radius at 5–10 kpc, which we call the *molecular front* (MF). A galaxy disk is clearly separated by this front: the inner molecular disk, and outer HI disk. The MF is determined by the interstellar pressure $P$, UV radiation field $U$, and metallicity $Z$. We constructed a model galaxy in which $P$, $U$ and $Z$ decrease exponentially as $\propto \exp(-r/r_e)$, with $r_e$ being the disk scale radius, and applied Elmegreen's theory. The theoretical $f_{\rm mol}$ variation reproduces well the observations, and a clear MF shows up. We find that the $Z$ (metal) gradient is most crucial for the formation of MF. The MF would, therefore, have crucial implications for understanding the chemical evolution of the ISM in galaxies. (References: Sofue, Y., Honma, M., Arimoto, N. 1995, A&A, 296, 33; Honma, Sofue, Y., M., Arimoto, N. 1995, A&A, in press.)

## CO-Line Tully-Fisher Relation: mm-Wave Cosmology

Y. Sofue (IoA, Univ. Tokyo; Kiso Obs., Univ. Tokyo), F. Schöniger (IoA, Univ. Tokyo), K. Wakamatsu (Dept. Physics, Gifu Univ.), T. Ichikawa (Kiso Obs., Univ. Tokyo), M. Honma (IoA, Univ. Tokyo), I. Kazes (Obs. Paris), J. Dickey (Dept. Astron., Univ. Minnesota)

The HI Tully-Fisher relation can reach galaxies with redshifts $cz \sim 10,000$ km s$^{-1}$. CO-line observations provide us with sharper beams and higher sensitivity to more distant galaxies and the atmospheric noise decreases at red-shifted CO frequencies. Astrophysically, molecular gas is better correlated with the stellar disk, so is less disturbed by tidal interactions than HI. Hence, we may expect that a CO line Tully-Fisher relation could reach farther distances than HI and better determination of $H_0$ would be possible. We have established the CO-line Tully-Fisher relation for the local calibrators and nearby galaxies from a sample of several tens of galaxies and demonstrated that CO can be used as an alternative to HI.

We have used the Nobeyama 45-m telescope to observe $^{12}$CO(J=1–0) line profiles of distant cluster galaxies at $cz \sim 10000-30000$ km s$^{-1}$. CO was detected in seven galaxies. We determined the distances to three galaxies from our CO sample by the Tully-Fisher relation using optical data from the Kiso Schmidt CCD images. A preliminary value of the Hubble constant for the galaxies is $H_0 \simeq 106 \pm 30$ km s$^{-1}$. (References: Dickey J., Kazes I., 1992, ApJ, 393, 530; Schöniger, F., Sofue, Y. 1993, A&A, 283, 21; Sofue Y., 1992, PASJ, 44, L231)

## $^{12}$CO ($J$=2–1) Emission in the Inner Galaxy

Kazuo Sorai, Tetsuo Hasegawa, Toshihiro Handa, Seiichi Sakamoto, Tomoharu Oka, Masumichi Seta, Kumiko Sato, Jun Ichi Morino (Institute of Astronomy, Univ. of Tokyo), Masahiko Hayashi (National Astronomical Observatory, Japan), Hiroko Shinnaga (Tokyo-Gakugei Univ.)

We have made a survey of the $^{12}$CO ($J = 2-1$) line emission from the inner Galaxy covering from 6 to 20 degrees in galactic longitude and from $-1$ to $+1$ degrees in galactic latitude with the University of Tokyo-Nobeyama Radio Observatory 60 cm radio telescope in Nobeyama. HPBW is $9'$ and sampling grid is $7.5'$.

The distribution of CO ($J = 2-1$) emission is similar to that of CO($J = 1-0$) shown by the Columbia survey (Dame *et al.* 1987). The intensity ratio, $R_{2-1/1-0}$, is $0.73 \pm 0.01$ averaged over the whole observed region. This value, taken superficially, corresponds to $n(\mathrm{H_2}) \lesssim 10^3$ cm$^{-3}$ and $T_\mathrm{k} \gtrsim$ 20 K in a framework of one-zone LVG analysis. The ratio shows a systematic gradient as a function of the galactocentric radius. Adding the data of 20 to 60 degrees in galactic longitude (Sakamoto *et al.* 1995) to our data, the ratio is about 0.5 in the solar vicinity and goes up to about 0.8 at 3 kpc from the galactic center. The ratio of molecular gas that appears to be associated with HII regions is 0.8 on the average markedly higher than average over the whole disk gas, but its contribution to the whole luminosity is small.

*References*: Dame,T.M. *et al.* 1987, ApJ, **322**, 706; Sakamoto,S. *et al.* 1995, ApJS, in press

**Circumstellar Molecular Gas of PV Cephei**

Karl Stapelfeldt (Jet Propulsion Laboratory, California Institute of Technology)

The peculiar young Ae star PV Cephei has been observed in the 1–0 lines of $^{13}CO$ and $C^{18}O$ and in the 2.7 mm continuum using the Owens Valley Millimeter Array. In all three tracers, circumstellar gas and dust are detected in the $3.9'' \times 3.2''$ beam. The dust continuum emission is spatially unresolved and centered exactly at the stellar position. The flux density ($38 \pm 4$ mJy at 110 GHz) suggests a circumstellar material mass of 0.3 $M_\odot$. CO isotope emission is also centered at the stellar position, but is spatially resolved with a deconvolved source size of 2600 AU at PA 120° and unresolved in the perpendicular direction. The $^{13}CO$ emission is seen over the VLSR velocity range 1.8–4.3 km $sec^{-1}$, but vanishes at the intermediate velocities 2.6–3.0 km $sec^{-1}$ because of self-absorption. The emission peak shifts from the NW side of the star at the lower velocities to the SE side of the star at higher velocities, suggesting rotation. These maps appear to confirm the existence of the circumstellar disk for PV Cephei which has been suggested by previous optical, infrared, and radio work. In contrast to the recent non-detections of other AeBe stars via millimeter interferometry, the results for PV Cephei show that Ae stars can possess massive disks at their earliest evolutionary stages associated with strong bipolar outflow.

**The Evolution of Detached Shells in AGB-stars**

Lars Stenholm (Uppsala Astronomical Observatory)

Detached shell AGB-stars have experienced a short period of intense mass loss. This massive shell is also faster than the previous normal wind. Observations of S Sct indicate that the shell is thin and to the first approximation spherical. There is, howerver, also indications of clumping along the shell. This project intends to identify the physical processes involved in the evolution of the shell. We used an 3D hydrodynamic code of SPH type to simulate the shell evolution.

## $CO^+$ in Photon-dominated Regions

Herbert Störzer (Universität zu Köln)

Recent observational studies have shown that molecular clouds in star-forming regions are inhomogeneous and clumpy, and that dense clumps appear to survive or form in clouds exposed to intense stellar far-ultraviolet radiation. These clumps may be studied directly by observing molecular species which are exclusively produced in dense PDRs. The molecular ion $CO^+$ is such a diagnostic probe.

We present the results of observational studies of $CO^+$ rotational emission from star-forming molecular clouds in Orion, M17, and S140. We give an interpretation of the results in the context of strong UV/high density PDR regions.

## Multiple Winds from the Carbon Star V Hydra: Studying Time-variations Using High-resolution Spectroscopy of the 4.6 $\mu$m CO Lines

Ben Sugerman (Occidental College), Raghvendra Sahai (JPL/Caltech), Kenneth Hinkle (KPNO/NOAO)

Observations of mm-wave CO lines have shown that the carbon star V Hydra possesses a slowly-expanding ($V_{\rm exp}\approx15$ km/s), largely spherically-symmetric "normal" circumstellar envelope, with a fast ($V_{\rm exp}\approx30$ km/s) secondary bipolar outflow. High-resolution spectra of the 4.6 $\mu$m P & R-branch CO lines over two epochs by Sahai and Wannier (1988, *Astr. Ap.*, **201**, L9) showed the presence of absorption components due to very fast moving ($\leq V_{\rm exp}\approx120$ km/s) gas. Spectral variations between these epochs implied changes in the high-velocity outflows over $\lesssim1$ yr. We have extended this study by obtaining higher quality data (KPNO 4-m FTS) over 7 epochs. We determine an expansion velocity of the slow component ($V_{\rm exp}\approx10\pm2$ km/s) which is less than that found from published mm-wave results. The high-velocity outflow consists of $\geq6$ kinematic components over a velocity range of $\sim70$ km/s. Analysis of integrated intensity ratios of the various components imply changes in the high-velocity structure over very short time-scales ($\sim2$ days) constraining this outflow to $\sim10^{16}$cm. This study will thus make an important contribution to our understanding of the physical processes which mediate the transformation of mass-losing stars like V Hydra into bipolar planetary nebulae.

## An Analysis of the Smoothness of Line Profiles

J.A. Tauber, M. Fridlund, G. Pilbratt (European Space Agency, Astrophysics Div.)

It is now generally agreed that molecular clouds are clumpy, i.e., they exhibit large point-to-point density fluctuations. Because individual clumps are observed to emit lines which are narrower than those emitted by their ensemble (the cloud), they must leave an imprint as velocity structure superimposed on ensemble profiles. Thus, a possible route to explore clumpiness is to observe emission lines with very high signal-to-noise ratios and very high velocity resolution, analyze the fluctuations in brightness temperature present on the line shape, and infer from them the properties of the clumps present in the beam. The advantage of this technique is that since a relatively large field of view is sampled, the statistical properties of clumps are obtained directly, without having to go through the process of detecting and mapping each clump as is done with interferometry.

This technique was developed recently and applied to the case of Orion (Tauber, Goldsmith and Dickman 1991), using CO and $^{13}$CO lines as tracers. However, in that paper the clumps were assumed to be "black disks", and the beam area filling factor was assumed to be small. We have now generalized the technique to include clumps with an arbitrary central opacity and radial distribution, using as a framework the clumpy cloud model of Martin, Sanders and Hills (1984). We show how with this model, the smoothness of the observed profiles can be used to constrain the parameter space of the unresolved clumps.

## Cometary Globules in the Orion Super-Bubble

David S. Theil, Ka Chun Yu, John Bally (Center for Astrophysics and Space Astronomy, University of Colorado, Boulder)

We present a preliminary search for cometary globules near the Orion OB associations using IRAS Sky Survey Atlas (ISSA) plates. Coincident $^{12}$CO scans of twenty-three of the objects are also presented.

The cometary globules were identified by examining ISSA plates at 25 $\mu$m, 60 $\mu$m, and 100 $\mu$m around the Orion region [R.A. range $7^h$ to $2^h$, and Dec. range $-24°$ to $+20°$]. The general morphologies (the locations and directions of "tails" in our sample) and kinematics (velocity information from the CO data) suggest that their interactions with the early-type stars in the Orion OB association and older associations in the region are dominated by ionizing radiation originating from a few distinct locations near the center of Barnard's Loop. Using CO observations made from the AT&T Bell Laboratories 7 m offset Cassegrain antenna, we obtain line velocities and line widths for these objects, as well as $^{12}$CO and $^{13}$CO maps for several of the globules.

Many of the globules with R.A. $> 5^h$ are moving towards the sun with $v \lesssim 25$ km s$^{-1}$ relative to the $V_{LSR}$ of $\sim 5$ km s$^{-1}$ for the Orion molecular cloud, which is consistent with numerical simulations of super-bubbles large enough to "blow out" of the galactic plane.

### The Distribution of Cold Gas in the Flocculent Spiral NGC 5055

Michele D. Thornley (University of Maryland)

BIMA and FCRAO CO $J$=1–0 observations are used to assess the properties of the molecular gas distribution in the nearby, flocculent spiral NGC 5055 (7.2 Mpc). As the weak spiral structure in flocculent galaxies may indicate the absence of spiral density wave activity, these observations can help determine the range of cloud sizes and masses that form in spiral galaxies without the influence of strong density waves. With an average surface density one-quarter that of the grand design spiral M51, NGC 5055 exhibits molecular structures with similar properties to the Giant Molecular Associations (GMA) detected in M51. The measured structures have masses between $6 \times 10^6$ and $1 \times 10^8$ M$_\odot$, and half of these have comparable virial masses. The distribution of CO cloud structures is also compared with the distribution of HI, H$\alpha$, and NIR peaks in order to examine the relationship between cold gas and star formation in NGC 5055. This comparison suggests structure organization on kiloparsec scales.

### Molecular Clouds in M51

Tomoka Tosaki (Nobeyama Radio Observatory), Yoshiaki Taniguchi (Tohoku University), Ryohei Kawabe (Nobeyama Radio Observatory)

We report results of higher spectral-resolution analyses (3.25 km s$^{-1}$) of $^{12}$CO($J = 1-0$) mapping data of the bright arm region in M51 obtained with the Nobeyama Millimeter Array (NMA). We identified 79 clouds as distinct components both spatially and kinematically. The less-massive clouds have the virial masses with a few $\times 10^5$ M$_\odot$, which is just comparable to a typical mass of the Galactic GMCs. We found that for almost all clouds the ratios of the mass estimated from CO flux using the Galactic CO-to-$H_2$ conversion factor to the virial mass estimated from the size and velocity width are much larger than 1. The mean ratio in all of the field is 5. This result suggests that the CO-to-$H_2$ conversion factor is smaller than the Galactic one, and thus implies that in M51 there are many clouds with smaller optical depth than those of the Galactic GMCs. The high temperature and the high metallicity abundance of molecular clouds in M51 may account for the small conversion factor, namely, small optical depth. To investigate the effect of the density wave on the nature of molecular clouds, we divided them into three types by their positions; upstream, on-arm, and downstream clouds. There is a clear difference in the brightness temperatures between the upstream and the downstream clouds. The average brightness temperature for the downstream clouds is high, 8.5 K, which is close to that for the on-arm clouds. This implies that the physical properties of the downstream clouds are closer to those of the on-arm clouds than those of the upstream clouds.

## CO(1–0), CO(3–2), and HCN(3–2) Observations of IRAS F10214+4724

Masato Tsuboi (Institute of Astrophysics and Planetary Science, Ibaraki University), Naomasa Nakai (Nobeyama Radio Observatory)

We have observed CO(1–0), CO(3–2) and HCN(3–2) emission from the high redshift galaxy IRAS F10214+4724 using the Nobeyama 45-m telescope. CO(1–0) and CO(3–2) were detected at $z$=2.286 with integrated intensities of 2.4±0.3 and 5.1±0.3 Jy km $s^{-1}$, respectively. The intrinsic brightness temperature ratio of CO(3–2) to CO(1–0) is 0.25. HCN(3–2) was not detected, although a weak $3\sigma$ emission line was recognized at $z$=2.286 ($1\sigma$=0.8 mK in $T_{\rm MB}$). The $5\sigma$ upper limit to the integrated HCN(3-2) intensity is 1.3 Jy km $s^{-1}$. The $5\sigma$ upper limit to the intrinsic brightness temperature ratio of HCN(3–2) to CO(3–2) is 0.2. In addition, we have searched for CO(3–2) from 4 points (east, west, north, and south) 15″ from the center position. There is no significant emission at these points. These results suggest there is a huge mass of low-density molecular gas at $n(H_2)$= $10^2$ $cm^{-3}$ in the object.

**The Discovery of Methanol Masers at 107 GHz**

I.E. Val'tts, A.M. Dzura, S.V. Kalenskii, V.I. Slysh (Astro Space Center of Lebedev Physical Institute, Russia), R.S. Booth and A. Winnberg (Onsala Space Observatory, Sweden)

A new methanol maser line was discovered in galactic sources in a transition between the K=0 and K=1 ladders at 107 GHz. The observations were carried out, using the radome-enclosed 20 m millimetre wave telescope of Onsala Space Observatory. A cryogenically cooled low-noise SIS mixer was used in the receiver; the single side-band receiver noise temperature was 80 K. The backend was a 64 MHz wide 256-channel filter spectrometer with a frequency resolution of 250 kHz (velocity resolution 0.7 km $s^{-1}$, at 107 GHz).

All detected sources are known Class II methanol masers. It is similar to the much stronger 6.7 GHz masers. The sources where the new maser line was observed are in general the same sources which are 6.7 GHz Class II line emitters. Toward Class I methanol maser sources, the 107 GHz line was found in absorption, or as rather weak quasi-thermal emission sources.

The detection of the new methanol maser line supports an assumption that the level population inversion is due to overpopulation of some K-ladders relative to neighboring ladders.

**Dust and CO in the T Tauri Circumstellar Matter**

Huib Jan van Langevelde (Joint Institute for VLBI in Europe, Dwingeloo and NRAO Socorro), Ewine F. van Dishoeck (Sterrewacht Leiden), Geoffrey A. Blake (Geological and Planetary Sciences, Caltech)

The circumstellar material around T Tau shows complicated structure at different scales and different observing frequencies. This is partly due to the fact that the optical star (T Tau N) is accompanied by a close infrared companion (T Tau S). We present sub-arcsecond resolution observations of gas and dust to try to discriminate between several components in the circumstellar material around T Tau. Observations with the Owens Valley Millimeter array at 267 GHz reveal continuum emission, as well as—on the short baselines—$HCO^+$ $J$=3→2 emission. The continuum emission indicates an unresolved component (70 AU) at the position of T Tau N, but the emission is at least a factor 4 weaker at the position of T Tau S. Submillimeter interferometer data between the CSO and JCMT taken at 357 GHz are consistent with a single component and provide clues about the spectral energy distribution of the dust around T Tau N. Comparison with single-dish data shows that considerable amounts of gas and dust are present on a larger, circumbinary scale of $\approx$ 1000 AU. Preliminary estimates indicate a relatively low CO-to-dust ratio in the circumbinary envelope.

## CO Observations of NGC 7674

L. Verdes-Montenegro (Instituto de Astrofísica de Andalucía, CSIC), J. Cernicharo (Instituto Geográfico Nacional), J. Perea, A. del Olmo, I. Márquez (Instituto de Astrofísica de Andalucía, CSIC)

We have mapped the CO(2–1) and (1–0) emission toward NGC 7674, 1st ranked galaxy of Hickson 96, with the IRAM 30 m radiotelescope (Granada, Spain). The beam sizes were $\sim$ 13″ and 22″, respectively. The molecular gas associated with NGC 7674 is overabundant and extended. Both CO transition maps show enhanced emission that seems to be associated with the spiral arms of the galaxy. We obtained a total mass of $2 \times 10^{10}$ $M_{\odot}$. A comparison of the masses of the interstellar gas in the molecular and atomic gas phases leads to a value of $M(H_2)/M(HI) \sim 2$ according to the interacting nature of the galaxy and its high infrared luminosity.

## CO Emission from QSO Absorption Line Systems

Constance E. Walker, Jill Bechtold, John H. Black (Steward Observ., U. Arizona)

By studying the physical conditions in galaxies at different redshifts, we can begin to understand the evolutionary process of star formation from early epochs to the present. Observations of the mm/submm CO lines in these galaxies provide a way to investigate abundances of, and conditions in, potential star-forming material. Here we report the detection of CO in emission toward 4 galaxies at redshifts of 0.005 to 0.05. Detections were made in two transitions toward each of the four sources: PKS 0038–019, S4 0248+43, 3C 232, and PKS 1327–206. From these observations, we computed the molecular gas mass to be $1.7\times10^8$ $M_{\odot}$ to $3\times10^9$ $M_{\odot}$. These estimates used the "standard conversion factor" for $H_2$ column density to integrated CO intensity ratio, $(N(H_2)/I_{co})$, and an $H_0$=75km/s/Mpc and $q_0 = 0.5$. We also report upper limits to the molecular gas mass in 33 other systems with redshifts ranging from 0.03 to 2.8 to be $< 4 \times 10^8$ $M_{\odot}$ to $< 5\times10^{11}$ $M_{\odot}$. All four of our sources detected in CO were selected to be metal line systems (with high HI column densities) and possess strong FIR fluxes detected by IRAS at 60 and/or 100 $\mu$m. The source possessing the brightest FIR emission among the four is the $z = 0.05$ 21 cm absorber toward S4 0248+43. The total observed FIR luminosity for this source is $L_{\rm FIR} = 3.0 \times 10^{11}$ $L_\odot$. Taking into account its luminosity and the additional correction for the integrated flux from 0.01 – 100 $\mu$m (Sanders et al. 1988), this object can be classified as a transition object between a normal luminous galaxy and an ultraluminous IR galaxy. This absorber galaxy may be the only known starburst in front of a quasar.

## An Experiment to Detect Linear Polarization of Millimeter-wave Emission Lines

Christopher K. Walker (Steward Observatory), Phillip R. Jewell (NRAO), Jason Glenn (Steward Observatory)

Using a sensitive, high efficiency polarimeter, we have used the NRAO 12-m to search for linear polarization in the $HCO^+$ $J = 1 \rightarrow 0$ transition (89.19 GHz) toward the redshifted molecular outflow lobe in DR21. $HCO^+$ was chosen as the molecular probe because of its high dipole moment (3.3 D). The conditions toward this outflow lobe satisfy many of the excitation requirements for linear polarization. After ~16 hrs of integration, linear polariztaion was *not* observed to the $\sim 0.07\%$ level. This sensitivity level is $\geq 10\times$ higher than what is required to detect line polarization using the most conservative theoretical estimates.

Why was linear polarization not observed? The most plausible explanation is that there are small-scale inhomogeneities in the velocity field along our line of sight that work to randomize the radiation field. Without an anisotropic radiation field, the Zeeman sublevels of a given transition are not differentially populated, and linear polarization is not produced. What is reponsible for the small-scale structure in the velocity field? Turbulence within the material entrained by the stellar wind is the most likely cause. Since no polarization was observed anywhere in the line profile, one is led to conclude turbulent gas is present along the entire line of sight. This result is consistent with a model in which the outflow lobe is filled with turbulent gas.

**Shocked Vibration - Rotation CO and $H_2$ in NGC 2071 IRS1**

D. M. Walther and T. R. Geballe (Joint Astronomy Centre/UKIRT)

Two lines of the fundamental vibration-rotation band of CO at 4.7 $\mu$m have been detected in emission in the vicinity of NGC 2071/IRS1. Velocity-resolved spectra of these two lines (P(3) and P(4)) as well as the neighboring $H_2$ 0–0 S(9) line are presented. The $H_2$ line has a velocity width of 110 km $s^{-1}$. The three lines have similar velocities of maximum line intensity and are broadened, suggesting that the CO emission, like that of $H_2$, occurs in shocked-heated gas. The CO lines are cut off on their blue-shifted sides by absorption by quiescent CO in the NGC 2071 cloud.

**CO Observations of Dust Lane Galaxies: NGC 5866 and NGC 3656**

Z. Wang (Smithsonian Astrophysical Observatory)

We present aperture synthesis CO (1–0) observations of early-type galaxies NGC 5866 and NGC 3656. The molecular gases are found to trace the dust lane as seen in optical images of these galaxies, but appear to be unsettled in the gravitational potential. Also, the rotation curves derived from the CO data are found to be rising faster than a simple kinematic model would predict. These results may shed some light on the origin of cold gas and dust in early-type galaxies.

**Photodissociation and Rotational Excitation of Interstellar CO**

Sophie Warin, Jean-Jacques Benayoun (Observatoire de Grenoble), Yves P. Viala (DEMIRM, Observatoire de Paris, Ecole Normale Supérieur)

We have developed a detailed interstellar cloud model that includes C and O chemistry, as well as $^{13}C$ and $^{18}O$ isotopic substitutes. All major processes that modify rotational population of $H_2$, CO and its isotopes are taken into account: selective photodissociation, UV pumping for $H_2$, chemical, collisional and radiative processes. Gas and grain temperature distribution can be obtained by solving the thermal equilibrium equation. We have computed a series of models for diffuse, translucent and dense dark clouds. For each type, we consider a "standard" model with a uniform density and temperature. Selective photodissociation of CO leads to a strong dependence on the rotational quantum number and a strong coupling between photodissociation and rotational excitation. In most of the cases, results show that the rotational populations of CO and its isotopes are generally sub-thermally excited with different excitation temperatures for each transition.

We carry out a study of CO abundances as a function of visual extinction to compare to observations. We ran models with uniform density and a distribution temperature computed by solving the thermal balance equation. We found a very good agreement between our model and Frerking et al. (1982) observations. Although we predict too low abundances of both $^{13}CO$ and $C^{18}O$, the column density ratio $R(^{13}CO/C^{18}O)$ normalized to the isotopic ratio matches pretty well the Lada et al. (1994) ratio for IC 5146.

## A High-resolution Multi-transition Study of CO, CI and Other Lines Towards the Galactic Centre Circumnuclear Disk: A Unique Example of a Photo-dissociation Region

G. J. White (QMW, University of London)

This paper reports the results of a major long-term programme on the JCMT to study the chemical structure and excitation gas in the unique circumnuclear molecular ring (CND) at our Galactic Centre. The CND provides us with our closest view of conditions that may be typical of many external galactic nuclei. The CND has been imaged by the JCMT with beams as small as 10″ in the CO $J$ = 2–1, 3–2 and 4–3 CO and their isotopic transitions, in atomic Carbon, CI, and in a variety of other strong lines and their isotopes including CS, CN, HCN, $HCO^+$ and $CH_3OH$. A well defined rotating ring structure is seen in the CO $J$ = 4–3 transition, and in maps of $^{13}CO$ and $C^{18}O$. However, the CO isotope line intensities are unusually weak, typically 5–10% those of the main line, suggesting they are surprisingly relatively optically thin. The CO structure is closely correlated to the structures seen in other high density tracers such as CS, CN, HCN, $HCO^+$ and $CH_3OH$, but weakly correlated with atomic carbon, CI, which has a distribution quite different from those of any of the molecules. The CO $J$ = 4–3 line intensities exceed 100 K at several points around the ring, as well as showing deep self-absorption features similar to those seen in CN and HCN. A spectral survey in the 200 and 300 GHz windows of two bright points on the ring reveals a simple chemistry, characteristic of photodissociation regions, and dominated by highly excited transitions of simple molecules. Characteristic values of the excitation temperatures and densities are 200 – 400 K and $10^5$ cm$^{-3}$ respectively. The material is highly clumped, its morphology and kinematics showing it is closely correlated to ionised gas making up the central 'mini-spiral'. The material at the edge of the CND is clearly being stripped of material, which is feeding along the arms of the mini-spiral into the Centre.

## Atomic Carbon, CO and Submillimetre Continuum Studies of the Orion Nebula and the Serpens Star Formation Regions

G. J. White (QMW, University of London)

This paper reports the results of details multi-transition studies of the Orion and Serpens molecular cores in CO and CI lines to study the relationship between atomic carbon and CO, and to attempt to measure the gas-to-dust ratios, and the behaviour of the isotopic abundance ratios of CO. The spatial distributions of the $^3P_1 -^3 P_0$ atomic fine structure of carbon (CI), the CO $J$=4–3, the CO $J$=2–1 transitions of CO, $^{13}$CO, C$^{18}$O and C$^{17}$O and the 790 $\mu$m continuum emission have been mapped towards the central region of the Orion molecular cloud (OMC1 cloud), and the Bright Bar ionisation front. The CO data are analysed in a consistent way, allowing the inter isotopomeric abundance ratios to be studied over a wide range of extinction values. The $^{13}$CO lines are optically thick; the $^{13}$CO abundance being enhanced because of strong isotopic fractionation near the Bright Bar, but less convincingly in the OMC1 cloud. The fractionation occurs mostly in the less opaque regions where N($^{13}$CO) may be enhanced by up to one order of magnitude, relative to the more shielded parts. No isotope selective enhancement of the other CO isotopomers was seen; C$^{18}$O may in fact show a slight depletion in more exposed material. The C$^{18}$O and C$^{17}$O lines are optically thin, and correlated with the 790 $\mu$m dust continuum emission. The CI emission comes from hot optically thin gas; the abundance ratios of [CI]/[CO] are typically 0.05–0.3, with the larger ratios towards the northern section of the Orion ridge. The CI abundance ratios are also high along the edge of the Bright Bar which is adjacent to the HII region (and the Trapezium cluster which excites it), but decrease in the dense shielded material behind the Bar. The Serpens Molecular cloud is a nearby low/intermediate mass star forming region that is in the final stages of forming a densely packed cluster of stars. Spectra and high angular resolution maps of the CO, $^{13}$CO, C$^{18}$O and C$^{17}$O $J$=2–1 and CO $J$= 4–3 transitions were obtained to study the distribution of molecular gas near the cluster. These are supplemented with data on the $J$= 4–3 HCO$^+$ line (to probe the denser gas), and $^3P_1 -^3 P_0$ line of atomic carbon. The mass of the region is estimated to be $\sim$ 1450 M$_\odot$, implying that the star formation efficiency in the region to date has been $\sim$ 2.5 percent. Several molecular outflows are visible in the maps; some are associated with compact objects visible in millimetre and submillimetre wave continuum maps, as well as more widespread diffuse high velocity gas that extends over much of the nebula. The mass and energy of material in the high velocity gas are relatively small, $\sim 0.3$ $M_\odot$ and $3 \times 10^{45}$ erg $s^{-1}$, consistent with the characteristics of outflows seen towards low mass star-formation regions. The directions of the overlapping outflow lobes do not, however, show a clear alignment with the cloud's large scale magnetic field as has been reported towards some other regions. The gas temperatures in the region are warmer ($\sim$ 30–40 K) than typical for dark clouds, suggesting that the Serpens cluster has interacted with, and heated the gas. Estimates of the CO isotopomeric abundance ratios from these data may be strongly affected by opacity and radiative transfer effects. In a comparison of the LTE and LVG techniques, the disagreement between the derived column densities is discussed - with particular relevance to the higher rotational transitions.

**CO in Selected Edge-on Galaxies**

R. Wielebinski (Max-Planck-Institut für Radioastronomie)

We have made extensive CO maps of numerous northern and southern edge-on galaxies in both the (1–0) and (2–1) line transitions. The observations were made both with the 30m Pico Veleta telescope and with the 15m SEST facility. The maps are made with receivers of good sensitivity so they represent the most extensive studies of these objects so far. Recently, high-angular resolutions studies in CO of nuclear areas of some of 'our' galaxies have been started with the Plateau de Bure interferometer. The main objective in this on-going project, which is being supported by a Max Planck-Forschungspreis (Australian partner John Whiteoak), is to study the dynamics of the galaxies. In particular, we wish to see if the combination of the various dynamic components (optical, HI, dust, molecular component, FIR, X-ray, etc.) will give us a better understanding of the whereabouts of the missing mass at the individual galaxy level. We collaborate with Michel Guélin (Grenoble) in the observation of the 1 mm dust. The HI observations with identical angular resolution come either from the VLA or the Australia Telescope. Optical observations are usually available in publications. We are working towards the adoption of various dynamical codes for galaxies to handle multi-component data (collaboration with Françoise Combes (Paris) and Jan Palous (Prague). The objects being studied (edge-on) are: NGC55, 253, 613, 891, 3034(M82), 3628, 4565,

4594(M104), 4631, 4945, 5907, and 7331. In addition we have observed: NGC1566, 1672, 1808, 2442, 3627, and M51.

**The Structure and Evolution of Giant Molecular Clouds**

Jonathan P. Williams (University of California at Berkeley)

The relationship between the properties of the molecular gas in a cloud and its star forming nature are investigated and interpreted in terms of an evolutionary scenario for molecular clouds. The internal clumpy structures in two clouds, the Rosette molecular cloud (RMC) and the Maddalena–Thaddeus molecular cloud (MTC) are compared and contrasted with their differing stellar content: the RMC contains the Rosette nebula, a very luminous OB association and, in addition, has several sites of ongoing star formation deeper into the cloud; but the MTC is very "infra-red cold", and there are no O stars, formed or forming, within its molecular boundaries.

Gradients of clump properties across the RMC are found that show gas density, temperature and star forming activity, $L_{\rm IR}/M$, all systematically increase toward the OB association. In the MTC, however, the gas properties are found to be remarkably similar to those in the RMC. The clump mass spectra, size-linewidth relations, clump profiles, *etc.*, share the same slope in the two clouds. Differences between the two are found not in the slopes of the relations, but in the offsets, that is, *not in structure, but in scale*, e.g., the virial mass scales as $M^{2/3}$ in both clouds, but whereas no clumps in the MTC are bound, all the star forming clumps in the RMC are.

The gradients that are seen in the Rosette suggest that star formation in a cloud may not be a purely stochastic process but is influenced by its environment. The structural similarities between the RMC and the MTC possibly reflect common cloud formation/evolutionary processes. Cloud evolution may proceed via clump agglomeration and dissipation in a way that preserves structure but leads to the formation of massive, bound, and ultimately star forming clumps.

**BIMA Array Observations of $C^{18}O$ and 2.7 mm Continuum Emission Toward NGC 2024**

T. L. Wilson (Max-Planck-Institut für Radioastronomie), D. M. Mehringer, and H. R. Dickel (University of Illinois)

NGC 2024 has been imaged in the J = 1–0 line of $C^{18}O$ and in the 2.7 mm continuum using the BIMA Array with an angular resolution of 10″. The continuum images show four sources which have been observed in previous studies. A 160 mJy point source coincides with the position of IRS 2. The spectral index of this source between 2.7 mm and 6 cm is 1.3. The other three 2.7 mm continuum sources correspond to the 1.3 mm sources FIR 3, 5, and 6 (Mezger et al. 1988, A&A, 191, 44; Mezger et al. 1992, A&A, 256, 637). The spectral indices of these sources between 2.7 and 1.3 mm are $\geq 4$. The other continuum sources, FIR 1, 2, 4, and 7, which are not detected in our data, have similar spectral indices. $C^{18}O$ emission is observed in three slightly extended regions. The two southern molecular clumps are offset from the FIR 5 and 6 continuum sources. The northern molecular clump is located $\leq 4$ from the FIR 2 continuum source. $H_2$ column densities derived from a viral analysis of the $C^{18}O$ clumps indicate that the CO relative abundance is only about 10% of the usual value. Depletion of CO in the gas phase may be due to "freezing out" of this molecule onto grain surfaces.

**Dynamical Models of Circumstellar Dust Shells around Long-period Variables: Spectrum, Lightcurves, Intensity Profiles, and Visibilities of IRC+10216**

Jan Martin Winters, Axel J. Fleischer, Erwin Sedlmayr (TU Berlin, Institut für Astronomie & Astrophysik, Sekr. PN 8-1, Hardenbergstr. 36, D-10623 Berlin, FRG), Andreas Gauger (Los Alamos National Laboratory, Theoretical Division)

Based on dynamical models for circumstellar dust shells (CDS) around long-period variables, including time-dependent hydrodynamics and a detailed treatment of carbon grain formation, continuum radiative transfer calculations have been carried out for a model which is capable of explaining several different observations of the prominent carbon star IRC+10216. It is shown that the resulting theoretical lightcurves exhibit the superposition of different oscillations: The oscillation with the period of the interior pulsation of the star is superposed by an oscillation with a longer period which is produced by the dynamical structure of the CDS. This dust shell model also causes a pronounced non-monotonic course of the spatial intensity distribution across the CDS, leading to the occurrence of wavelength-dependent internal structures of the calculated visibility functions. A quantitative comparison with specific observations obtained for IRC+10216 shows the layered distribution of dust across the CDS might be a natural physical explanation of the multiperiodic photometric lightcurves, the step-like behaviour of the brightness profile, and the internal structures present in the measured visibilities. **Acknowledgements:** This work has been supported by the BMBF, the DFG, and the DOE.

### The Monoceros OB1 Dark Cloud: A Comparison of $^{13}$CO, C$^{18}$O, and CS Observations of Outflow Regions

Grace A. Wolf-Chase (NASA/Ames Research Center)

We compare $^{13}$CO and C$^{18}$O J=1$\rightarrow$0 observations of four massive outflows in the Mon OB1 dark cloud with CS J=2$\rightarrow$1 (Wolf-Chase, Walker, & Lada 1995; Wolf-Chase & Walker 1995) and CO J=1$\rightarrow$0 (Margulis, Lada, & Snell 1988) observations. We find the $^{13}$CO J=1$\rightarrow$0 transition to be optically thin almost everywhere and derive a $^{13}$C$^{16}$O/$^{12}$C$^{18}$O abundance ratio of $\sim$ 8.1. Comparison of $^{13}$CO and CS column densities in the line cores indicates that the relative CS abundance may be enhanced by a factor of two to four over the typical abundance in quiescent clouds. However, comparison of the column densities derived from the outflowing gas in the line wings indicates the CS abundance may be enhanced by one or two orders of magnitude in these regions. The CS abundance is correlated with the mechanical luminosity of the outflows, suggesting a possible correlation of CS abundance with the strength of the shocks where the wind impacts the ambient cloud material. The CS abundance is about an order of magnitude higher in the redshifted outflowing gas than in the blueshifted outflowing gas. We speculate this may be due to the higher velocities observed in the redshifted gas associated with these outflows.

### Ammonia and Formaldehyde Gas Temperatures and Kinematics in Protostellar Cores

Alwyn Wootten (NRAO), Jeffrey G. Mangum (SMT)

Protostellar cores can be distinguished by possession of accompanying dust concentrations equalling or exceeding in mass the stellar mass inferred from the observed luminosity of the source. We have imaged a number of these cores to compare dust and gas distributions on scales of several thousand AU. Neither thermal nor kinematic properties of the near-protostellar region can be easily measured using dust emission measurements alone. We have employed radiation from symmetric ($NH_3$) or slightly asymmetric ($H_2CO$) top molecules to measure temperature variations on these scales and to explore core velocity structure. In more luminous sources, (e.g., DR21(OH) or S106FIR) there appears to be a close correlation between properties derived from all of these constituents of the core. In less luminous cores (e.g., L1448-IRS3 or IRAS16293), ammonia emission correlates poorly with properties deduced from dust or formaldehyde emission. In some regions (e.g., L1448-IRS3 or VLA1623), dust emission peaks at saddle points in the ammonia distribution, possibly indicating that ammonia is processed into other forms in regions of sufficiently high density and low temperatures. We suggest that gas temperatures measured from formaldehyde more securely probe the region near low luminosity protostars than either blackbody-derived dust temperatures or modeled ammonia temperatures.

## Ammonia Observations for Dense Cores of High Velocity Molecular Outflow

Yuefang Wu (Geophysics Department, Peking University), M. Miller (I. Physikalisches Institut, Universität zu Köln), G. Winnewisser (I. Physikalisches Institut, Universität zu Köln)

Observations with the 100 m telescope at Effelsberg in the $NH_3$ (1,1) and (2,2) lines towards eight molecular outflows were reported. All sources were mapped with the two lines except L1582B in which only weak $NH_3$(1,1) emission was detected. The $NH_3$ cores detected are coincident with the corresponding flow center sources. In the previous known source IRAS 00338+6312, our data agree with the previous results. In L1535, L483, and possibly in B361, the condensations are elongated in the direction perpendicular to the outflow axis. In other sources, the cores are between the two lobes or overlapped with the lobes. In L1535, L483, and L1251A, there are velocity gradients along the major axis of the condensation, showing evidence that the outflow material may be accelerated and collimated by rotation of the disk. The $NH_3$(1,1) line widths in these cores are wider than those of the cores without outflows on the whole. Our analysis also shows that the gas in the cores with high luminosity sources has high kinetic temperature and vice versa. The line width and gas temperature comparisons suggest that the perturbation of the dense gas produced by moment coupling with strong winds may be more than by the radiation of the center source. Core masses were derived with different methods. The stability of the cores is analysed.

**Observational Studies of Molecular Outflows in Star Forming Regions**

Yuefang Wu (Geophysics Department, Peking University)

Molecular outflows were searched widely towards young stellar objects. With surveys which were in unbiased (e.g., Margulis and Lada 1986; Fukui 1989) or guided by IRAS data (e.g., Levreault 1985; Morgan and Bally 1991) or other star forming activities (e.g., Evans et al. 1994), many new outflows were found. According to Wu, Huang and He (1995), there were 252 outflows observed until May of 1995. Resulting from high sensitivity and spatial resolution measurements, weak outflows were detected and polar or more collimated flows increase compared with those in the early period of the outflow discoveries. The velocity range, emission size and mass of EHV gas were investigated. Physical parameter ranges are summarized: $\Delta$V: 3–320 km $s^{-1}$; Size: 0.01 – 2 pc; mass: 0.008 – 600 $M_\odot$; energy: $10^{41} - 10^{47}$ erg; mass loss: $2 \times 10^{-9} - 10^{-5}$ $M_\odot$/yr; $L_{mech}$: 0.0002 – 2600 $L_\odot$; $\tau : 1 \times 10^3 - 4.6 \times 10^5$ yr; $L : 0.2 - 10^5$ $L_\odot$. The ranges tend to be wider than the earlier sources on the whole (Lada 1985). And some statistics based on the large number of samples may be meaningful. There were 80 molecular outflows associated with water masers and 59 flows associated with HH objects. The correlations between the outflows with these associated objects and the source bolometric luminosities were analysed. New observational advances in infall or inflow detection was presented. The flow region structures which show evidence for theoretical driving models are also discussed.

**Water Vapor Masers in Envelopes of Late Type Stars**

Yuefang Wu (United Radio Labs, Academic Sinica & Peking Univ.), Jing Lu (Purple Mountain Obs.), Huawei Zhang (Peking Univ.)

A survey was made of 98 sources to study the occurrence of water maser emission in the circumstellar shells of late-type stars. The sample was selected from the IRAS PSC with new colour criteria resulting from analyses of known water masers: [12]–[25]: –1.5 – 0.5, [25]–[60]: –2.3 – (–1.0); here

$[\lambda_1]$–$[\lambda_2]$= $2.5\times[\log(f_{\lambda 2}\ /f_{\lambda 1})]$. The survey was carried out with the 13.7m telescope at De Lin Ha, Qinghai province of China during August of 1993 to January of 1994. We detected weak water maser emission with S/N>4 or $\simeq$4 in 33 sources for the first time. The detection rate was 34%. Considering that a number of $H_2O$ masers detected previously were excluded from the sample, one can see that the water maser occurrence rate is rather high in this two color region. Maser emission was also detected in IRAS LRS class 4n sources, which exhibit the 11$\mu$m SiC future. Combining the known water masers, the dependence of the detection rate on IRAS LRS class, variability and colour index is analysed. Results show that stellar $H_2O$ masers occurrence most frequently in thick oxygen-rich envelopes or those with a variable star. However water masers were detected in IRAS LRS 1n and 4n objects. It may suggest that water masers can also occur in thin oxygen-rich envelopes which may be forming and diffusing. Since the S/N of our results is low, more sensitive observations are needed.

## Water Maser Search in Candidate HII Regions

Yuefang Wu (United Radio Lab., Academic Sinica & Dept. of Geophysics, Peking Univ.), Ruiqing Mao (Purple Mountain Obs. Qinghai Station)

Water maser emission has been searched for in 63 sources chosen from the IRAS PSC. The infrared index of the sample satisfies the criteria suggested by Wood and Churchwell (1989) for being associated with ultracompact HII regions. We used $f(60\mu m) > 150$ Jy to limit number the number of objects observed and to choose objects with a high expected detection rate. We also limited the declination to $\geq -25°$ so the source elevation was high enough. The observations were made with the 13.7m telescope of the Purple Mountain Observatory at De Lin Ha in Qinghai Province from June to December of 1993. The pointing accuracy is $20''$. The efficiency of the antenna is 47%. All the spectra were taken with a 1024 channel acousto-optical spectrometer. Its total bandwidth is 12.38 MHz. The system noise temperature was 320-350 K. For most of the observations, frequency switching was used. We observed 16 sources with signal-to-noise ratios $\geq 4$, including IRAS 18110–1854, which was also detected by Brand et al. (1993). The primary result is a 25% detection rate. It may be not typical that some sources satisfying the WC criteria were found with water masers already (Palla et al. 1991). Combining known maser statistics shown in sources with high LRS and VAR index, masers may occur rather frequently but that is not the necessary condition. And water maser emission may be easily excited around strong infrared sources.

## CO Observations for High Velocity Gas in Molecular Clouds

Yuefang Wu (Dept. of Geophysics, Peking Univ.), Ruiqing Mao (Purple Mountain Obs., Qinghai Station), M. Miller (I. Physikalisches Institut, Universität zu Köln)

Searches for high velocity gas were made in six molecular clouds with the CO $J$=2–1 transition line. One (S64) of the observed sources is in a high mass star formation region and five are low mass cores in nearby dark clouds. The core TMC2A is an IRAS source, but the others (L1512, L1523, L1582A, and L1681A) are without associated infrared sources (Beichman et al. (1986). The observations were carried out with the 3m KOSMA telescope of the Gornergrat Observatory of I. Physics Institute, Universität zu Köln, near Zermatt from February to September of 1992. Both blue and red wings were detected in L1582A and L1681A. The maps indicate the wings are probably bipolar outflows, showing star forming activity. There may be weak infrared sources in these clouds that need to be investigated further. There is also a bipolar outflow in the low mass source TMC2A. The $V_{\rm lsr}$ distribution in this core shows there is rotation with the axis along the northeast to southwest direction. Strong red wings were measured in S64. The observed area is $12' \times 12'$, which shows there is a red monopolar outflow in the cloud. It suggests the infrared source is located near the surface of the molecular region. The result confirms the layer structure consisting of an optical HII region, a CII radiation region, and a molecular region (Sharpless 1959; Vallee 1987; Zelik & Lada 1978). Since our maps are not complete, more observations are needed for further analysis.

## The Giant Molecular Cloud Monoceros R2: A Large "Wok-Shaped" Shell Possibly Resulting from a Champagne Flow

Taoling Xie (University of Maryland), P. F. Goldsmith (Cornell University), J. Bally (University of Colorado), W. D. Langer (Jet Propulsion Lab), R. W. Wilson (Bell Labs)

We present abundant data at millimeter wavelengths which show that the nearby Giant Molecular Cloud Monoceros R2 is largely a giant, massive expanding "wok" shell with its open side pointing away from us. This type of structure is expected in the "champagne flow" phenomenon when stars form on one side of a molecular cloud. Of particular interest is the large scale (over 30 pc) and the apparent longevity (over a few millions years) of this structure. Our observations show that the wok shell incorporated almost all the available molecular gas in this GMC ($\sim 40,000\ M_\odot$). The age of the structure is consistent with the age of a cluster of main sequence stars in this GMC. The existence of numerous younger stellar objects with an age on the order of $10^5$ years or younger on the shell strongly suggest the occurrence of triggered star formation. Overall, it seems that existing theories on the development of molecular shells and the subsequent triggered star formation contain essential physical elements to confront the wealth of molecular data on a large scale, but some discrepancies in details between theories and observations remain to be explained.

Xie, T., & Goldsmith, P.F. 1994, *ApJ*, *430*, 252

## Molecular Line Surveys of Dense Cores in Regions of High Mass Star Formation

Igor Zinchenko (Institute of Applied Physics, Russian Academy of Sciences)

We analyze further the observational data for dense cores in regions of high mass star formation (HMSF) obtained recently with the SEST and Metsähovi radio telescopes. The statistical distributions of the core sizes, masses, mean densities, CS line widths and CO temperatures are constructed.

The galacto-centric dependence of these parameters in the range $R \approx 7 - 11$ kpc is investigated. There is a trend to a systematic decrease of the mean density of the cores with increasing galacto-centric distance.

We have attempted to investigate the radial dependence of some physical parameters in several cores which have a simple structure. The CS line areas drop rather quickly with $r$, corresponding to $r^{-1.5}$ to $r^{-2}$. The CS line widths decrease towards the core edges as well, sometimes below the $C^{34}S$ line width (measured at the CS peak position). The ratios of the CS and $C^{34}S$ line widths and intensities are described preferably by the models with density inhomogeneities in the cores.

We investigate the $L - \Delta V$ and $L - \bar{n}$ relationships for the CS cores. Although there is a trend for increasing line width with increasing size, the correlation is very weak. There is, in general, a higher velocity dispersion in HMSF cores than in CO clouds and dense cores in dark clouds.

# AUTHOR INDEX

# OBJECT INDEX

R



S



T



U



V



W



X

Y

Z

# SUBJECT INDEX

# PARTICIPANTS

Aalto, Susanne
Caltech
sab@astro.caltech.edu

Alten, Victoria
APAS, U Colorado
alten@casa.colorado.edu

Apponi, Aldo
Arizona State University
aldo@quasar.la.asu.edu

Armağan, Demet S.
Marmara Research Centre, Turkey

Aspin, Colin
JACH
caa@jach.hawaii.edu

Baars, Jaap
SMTO/MPIfR
jbaars@as.arizona.edu

Bally, John
APAS, U Colorado
bally@janos.colorado.edu

Barnes, Peter
Haystack Observatory
peterb@wells.haystack.edu

Barsony, Mary
U California, Riverside
fun@nusun.ucr.edu

Barvainis, Richard
Haystack Observatory
reb@wells.haystack.edu

Benford, Dominic
Caltech
dbenford@tacos.caltech.edu

Bergin, Edwin
FCRAO, U Massachusetts
eab@fcrao1.phast.umass.edu

Bergman, Per
Caltech
bergman@csardas.caltech.edu

Bieging, John
Steward Observatory, U Arizona
jbieging@as.arizona.edu

Billawala, Youssef
CASA, U Colorado
billawal@janos.colorado.edu

Black, John
Steward Observatory, U Arizona
black@as.arizona.edu

Blitz, Leo
Astronomy Program, U Maryland
blitz@astro.umd.edu

Bloemen, Hans
SRON, Utrecht
bloemen@hea.sron.ruu.nl

Boland, Wilfried
NFRA
boland@nfra.nl

Boogert, Adwin
Kapteyn Observatory

Booth, Roy
Onsala Space Observatory
roy@oso.chalmers.se

Bourke, Tyler
CfA/Harvard-Smithsonian
bourke@cfara1.harvard.edu

Brinks, Elias
NRAO, Socorro
ebrinks@aoc.nrao.edu

Bronfman, Leonardo
U Chile
leo@das.uchile.cl

Brown, Robert
NRAO Charlottesville
rbrown@nrao.edu

Bryant, Peter
Caltech
pmb@astro.caltech.edu

Buckley, Henry
ROE
h.buckley@roe.ac.uk

Caldwell, Doug
NRAO/Rensselaer
dcaldwel@nrao.edu

Campbell, Clara
U Colorado
campbelc@bogart.colorado.edu

Carpenter, John
IFA, U Hawaii
carp@ifa.hawaii.edu

Casoli, Fabienne
Observatoire de Paris
casoli@obspm.fr

Castets, Alain
Observatoire de Grenoble
castets@gag.observ-gr.fr

Chernin, Lawrence
U California, Berkeley
lchernin@astron.berkeley.edu

Clarke, Jeff
SMTO
jclarke@as.arizona.edu

Comeron, Fernando
U Barcelona
fcomeron@mizar.am.ub.es

Cunningham, Charles
Herzberg Inst. of Astrophysics
ctc@hiaras.hia.nrc.ca

Dame, Thomas M.
CfA/Harvard-Smithsonian
tdame@cfa.harvard.edu

Deane, James R.
IFA, U Hawaii
deane@gmc.ifa.hawaii.edu

Devereux, N.
New Mexico State U
devereux@nmsu.edu

Dickel, Hélène R.
U Illinois
lanie@astro.uiuc.edu

Digel, Seth
NASA/GSFC
digel@egret.gsfc.nasa.gov

Dorfi, Ernst
Inst. für Astronomie, Vienna
ead@astro.ast.univie.ac.at

Durouchoux, Philippe
CE-SACLAY-DAPNIA-SAP
durouchoux@sapvxg.saclay.cea.fr

Eckart, Andreas
M-P-I für Extraterrestrische Physik
eckart%mpeis1@mpe.mpe-garching.
mpg.de

El-Nawawy, M. S.
United Arab Emerites U

Emerson, Darrel
NRAO Tucson
demerson@nrao.edu

Encrenaz, Therese
DESPA, Obs. de Paris
therese.encrenaz@obspm.fr

Engelbracht, Chad
Steward Observatory, U Arizona

Evans, Aaron
IFA, U Hawaii
evans@galileo.ifa.hawaii.edu

Falgarone, Edith
Ecole Normale Supérieure
falgarone@ensapa.ens.fr

Falstreau, Ron
S.A.R.A., Texas

Fich, Mike
U Waterloo, Canada
fich@uwaterloo.ca

Fleischer, Axel J.
Tech. Universität, Berlin
fleischer@physik.tu-berlin.de

Frayer, David
NRAO/U Virginia
dfrayer@nrao.edu

Freund, Robert
NRAO
rfreund@nrao.edu

Friberg, Per
JACH, Hawaii
friberg@jach.hawaii.edu

Gao, Yu
SUNY, Stony Brook
ygao@astro.sunysb.edu

Gauger, Andreas
Los Alamos National Laboratory
ahg@t4.lanl.gov

Ge, Jian
Steward Observatory, U Arizona

Genzel, Reinhard
MPI für Physik und Astrophysik
genzel%mpeis1@mpe-garching.mpg.de

Giannakopoulou, Jean
U Waterloo, Canada
jean@astro.uwaterloo.ca

Glenn, Jason
Steward Observatory, U Arizona

Goldsmith, Paul
NAIC-Cornell U
pfg@astrosun.tn.cornell.edu

Goodman, Alyssa
Harvard U
agoodman@cfagood.harvard.edu

Gözel, Isfendiyar
Marmara Research Centre, Turkey

Grenier, Isabella
U Paris/DAPNIA/Centre d'Etudes, Saclay
grenier@sapvxg.saclay.cea.fr

Guélin, Michel
IRAM, Grenoble
guelin@iram.grenet.fr

Haikalä, Lauri
U Helsinki
haikala@hylk.helsinki.fi

Handa, Toshihiro
U Tokyo
handa@ghz.mtk.ioa.s.u-tokyo.ac.jp

Harris, Andrew
FCRAO/U Massachusetts
harris@fcrao1.phast.umass.edu

Harrison, Andrew
ROE
aph@roe.ac.uk

Hartmann, Dap
CfA/Harvard-Smithsonian
dap@abitibi.harvard.edu

Hasegawa, Tetsuo
Tokyo U
tetsuo@ghz.mtk.nao.ac.jp

Heikkilä, Arto
Onsala Space Observatory
arto@oso.chalmers.se

Heithausen, Andreas
Physikalisches Inst./U Köln
heithausen@ph1.uni-koeln.de

Helfer, Tamara
U Maryland
thelfer@astro.berkeley.edu

Herbst, Eric
Ohio State U
herbst@ohstpy.mps.ohio-state.edu

Heyer, Mark
FCRAO/UMass
heyer@fcrao1.phast.umass.edu

Hinkle, Kenneth
KPNO/NOAO
hinkle@noao.edu

Hjalmarson, Ake
Onsala Space Observatory
hjalmar@oso.chalmers.se

Hogg, Dave
NRAO Charlottesville
dhogg@nrao.edu

Horellou, Cathy
Onsala Space Observatory
horellou@oso.chalmers.se

Hron, Josef
Inst. für Astronomie, U. Vienna
hron@astro1.ast.univie.ac.at

Hunter, Deidre
Lowell Observatory
dah@lowell.edu

Hurt, Robert
U California, Riverside
rhurt@fun.ucr.edu

Inoue, Motoko
Nobeyama Radio Observatory
motoko@nro.nao.ac.jp

Irvine, William
FCRAO, U Massachusetts
irvine@fcrao1.phast.umass.edu

Irwin, Judith
Queens U, Canada
irwin@astro.queensu.ca

Ishiguro, Masato
Nobeyama Radio Observatory
ishiguro@nro.nao.ac.jp

Ishizuki, Sumio
Tohoku Univ.
ishizuki@astroa.astr.tohoku.ac.jp

Ivanov, Valentin
Steward Observatory, U Arizona

Jewell, Phil
Joint Astronomy Centre, Hawaii
pjewell@jach.hawaii.edu

Justtanont, Kay
NASA/Ames
kj@prometheus.arc.nasa.gov

Kawabe, Ryohei
Nobeyama Radio Observatory
kawabe@nro.nao.ac. jp

Keady, John
Los Alamos Nationall Laboratory

Keene, Jocelyn
Caltech
jbk@tacos.caltech.edu

Knee, Lewis B. G.
Swedish-ESO Submm Telescope/ESO
lknee@eso.org

Kobulnicky, H. "Chip"
U Minnesota
chip@astro.spa.umn.edu

Koerner, David W.
Jet Propulsion Laboratory
davidk@coma.jpl.nasa.gov

Kono, Kotaro
Nobeyama Radio Observatory
kotaro@nro.nao.ac.jp

Kramer, Carsten
IRAM Granada
kramer@iram.es

Kutner, Marc
RPI/NRAO
mkutner@nrao.edu

Kwok, Sun
U Calgary
kwok@iras.ucalgary.ca

Lada, Charles
Smithsonian Astrophysical Obs.
clada@cfa.harvard.edu

Ladd, Ned
FCRAO, U Massachusetts
ladd@fcrao1.phast.umass.edu

Langer, William
Jet Propulsion Laboratory
langer@langer.jpl.nasa.gov

Larsen, Fredrik
Stockholm Observatory
fredrik@astro.su.se

Latter, William
NASA/Ames
latter@dusty.arc.nasa.gov

Lebzelter, Thomas
Inst. für Astronomie, U Vienna
lebzelter@astro.ast.univie.ac.at

Lee, Terry J.
ROE
tjl@roe.ac.uk

Lefloch, Bertrand
IRAM Granada
lefloch@iramed.iram.es

Lis, Darek
Caltech
dcl@tacos.caltech.edu

Liszt, Harvey
NRAO Charlottesville
hliszt@nrao.edu

Lockman, Felix J.
NRAO Green Bank
jlockman@nrao.edu

Loinard, Laurent
CfA/Harvard-Smithsonian
lloinard@cfa.harvard.edu

Lucas, Robert
IRAM Grenoble
lucas@iram.fr

Maddalena, Ron
NRAO Green Bank
rmaddale@nrao.edu

Magnani, Loris
U Georgia
loris@zeus.physast.uga.edu

Malhotra, Sangeeta
Princeton U
san@astro.princeton.edu

Maloney, Phil
JILA/CASA, U Colorado
maloney@shapley.colorado.edu

Mangum, Jeff
SMTO
jmangum@as.arizona.edu

Martin, Robert
SMTO
rmartin@as.arizona.edu

Mauersberger, Rainer
Steward Obs / MPIfR
p373mau@mpifr-bonn.mpg.de

May, Jorge
U Chile
jmay@das.uchile.cl

McKee, Chris
U California, Berkeley
cmckee@mckee.berkeley.edu

McMullin, Joseph
SMTO
jmcmulli@maisel.as.arizona.edu

Mead, Kathryn
NRAO/Union College
kmead@nrao.edu

Minchin, Nigel
Q. Mary & Westfield, U London
n.r.minchin@qmw.ac.uk

Mitchell, George
St. Marys U, Halifax NS
mitchell@husky1.stmarys.ca

Miyawaki, Ryosuke
Fukuoka Univ.
miyawaki@fueipc.fukuoka-edu.ac.jp

Moorey, Graham
ATNF/CSIRO
gmoorey@atnf.csiro.au

Morino, Junichi
Nobeyama Radio Observatory

Morris, Mark
U California, Los Angeles
morris@osprey.astro.ucla.edu

Myers, Philip
CfA/Harvard-Smithsonian
myers@cfa.harvard.edu

Nakai, Naomasa
Nobeyama Radio Observatory
nakai@nro.nao.ac.jp

Narayanan, Gopal
Steward Observatory, U Arizona
gopal@as.arizona.edu

Neininger, Nikolaus
IRAM, Grenoble
neininge@iram.fr

Nummelin, Albert
Onsala Space Observatory
albert@oso.chalmers.se

Oka, Tomoharu
U Toyko
tomo@sgr.nro.nao.ac.jp

Olofsson, Goran
Stockholm U
olofsson@astro.su.se

Osterberg, Juergen
Radioastronomisches Inst., U Bonn
josterb@astro.uni-bonn.de

Padgett, Deborah
Caltech/IPAC
dlp@ipac.caltech.edu

Papadopoulos, Peter
U Toronto
papadop@radio.astro.utoronto.ca

Patel, Nimesh
Smithsonian Astrophysical Obs.
npatel@cfa.harvard.edu

Payne, John
NRAO Tucson
jpayne@nrao.edu

Peters, William
Steward Observatory, U Arizona
peters@as.arizona.edu

Pound, Marc W.
U California, Berkeley
pound@ral.berkeley.edu

Pratap, Preethi
FCRAO/U Massachusetts
preethi@fcrao1.phast.umass.edu

Radford, Simon
NRAO Tucson
sradford@nrao.edu

Rand, Richard J.
U Maryland
rjr@astro.umd.edu

Reipurth, Bo
ESO
breipurt@eso.org

Remy, Gilles
Centre d'Etudes, Saclay
gremy@mesioa.obspm.fr

Richer, John
MRAO, Cambridge U
j.s.richer@mrao.cam.ac.uk

Richter, Matthew J.
U California, Berkeley
richter@triscuit.berkeley.edu

Robson, Ian
JACH
eir@jach.hawaii.edu

Rubio, Monica
U Chile
monica@das.uchile.cl

Ruiz, Milagros
Steward Observatory, U Arizona

Sage, Leslie
Nature Magazine
lsage@naturedc.com

Saito, Masao
Nobeyama Radio Observatory
masao@nro.nao.ac.jp

Sakamoto, Kazushi
Nobeyama Radio Observatory
Sakamoto@nro.nao.ac.jp

Sakamoto, Seiichi
Nobeyama Radio Observatory
seiichi@sgr.nro.nao.ac.jp

Sanders, Dave
IFA, U Hawaii
sanders@galileo.ifa.hawaii.edu

Sargent, Anneila
Caltech
afs@astro.caltech.edu

Sato, Kumiko
U Tokyo
satok@ghz.mtk.ioa.s.u-tokyo.ac.jp

Saucedo-Morales, Julio
Steward Observatory

Schilke, Peter
Caltech
schilke@medoc.caltech.edu

Schloerb, Peter
FCRAO, U Massachusetts
schloerb@fcrao1.phast.umass.edu

Schneider, Nicola
I. Physik. Institut, U Köln
schneider@ph1.uni-koeln.de

Scoville, Nick
Caltech
nzs@radio.caltech.edu

Serabyn, Eugene
Caltech
serabyn@tacos.caltech.edu

Seta, Masumichi
U Tokyo
seta@ghz.mtk.ioa.s.u-tokyo.ac.jp

Shaver, Peter
ESO
pshaver@eso.org

Shen, Jianjun
U Illinois
shenj@astro.uiuc.edu

Shinnaga, Hiroko
U Tokyo
shinnaga@ghz.mtk.ioa.s.u-tokyo.ac.jp

Snell, Ronald
FCRAO/UMass
snell@fcrao1.phast.umass.edu

Snyder, Lewis
U Illinois
snyder@sirius.astro.uiuc.edu

Sofue, Yoshiaki
U Tokyo
sofue@mtk.ioa.s.u-tokyo.ac.jp

Solomon, Phil
SUNY, Stony Brook
psolomon@sbast7.ess.sunysb.edu

Sorai, Kazuo
U Tokyo
sorai@ghz.mtk.ioa.s.u-tokyo.ac.jp

Stapelfeldt, Karl
Jet Propulsion Laboratory
krs@wfpc2-mail.jpl.nasa.gov

Störzer, Juergen
I. Physik. Institut, U Köln
stoerzer@ph1.uni-koeln.de

Strom, Stephen
FCRAO
sstrom@donald.phast.umass.edu

Stutzki, Juergen
I. Physik. Institut, U Köln
stutzki@ph1.uni-koeln.de

Sugerman, Ben
JPL/Occidental College
t42@oxy.edu

Tacconi, Linda
MPI für Ext. Physik
linda%mpeis1@mpe.mpe-garching.
mpg.de

Tauber, J.
European Space Agency
jtauber@astro.estec.esa.nl

Taylor, Christopher
U Minnesota/NRAO
ctaylor@nrao.edu

Thaddeus, Patrick
CfA/Harvard-Smithsonian
thaddeus%cfaco1.decnet@cfa.
harvard.edu

Theil, David
CASA, U Colorado
theil@casa.colorado.edu

Thornley, Michele
U Maryland
michele@astro.umd.edu

Tilanus, Remo
JACH
rpt@jach.hawaii.edu

Tosaki, Tomoka
Nobeyama Radio Observatory
tomoka@nro.nao.ac.jp

Tsuboi, Masato
Ibaraki U, Japan
tsuboi@masato.sci.ibaraki.ac.jp

Turner, Jean
UCLA
turner@bonnie.astro.ucla.edu

Ungerechts, Hans
IRAM, Granada
ungerechts@iram.es

Vanden Bout, Paul
NRAO Charlottesville
pvandenb@nrao.edu

van Langevelde, Huib
Joint Inst. for VLBI-Europe/NRAO
hvanlang@nrao.edu

Verdes-Montenegro, Lourdes
CSIC-Spain
lourdes@iaa.es

Walker, Chris
Steward Observatory, U Arizona
cwalker@as.arizona.edu

Walker, Connie
Steward Observatory, U Arizona
Connie@as.arizona.edu

Wall, William
INAOE, Mexico
wwall@tonali.inaoep.mx

Walmsley, C. Malcolm
U Köln
walmsley@apollo.ph1.uni-koeln.de

Walther, Dolores
JACH
dolores@jach.hawaii.edu

Wang, Zhong
Smithsonian Astrophys. Obs
ysw@mrao.cam.ac.uk

Warin, Sophie
Obs. Grenoble
warin@gag.observ-gr.fr

Welch, Jack
U California, Berkeley
welch@bkyast.berkeley.edu

White, Glenn J.
Qu. Mary & Westfield, U London
g.j.white@qmw.ac.uk

Whiteoak, John
ATNF/CSIRO
jwhiteoa@rp.csiro.au

Wielebinski, Richard
MPIfR
gabi@mpifr-bonn.mpg.de

Wild, Wolfgang
IRAM Granada
wild@iram.es

Willacy, Karen
Edinburgh U
kaw@roe.ac.uk

Williams, Jonathan
UC Berkeley
jwilliams@astro.berkeley.edu

Wilson, Robert
Harvard University
rww@big.att.com

Winters, Jan Martin
Inst. Astron/Astrophys, Berlin
winters@physik.tu-berlin.de

Wolf-Chase, Grace A.
NASA-Ames
wolf@ssa1.arc.nasa.gov

Wootten, Al
NRAO Charlottesville
awootten@nrao.edu

Wu, Yuefang
Peking U
gao@bao01.bao.ac.cn

Xie, Taoling
U Maryland
tao@astro.umd.edu

Young, Judith
FCRAO/UMassachusetts
young@fcrao1.phast.umass.edu

Young, Ken
CSO, Hilo
rtm@tacos.caltech.edu

Yu, Ka Chun
CASA, U. Colorado
kachun@mensae.colorado.edu

Yun, Min Su
Caltech
msy@astro.caltech.edu

Zinchenko, Igor
Russian Acad. Sciences
zin@appl.nnov.su

Ziurys, Lucy
Steward Obs U Arizona/ASU
lziurys@as.arizona.edu

Zmuidzinas, Jonas
Caltech
jonas@tacos.caltech.edu